Research Reports in Physics

Research Reports in Physics

Nuclear Structure of the Zirconium Region
Editors: J. Eberth, R. A. Meyer, and K. Sistemich

Ecodynamics Contributions to Theoretical Ecology
Editors: W. Wolff, C.-J. Soeder, and F. R. Drepper

J. Eberth R. A. Meyer K. Sistemich (Eds.)

Nuclear Structure of the Zirconium Region

Proceedings of the International Workshop,
Bad Honnef, Fed. Rep. of Germany, April 24-28, 1988

Organized by
the Kernforschungsanlage Jülich, Universität zu Köln,
and Lawrence Livermore National Laboratory

With 232 Figures

Springer-Verlag Berlin Heidelberg New York
London Paris Tokyo

Dr. Jürgen Eberth
Institut für Kernphysik, Universität zu Köln, Zülpicher Straße 77,
D-5000 Köln 41, Fed. Rep. of Germany

Dr. Richard A. Meyer
Nuclear Chemistry Division, Lawrence Livermore National Laboratory,
Livermore, CA 94550, USA

Professor Dr. Kornelius Sistemich
Institut für Kernphysik, Kernforschungsanlage Jülich, Postfach 1913,
D-5170 Jülich, Fed. Rep. of Germany

ISBN 3-540-50120-7 Springer-Verlag Berlin Heidelberg New York
ISBN 0-387-50120-7 Springer-Verlag New York Berlin Heidelberg

This work is subject to copyright. All rights are reserved, whether the whole or part of the material is concerned, specifically the rights of translation, reprinting, reuse of illustrations, recitation, broad-casting, reproduction on microfilms or in other ways, and storage in data banks. Duplication of this publication or parts thereof is only permitted under the provisions of the German Copyright Law of September 9, 1965, in its version of June 24, 1985, and a copyright fee must always be paid. Violations fall under the prosecution act of the German Copyright Law.

© Springer-Verlag Berlin Heidelberg 1988
Printed in Germany

The use of registered names, trademarks, etc. in this publication does not imply, even in the absence of a specific statement, that such names are exempt from the relevant protective laws and regulations and therefore free for general use.

Offset printing and bookbinding: Weihert-Druck GmbH, D-6100 Darmstadt
2157/3150-543210 – Printed on acid-free paper

Preface

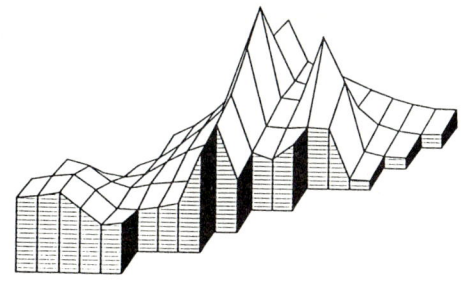

The International Workshop on "Nuclear Structure of the Zirconium Region" was organized in order to explore the special nuclear physics information which can be obtained through the study of medium-mass nuclei. The regions of proton-rich nuclei with mass numbers around 80 and of neutron-rich isotopes with $A \sim 100$ are characterized by the occurrence of very different structures. Thus, ^{90}Zr and ^{96}Zr are magic nuclei whereas strong ground-state deformations are observed for the isotopes both at $A \sim 80$ and at $A \sim 100$. Remarkable features have been discovered such as moments of inertia close to the rigid-rotor values, shape coexistence and a strong influence of the proton-neutron interaction in selected orbitals on the nature of the nuclei. Therefore, the zirconium region offers unique possibilities of obtaining insight into a large variety of nuclear structure phenomena which depend sensitively on the number of nucleons. Because of this, intensive experimental and theoretical work has been invested in the study of this nuclear region.

It was the aim of the workshop to bring together experts on both the $A \sim 80$ and $A \sim 100$ regions in order to initiate art-in-depth discussions on the emerging physical picture. It was expected that especially a joint workshop would have a productive influence on future studies since both mass regions complement each other to a large extent. The workshop was held at the "Physikzentrum" of the German Physical Society, April 24–29, 1988. About 70 scientists convened, attended about 50 presentations, and took part in lively discussions. It was a pleasure to note in particular the participation and active interest of younger scientists in the field. These proceedings contain all the presentations by the attendees. There are several review papers which put the knowledge about the zirconium region into general perspective. Most of the shorter papers present very new results. Thus, the book provides a useful compilation of nuclear structure information, which is focussed on this region, and stimulates further studies.

In organizing the workshop we experienced valuable help from the International Advisory Committee consisting of P. von Brentano (Cologne), R.F. Casten (Brookhaven), K. Heyde (Ghent), K.-L. Kratz (Mainz), K.-P. Lieb (Göttingen), I. Ragnarsson (Lund), and O.W.B. Schult (Jülich). Their extremely constructive contributions to the selection of the conference topics and to the program is gratefully acknowledged.

The workshop could only be realized thanks to the sponsorship by the Volkswagenstiftung, the Kernforschungsanlage Jülich and the Deutsche Physikalische Gesellschaft, to whom the organizers are very grateful.

Thanks are also due to several members of the Kernforschungsanlage Jülich and especially to the conference secretary, Ms. C. Bong, for their efficient assistance in preparing and running the meeting. It was a pleasure to enjoy the hospitality of Dr. Debrus, Mrs. Kluth, Mrs. Offerzier, and their coworkers at the Physikzentrum.

Köln	*J. Eberth*
Livermore	*R.A Meyer*
Jülich, June 1988	*K. Sistemich*

Contents

Part I   Introduction

Nuclear Structure in the Zr Region: Some Introductory Remarks
By K. Heyde (With 9 Figures) 3

Part II   Shape Transitions and Shape Coexistence

Some Results from Potential Energy Surface Calculations for Nuclei
in the Mass 70–80 Region
By R. Bengtsson (With 9 Figures) 17

Collectivity of Neutron Deficient Odd Strontium and Yttrium
Isotopes
By D. Bucurescu, G. Cǎta, D. Cutoiu, M. Ivaşcu, C.F. Liang,
P. Paris, and N.V. Zamfir (With 2 Figures) 26

Microscopic Description of Even Ge and Se Nuclei
By F. Grümmer, A. Petrovici, and K.W. Schmid (With 3 Figures) . 32

Coexistence of Shapes and Structure of Nuclei in the A = 80 Region
By R. Sahu and S.P. Pandya (With 4 Figures) 39

Competition of Oblate and Prolate Deformation in $^{69,70,71,72}_{34}$Se
By T. Mylaeus, J. Busch, P. von Brentano, J. Eberth, M. Liebchen,
N. Schmal, R. Sefzig, S. Skoda, W. Teichert, M. Wiosna,
and W. Nazarewicz (With 5 Figures) 45

Moments and Radii of $^{78-100}$Sr
By R. Neugart, E. Arnold, W. Neu, K. Wendt, P. Lievens,
R.E. Silverans, L. Vermeeren, F. Buchinger, E.B. Ramsay, G. Ulm,
and The ISOLDE Collaboration (With 4 Figures) 52

Experimental Evidence for Shape Coexistence in ^{97}Sr$_{59}$ and
Implications for the Structure of the Odd-Odd Isotone ^{98}Y
By G. Lhersonneau, K.-L. Kratz, H. Ohm, B. Pfeiffer,
and K. Sistemich (With 2 Figures) 58

Shape Coexistence and Mixing of Spherical and Deformed Shapes in the N = 60 Isotones
By J.C. Hill, F.K. Wohn, R.F. Petry, R.L. Gill, H. Mach,
and M. Moszynski (With 3 Figures) 64

Transition Probabilities and Static Moments in Transitional Nuclei
By A. Wolf and R.F. Casten (With 3 Figures) 70

Evidence for Shape Coexistence in Neutron-Rich Rh and Ag Nuclei
By J. Rogowski, N. Kaffrell, H. Tetzlaff, N. Trautmann, D. De Frenne,
K. Heyde, E. Jacobs, G. Skarnemark, J. Alstad, M.N. Harakeh,
J.M. Schippers, S.Y. van der Werf, W.R. Daniels, and K. Wolfsberg
(With 4 Figures) 76

Shape Coexistence and the Structure of Even-Even Se, Pd, and Ru Nuclei
By R.A. Meyer, D.F. Kusnezov, M.A. Stoyer, and R.P. Yaffe 82

Monopole Degrees of Freedom in the Odd-Mass Neighbours of Even-Even Nuclei with Low-Lying 0^+ Levels
By E.F. Zganjar and J.L. Wood (With 5 Figures) 88

Part III   High Spin Effects

Techniques for the Study of the $A \sim 80$ and $A \sim 100$ Nuclei Far from Stability
By W. Gelletly, Y. Abdelrahman, A.A. Chishti, J.L. Durell,
J. Fitzgerald, C.J. Lister, J.H. McNeill, W.R. Phillips,
and B.J. Varley (With 18 Figures) 101

M1 and E2 Transition Probabilities in Kr Nuclei
By L. Funke and G. Winter 115

Lifetimes and Sidefeeding Times in $A = 70$–83 Nuclei
By K.P. Lieb, F. Cristancho, W. Fieber, C. Gross, J. Heese,
T. Osipowicz, S. Ulbig, and B. Wörmann (With 6 Figures) 120

The Moment of Inertia of ^{78}Sr
By C.J. Gross, J. Heese, K.P. Lieb, C.J. Lister, B.J. Varley,
A.A. Chishti, and W. Gelletly (With 3 Figures) 127

g-Factors Near the First Backbend in ^{82}Sr and ^{84}Sr
By A.I. Kucharska, J. Billowes, and C.J. Lister (With 1 Figure) ... 131

Core-Related Effects on Electromagnetic Transition Strengths in Medium-Mass Nuclei
By W. Andrejtscheff, L.K. Kostov, P. Petkov, and Y. Sy Savane
(With 3 Figures) 137

Microscopic Description of the Ground-State and High-Spin Properties of the Light Strontiums
By H.C. Flocard (With 11 Figures) 143

Part IV   Complete Spectroscopy

Complete Spectroscopy
By P. von Brentano, A. Dewald, W. Lieberz, R. Reinhardt, K.O. Zell,
and V. Zipper (With 8 Figures) 157

^{73}Se Investigated by the $(\alpha,n\gamma)$ Reaction
By F. Seiffert, R. Wrzal, K.O. Zell, K.P. Schmittgen, R. Reinhardt,
W. Lieberz, A. Dewald, A. Gelberg, and P. von Bretano
(With 4 Figures) 172

Complete Spectroscopy of ^{87,88,89}Sr with (n,γ) and (d,p) Reactions?
By C. Winter, B. Krusche, K.P. Lieb, G. Hlawatsch, T. v. Egidy,
F. Hoyler, and R.F. Casten (With 4 Figures) 176

Quantum Chaos and the Boson-Fermion Approach to $A \sim 100$ Region
By V. Paar, D. Vorkapić, S. Brant, H. Seyfarth, and V. Lopac
(With 3 Figures) 184

Part V   Shell Effects

The Approach to 2:1 Deformations Induced by Particle Numbers
Around 40 and 60
By I. Ragnarsson and T. Bengtsson (With 5 Figures) 193

IBMF and IBFFM Approach to Nuclei in the $A \simeq 100$ Region
By S. Brant, K. Sistemich, H. Seyfarth, H. Ohm, M.L. Stolzenwald,
V. Paar, D. Vretenar, D. Vorkapić, V. Lopac, R.A. Meyer,
G. Lhersonneau, K.-L. Kratz, and B. Pfeiffer (With 8 Figures) 199

Single-Particle Excitations and Collective Vibrational Modes in ^{96}Zr
By G. Molnár, T. Belgya, B. Fazekas, A. Veres, S.W. Yates,
E.W. Kleppinger, R.A. Gatenby, H. Mach, R. Julin, J. Kumpulainen,
A. Passoja, and E. Verho (With 5 Figures) 215

Doppler-Shift Lifetime Measurements in ^{96}Zr
By T. Belgya, G. Molnár, B. Fazekas, A. Veres, S.W. Yates,
and R.A. Gatenby (With 2 Figures) 227

In-Beam Studies of ^{96}Zr and ^{98}Zr: Collective Excitations
By E.A. Henry, R.A. Meyer, A. Aprahamian, K.H. Maier, L.G. Mann,
and N. Roy 233

Band Structures in ^{96}Zr and ^{98}Zr Studied Through the High-Spin
Beta Decays of the Y Parents
By M.L. Stolzenwald, S. Brant, H. Ohm, K. Sistemich,
and G. Lhersonneau (With 5 Figures) 239

Possible Evidence for Subshell Closures at N = 38, 40 and 56
By N. Severijns, E. van Walle, D. Vandeplassche, J. Wouters,
J. Vanhaverbeke, W. Vanderpoorten, and L. Vanneste
(With 2 Figures) 245

Subshell Closure at N = 56 in Very Neutron-Rich Bromine Isotopes
By M. Graefenstedt, U. Keyser, F. Münnich, F. Schreiber,
The ISOLDE, LOHENGRIN, and OSTIS Collaboration
(With 3 Figures) 251

Recent Results of Beta-Decay-Energies of Very Neutron-Rich Nuclei
Around Mass Number A = 100
By M. Graefenstedt, U. Keyser, F. Münnich, F. Schreiber,
The ISOLDE, LOHENGRIN, and OSTIS Collaboration
(With 4 Figures) 257

Level Structure of $^{89}_{39}$Y, N = 50 Isotone
By T. Batsch, J. Kownacki, Z. Zelazny, M. Guttormsen, T. Ramsøy,
and J. Rekstad (With 2 Figures) 263

Nuclear Structure Close to ^{100}Sn: Excitations of the Shell Model Core
By H. Grawe, D. Alber, H. Haas, H. Kluge, K.H. Maier,
B. Spellmeyer, and X. Sun (With 4 Figures) 269

Part VI Deformations

$A \approx 80$ and $A \approx 100$ Nuclei Studied Within the Mean Field Approach
By W. Nazarewicz and T. Werner (With 19 Figures) 277

On the Solution to the Large Amplitude Collective Motion of Finite
Interacting Fermi Systems
By F. Dönau, Jing-ye Zhang, and L.L. Riedinger 293

Nuclear Structure Effects along the N = Z Line
By C.J. Lister, A.A. Chishti, B.J. Varley, W. Gelletly,
and A.N. James (With 2 Figures) 298

Core Polarization Effects in Odd $A \approx 80$ Nuclei
By S.L. Tabor, P.D. Cottle, C.J. Gross, D.M. Headley,
U.J. Hüttmeier, E.F. Moore, and W. Nazarewicz (With 6 Figures)   303

Deformation of Light Br Isotopes: New Ground State Spin and
Moment Measurements
By N.J. Stone, C.J. Ashworth, I.S. Grant, A.G. Griffiths, S. Ohya,
J. Rikovska, and P.M. Walker (With 2 Figures) 309

Anomalous Behaviour of Transition Probabilities in ^{75}Kr
By S. Skoda, J. Busch, J. Eberth, M. Liebchen, T. Myläus,
N. Schmal, R. Sefzig, W. Teichert, and M. Wiosna (With 4 Figures)   315

Backbending Behaviour in the Odd Neutron Nuclei ^{75}Kr and ^{79}Sr
By A.A. Chishti, M. Campbell, W. Gelletly, L. Goettig, A.N. James,
C.J. Lister, D.J.G. Love, J.H. McNeill, R. Moscrop, O. Skeppstedt,
and B.J. Varley (With 6 Figures) 320

Proton-Neutron Interaction and the 1 g Orbital in the $33 \leq Z \leq 50$
Region. By A.G. Griffiths and J. Rikovska (With 3 Figures) 326

Particle-Rotor Calculations in the A = 100 Region
By P.B. Semmes (With 5 Figures) 332

Large Inertial Moments, Pairing Reduction and the Effects of Coriolis
Mixing on Rotational Bands in Odd-Z Deformed $A \simeq 100$ Nuclei
By F.K. Wohn, J.C. Hill, R.F. Petry, and R.L. Gill (With 3 Figures)   343

The Nuclear Structure of ^{101,103}Nb
By T. Seo, R.A. Meyer, and K. Sistemich (With 4 Figures) 349

Part VII   Proton-Neutron Interactions and Pairing Reductions

Influence of the Proton-Neutron Interaction in the $A \simeq 100$ Region
By P. Federman (With 1 Figure) 357

The Proton-Neutron Interaction in Neutron-Rich $A \simeq 100$ Nuclei
By B. Pfeiffer and K.-L. Kratz (With 3 Figures) 368

The P-Factor and Atomic Mass Systematics: Application to Medium
Mass Nuclei
By D.S. Brenner, P.E. Haustein, and R.F. Casten (With 5 Figures) . 376

The Spin Dipole Interaction in Odd-Odd Nuclides
By W.B. Walters (With 1 Figure) 382

Part VIII   Beta Decay at A~100

Fast First-Forbidden Transitions and Subshell Closures in the Region
of ^{96}Zr
By H. Mach, E.K. Warburton, R.L. Gill, R.F. Casten, A. Wolf,
Z. Berant, J.A. Winger, K. Sistemich, G. Molnár, and S.W. Yates
(With 1 Figure) 391

Structure of 1$^+$ States in Neutron Deficient Odd-Odd Indium
Isotopes
By L. Kalinowski (With 1 Figure) 397

Gamow-Teller Beta Decay of Even Nuclides Near ^{100}Sn
By R. Barden, H. Gabelmann, I.S. Grant, R. Kirchner, O. Klepper,
G. Nyman, A. Plochocki, G.-E. Rathke, E. Roeckl, R. Rykaczewski,
D. Schardt, J. Szerypo, and J. Zylicz (With 4 Figures) 403

Gamow-Teller Beta Decay of Neutron Rich Tc, Ru, Rh and Pd Isotopes
By J. Äystö, P. Jauho, V. Koponen, H. Penttilä, K. Rykaczewski,
P. Taskinen, and J. Zylicz (With 5 Figures) 409

Giant Gamow-Teller Excitations in ^{104}Cd
By M. Huyse, V.R. Bom, P. Dendooven, R.W. Hollander,
P. Van Duppen, and J. Vanhorenbeeck (With 3 Figures) 415

Subject Index 421

Index of Contributors 423

Part I

Introduction

Nuclear Structure in the Zr Region: Some Introductory Remarks

K. Heyde

Institute for Nuclear Physics, Proeftuinstraat 86,
B-9000 Gent, Belgium

The study of the A≈100 mass region, or better of the Zr region, is a very specific choice that was made by the organisers. Therefore, all of you present here are specialists in one or another aspect of this interesting mass region and will talk and discuss about it. In that respect, my contribution will be more of setting the scene, giving a number of introductory remarks and some specific thoughts of mine that are definitely somewhat biased. I would like to concentrate on the present state-of-the art of our understanding concerning the nuclei in the Zr region and point out a number of severe problems that are still around. Problems that hopefully will get an answer by the next time we could meet around such a topical mass region workshop.

I have taken some freedom in presenting these things. I could have given you just a list of one or two transparancies with the status and open problems and leave you with that: I expect that this is not what the organisers have ment. I will therefore try to bring our present day of understanding together in some coherent framework.

The mass region around Zr(Sr,Mo,...), starting from the double subshell closed nucleus ^{76}Sr up to the neutron rich side at ^{96}Sr, ^{98}Zr, ^{100}Mo, ... where clear indications of strong prolate deformed shapes do show up, is a rich region where nuclear excitations are observed that otherwise, for the strongly deformed rare-earth region, need a much larger mass region to be studied. This concentration of rapid evolving shapes and nuclear excitations make the Zr region intruiging but also very difficult to comprehend in a coherent way. I do not mean using a different model for every single nucleus around. In this region, the interplay of single-particle and collective excitations is sometimes simple (strongly deformed bands in odd-A nuclei), sometimes highly complicated (the odd-mass and even-even nuclei in the light N=Z region around mass A=80).

Because of my education at the Utrecht shell-model school, I still feel that the nuclear shell-model is one of the best starting points from which to start in describing nuclear excitations. Thereby one remains on firm ground and can calculate nuclear properties in a well-defined framework. If we take the best inert core around, then the Z=50,N=50 nucleus is probably the best choice, clearly much better than the Z=38,N=50 nucleus is.

Basically, Hartree-Fock theory gives a prescription on how to calculate the single-particle energies and the mean field starting from a given two-body nucleon-nucleon interaction. Also, the variation for the above quantities with mass number can be obtained. Since the Hartree-Fock single-particle energy can be split into a part coming from filled orbitals and from partially filled orbitals β, one can split the energy into two parts

$$\tilde{\epsilon}_\alpha = \epsilon_\alpha + \mu_\alpha \quad , \tag{1}$$

with

$$\mu_\alpha = \sum_\beta <\alpha\beta|V|\alpha\beta> v_\beta^2 \tag{2}$$

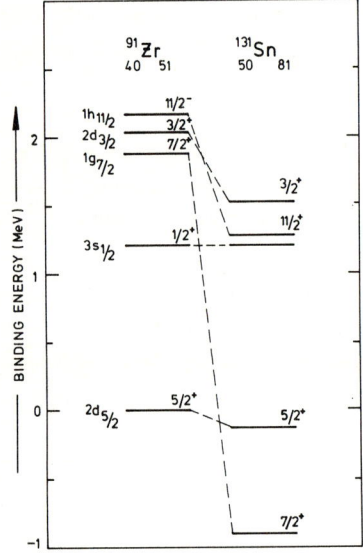

Fig.1 Variation of the neutron single-particle energies in the $50 \leq N \leq 82$ shell, as a function of the filling of the proton $1g_{9/2}$ orbital between Z=40 and Z=50. The $1/2^+$ state is taken as a reference state.

If the ϵ_α are determined for a given core, then the μ_α give the variation over a number of isotones(for ϵ_ν) of for a number of isotopes (for ϵ_π). A typical illustration is given in fig.1 where the variation for the neutron single-particle energy between (Z=40,N=51) and (Z=50,N=81) is given [1]. Because of the filling of the proton $1g_{9/2}$ proton orbital, the $1g_{7/2}$ neutron orbital gains a lot of extra binding energy via the $\mu_{1g7/2}$ term ,relative to the $2d_{5/2}$, $2d_{3/2}$, $3s_{1/2}$ and $1h_{11/2}$ orbitals. This specific idea of radial overlap between the proton and neutron orbitals (which as you now all see,is coming from Hartree-Fock theory , the basics of the nuclear shell-model) has been emphasized very much over the years in this particular mass region, first by de-Shalit and Goldhaber [2], by Federman and Pittel [3-7] and later by many others. Since the particular single-particle spacings do determine where gaps (major of subshell gaps) in the single-particle spectrum appear, this idea of proton-neutron residual interaction is playing a central role in order to determine the conditions for which typical shell-model excitations or aspects of nuclear deformation will dominate the nuclear structure at low excitation energy. Of course, there is a mutual interplay so that the filling of the proton $1g_{9/2}$ orbital modifies the neutron single-particle spectrum and filling these neutron orbitals will again modify the proton $1g_{9/2}$- $2p_{1/2}$ energy separation. This mutual influence has to be considered in a self-consistent way.

The shell-model configuration space can now become prohibitively large when using the Z=50,N=50 closed shell nucleus as a an inert core in order to study the region of Zr nuclei. Therefore, in many cases, the ^{88}Sr core (Z=38,N=50) has been used as a closed core [5,8] to start shell-model calculations. This choice is not very good since particle-hole excitations across the Z=38 shell do occur in nearby nuclei at rather low energy ($2^+,3^-,4^-,5^-$) because of the $2p_{3/2}^{-1}$-$1g_{9/2}$, $2p_{3/2}^{-1}$-$2p_{1/2}$, $1f_{5/2}^{-1}$-$2p_{1/2}$ excitations. Still, a number of interesting results, in particular for the low-lying excited 0^+ states were obtained signaling that states with proton excitations into the $1g_{9/2}$ shell and neutron excitations into the $1g_{7/2}$ shell can give a large binding energy gain through the proton-neutron interactions [5]. So, we come into a picture where exciting particles into higher-lying orbitals, thereby partially reducing the pairing correlations, can still result into a large gain in energy through the residual proton-neutron interaction energy, in particular when the radial overlap between the individual proton and neutron orbitals is maximal [9,10]. This picture of **loss-versus-gain** of energy has come to play a major role in our present day understanding of the nuclear structure around mass A≈100 for the Sr,Zr,Mo,.. nuclei. It is expressed e.g. for α-like correlations [11,12] through the energy for such α-like particle hole excitations given as

$$E_x(\alpha) = 2BE(A,Z) - BE(A-4,Z-2) - BE(A+4,Z+2) . \qquad (3)$$

A contour plot of this quantity is given in figure 2 where indeed, near Z=40,N=56 and Z=42,N=58 very low values of $E_x(\alpha)$ occur. The more specific variation for the Zr nuclei, where also proton 2p-2h and neutron 2p-2h unperturbed excitation energies are given is presented in fig.3. Here, one observes that near the double subshell closure, the α-like configuration comes lowest in energy [11]. Recently, evidence for such excitations was discussed in ^{96}Zr by Molnar et al. [13-15],

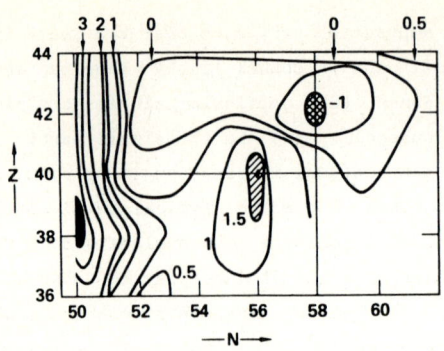

Fig.2 Alpha-correlation energy $E_x(\alpha)$ (according to eq.3) for nuclei with $49<N<63$ and $36 \leq Z \leq 44$.

Mach et al. [16] and in ^{98}Zr by Meyer et al. [12] using various methods (β-decay,...). Excited 0^+ states are also strongly excited in α-transfer reactions e.g. (d,^{6}Li) [17,18]. A more detailed mapping of not only these 4p-4h states, but also of the other 2p-2h excitations would be most interesting in order to find out if these ideas describe the actual nuclear structure phenomena rather well.

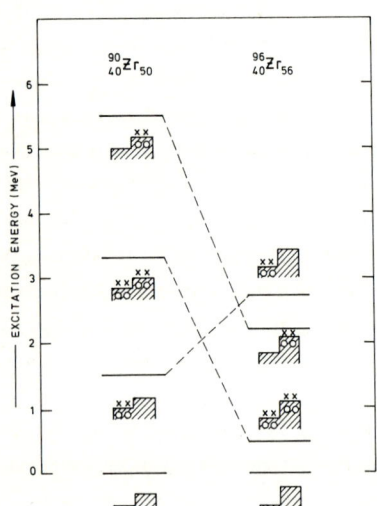

Fig.3 Energy systematics of the proton- and neutron pairing and α- vibrational modes in ^{90}Zr and ^{96}Zr. The energies were calculated according to the expressions
$E_{2p-2h}=BE(^{A-2}Zr)+BE(^{A+2}Zr)-2BE(^AZr)$ (neutrons)
and $BE(^{A-2}Sr)+BE(^{A+2}Mo)-2BE(^AZr)+\Delta E$ (protons)
and eq.(3), also corrected for the Coulomb energy. Here, $\Delta E \simeq 4\Delta E_c$ with ΔE_c the attractive energy arising from the proton particle-hole attraction energy (see ref.11).

If we would allow the spherical field to deform, this picture of large binding energy for optimal overlapping orbitals becomes also well established but expressed in a slightly different language. The binding energy resulting through the proton-neutron matrix element for given deformed orbitals $|\Omega_p>$, $|\Omega_n>$, using a quadrupole-quadrupole force, becomes

$$<\Omega_p \Omega_n | -\kappa Q_p \cdot Q_n | \Omega_p \Omega_n> = -\kappa <\Omega_p|Q_p|\Omega_p> \cdot <\Omega_n|Q_n|\Omega_n> , \qquad (4)$$

where the quadrupole single-particle matrix elements are given by the intrinsic quadrupole moments for the particular orbitals. This product is maximal if both the proton and neutron are moving in orbitals which are both down-sloping or

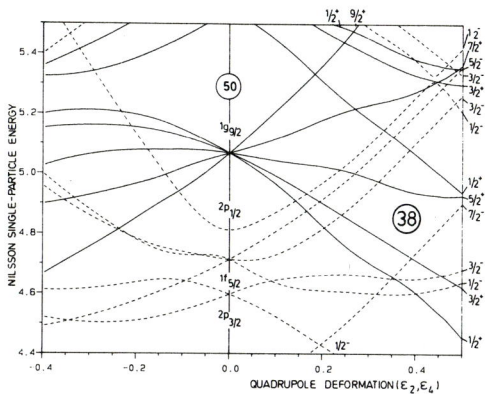

Fig.4.a Part of the proton Nilsson single-particle spectrum for ^{80}Zr, as a function of the quadrupole deformation. On the horizontal axis, we, however, indicate (ϵ_2,ϵ_4) since we have calculated along a line, minimizing the total potential energy surface in the hexadecapole deformation ϵ_4. The parametes of the Nilsson model are given in ref.29. Moreover, the prolate gap at $Z=38$ is also indicated. The single-particle levels are identified with Ω^π.

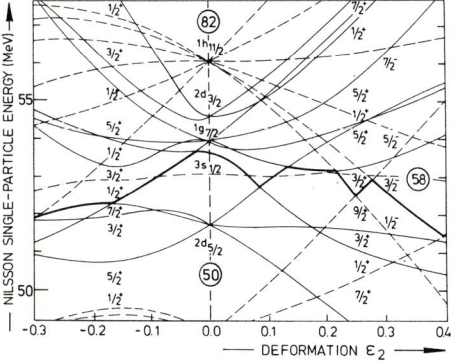

Fig.4.b Part of the neutron Nilsson single-particle spectrum for ^{98}Zr (using the parameters $\kappa_n=0.060, \mu_n=0.22$ for $N=4$ orbitals and $\kappa_n=0.066, \mu_n=0.35$ for $N=5$ orbitals). The thick line gives the Fermi level for a sharp distribution at $N=58$. The single-particle levels are indentified with Ω^π. The prolate gap at $N=58$ is also indicated.

up-sloping and moreover result in nearby Ω- values [19]. This is expressed for the A≈100 Zr region in figure 4 where exciting protons into the down-sloping $\Omega=1/2^+$ ($1g_{9/2}$) and neutrons in the down-sloping $\Omega=1/2^+, 3/2^+$ ($1g_{7/2}$) and $\Omega=1/2^-, 3/2^-$ ($1h_{11/2}$) orbitals gives a large binding energy. Moreover, a clear picture can be formed of this type of proton-neutron interaction in a strong-coupling scheme. In this figure 4, besides excerpts of the proton and neutron Nilsson single-particle spectra, the intrinsic quadrupole moments for

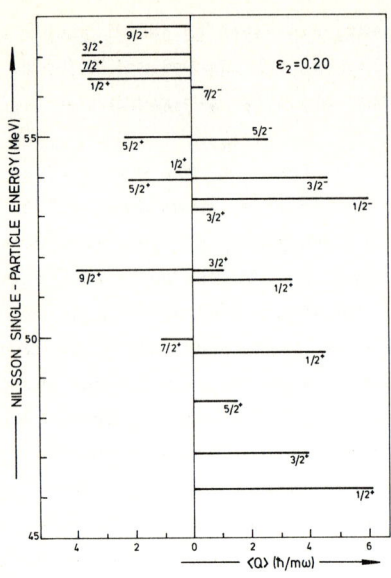

Fig.4.c The neutron intrinsic quadrupole moments $\langle\Omega|\sqrt{16\pi/5}\cdot Y_{2,0}|\Omega\rangle$ occuring in eq.4, for the orbitals corresponding to ^{98}Zr of fig.4.b at $\epsilon_2=0.20$. The intrinsic quadrupole moments for the lowest neutron $1/2^+, 3/2^+, 5/2^+, 7/2^+, 9/2^+$ levels are very similar to the proton intrinsic quadrupole moments for the analogous levels in fig.4.a at the same deformation.

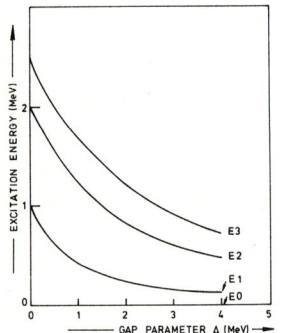

Fig.5 The one-quasi particle energies, $E_i=[(\epsilon_i-\lambda)^2+\Delta^2]^{1/2}$ for a set of levels $\epsilon_0(=\lambda), \epsilon_1=1$ MeV, $\epsilon_2=2$ MeV, $\epsilon_3=2.5$ MeV, as a function of the gap parameter Δ. The one quasi-particle energies are drawn relative to the lowest level for which we take $\epsilon_0=\lambda$, independent of the value of Δ.

excitations at deformation $\epsilon_2=0.20$ are given. I will come back to this point of deformation later.

The question of pairing reduction when forming particle-hole excited states has been coming up as a point of debate in the Zr,Sr,... region, recently [20,21]. If the protons and neutrons can profit of going into largely overlapping orbitals, pairing has to be reduced somehow. This reduction should most clearly show up in the density of excited states in the spectra of odd-mass and doubly-odd nuclei, in particular for the single-particle excitations. This point has definitely not been solved to a good end as yet and we will hear more about this during the workshop. I just like to mention that in BCS calculations, a reduction of pairing means going closer to the unperturbed single-particle spectrum without interactions and thus to a spectrum with a decreasing level density as illustrated in figure 5 for a simple case.

Experimental evidence for the border regions of major shells has abundantly been given. For subshells, on the contrary, as shown before, a strong dependence on mass number shows up. These subshells can be mapped by measuring e.g. S_{2n}, S_{2p} values. Suppose we have a somewhat idealized situation of a closed

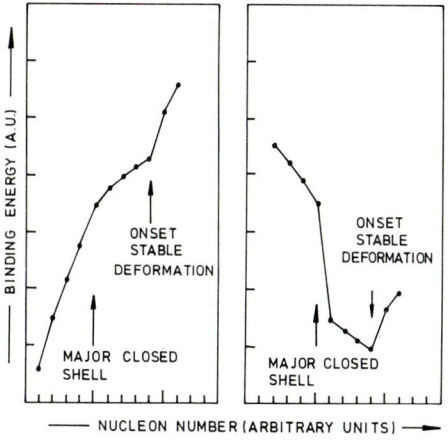

Fig.6 Schematic figure of the variation (in arbitrary units) for the binding energy and the related S_{2n} values (also in arbitrary units) when (i) a major closed shell is passed, (ii) a small shell (e.g. the $2d_{5/2}$ orbital) is filled afterwards, and (iii) stable, deformed shapes set in.

shell (take N=50) with the $2d_{5/2}$ orbital to be filled above this shell closure (see fig.6). The variations one could expect on the basis of binding energy, and thus also in the S_{2n} values, reflects in a very detailed way the values of gaps and moreover how gaps move with changing proton and neutron number. Also, regions where due to the onset of stable deformed (quadrupole, octupole?,..)shapes a strong increase in binding energy and thus of a change in slope of the S_{2n}, S_{2p} curves results, can easily be observed and followed as a function of Z and N

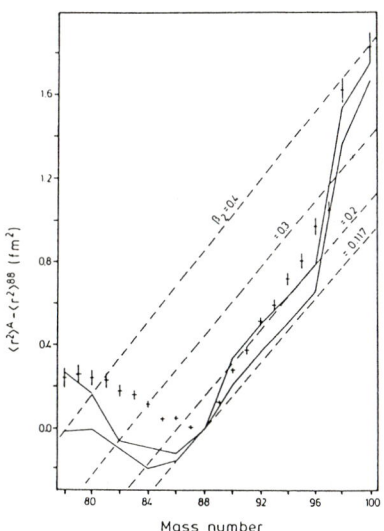

Fig.7 Variation of the nuclear radius for the Sr isotopes. Dashed lines correspond with fixed deformation in the droplet model. The full lines correspond to the droplet model, but using β values deduced from B(E2) rates.

(see [22]). A few comments are in place here: any subshell closure will induce a change in the BE(A,Z) surface and thus in the $S_{2n}(A,Z)$ surface. Just above a closed shell, a valence pair is less bound compared with the closed shell itself and a subsequent drop in S_{2n} will occur.

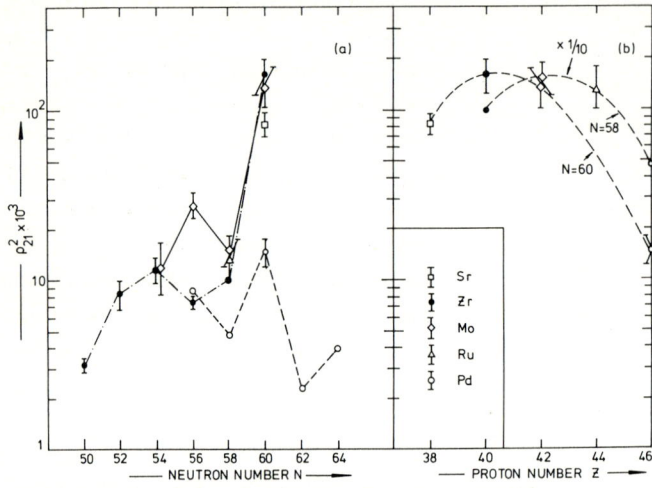

Fig.8 (a) Detailed results for ρ_{21}^2 in the mass A≈100 region. The data are taken from references discussed in [26] and (b) the monopole strength for the N=58 and N=60 isotones. The values for the N=58 isotones have been multiplied by a factor of 10 for comparison.

Another type of experiments that give unambigous evidence for the nuclear shapes and shape changes is the measurement of the isotopic shift $\Delta\langle r^2\rangle$. This quantity has been measured over large sequences of isotones of Sr more recently. Thereby, the onset of deformation, or better, a large increase in the nuclear radius is observed near N=38 and near N=58. I refer to the results obtained by Eastham et al. [23] and Buchinger et al. [24] as a very nice example of what actually happpens to nuclei in their ground state and what can be reached experimentally (fig.7). We hope for more such measurements in the future.

A most peculiar behaviour in the E0 transition rates has been observed for transitions connecting the lowest excited 0^+ to the ground state(see fig.8). These measurements give an independent indication for the existence of low-lying particle-hole configurations and on their mixing with the ground state itself. The expectation value of the E0 operator, within a given state, gives a measure of the nuclear radius. Since the wave functions, corresponding to 0^+ states with largely different deformation are poorly overlapping, only that overlap region is relevant in determining the E0 matrix element. In this light, the very large ρ^2 values near N=60, the specific neutron number at which the excited 0^+ states come lowest in energy where unexpected at first [25,26]. The correlation between the ρ^2 maximum and the $E_x(0_2^+)$ minimum, however, is too explicit to be without meaning: we still have to interpret these very interesting results in a correct way. More systematic E0 measurements all through the Sr,Zr region are definitely needed but also, difficult to be carried out.

In the above discussion, we have emphasized that due to the rapid variations in the nuclear shell-model single-particle energies, shell-model calculations have a hard time in order to cover the many properties of nuclei in the Zr,Sr,... A≈100 region. Near N=58 deformation sets in and also in the Z≈N region of very neutron deficient nuclei, very large deformation effects have been observed [27-29]. In fig.9, we show as a general overview, a contour plot of the results obtained by Møller and Nix [30] concerning the ground state quadrupole moments, given by

$$Q_2 = 2 \left(\frac{3Z}{4\pi r_0^3 A} \right) \int r^2 P_2(\cos\theta) d\vec{r} \quad , \tag{5}$$

with r_0=1.6fm and contour lines (obtained from ref.30 and using an interpolation program)with a distance of 0.5 eb. between adjacent lines. The black region corresponds to Q_2>3.0 eb. In the two extreme regions with many valence nucleons outside of the closed Z=50,N=50 core, the spherical shell-model is definitely not

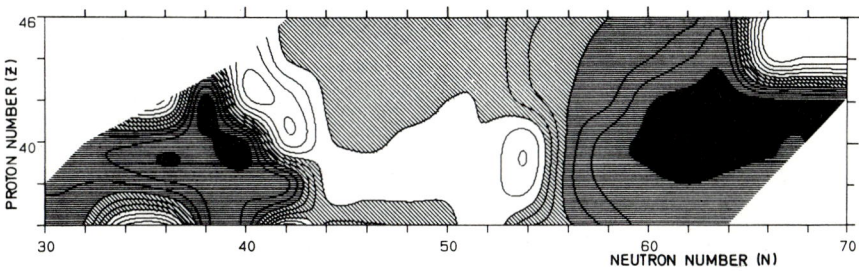

Fig.9 Contour map, indicating the ground state quadrupole moments, defined in eq.5 for the region 36≤Z≤46 and 30≤N≤70. The calculations are from ref.30.
The contour lines are explained in the text.

the best approach any longer. Deformed shapes and the corresponding deformed potential offer a more appropriate way of expressing essentially the same results, but using a better zero-order Hamiltonian.

The best way to start will again be, as was used for the spherical shell-model,to determine the optimal average field from a given nucleon-nucleon force as a function of a number of constraining operators: axial quadrupole deformation $<Q_{2,0}>$, axial octupole deformation $<Q_{3,0}>$, triaxial quadrupole deformation $<Q_{2,\pm2}>$,.. For deformed nuclei ,Bonche et al. [31] have made efforts to solve the Hartree-Fock equations immediately in coordinate space.These self-consistent calculations have been performed using the Skyrme interaction SIII ,already successfully used in order to describe properties of nuclei in several mass regions of the periodic table. I just show some results that were obtained for the neutron rich Kr,Sr,Zr,Mo nuclei with deformation energy surfaces in the (β,γ) plane ,showing large, prolate deformation (see ref.30). I hope we will hear more about the Hartree-Fock calculations , not just for the ground-state properties but also for higher-lying excited and eventually high-spin states. Of course, aspects of nuclear deformed shapes have also been studied along other ways, using various degrees of sophistication: the Nilsson model, Woods-Saxon field, folded Yukawa,.. [28-31]. Here too, many new results were recently obtained. I hope we will learn about them in the workshop and have discussion on questions like: does superdeformation (using the more recent definition of "super") also occur in the lighter $Z \approx N$ $A \approx 76$ regions.

Of course, there is still a way to go ,even for the Hartree-Fock calculations in order to supply not just the potential energy surfaces but also the kinetic energy surfaces and thus to be able of constructing a microscopic Bohr-Hamiltonian and its solutions. Going beyond the static part of the Hartree-Fock could also be pursued by constructing besides the ground-state Slater determinant wave function Φ, the low-lying excited Slater determinants $\Phi',\Phi'',..$ that can span a basis for a multi-configurational calculation. It is in precisely this way that the MONSTER,VAMPIR (and what more to come!) calculations from Tubingen have to be situated [32].

It is not the place here ,just at the beginning of this workshop to start already summarizing. I hope that the present introductory remarks have opened for the different roads to be taken in order to address both experimental and theoretical topics for the days to come but in a lot of more detail. I expect that we still can keep on a general track to try to get convergence of ideas and prospects by the end of the workshop when somebody else is going to take over my task by summarizing what has been said and accomplished.

The author is particularly indebted to P.von Brentano,P.Federman, W.Gelletly,M.Huyse,K.L.Kratz,R.A.Meyer,G.Molnar,V.Paar,K.Sistemich, D.Warner and J.Wood for many discussions on theoretical and experimental aspects of this Zr mass region. Part of this work was carried out with support from the IIKW, the NFWO and a research grant NATO 0452/86

References

1. J.L.Wood, private communication
2. A.de-Shalit and M.Goldhaber, Phys.Rev.C92(1953),1211
3. P.Federman and S.Pittel, Phys.Lett.69B(1977),385
4. P.Federman and S.Pittel, Phys.Lett.77B(1978),29
5. P.Federman and S.Pittel, Phys.Rev.C20(1978),820
6. P.Federman,S.Pittel and R.Campos, Phys.Lett.82B(1979),9
7. P.Federman,S.Pittel and A.Etchegoyen, Phys.Lett.140B(1984),269
8. J.Blomqvist and L.Rydstrom, Phys.Scripta 31(1985),31
9. K.Heyde et al.,Nucl.Phys. A466(1987),189
10. K.Heyde,P.Van Isacker,M.Waroquier,J.L.Wood and R.A.Meyer, Phys.Reports 102(1983),291
11. F.Catara,L.Ferreira, A.Insola,A.Vitturi and R.A.Broglia, Nucl.Phys. A372(1981),237
12. R.A.Meyer,E.A.Henry,L.G.Mann and K.Heyde, Phys.Lett. 177B(1986),271
13. G.Molnar et al.in : Symmetries and Nuclear Structure, eds. R.A.Meyer and V.Paar (Harwood Acad.Publ., 1987), p 237
14. G.Molnar et al.,preprint
15. G.Molnar,S.W.Yates and R.A.Meyer, Phys.Rev.C33(1986),1843
16. H.Mach et al., Phys.Rev.C37(1988),254
17. A.Saha,G.D.Jones,L.W.Put and R.H.Siemssen, Phys.Lett. 82B(1979),208
18. A.M.van den Berg and R.H.Siemssen,Il Nuovo Cim. 81A(1984),318
    A.M.van den Berg, Ph.D. Thesis, University of Groningen,1983, unpubl.
19. R.F.Casten,D.D.Warner,D.S.Brenner and R.L.Gill, Phys.Rev.Lett.47(1981), 1443
20. B.Pfeiffer et al.,in: Nuclei far from Stability,AIP Conference Proceedings 164,ed. I.S.Towner(AIP,New-York,1988),403
21. L.K.Peker,J.H.Hamilton and P.G.Hansen, Phys.Lett.167B(1986),203
22. M.Graefenstedt et al., Z.Phys.A327(1987),383 and refs. therein
23. D.A.Eastham et al.,Phys.Rev.C36(1987),1583
24. F.Buchinger et al.,in:Nuclei far from Stability,AIP Conference Proceedings 164,ed. I.S.Towner(AIP,New-York,1988),197 and P.Lievens, L.Vermeeren and R.E.Silverans, in : Progress Report 1986/1987.IKS, Leuven; pg.13.
25. R.J.Estep et al., Phys.Rev.C35(1987),1485
26 K.Heyde and R.A.Meyer, Phys.Rev.,to be publ.
27. R.Bengtson,P.Møller,J.R.Nix and Jing-ye Zhang,Phys.Scripta 29(1984),402
28. I.Ragnarsson and R.K.Sheline, Phys.Scripta 29(1984),385
29. K.Heyde,J.Moreau and M.Waroquier, Phys.Rev.C29(1984),1859
30. P.Moller and J.R.Nix, At.Nucl.Data Tables 26(1981),165
31. P.Bonche et al., Nucl.Phys.A443(1985),39
32. K.W.Schmid and F.Grummer, Repts. on Progr. in Physics, 50(1987),731

Part II

Shape Transitions and Shape Coexistence

Some Results from Potential Energy Surface Calculations for Nuclei in the Mass 70–80 Region

R. Bengtsson

Dept. of Mathematical Physics, Lund Inst. of Technology,
P.O. Box 118, S-22100 Lund, Sweden

Abstract: Low spin properties of even-even nuclei in the A=70-80 mass region are interpreted in terms of shape coexistence and shape changes.

Near the ground state nuclei in the A=70-80 mass region show very irregular features. This is illustrated in fig. 1, which shows the energy of lowest 2^+ state in the even-even nuclei. If instead the $6^+ \to 4^+$ transition energy is considered a much more regular pattern appears, as shown in fig. 2. There the equivalent 2^+ energy is calculated as $E'_{2^+} = 3[E(6^+) - E(4^+)]/11$, which means that we have assumed that the 4^+ and 6^+ states are members of a regular rotational band with a constant moment of inertia. The energies used to construct fig. 2 decrease in a relatively smooth way towards the center of the deformed region in a way similar to the deformed rare earth region, where, however, also the 2^+ energy itself shows this regular pattern. It should also be observed that the energies in fig. 2 are considerably smaller than those in fig. 1, indicating that the moment of inertia associated with the $6^+ \to 4^+$ transition is much larger than that associated with the $2^+ \to 0^+$ transition.

Theoretically calculated potential energy surfaces (using the Woods-Saxon potential and monopole pairing[1,2]) give a clue as to how the data presented in figs. 1 and 2 should be interpreted. The energy surfaces show a great variety of different patterns as illustrated in fig. 3. They may have a single well-developed minimum at prolate, oblate or triaxial shape, they may have coexisting prolate and oblate or sometimes spherical minima or they may be very soft with no well-defined minimum. Due to the great complexity it is hard to summarize the information contained in the energy surfaces in a condensed way, although an attempt is made in fig. 4. In a first approximation four groups of nuclei can be distinguished: i) Nuclei with a large prolate ground state deformation centered around N=Z=38-40. These particle numbers correspond to gaps in the single-particle energy spectrum at large prolate shapes, cf. fig. 5. ii) Nuclei with a large oblate deformation centered around N=Z=34-36. Gaps in the single-particle spectrum appear at oblate shape for these particle numbers. iii) Nuclei with spherical or near spherical shape

---

\# The results reported here summarize parts of a longer paper, presently being prepared for publication in collaboration with W. Nazarewicz.

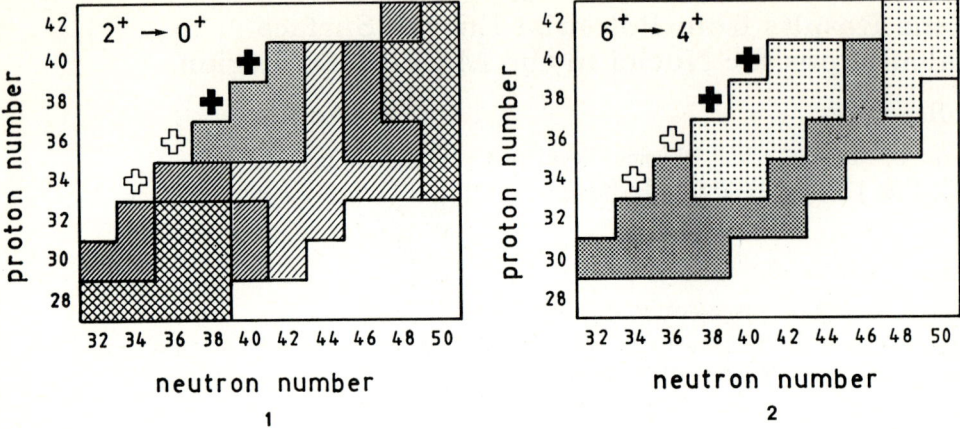

Figs. 1 and 2 The energy of the experimental 2^+ state (1), and the equivalent 2^+ energy derived from the experimental 4^+ and 6^+ states (see the text for details). The various shadings indicate the 2^+ energy: Sparsely dotted if $E(2^+) < 0.25$ MeV, densely dotted if 0.25 MeV $\leqslant E(2^+) < 0.50$ MeV, sparsely lined if 0.50 MeV $\leqslant E(2^+) < 0.75$ MeV, densely lined if 0.75 MeV $\leqslant E(2^+) < 1.00$ MeV and cross-hatched if $E(2^+) \geqslant 1.00$ MeV.

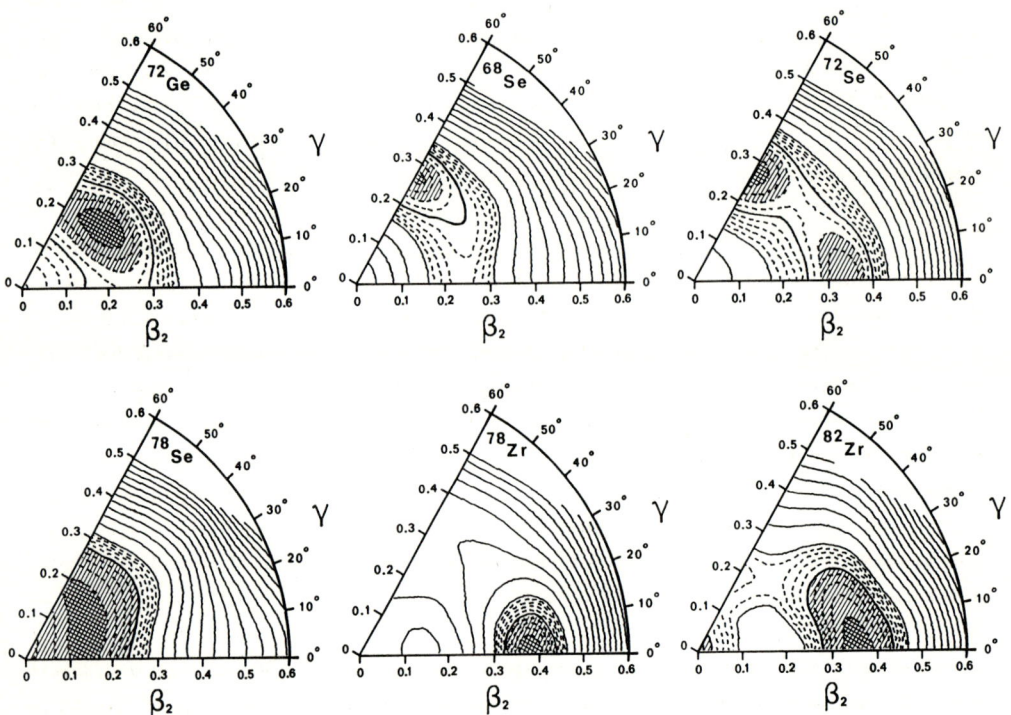

Fig. 3 Selected examples of calculated ground state potential energy surfaces. The separation between contour lines is 0.1 MeV. Minima are indicated by shading.

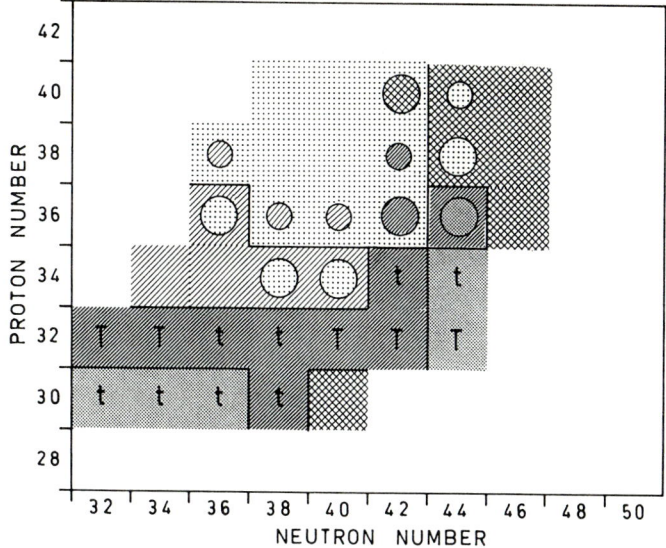

Fig. 4 A summary of the deformations calculated for minima in the ground state potential energy surfaces: sparsely dotted regions correspond to prolate nuclei with $\beta_2 > 0.25$, densely dotted to prolate nuclei with $0.10 \leq \beta_2 \leq 0.25$, sparsely lined to oblate nuclei with $|\beta_2| > 0.25$, densely lined to oblate nuclei with $0.10 \leq |\beta_2| \leq 0.25$ and cross-hatched to spherical or near sphericl nuclei with $|\beta_2| < 0.10$. The corresponding information for coexisting excited minima is given inside the circles. A large circle is used if the excitation energy is less than 0.5 MeV and a small circle if it is between 0.5 and 1.0 MeV. A t is used to indicate γ-softness (the oblate-prolate energy difference is then smaller than 0.5 MeV). A T indicates triaxial deformation ($15° \leq \gamma \leq 45°$). In such cases the shading shows whether the minimum is closer to the oblate or prolate axis.

which appear when N approaches the magic number 50. The tendency towards sphericity is most pronounced for Z=40, which can be considered as a semi-magic closed shell at spherical shape, cf. fig. 5. iv) Triaxially deformed or very γ-soft nuclei with in most cases moderate β_2-deformations. This group of nuclei mainly contain the Z=30 and Z=32 isotopes. The former have a slight preference for prolate shapes while the latter have a similar preference for oblate shapes. Another interesting feature of fig. 4 is that almost all the nuclei along the border line of the region of nuclei with a large prolate deformation have coexisting minima in the potential energy surface.

Nuclei with complex potential energy surfaces may respond to a rotation in a number of different ways. However, in most cases the behaviour can quite easily be predicted. The main rule is that the nucleus with increasing angular momentum tries to exploit deformations associated with higher moment of inertia, since this will lower the rotational energy. The expected behaviour is illustrated for a few typical situations in fig. 6. Corresponding examples from real nuclei are shown in fig. 7.

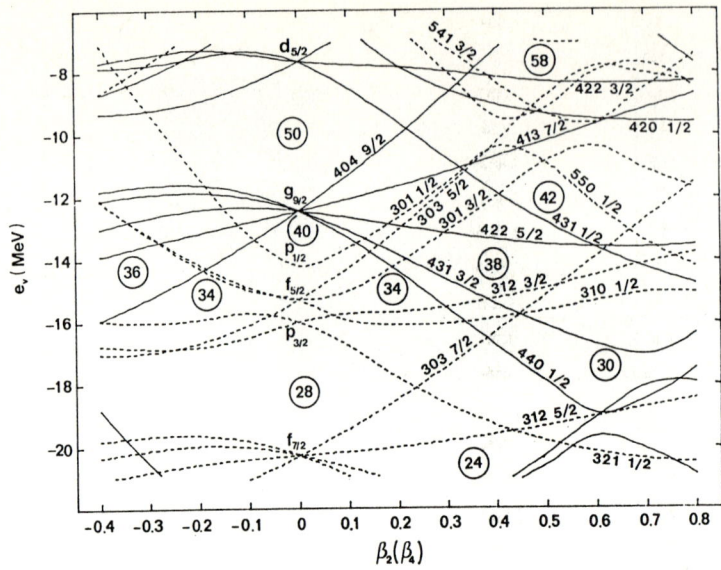

Fig. 5 Neutron single-particle levels in the Woods-Saxon potential for ^{80}Sr as functions of β_2. For each β_2 the value of β_4 that minimizes the liquid drop energy has been used. The proton levels are very similar.

In fig. 6 P denotes a large deformation, which in realistic cases usually is prolate, while O denotes a smaller deformation, which could be oblate, prolate or spherical depending on what nucleus that is considered. If the deformation is very stable a nearly perfect rotational band is expected (fig. 6b). In real nuclei this situation is rare, although ^{102}Zr gives a good example (fig. 7b). Much more common is that the deformation increases slightly with increasing angular momentum (fig. 6e). Simultaneously γ usually changes slightly in direction towards -30°, since this is the value of γ that gives the largest moment of inertia (for a given value of β_2). The result of a deformation change of this type is that the 0^+ state gets a lower energy than could be expected from the 6^+ and 4^+ states assuming a constant moment of inertia (cf. the full and the dashed lines in fig. 6e). ^{78}Sr is a good example of a nucleus of this kind. In softer nuclei the deformation change may be much larger and the influence on the ground band energies so strong that this band no longer looks like a typical rotational band (fig. 6d). A ground band spectrum showing these features is found e.g. in ^{80}Kr.

In nuclei with shape coexisting minima, each one associated with a rotational band, two different situations can be distinguished. If the ground state minimum has a smaller deformation than the excited minimum, the ground state band will be crossed by the band built on the excited minimum as shown in fig. 6c. Several examples of crossings between shape coexisting rotational bands of this kind are known, one example being ^{72}Se. If the ground state minimum has a larger deformation than the excited minimum the corresponding rotational bands will not cross each

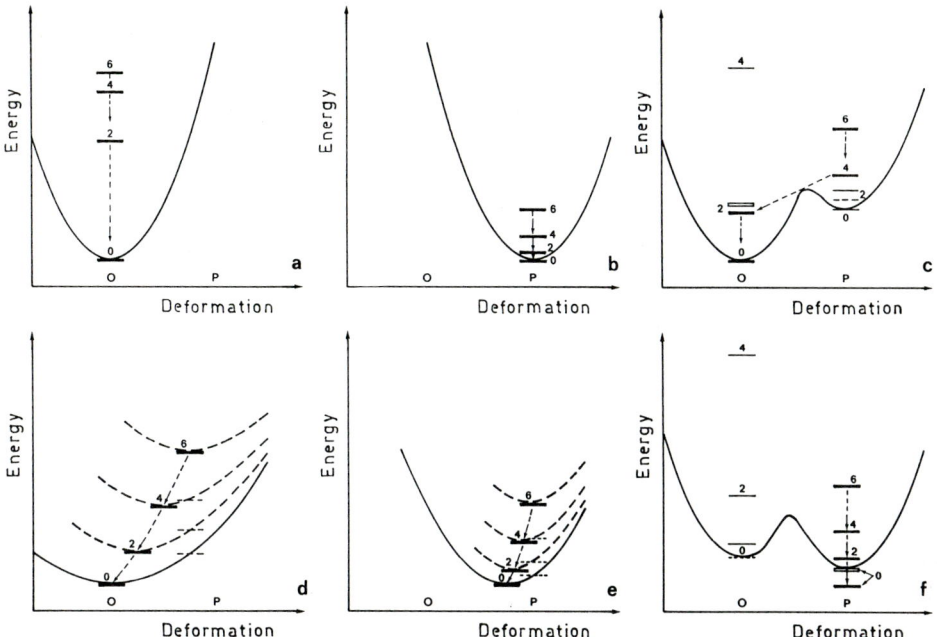

Fig. 6 A schematic illustration of different mechanisms which may cause deviations from a pure rotational spectrum. Potential energy curves are drawn with full lines(0^+ state) or dashed lines (excited states). In the excitation spectra thick lines are used for yrast states and thin lines for excited states. These may be perturbed near a bandcrossing. The unperturbed states are in such cases shown as open thick lines and dashed lines respectively (in c and f). In d and e the dashed spectrum is the one to be expected if the deformation was constant, equal to that of the I=6 state.

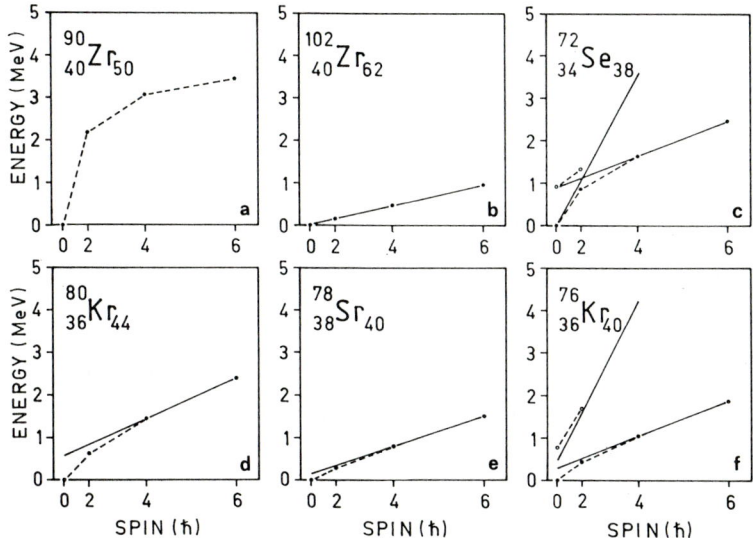

Fig. 7 Examples of experimental spectra, which can be related to the various situation, which were schematically illustrated in fig. 6.

other. However, if the energy of the two minima is very similar it is still possible that the ground state (0^+) and possibly also the 2^+ state of the ground band may be pushed down in energy due to a residual interaction between the two bands, cf. fig. 6f. The probably best example of this kind is ^{76}Kr (fig. 7f). As can be easily seen from fig. 7 it may in practice be very hard to distinguish the situations illustrated in figs. 6c and 6d or those in 6e and 6f, in particular if the knowledge of the excited 0^+ and 2^+ states is incomplete.

In the considered mass region there are also a number of spherical or near spherical nuclei centered around the semi-magic nucleus ^{90}Zr, which do not have any rotational band structures (cf. figs. 6a and 7a).

It should also be mentioned that many of the γ-soft nuclei with Z=30 and 32 develop separated shape coexisting minima already at quite low rotational frequencies. In most cases the coexisting states have configurations with different numbers of $g_{9/2}$ intruder orbitals occupied and can therefore easily be distinguished. Structural differences of this kind are in fact present also in the ground state potential energy surface, although no distinguishable minima can be seen, since the separating barrier has been washed out by the pairing interaction. (For a more detailed discussion of this topic, see ref[3].)

All the deformation effects discussed in connection to figs. 6 and 7 result in a lowering of the ground state (0^+) energy relative to what could be expected from the 6^+ and 4^+ states assuming a rotational band with a constant moment of inertia. The lowering can be calculated from the experimental energies as

$$\delta E_{0^+} = [21E(4^+) - 10E(6^+)]/11 - E(0^+)$$

A small value obviously implies that the nucleus is a good rotor while a large value can be given different explanations, which is evident from fig. 6.

The experimental values of δE_{0^+} are sumarized in fig. 8. Combined with theoretical calculations (cf. figs. 3 and 4) and additional experimental information it is possible to make a rough characterization of the yrast bands of even-even nuclei in the A=70-80 mass region at low spins. Nuclei with Z≥36 and N≤42 are the best rotors. They are all expected to have a prolate shape. The nuclei near ^{90}Zr with the largest values of δE_{0^+} are spherical or near spherical and should not be interpreted in terms of rotational band structures. For the remaining nuclei in this mass region which have intermediate values of δE_{0^+} alternative interpretations are possible. However, for the N=38 and 40 isotones of the elements with Z≤34 shape coexisting 0^+ states are experimentally well established and the low spin part of the yrast line can be understood in terms of a crossing between two shape coexisting

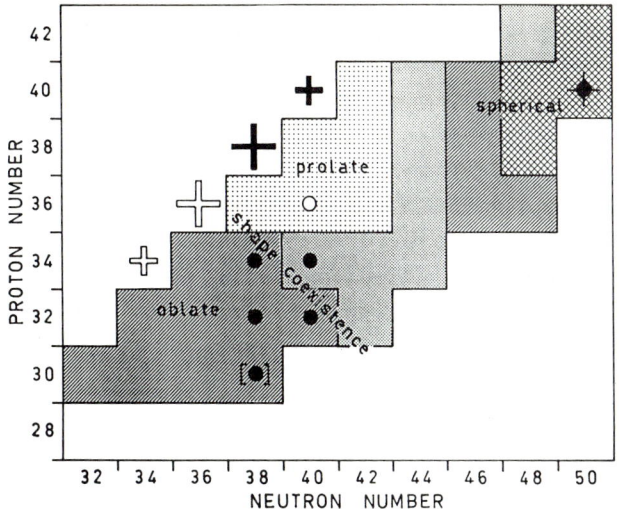

Fig. 8 The energy δE_{0^+} ground state deviates from the 0^+ energy of a hypothetical rotational band with constant moment of inertia passing through the 6^+ and 4^+ yrast states. Sparsely dotted regions have $\delta E_{0^+} < 0.40$ MeV, densely dotted regions have 0.40 MeV $\leqslant \delta E_{0^+} < 0.75$ MeV, lined regions have 0.75 MeV $\leqslant \delta E_{0^+} < 1.5$ MeV and cross-hatched regions have $\delta E_{0^+} \geqslant 1.5$ MeV. Nuclei with experimentally observed shape coexisting low-lying 0^+ states are indicated with circles, which are filled if the ground state has the smaller deformation.

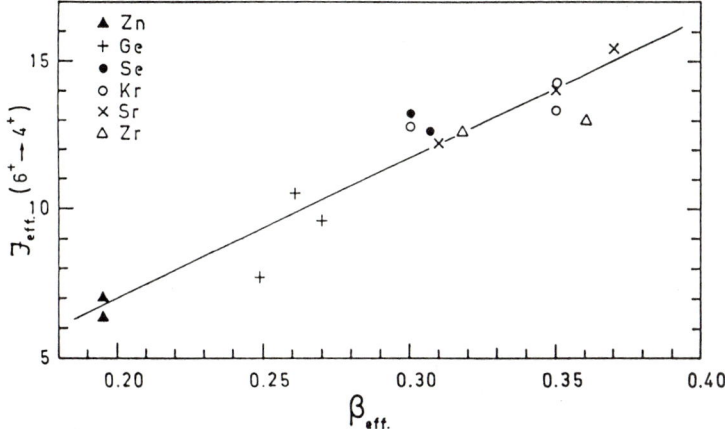

Fig. 9 The experimental moment of inertia for the 6^+ and 4^+ transition plotted versus the theoretical deformation parameter β_{eff} defined in the text. The line is just to guide the eye.

rotational bands, cf. fig. 7. The calculations suggest that shape coexistence may also appear in the N=44 isotones. However, the experimental excited 0^+ states lie too high and a rotational band built on them would hardly influence the yrast states through a residual interaction. We would therefore rather suggest that the N=44 isotones are soft rotors which undergo substantial deformation changes along the yrast line. Prolate or near prolate shapes are generally expected. Most of the light nuclei with Z≤34 and N≤36 may also be interpreted as soft rotors. However, it cannot be excluded that the yrast states are perturbed by interactions with higher lying shape coexisting states. It is also hard to make any prediction about the shape of these nuclei, since the calculations give very γ-soft potential energy surfaces for most of them. A preference for oblate shapes is however present in particular for Z=34 (note the gap at oblate shape at this particle number in fig. 5) and to a lesser extent for Z=32.

We have so far only made very qualitative comparisons between theory and experiment. On this level of comparison the main features of figs. 1 and 2 can be understood in terms of crossings between shape coexisting bands and in terms of larger or smaller deformation changes within a single rotational band. For a more quantitative comparison these deformation changes must be calculated. However, if we exclude the softest nuclei and make use of the observation that in the nuclei with shape coexisting bands the band with the larger deformation usually is yrast at $I=4^+$ and $I=6^+$, we would expect to find a close correlation between the moment of inertia, defined as $\mathcal{J}_{eff}=I_x/\omega$ calculated for the $6^+ \to 4^+$ transition and the deformation at the energy minimum in the ground state potential energy surface. If coexisting minima exist the one with the larger deformation should be used. To correct for deviations from axially symmetric shapes an effective deformation (cf. ref.[4]) $\beta_{eff} = \beta_2 \cdot \cos^2(\gamma+30°)/\cos^2(30°)$ is used, where the value of γ should be chosen negative. The correlation between $\mathcal{J}_{eff}$ and β_{eff} is shown in fig. 9. Due to the presence of pairing correlations a linear dependence is expected in a lowest order approximation over the considered deformation range. A correlation of this kind is clearly present, showing that the relative magnitude of the calculated deformations indeed is correct. To varify if the absolute values are correct would require that the moment of inertia is explicitly calculated. A similar comparison for the 0^+ and 2^+ states cannot be made since the energy of these states is in many nuclei perturbed by interactions between shape coexisting bands. Neither can the 8^+ or higher states be used because of the onset of alignment of $g_{9/2}$ particles.

In conclusion, potential energy surfaces calculated with the Woods-Saxon potential with monopole pairing allow us to qualitatively understand the low spin properties of even-even nuclei in the A≈ 70-80 mass region.

References:

1. W. Nazarewicz et al., Nucl. Phys. A435 (1985) 397.

2. R. Bengtsson and W. Nazarewicz, Proc. XIX Winter School, Zakopane, Poland, 1984 (Report IFY No 1268lPL), p. 171.

3. R. Bengtsson and W. Nazarewicz, Lund preprint MPh-87/08, to appear in Nucl. Phys. A.

4. S. Frauendorf and F.R. May, Phys. Lett, 125B (1983) 245.

Collectivity of Neutron Deficient Odd Strontium and Yttrium Isotopes

D. Bucurescu[1], G. Cãta[1], D. Cutoiu[1], M. Ivaşcu[1], C.F. Liang[2], P. Paris[2], and N.V. Zamfir[1]

[1]Central Institute of Physics, Bucharest, Romania
[2]C.S.N.S.M. Orsay, F-91406 Orsay Cedex, France

Abstract: *IBFM-1 calculations are reported for the Yttrium (79 to 85) and Strontium (77 to 83) isotopes, which describe well the shape transition from the A ≈ 80 nuclei. Possible systematics of the IBFM parameters in this region are discussed.*

The existence of nuclei with large permanent deformation around $Z,N \approx 38$ is now well established experimentally. The transition from spherical nuclei ($N \approx 50$) towards deformed ones has been rather systematically studied along the even isotope chains (Kr, Sr, Zr), the IBA model proving able to accomodate in a natural way the occurence and development of the collectivity. By comparison, odd nuclei in the same region are less systematically studied.

In this work we present a study of the low energy level schemes in the light odd isotopes of Strontium (A=77 to 83) and Yttrium (A=79 to 85) in the framework of the interacting boson-fermion model (IBFM). There are a few IBFM calculations in this mass region, for some Rb [1], Sr and Y [2] isotopes, but the number of nuclei studied is still reduced and a systematics of the model parameters is missing.

Our study will be centered on the positive parity states which are better studied experimentally. Some references to the negative parity states will be also made.

Most of the experimental information on the $\pi = +$ states is displayed in Fig. 1. Distinctive features are the occurence of well developped rotational bands in the lightest isotopes, and the strong competition (for the lowest energies) in the transitional region (N=44 and N=42), between the state J=9/2 and the states J-1 and J-2. A similar structure evolution appears in the light Kr and Rb nuclei which are isotones of the nuclei discussed here.

The spin sequence 5/2, 7/2 and 9/2 for the lowest $\pi = +$ states in

A≈80 nuclei with large neutron deficit has been mainly interpreted in terms of models which couple a particle to a (prolate) deformed core (see, e.g., refs. [3]).

Our approach is based on the IBFM-1 formalism. Y isotopes are regarded as a proton coupled to even Sr cores, while the Sr one as a neutron hole coupled to the same cores. The Sr cores (A=78 to 84) are described by the IBA-1 model, according to ref. [2]. The code ODDA [4] was used for the diagonalization of the well known IBFM-1 Hamiltonian [5]. The following parametrizations have been used for the strengths of the three usual terms of the boson-fermion interaction (the monopole-monopole, quadrupole-quadrupole and exchange) [6]:

$$A_j = \sqrt{5(2j+1)} A_o; \quad \Gamma_{jj'} = \Gamma_o(u_j u_{j'} - v_j v_{j'})$$

$$\Lambda_{jj'}^{j''} = -\sqrt{5}\Lambda_o[(u_{j'}v_{j''}+v_{j'}u_{j''})Q_{j'j''}\beta_{j''j'} + (u_j v_{j''}+v_j u_{j''})Q_{j''j}\beta_{j'j''}]/\sqrt{2j''+1} \quad (1)$$

with

$$Q_{jj'} = <j||Y^{(2)}||j'>$$

$$\beta_{jj'} = (u_j v_{j'} + v_j u_{j'})Q_{jj'}/(E_j + E_{j'} - \hbar\omega) \quad (2)$$

The standard notations are employed. The odd fermion was allowed to occupy the shell model orbitals $1g_{9/2}$ and $2d_{5/2}$ (for the $\pi = +$ state calculations) and $2p_{3/2}, 2p_{1/2}, 1f_{5/2}$ (for the $\pi = -$ states). The occupation probabilities v_j^2 and the qp energies E_j were provided by a BCS calculation based on sp levels from refs. [7,8]. The x-value in the boson quadrupole operator was chosen as $\chi_{\nu(\pi)}$ from the IBA-2 description of the core, when the odd fermion was of the proton (neutron) type. The energy difference between the D and S state was chosen $\hbar\omega = 1.5$ MeV. The remaining parameters A_o, Γ_o and Λ_o were determined by reproducing the experimental level ordering and spacings. For both Y and Sr we found that little variation of A_o and Γ_o is required along the isotope chain, whereas Λ_o had to vary with N in order to be able to reproduce the essential features of the data. The adopted parameters are:

Y: $A_o=-0.23$ MeV; $\Gamma_o=0.45$ MeV; $\Lambda_o = 5.1, 7.4, 10.6$ and 12.4 MeV^2
(for A = 85, 83, 81 and 79, respectively).

Sr: $A_o=-0.10$ MeV; $\Gamma_o=0.30$ MeV; $\Lambda_o=3, 3.4, 8$ and 19 MeV^2
(for A = 83, 81, 79 and 77, respectively).

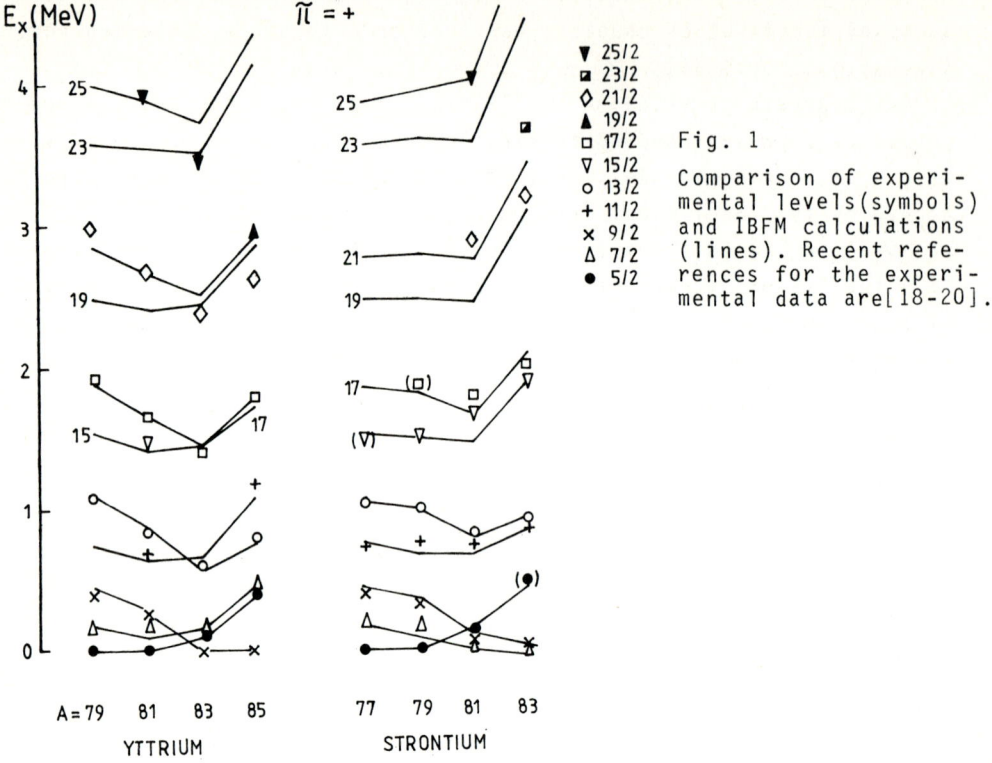

Fig. 1 Comparison of experimental levels (symbols) and IBFM calculations (lines). Recent references for the experimental data are [18-20].

Fig.1 shows the comparison of the calculations with the experimental levels.

Negative parity states (experimentally known in ^{83,85}Y and ^{79,81}Sr) were calculated also with the above parameters. A fair agreement with the data was obtained, which could be easily improved by slightly readjusting the parameters. For example in ^{83}Y a good description of the $\pi=-$ states requires a lower Λ_0 (about 6 MeV^2 instead of 7.4 MeV^2) which still provides a reasonable description of the $\pi = +$ states. Thus, it appears that in this mass region one can describe the states of both parities with the same Hamiltonian. Since there is no information on $\pi=-$ states in all studied isotopes, we discuss, for consistency, the parameters deduced on the basis of the $\pi = +$ states.

We consider first the exchange force, which plays such an important role in these nuclei. Since the Λ_0 value depends on the choice of $\hbar\omega$ (eqs. (1) and (2)), we consider one of the elements $\Lambda_{jj'}^{j''}$, and namely that with $j'' = j = j' = 9/2$, denoted by Λ_{99}^{9}; for a given qp scheme, the other elements are in a well-defined relationship with Λ_{99}^{9}. This procedure will allow us to make comparisons with other calculations where different parametrizations of $\Lambda_{jj}^{j''}$, have been used. We also use the recent observation of Tabor [9] that data characterizing both the even and odd nuclei in the A≈80 region can be convenably parametrized in terms of $N_p N_n$, where $N_{p(n)}$ are the number of active protons (neutrons) counted with respect to the shell numbers 28 and 50. For the even nuclei, the $N_p N_n$ scheme of Casten [10] led, as a consequence, to a sinmple parametrization of the IBA-1 parameters [11]. Along these ideas, we have tried to view the quantity Λ_{99}^{9} not as a function of N or A, but of the product $N_p N_n$, and also on $P = N_p N_n/(N_p + N_n)$ recently considered by Casten [12]. The two representations are shown in Fig.2. Besides the Sr and Y values, we have included those for ^{81,83}Rb from ref. [1], and for $^{97-105}$Rh [13], although in both these cases formulas slightly different from (1) and (2) were used for Λ. The Rh nuclei were included based on the fact that their energy difference $E_{13/2^+} - E_{9/2^+}$ follows the systematics of Tabor [9] for the odd Br, Rb and Y nuclei. From Fig.2 one sees that in the $N_p N_n$ scheme the Λ values are somewhat scattered, whereas in the P scheme an improvement is obtained, all isotope families considered following rather closely a unique curve.

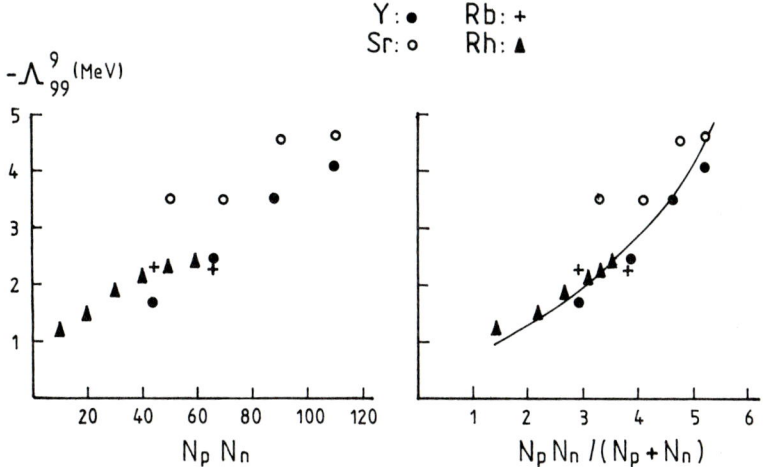

Fig. 2 Two different representations of the exchange force strength elements $\Lambda_{9/2,9/2}^{9/2}$ (see text for details). The continuous line is drawn to guide the eye.

One should remark that Λ values in Fig.2 correspond to a good description of the level scheme of each nucleus (Fig.1). A better coalescence of the points in Fig.2, which would require, in average, less that 15%

variation of Λ_0 in several cases, would still preserve the reproduction of the general trend of the level schemes along an isotope chain.

As remarked in [12], the value of P can be viewed as the average number of interactions of each valence nucleon with those of the other type, and it also characterizes the relative integrated strengths of the p-n and like-nucleon (pairing) interactions. Qualitatively, one can assign the empirically observed dependence of the exchange force strength on P to the microscopic origin of the exchange term which lies in both the n-p quadrupole interaction [14,15] and the like-particle interaction [16].

One should remark, nevertheless, that microscopic derivations of the boson-fermion interaction, based only on the n-p quadrupole interaction lead to the prescription that $\Lambda_0 \sim P$ and Γ_0/Λ_0 = ct for an isotope chain [14]. Only a qualitative agreement of the empirical values was found with this prescription, usually the experimental Γ_0/Λ_0 ratio being found significantly different from that estimated theoretically (see, e.g., [13,17]); a possible cause of this discrepancy was thought to be the neglect of the like particle interactions [17].

It does not appear as simple to treat similarly the strength of the quadrupole-quadrupole interaction. In the Y isotopes this interaction varies mainly due to the variation of χ, while in Sr mainly due to the variation of the occupation probabilities. Thus, on the whole, an appropriate quantity to follow would be $\chi\Gamma_0$(or $\chi\Gamma_{jj'}$), but from the available data it seems that for n-odd and p-odd nuclei it follows different trends, so that more nuclei should be studied to evidence a possible systematic behaviour.

In conclusion, from a phenomenological study of two isotope chains, Y and Sr, with A≈80, we have evidence that the IBFM-1 parameters for these nuclei might follow a simple, systematic behaviour which would allow the description of many nuclei in this region with a small number of parameters. First of all, we consider such systematic calculations for other A≈80 nuclei. The predictions of the present calculations for other spectroscopic quantities will be given in a more extended paper.

REFERENCES

[1]. U.Kaup *et al*, Phys.Rev. C22(1980) 1738; J.Panqueva *et al*, Phys. Lett. 98B (1981) 248; Nucl.Phys. A389 (1982) 424.
[2]. D.Bucurescu *et al*, Nucl.Phys. A401 (1983) 22.

[3]. G.Garcia-Bermudez *et al*, Phys.Rev. <u>C30</u> (1984) 1208; S.E.Arnell *et al*, J.Phys. <u>G9</u> (1983) 1217; C.J.Lister *et al*, J.Phys. <u>G11</u> (1985) 969; B.Wörmann *et al*, Nucl.Phys. <u>A431</u> (1984) 170.

[4]. O.Scholten, Computer code ODDA, Internal Report KVI 252 (Groningen).

[5]. F.Iachello, O.Scholten, Phys.Rev.Lett. <u>43</u> (1979) 679.

[6]. R.Bijker and A.E.L.Dieperink, Nucl.Phys. <u>A379</u> (1982) 221.

[7]. B.S.Reehal and R.A.Sorensen, Phys.Rev. <u>C2</u> (1970) 819.

[8]. D.Galeriu, D.Bucurescu, M.Ivașcu, J.Phys. <u>G12</u> (1986) 329.

[9]. S.L.Tabor, Phys.Rev. <u>C34</u> (1986) 311.

[10]. R.F.Casten, Phys.Lett. <u>152B</u> (1985) 145; Phys.Rev.Lett. <u>54</u> (1985) 1991.

[11]. R.F.Casten *et al*, Nucl.Phys. <u>A444</u> (1985) 133.

[12]. R.F.Casten, Phys.Rev.Lett. <u>58</u> (1987) 658.

[13]. D.Bucurescu *et al*, Nucl.Phys. <u>A443</u> (1985) 217.

[14]. O.Scholten, Ph.D. Thesis, Univ. of Groningen, 1980.

[15]. I.Talmi, in *Interacting Bose-Fermi systems in nuclei*, ed. F.Iachello (Plenum, 1981) p. 329.

[16]. U.Kaup, Progr.Part.Nucl.Phys. <u>9</u> (1983) 561; T.Otsuka *et al*, Phys.Rev. <u>C35</u> (1987) 328.

[17]. O.Scholten, T.Ozzello, Nucl.Phys. <u>A424</u> (1984) 221.

[18]. C.J.Lister *et al*, J.Phys. <u>G11</u> (1985) 969.

[19]. M.S.Rapaport, C.F.Liang, P.Paris, Phys.Rev. <u>C36</u> (1987) 303.

[20]. U.Lenz *et al* (München), communication at the European study conference on nuclear shapes, Crete (1987) and private communication.

Microscopic Description of Even Ge and Se Nuclei

F. Grümmer[1], *A. Petrovici*[2], *and K.W. Schmid*[3]

[1]Institut für Kernphysik, Kernforschungsanlage Jülich,
 Postfach 1913, D-5170 Jülich, Fed. Rep. of Germany
[2]Central Institute for Nuclear Physics and Engineering,
 Bucharest-Maguerele, Romania
[3]Institut für Theoretische Physik, Universität Tübingen,
 Auf der Morgenstelle, D-7100 Tübingen, Fed. Rep. of Germany

1. Introduction

In the last years a family of microscopic nuclear structure models based on Hartree-Fock-Bogoliubov mean fields has been developed by the authors, which are known as the MONSTER and the VAMPIR approaches[1]. These models are supposed to be an alternative to the 'exact' shell model configuration mixing approach, which is limited to too small single particle basis spaces in order to be useful for realistic applications in heavy nuclei. We will briefly describe here the theoretical background and the numerical implementation of these models.

The main part of this talk, however, will consist in the presentation of results we have obtained with the EXCITED VAMPIR approach[2] in the $A \sim 70$ region[3], especially for some Ge and Se isotopes. The experimental data in these nuclei suggest that shape coexistence phenomena play an important role. This work demonstrates the principal potential of the EXCITED VAMPIR method to describe such nuclei. Looking at low lying states of small as well as high angular momentum we find not only qualitative agreement with the experimental data, but in many cases a surprisingly good quantitative agreement. It turns out that a proper renormalization of the effective Hamiltonian describing the nuclei under consideration is tremendously important in order to get satisfatory results.

2. Mean field models and SCM

We are starting from the well justified assumption that for the description of most low energy nuclear phenomena we may consider the 'classical' nuclear many body problem. This means that our aim is the construction of microscopic, quantummechanical A-nucleon wave functions for the ground and excited states of nuclei. The model we are using for this purpose should be universal, i.e. it should be valid for all nuclei in all mass regions.

The essential assumptions we have to make in order to be able to construct well defined models are:

- We assume that the nucleus consists only of nucleons without internal degrees of freedom, which may be treated nonrelativistically.

- We assume that for the description of the low lying states a finite single particle model space may be used. In realistic applications this finite model space can actually be quite large.

- We assume that an appropriate "effective" many body Hamiltonian is known. This is actually not the case in large basis spaces, because up to now it was not possible to perform microscopic nuclear structure calculations in such large modelspaces. Thus it is one of the major issues of future research to determine such effective Hamiltonians for various mass regions.

In practice we start with a finite single particle basis like e.g. a harmonic oscillator basis of M states $\{c_i^+, c_j^+, \ldots, c_M^+\}$, where the index i denotes the quantumnumbers of the state $i = \{\tau_i, \pi_i, n_i, \ell_i, j_i, m_i\}$. For this finite basis we may specify a many body Hamiltonian consisting of a one- and a two-body part.

$$\hat{H} = \sum_{i,j} t(ij) c_i^+ c_j + \sum_{i,j,r,s} v_{eff}^{a.s.}(ijrs) c_i^+ c_j^+ c_s c_r$$

The many body problem formulated in this fashion can in principle be solved exactly using the shell model configuration mixing (SCM) method. This means that one has to construct all possible A-nucleon Slater determinants and diagonalize the Hamiltonian in the space of these wavefunctions. One sees immediately, however, that this method is limited to very small single particle basis spaces, since the number of configurations increases dramatically with the number of basis states.

Therefore, we proposed the more physical solution to put as many as possible of the correlations in an optimal mean field, so that it will be a well justified truncation scheme to decribe additonal correlations by only few n particle – n hole (or equivalently 2n quasiparticle) configurations.

The starting point of all microscopic models based on this mean field concept is the Hartree-Fock-Bogoliubov transformation, which connects the original single particle basis with a new basis of independent quasiparticles.

$$a_\alpha^+ = \sum_{i=1}^{M} \{A_{i\alpha} c_i^+ + B_{i\alpha} c_i\}$$

$$a_\alpha = \sum_{i=1}^{M} \{B_{i\alpha}^* c_i^+ + B_{i\alpha}^* c_i\}$$

With the help of this transformation we can now define an intrinsic many body wave function in the form of a quasiparticle determinant.

$$|F\rangle = \prod_\alpha a_\alpha |0\rangle$$

Since the transformation matrices A and B usually mix all the good quantum numbers of the original basis states the resulting wave function $|F\rangle$ will not have the required symmetries like being an eigenstate of the angular momentum, particle number, parity, and so on. Thus one has to restore the correct symmetries with the help of projection operators. This leads us to physical many body wave functions in the form of projected quasiparticle determinants.

$$|I^\pi M; A, T_z(F)\rangle = \sum_K \hat{P}_{MK}^I \hat{P}^\pi \hat{Q}_A \hat{Q}_{T_z} |F\rangle$$

If we now vary the energy of this state (i.e. the expectationvalue of the Hamiltonian using the above state) with respect to the mean field transformation coefficients A and B characterizing the mean field $|F\rangle$, one is able to determine an optimal mean field for the lowest state with a given set of quantum numbers. This method has been given the name "VAMPIR"[4].

An excited state with the same quantumnumbers may now be described as a superposition of the ground state with a second still unknown projected HFB wave function. The transformation matrices A and B of this second transformation are then varied in order to obtain a minimal energy of the first excited state. The mixing coefficient of the two projected determinants is fixed by the requirement that the first excited state has to be orthogonal to the ground state. This procedure may be extended to more excited states and will be referenced as the EXCITED VAMPIR procedure[2].

3. Application to the $A \sim 70$ region

Looking at the Nilsson diagram for the $A \sim 70$ region one notices gaps at various prolate and oblate deformations, which suggest the occurence of shape coexistence phenomena. Thus the nuclei in this region should provide a rather challenging testing ground for the EXCITED VAMPIR approach.

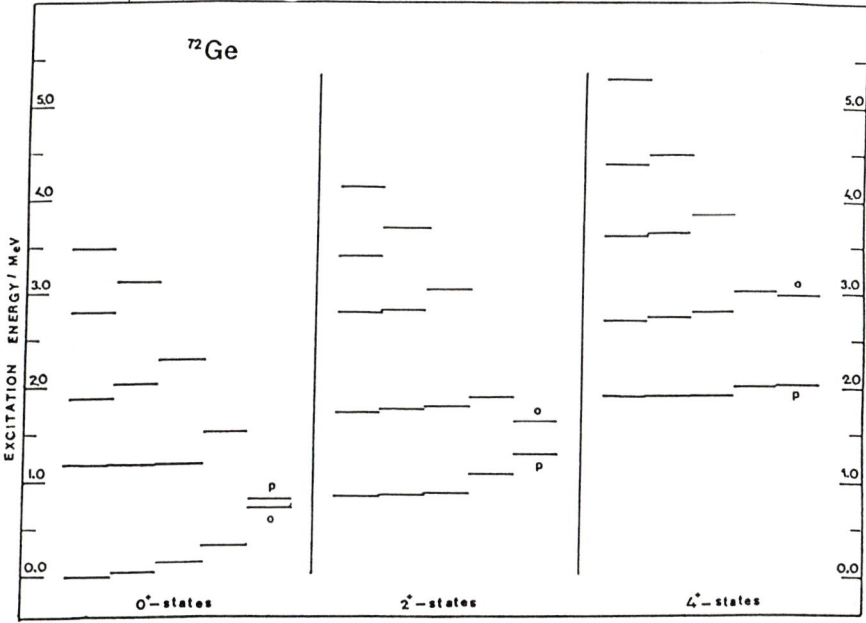

Fig. 1: 0^+-, 2^+- and 4^+- states generated with the EXCITED VAMPIR in ^{72}Ge. The states labeled with o and p are the oblate and prolate solutions found using only one quasiparticle determinant. From right to left always one more mean field wavefunction is added to the many body basis.

We use here a single particle basis consisting of the $N = 3$ and $N = 4$ major shells for both protons and neutrons together with the $0h_{11/2}$ orbit for the neutrons. The single particle energies of these states are essentially taken from spectra of odd nuclei in the $A \sim 50$ and $A \sim 80$ regions. For the construction of an appropriate effective interaction with as few as possible free parameters we start from a nuclear matter G-matrix based on the Bonn OBEP potential[5]. This parameter free interaction has, however, to be renormalized slightly in order to reproduce pairing and deformation properties in the mass region under consideration correctly. Therefore we add two short range ($\mu = 0.7 fm$) gaussians in the $T = 1$ channel with strenghts of $V_{pp} = -50 MeV$ and $V_{nn} = -40 MeV$. In addition the $\langle fg_{9/2}|G|fg_{9/2}\rangle^{T=0}$ matrix elements are shifted by $-0.115 MeV$ in order to reproduce correct ground state deformations. The so defined effective Hamiltonian is then used for all subsequent calculations.

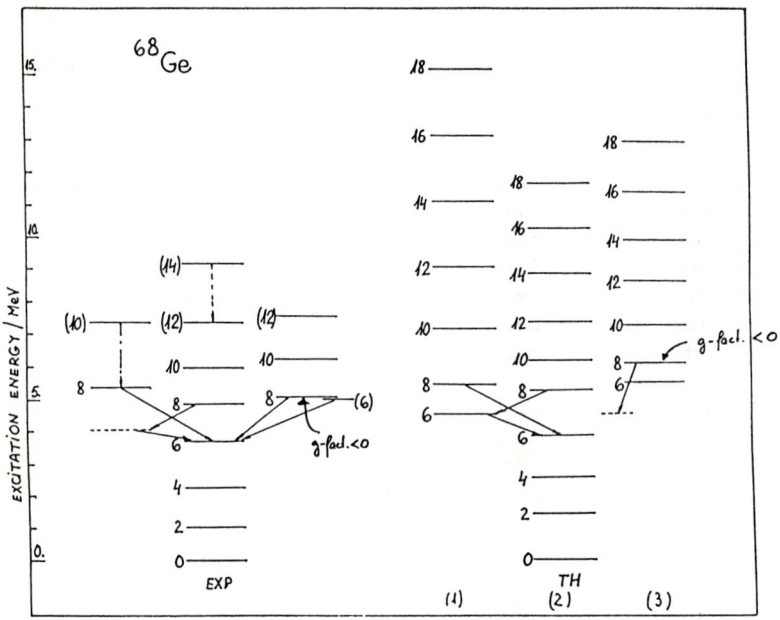

Fig. 2: *Some experimentally known bands in ^{68}Ge are compared to the theoretical results.*

Fig.1 shows the results of EXCITED VAMPIR calculations for the lowest 0^+, 2^+ and 4^+ states in ^{72}Ge. We find that the projected states based on a prolate and on an oblate mean field are energetically nearly identical. Mixing these two states gives us roughly the correct energy splitting of the two lowest 0^+ states. We see also that this mixing effect becomes less important for the higher angular momenta. If we add projected determinats in order to describe higher excited states, we find that they do not influence the lowest states much any more.

In all considered nuclei in the $A \sim 70$ region we find several low lying 0^+ states below $5 MeV$, a fact, which in itself is already a success of the EXCITED VAMPIR approach. Shape coexistence plays an important role in the description of these states.

An even more crucial test of the model and the effective Hamiltonian is the description of the high spin bands. Thus we calculated states with angular momenta up to 22^+ in the nuclei ^{68}Ge and ^{72}Se and tried to group them into bands according to the structure of their wave functions and according to the $B(E2)$-transitions between them. The results of this feature can be seen in figs.2 and 3, where parts of the experimentally known spectra are compared with our theoretical spectra[6-7]. One finds a very good agreement, although in some cases, when one finds several states of the same angular momentum in a range of a few hundred keV, the ordering of these states is not always reproduced correctly.

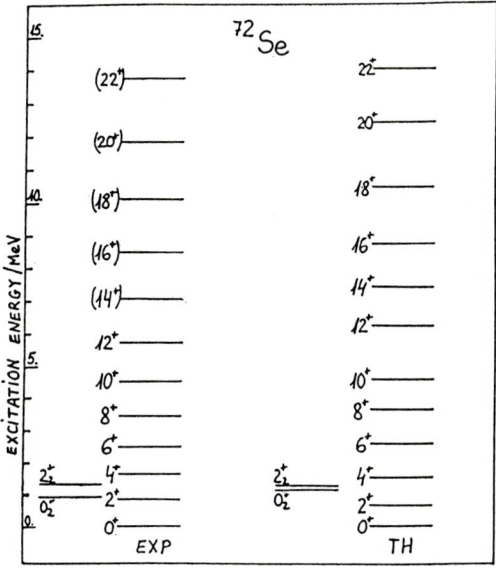

Fig. 3 : *Some experimentally known states in ^{72}Se are compared to the theoretical results.*

4. Summary

With the EXCITED VAMPIR we have developed a microscopic nuclear structure model, which allows us to describe all low lying nuclear states on the basis of only few optimally chosen quasiparticle determinants. All kind of correlations, whether of collective or of single particle nature are automatically taken into account by the variational principle. We applied this model for several nuclei in the $A \sim 70$ region. Using a modified nuclear matter G-matrix as effective interaction we obtain reasonable agreement with experimental data in Ge and Se nuclei. We see the effects of shape coexistence in low spin states as well as in high spin states. The moste crucial part for the quantitative understanding of the data turns out to be an appropriate effective interaction, i.e. single particle energies and the renormalization of certain two body matrixelements. In the future we hope to obtain some more insight in these dependencies in order to be able to find effective interactions also in other mass regions. Finally, we have been restricted up to now to the description of states with even angular momentum and positive parity in evev-even nuclei. With the newly developed COMPLEX EXCITED VAMPIR[8] we will in the future be able to go beyond this restriction and extend the class of accesible states.

REFERENCES

1. K.W.Schmid and F.Grümmer, *Rep. Prog. Phys.* **50** (1987) 731–781
2. K.W.Schmid, F.Grümmer, M.Kyotoku and A.Faessler, *Nucl. Phys.* **A 452** (1986) 493
3. A.Petrovici, K.W.Schmid, F.Grümmer and A.Faessler, *Nucl. Phys.* **A**, in press
4. K.W.Schmid, F.Grümmer and A.Faessler, *Nucl. Phys.* **A 431** (1984) 205
5. K.Holinde, K.Erkelenz, R.Alzetta, *Nucl. Phys.* **A 194** (1972) 161
6. J.Heese, K.P.Lieb, L.Lühmann, F,Raether, B.Wörmann, D.Alber, H.Grawe, J.Eberth, T.Mylaeus, *Z. Phys.* **A 325** (1986) 45
7. J.H.Hamilton in Treatise on Heavy Ion Science 8, D.A.Bromley Ed.,"Nuclei far from Stability",Plenum Press
8. K.W.Schmid, F.Grümmer and A.Faessler, *Ann. Phys.* **180** (1987) 1

Coexistence of Shapes and Structure of Nuclei in the A = 80 Region

R. Sahu[1] and S.P. Pandya[2]

[1] Physics Dept., Berhampur University, Berhampur-760007,
Ganjam, Orissa, India
[2] Physical Research Laboratory, Ahmedabad-380009, India

In recent years, the nuclei in the mass range A=60-90 have been studied extensively by the experimentalists. This has led to exciting discoveries of large deformations, co-existence of shapes, rapid variation of structure with changes in neutron and proton number etc. It has been shown in particular that as the neutron or proton number approaches 38-40, nuclei show strong deformations and rapid variation of structure. In this development studies of coexistence of different shapes in Se nuclei and the structure of the heaviest N = Z isotope ^{80}Zr would be quite interesting.

Theoretically this region has not been amenable to detailed microscopic calculations. Even with the assumption of an inert core of ^{56}Ni, the number of active nucleons to be treated is quite large and the configuration space must include at least four active single particle orbits: $p_{3/2}$, $f_{5/2}$, $p_{1/2}$ and $g_{9/2}$. Few such calculations have been carried out and then only for simple systems and with drastic approximations. Most of the theoretical investigations in this region employ vibrational models, coupling of quasiparticles to vibrations, rotational models, the dynamic deformation model, interacting boson model etc. It would be very useful to have a detailed microscopic model which can track the rapid changes of nuclear structure over the entire sequence of nuclei and give an insight into the role of nuclear interactions as well as the dynamics of the single particle motions.

In our microscopic model, we take ^{56}Ni to be the closed inert core with $p_{3/2}$, $f_{5/2}$, $p_{1/2}$ and $g_{9/2}$ as the active single particle orbits. The single particle energies for the first three orbits are taken as 0.0 MeV, 0.78 MeV, 1.08 MeV from ^{57}Ni data. The g-orbit is placed at 4.75 MeV. An effective interaction, generated for this space by Kuo (Private Communication) and partially modified by Bhatt[1], is used.

Shape coexistence in Se isotopes:

The shape coexistence in ^{72}Se was discovered by Hamilton et al.[2] almost simultaneously with its discovering in mercury isotopes. For this nucleus the lowest states 0^+, 2^+ appear to conform to a vibrating spherical nucleus, the states with $J = 4^+$ and above belong to a band based on a strongly deformed intrinsic state with $K = 0^+$. A plot of the moment of inertia against the rotational frequency confirmed this coexistence of shapes. Such a property was totally unexpected since, for a long time, the nuclei in this mass region were considered to be nearly spherical.

In our calculation, we carry out axially symmetric Hartree-Fock calculations which give lowest energy prolate and oblate solutions. A variety of configurations (deformed intrinsic states) can now be constructed and for each of these we redo a variational constrained calculation. Standard projection methods are used to project out states of good angular momentum from each of these intrinsic states. Finally the hamiltonian operator is diagonalized in the space of all states with the same J. Mixing several $J = 0$ states effectively takes into account pairing.

Figure 1 shows the single particle spectra corresponding to the lowest prolate solutions. There are lowlying oblate solutions also. But we do not include them in our calculation[1].

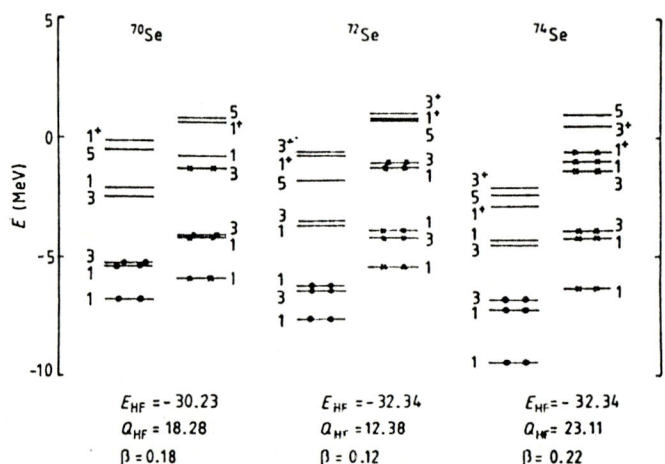

Figure 1. The spectra of HF orbits for the lowest energy prolate states for ^{70}Se, ^{72}Se, ^{74}Se. The protons are represented by circles and the neutrons by crosses. The numbers represent 2k values.

For ^{72}Se, the lowest intrinsic state has a rather small deformation. The protons as well as the neutrons are entirely distributed in p-f orbits. However the K = $1/2^+$ member of the $g_{9/2}$ multiplet is only a couple of MeV above the highest occupied state of the neutrons. One can easily generate a state with much larger quadrupole moment (and a larger intrinsic deformation) by exciting two neutrons from the Fermi state to the K = $1/2^+$ orbit. Indeed a tagged HF calculation with two neutrons in the K = $1/2^+$ orbit gives an intrinsic state with a quadrupole moment about two and a half times that of the lowest state ! In our calculation we have considered these two intrinsic states and carried out a band mixing calculation. It is also clear that proton excitation will not generate more deformed states and exciting two protons to a corresponding K = $1/2^+$ orbit would place such an intrinsic state at almost 10 MeV above the loswest intrinsic state. Since our interest is only in understanding broad features of the spectral structure, we have not considered more band mixing. A look at the HF single particle spectrum of ^{72}Se suggests immediately that ^{70}Se should show a similar phenomenon. In the lowest intrinsic state, the highest states occupied by neutrons in ^{72}Se are K = $1/2^-$ and $3/2^-$. Exciting a pair of these to the K=$1/2^+$ state generates a highly deformed intrinsic state. ^{70}Se has two neutrons less, but the last two neutrons would occupy the K = $3/2^-$ or K = $1/2^-$ state. These are nearly degenerate in energy and hence even in ^{70}Se to excite the two neutrons from the Fermi state to K = $1/2^+$ state would require almost the same amount of energy. So like ^{72}Se, ^{70}Se, should also have a lowlying highly deformed intrinsic state and should thus show a coexistence phenomenon. In fact we have been led to include ^{70}Se in our work only after seeing the structure of the HF states of ^{72}Se. This is an insight only revealed by a microscopic calculation.

In the case of ^{74}Se, one may expect a spherical intrinsic state where the twelve neutrons completely fill up the $p_{3/2}$, $p_{1/2}$ and $f_{5/2}$ orbits and the protons fully occupy the $p_{3/2}$ and $p_{1/2}$ orbits. However, energetically it should be very easy to excite two neutrons or even four neutrons to K = $1/2^+$ and $3/2^+$ in $g_{9/2}$ multiplet and obtain intrinsic states of large deformations. Thus tagged HF calculations give a state with two neutrons in K = $1/2^+$ orbit (with a large quadrupole moment) almost degenerate in energy with spherical HF state. Further, we find, only 1 MeV higher, another intrinsic state with four neutrons in K = $1/2^+$ and $3/2^+$ orbits and an even larger intrinsic quadrupole moment. For the purpose of our calculation, we have projected good J states from these three intrinsic states and carried out a band mixing calculation.

It is clear from the above discussion that it is the easy access to the $g_{9/2}$ orbit for neutrons in the topmost occupied members of pf orbits that produces, at relatively low excitation energies, strongly deformed intrinsic states. This effect will occur easily for neutron number 36-40.

Figure 2 shows the spectra obtained for ^{70}Se, ^{72}Se and ^{74}Se.

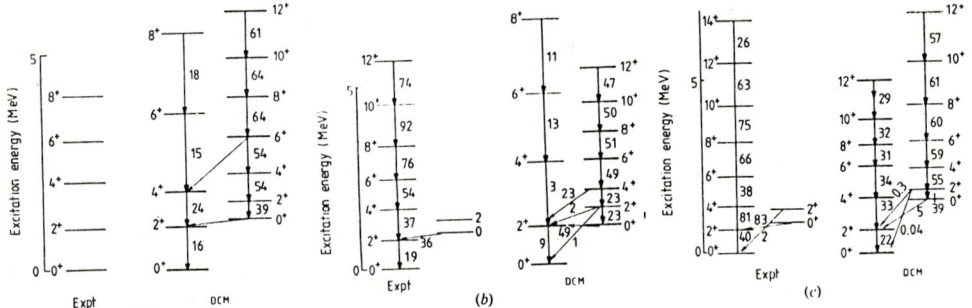

Figure 2. A comparison between experimental and calculated yrast spectra. The quantities near the arrows represent B(E2) values in Wu. (a) ^{70}Se (b) ^{72}Se (c) ^{74}Se.

The yrast sequence in ^{72}Se consists of J = 0^+ and 2^+ states of the nearly spherical ground state band and J = 4^+, 6^+ and higher J states belong to the strongly deformed excited band. The level structure of ^{70}Se show a similar architecture. In this case 0^+, 2^+ and 4^+ yrast states belong to the less deformed band whereas 6^+ and higher J yrast states belong to the more deformed band. However, for ^{74}Se, our calculation does not show any band crossing.

Structure of ^{80}Zr :

In a recent investigation, lister et al.[3] have observed the heaviest N = Z isotope, ^{80}Zr and measured some of its properties. These results are very important in view of the competition of prolate, oblate and spherical shell gaps at different values of N and Z. It has been reported that ^{80}Zr has a large quadrupole deformation, β = 0.4. Lowlying states at 290 and 828 KeV have also been reported. These two states are considered to have spin 2^+ and 4^+, respectively. We study the structure of this nucleus in the frame work of the microscopic self-consistent Hartree-Fock formalism.

As has been discussed earlier, the energies of the single particle states $p_{3/2}$, $f_{5/2}$ and $p_{1/2}$ are taken from ^{57}Ni data. The position of the $g_{9/2}$ orbit is not defined experimentally. For lighter nuclei its energy was taken to be around 5 MeV. However, for $A \approx 80$, it appears more appropriate to take this energy at about 3 MeV.

In figure 3, we plot the HF spectrum for this nucleus taking $\epsilon(g_{9/2}) = 3.0$ MeV and $\epsilon(g_{9/2}) = 4.0$ MeV. In both the cases the spherical solutions occur at energies higher than those for the deformed prolate solutions. The difference in energies is as large as 6 MeV for $\epsilon(g_{9/2}) = 3$ MeV but goes down to about 1.5 MeV for $\epsilon(g_{9/2}) = 4$ MeV. It is clear that the energy of the $g_{9/2}$ state is very crucial for generating deformations and shape coexistence phenomena. The values of β extracted from the intrinsic quadrupole moments are 0.34 and 0.19 corresponding to $\epsilon(g_{9/2}) = 3.0$ MeV and $\epsilon(g_{9/2}) = 4.0$ MeV. The energy of the $g_{9/2}$ state strongly controls the deformation through the occupation of the $g_{9/2}$ state. For $\epsilon(g_{9/2}) = 4$ MeV, there are two neutrons and two protons in the $g_{9/2}$ orbital while the corresponding numbers are four protons and four neutrons for $\epsilon(g_{9/2}) = 3$ MeV. We should emphasize that these details are also quite sensitive to the nature of the effective interaction. If further experimental data on occupancies or

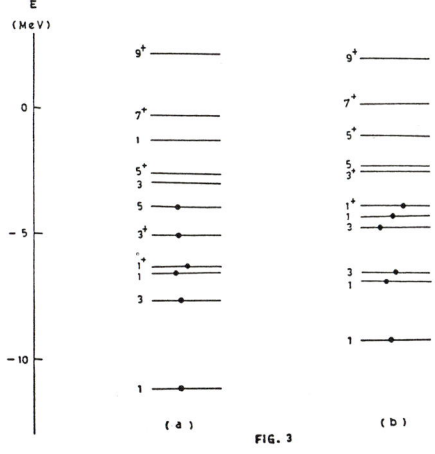

FIG. 3

Figure 3. Lowest energy HF states for ^{80}Zr. (a) $\epsilon(g_{9/2}) = 3.0$ MeV (b) $\epsilon(g_{9/2}) = 4.0$ MeV. Each occupied level is four fold degenerate. The occupancy is represented by a solid circle.

on the energy of spherical 0^+ state become available, we would have a powerful probe in the nature of the effective interaction.

The spectrum projected from the above HF states are given in figure 4.

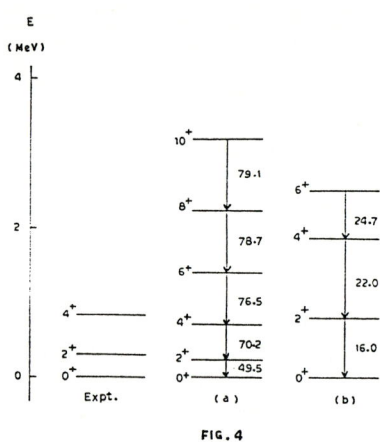

FIG. 4

Figure 4. comparison of experimental and calculated spectrum (a) $\epsilon(g_{9/2}) = 3.0$ MeV (b) $\epsilon(g_{9/2}) = 4.0$ MeV.

In summary, a simple microscopic calculation has been carried out to predict correctly in a self consistent method the deformations of ^{80}Zr and the spectrum of excited states. The treatment brings out clearly the role of $g_{9/2}$ state and its energy in breaking the expected shell closure at N = Z = 40 and producing large deformation.

References :

1. D.P.Ahalpara, K.H.Bhatt and R.Sahu, J.Phys.G 11(1985)735
2. J.H.Hamilton et al., Phys.Rev.Lett.32(1974)239
3. C.J.Lister et al., Phys.Rev.Lett.59(1987)1270

Competition of Oblate and Prolate Deformation in $^{69,70,71,72}_{34}$Se*

T. Mylaeus[1], J. Busch[1], P. von Brentano[1], J. Eberth[1], M. Liebchen[1],
N. Schmal[1], R. Sefzig[1], S. Skoda[1], W. Teichert[1], M. Wiosna[1],
and W. Nazarewicz[2]

[1]Institut für Kernphysik der Universität zu Köln,
Zülpicher Str. 77, D-5000 Köln 41, Fed. Rep. of Germany
[2]Institute of Physics, Warsaw Institute of Technology,
ul. Koszykowa 75, 00662 Warsaw, Poland

In the course of our systematic nuclear structure investigations of the $A \approx 70$ mass region, our special interest focused on the light Se isotopes, which are very good examples for shape coexistence phenomena at low spins. In these nuclei, the ground state rotational band is crossed at $I \approx 4-6$ by a more deformed configuration [1,2]. Due to the large band interaction (of the order of a few hundred keV) both bands are strongly mixed in the 0^+, 2^+ and 4^+ states. Calculations employing the deformed shell model and the shell correction procedure explain shape coexistence with two minima in the potential energy surface, one at small oblate, the other at large prolate deformation [3]. A rotational band built on the excited 0_2^+ state was known since long up to spin 10^+ in ^{72}Se (see e.g. the contribution of R.Bengtsson in this book). The microscopic reason for the prolate oblate shape coexistence in Se isotopes can be easily understood from the single particle Nilsson diagram (see e.g. refs [3,4]). The large (about 1.5 MeV) Z=34 gaps appear both at prolate and oblate shapes, at N=38 the gap opens at large prolate shapes while oblate configurations in Se are stabilized by the N=36 shell gap.

However, firm experimental evidence for oblate quadrupole deformation in this mass region was missing until now due to the strong mixing of the different intrinsic structures in even even nuclei studied so far. In odd isotopes the sign of the quadrupole moment can be determined from the sign of the mixing ratio $\delta = \langle I|E2|F\rangle/\langle I|M1|F\rangle$ of $\Delta I=1$ transitions between favoured and unfavoured bands and the signature splitting of these bands. However, odd nuclei studied so far: ^{73}Se[5] ^{75,77}Kr[6,7], ^{73,75}Br[8,9] have neutron numbers close to 38 and in those nuclei the odd particle stabilizes large prolate deformations ($\beta_2 \approx 0.35-0.40$) in agreement with the theoretical predictions. The most favoured nuclei for a pronounced oblate shape should be ^{69}Se, ^{70,71}Br, ^{72,73}Kr. None of these nuclei was studied with in-beam spectroscopy until now due to their very weak population in fusion-evaporation reactions.

The investigation of the even Se isotopes was carried out with the **OSIRIS** [14] Anticompton spectrometer, using the reaction ^{40}Ca (^{36}Ar, [4p14p2n]) ^{70,72}Se, E(^{36}Ar)= 145 MeV. The beam was provided by the VICKSI accelerator of the HMI in Berlin. ^{72}Se is produced with a relative cross section of 13.2% as strongest exit channel at this beam energy, ^{70}Se with σ_{rel}=2.26% of σ_{total}= 849 mb, according to CASCADE [16] calculations. The maximum angular momentum brought into the system was approximately 45ℏ. The program of measurements included (i) $\gamma\gamma$ coincidences, (ii) $\gamma\gamma$ angular correlations, (iii) correlation of γ rays with cyclotron frequency (time spectra), and (iv) different geometrical detector setups (besides a 90° setup, also a 30°/90° **OSIRIS** setup was used to measure fully Doppler shifted γ ray lines). This procedure enabled us to extract DCO ratios for the transitions in the yrast bands. The theoretical ratio for stretched E2 transitions is unity. The DCO ratios extracted for all the transitions in the yrast bands are unity within the errors and thus consistent with an E2 character of the radiation. Two cross bombardement experiments were carried out at the Cologne accelerator facility **KöBes** with OSIRIS to confirm the assignment of new γ-rays: ^{40}Ca(^{32}S,2p)^{70}Se at 100 MeV and ^{58}Ni(^{16}O,[αl2p])^{70,72}Se at 65 MeV.

Out of the complex levelschemes, which were notably extended, the Yrast bands are shown in fig. 1a,b together with plots of the kinematical and dynamical moments of inertia vs. rotational frequency (fig.1,c) and B(E2) values, taken from ref. 9, vs. spin (fig.1,d). It is very interesting to compare both isotopes. While for low spin states ($I \leq 8^+$) ^{70}Se shows a second discontinuity, for high spins with $I \geq 10^+$ both

* Work supported by Bundesministerium für Forschung und Technologie under contract Nr.: 060 K272

nuclei have a smooth increase of $J^{(1)}$ over a very wide range of rotational frequency ($0.45 \leq \hbar\omega \leq 1.2$). Although the Yrast band of ^{70}Se could only be established up to $I^\pi = 16^+$, the trend indicates a similar behavior as for ^{72}Se for the higher spins. The dynamical moments of inertia are very close to each other and close to the rigid body value calculated for the near prolate shape (see below). For ^{72}Se, a backbending is seen at $\hbar\omega \approx 0.45$ MeV while in ^{70}Se the irregularity is observed at 0.5 MeV. The difference at low spins however is pointed out even more drastically by the B(E2) values (fig 1,d). In ^{72}Se, the B(E2) values increase steadily – except for the dip at the 6^+ to 4^+ transition, where the rotational band merges into the yrast band. They can be estimated not to drop above spin 10^+ from Doppler broadened line shapes of the high spin transitions. In ^{70}Se they drop from the very beginning down to a single particle unit.

Besides many quasi particle states found in ^{70}Se, a high spin Isomer with $\tau = 2.3 +/- 0.4$ ns was found which feeds into the 12^+ yrast state. On the basis of a Weisskopf estimate, a spin value of (14^-) is proposed for this state.

In order to understand the structure of ^{70,72}Se, cranking calculations based on the deformed Woods Saxon potential have been performed. Details of these calculations are given in ref [4]. The Total Routhian Surfaces (TRS) were calculated [3] with pairing and were minimized with respect to the hexadecupole deformation β_4 at each (β_4,γ) mesh point.

The TRS for ^{70,72}Se shown in fig.2 predict these nuclei to be very γ-soft at the ground state and at low spins ($\hbar\omega \leq 0.3$ MeV). Three minima in the TRS are seen: two oblate minima with $\beta_2 = 0.25$ ($60°$ and $-60°$, they correspond to the same oblate configuration at $\omega = 0$), and the excited collective prolate minimum at $\beta_2 = 0.32$, $\gamma = 0°$ for ^{72}Se, $\beta_2 = 0.25$, $\gamma = -5°$ for ^{70}Se. The oblate and prolate minima are fairly well separated by a potential barrier of about 330 keV. This result supports the interpretation of the low spin states in terms of shape coexistence suggesting the ground state to be of smaller oblate deformation. At higher frequencies, the prolate minimum stabilizes at $\beta_2 = 0.33$, $\gamma = -5$ for ^{72}Se (see fig. 2c,d, $\hbar\omega = 0.5$) and it does not change in the β-γ plane up to 1.2 MeV. Calculations for ^{70}Se (see fig. 2g,h) indicate a very similar deformation pattern. However, the oblate minimum persists up to $\hbar\omega \approx 0.4$, and the well deformed configuration which stabilizes above 0.5 MeV corresponds to the near prolate shape with $\gamma = 15$, $\beta_2 = 0.27$. At $\hbar\omega = 0.4$ MeV an excited non collective oblate minimum appears.

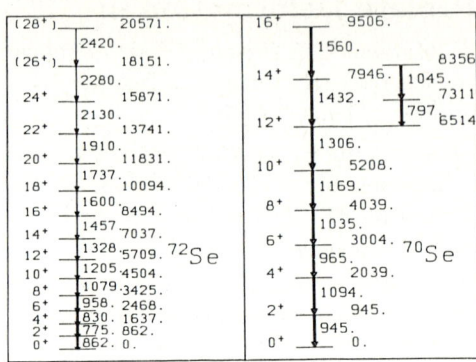

Fig.1a: Yrast band of ^{72}Se **1b:** Yrast band of ^{70}Se

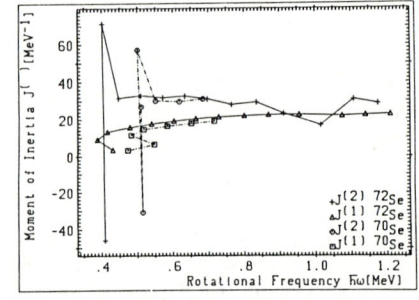

Fig.1c: Kinematical (J^1) and dynamical (J^2) moments of inertia of ^{70,72}Se.

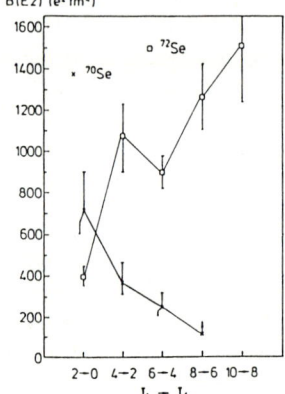

Fig.1d: Comparison of B(E2) for ^{70,72}Se, taken from [9]

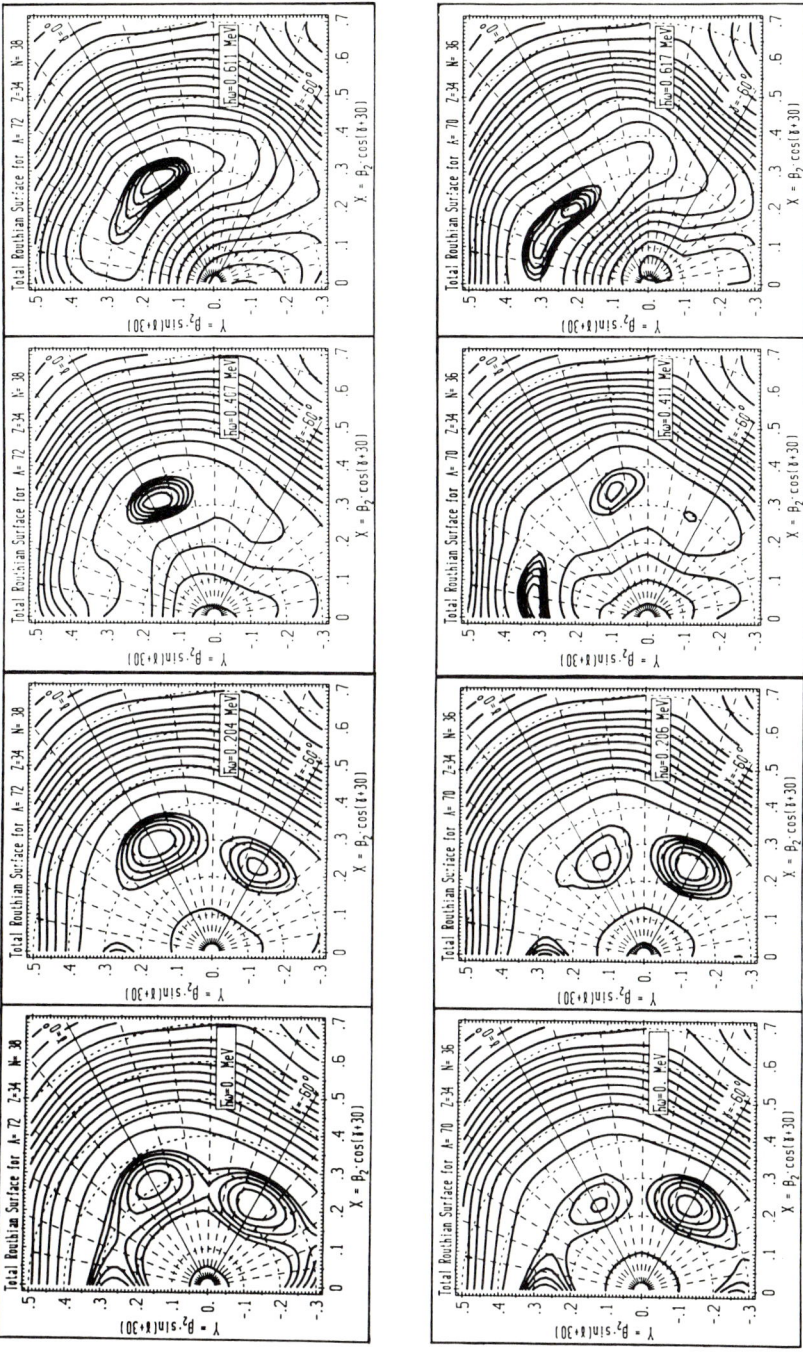

Fig.2: *Total Routhian surfaces of ^{70,72}Se, calculated at four different frequencies from $\hbar\omega=0$ to 0.6 MeV. For details see text and ref. 3.*

The quasiparticle routhians representative for the prolate deformation in ^{72}Se are shown in fig.3a (protons) and fig.3b (neutrons). In both cases the first $g_{9/2}$ crossing is predicted to appear at low rotational frequencies:
$\hbar\omega_c(p) = 0.45$ MeV, $\hbar\omega_c(n) = 0.5$ MeV. Such a dramatic decrease in predicted values of the $g_{9/2}$ crossing frequencies, when going down from the well deformed Sr and Kr isotopes towards $Z = 34$, $N = 38$, is in fact consistent with experimental systematics of proton and neutron crossings in the whole mass region. For example, the first proton crossing in the neighbouring ^{75}Br is seen [8] at 0.35 MeV. The yrast yrare interaction in ^{72}Se is predicted to be very large for protons and neutrons: about 0.35 MeV. Therefore no sharp backbendings but rather gradual alignments are expected. For ^{70}Se both, proton and neutron crossings, are predicted at $\hbar\omega_c = 0.48$ MeV. The neutron band interaction is considerably lower (see fig. 3d), but it is difficult to say, at the moment, whether this result can be correlated with the apparent strong band crossing seen in the ^{70}Se data between spins 4^+ and 8^+.

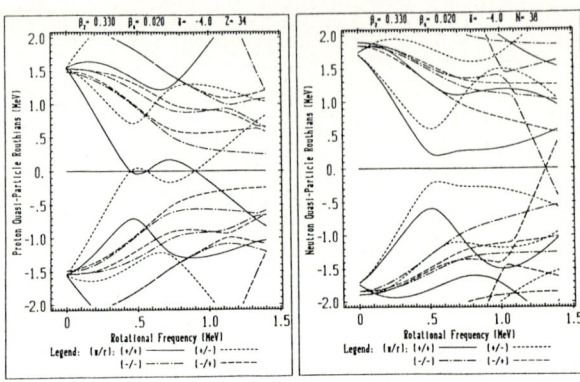

Fig.3: *Quasiparticles Routhians for ^{72}Se,*
(a) *protons*   (b) *neutrons*

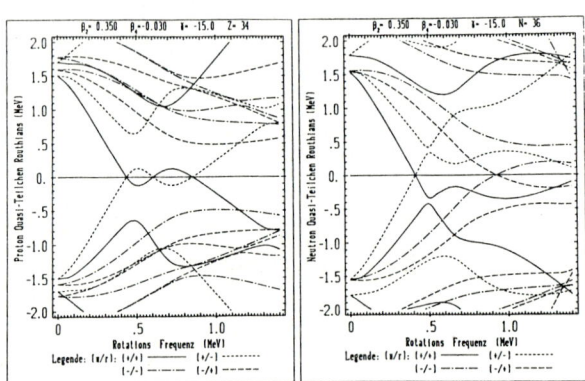

Fig.3: *Quasiparticles Routhians for ^{70}Se,*
(c) *protons*   (d) *neutrons*

Based on the results shown in fig.3 two conclusions can be drawn:
(i) The irregularities shown at low spins in ^{70,72}Se spectra are caused both by the interaction between configurations of different shape and by the alignment of $g_{9/2}$ quasiprotons followed by the $g_{9/2}$ neutron crossing. Above $I \approx 10^+$, yrast configurations have a dominating four quasiparticle component. For ^{70}Se, this interpretation is supported by the observation of the (14^-) isomer, which must be a 4 quasiparticles state according to its high spin.
(ii) The second proton and neutron crossings are predicted at much higher frequencies, above 1 MeV (see fig.3). This explains why there are no strong irregularities observed in the high spin part of the experimental data. The "kink" seen in the $J^{(2)}$ plot for ^{72}Se at 1.1 MeV may be explained in a similar way as in the case of ^{84}Zr [12], i.e. may be caused by the collapse of the static neutron pairing correlations. Such a scenario is fully consistent with the interpretation of ^{74}Se data given in ref.[13].

In summary, the yrast sequences of ^{70,72}Se have been observed up to 16^+ and (28^+), respectively. At low spins, the calculations predict an oblate to prolate shape transition mixed with the gradual alignment of $g_{9/2}$ protons and neutrons. Therefore the medium and high spin part of the yrast band in both nuclei has a four quasiparticle structure.

Now we report on the identification and investigation of ^{69}Se (see also ref.10) using **OSIRIS** in combination with an 8-fold neutron multiplicity filter as a trigger for the evaporated neutrons, described in detail in ref. 15.

The measurements performed at the Cologne FN Tandem accelerator included: (i) Prompt and delayed γ-ray singles spectra in coincidence with the evaporated neutrons, (ii) γ and nγ-excitation functions, (iii) γγ-coincidences, (iv) prompt and delayed conversion electron spectra (e and en), (v) n-gated γ-angular distributions. In these experiments ^{69}Se was populated via the reaction ^{40}Ca(^{32}S,2pn)^{69}Se at E(^{32}S) = 100 MeV using enriched (99.8%) ^{40}Ca-targets backed by a Bi-layer to stop the beam and the recoil products.

γ-rays were assigned to ^{69}Se populated in the 2pn channel from the following arguments: (i) The neutron multiplicity of the reaction channel was determined to be one, (ii) residual nuclei from oxygen impurities of the ^{40}Ca target could be excluded by comparing targets with very different oxygen concentrations: γ-rays assigned to ^{69}Se were not affected, (iii) in cross bombardements using the reactions ^{16}O + ^{58}Ni, ^{14}N + ^{58}Ni and ^{36}Ar + ^{40}Ca (^{36}Ar beam provided by VICKSI in Berlin) ^{69}Se was populated with the predicted cross section and neutron multiplicity, (iv) the experimental cross section of 2.4 % of the total fusion cross section for the reaction ^{40}Ca(^{32}S,2pn)^{69}Se compares well with the calculated values from the fusion-evaporation codes CASCADE [16] (2.4 %) and JULIAN/PACE [17] (2.7 %).

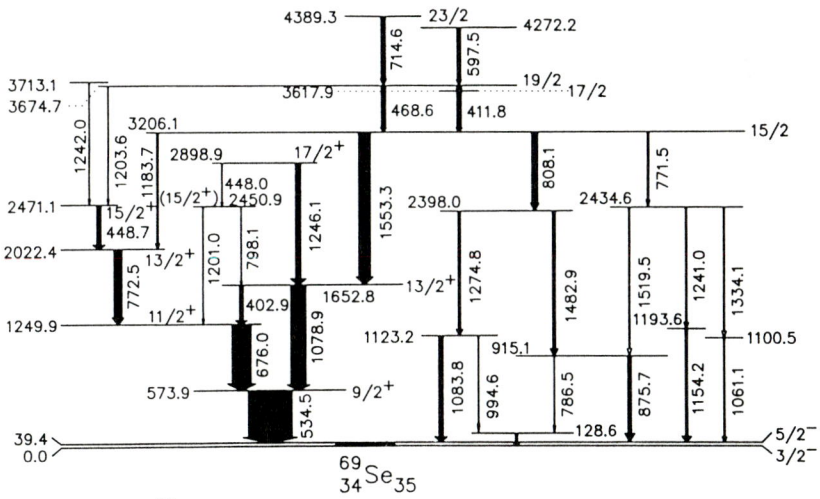

Fig.4a: Levelscheme of ^{69}Se.

The results of these measurements are comprised in fig. 4a. The groundstate spin of $J^\pi = 3/2^-$ could be determined from the analysis of the β^+-decay [18] to low lying states in ^{69}As using a recent spin assignment of $J^\pi = 3/2^-$ to the 98 keV level in ^{69}As[19]. For the 534.5 keV transition depopulating the level at E_x = 573.9 keV conversion electrons were measured. The experimental conversion coefficient α_k = 0.0042±0.0007 establishes the M2 character of the 534.5 keV transition. From this M2 assignment, from the shape of the yield function and from the nonexistence of a decay branch of the 573.9 keV level to the groundstate, spins and parities of $9/2^+$ and $5/2^-$ were assigned to the levels at 573.9 keV and 34.9 keV, respectively. Any other assignment contradicts our experimental data. From delayed nγ-coincidences a lifetime of τ = 0.4(+0.5,-0.2) sec was determined for the 573.9 keV level supporting the M2-assignement of this transition. Other spin and parity assignments given in the level scheme were extracted from n-gated γ-angular distributions, yield functions for the γ-transitions and from the fact, that no other long lived states were observed. Our results for the nucleus ^{71}Se were reported previously[20]. From a reinvestigation of the reaction ^{58}Ni(^{16}O,2pn)^{71}Se at E(^{16}O)=65 MeV using the **OSIRIS** spectrometer we identified the unfavoured $g_{9/2}$ band from nγ-coincidences. Spin and parity assignments for the new states (fig. 4c) are based on angular distributions, excitation functions and lifetime arguments[20].

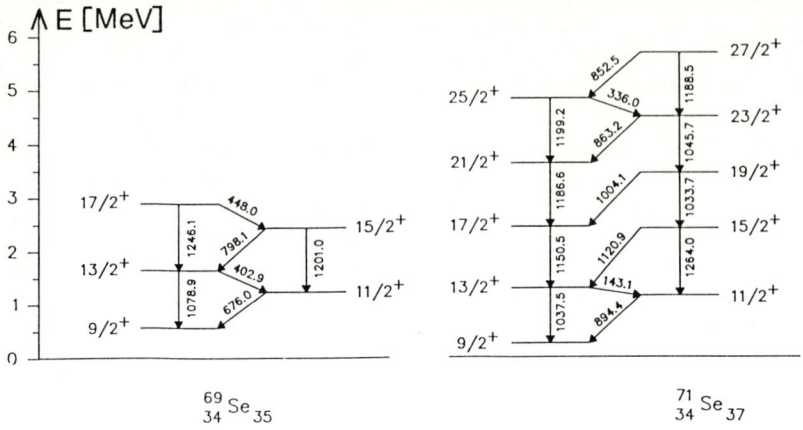

Fig.4b: Partial levelscheme of ^{69}Se **Fig.4c**: Partial levelscheme of ^{71}Se: the $g_{9/2}$ bands.

The results of calculations in the asymmetric rotor + particle model for the positive parity states in ^{69,71}Se using the code AROVM2 [21] are shown in fig. 5. For both nuclei the $g_{9/2}$ bands can only be reproduced by the model if an oblate deformation with γ close to 60° is assumed (notation used: $\gamma=0°$: prolate, $\gamma=60°$: oblate) but not for triaxial or prolate shapes.

The calculated oblate shape agrees with the observed small signature splitting especially in ^{69}Se indicating small Coriolis interaction and – in consequence – a high K-value of the involved Nilsson configuration. But high K-Nilsson orbitals, namely the [404]9/2 and [413]7/2 orbits imply an oblate deformation of this band whereas in the case of prolate deformation only low K-orbits are involved leading to decoupled bands (See e.g. ref. 4).

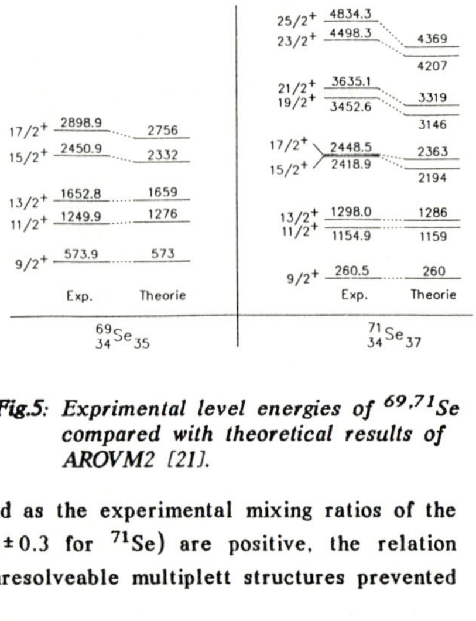

In addition, the oblate structure of the $g_{9/2}$-bands in both nuclei could also be derived from the sign of the mixing ratios of the 11/2$^+$ to 9/2$^+$ intraband transitions. Calculations in a Nilsson model code [11] give $-0.78 < g_K - g_R < -0.66$ for the [404]9/2 and [413]9/2 Nilsson configuration using a quenched $g_s = 0.7 g_{free}$ and $g_R = Z/A$. In these calculations a deformation range of $-0.30 < \beta_2 < -0.20$ and $-0.01 < \beta_4 < 0.01$

Fig.5: Exprimental level energies of $^{69,71}Se$ compared with theoretical results of AROVM2 [21].

was taken into account. As $g_K - g_R$ is negative in all cases and as the experimental mixing ratios of the 11/2$^+$ to 9/2$^+$ transitions ($\delta = 3.3 \pm 0.4$ for ^{69}Se and $\delta = 1.6 \pm 0.3$ for ^{71}Se) are positive, the relation sign(δ) = sign(($g_K - g_R$)/Q_0) gives a negative value of Q_0. Unresolveable multiplett structures prevented angular distribution analysis of other intraband transitions.

The spectra of $^{69-72}$Se are governed by the $g_{9/2}$ intruder states. In the even isotopes, proton and neutron crossings are very low ($\hbar\omega \leq 0.5$ MeV), which reflects the fact, that low K subshells are the intruders on the prolate side. In the odd isotopes, rotational bands are observed also in the oblate minima of the potential, but they do not reach as high spins as they do at prolate shape and show no backbending. This fact probably reflects the small Coriolis interaction in the j=9/2, high K subshells. At frequencies of about 0.4-0.5 MeV, these bands drain out and non collective oblate states are observed at high spins. The systematic trend observed in the light Se isotopes – when going from heavier to lighter isotopes – is that oblate shapes dominate nuclear spectra more and more.

References

1. J.H.Hamilton et.al., Phys. Rev. Lett. 32, 239, (1974); and
   A.V.Ramayya et.al., Phys. Rev. C12, 1360, (1975)
2. J.H.Hamilton, Lecture Notes in Physics 168,
   Heavy Ion Collision (Springer Verlag), 1982, p.287; and
   R.B.Piercey et.al., Phys.Rev.Lett. 47, 1514 (1981)
3. R.Bengtsson, W.Nazarewicz, Proc. XIX Winter School on Physics, ed. Z.Stachura,
   Report IFJ No. 1268 (1984), p.171
4. W.Nazarewicz, J.Dudek, R.Bengtsson, T.Bengtsson and I.Ragnarsson,
   Nucl.Phys. A435, 397 (1985)
5. K.O.Zell, B.Heits, W.Gast, D.Hippe, W.Schuh and P.von Brentano,
   Z.Phys.A, 279, 373 (1976); and
   Li Guangsheng, L.Cleemann, J.Eberth, T.Heck, W.Neumann, M.Nolte, J.Roth,
   Chinese Jou. Nucl. Phys. 5, 217 (1983)
6. M.A.Herath-Banda et. al., J.Phys.G: Nucl.Phys., 13, 43 (1987)
7. B.Wörmann, K.P.Lieb,R.Diller, L.Lühmann, J.Keinonen, L.Cleemann, J.Eberth,
   Nucl. Phys. A431, 170 (1984)
8. L.Lühmann et.al., Phys.Rev.C, 31, 828 (1985)
9. B.Wörmann et.al., Z. Phys.A, 322, 171 (1985)
10. M.Wiosna, J.Busch, J.Eberth, M.Liebchen, T.Mylaeus, N.Schmal, R.Sefzig, S.Skoda,
    W.Teichert, Phys. Lett. 200 B, 255 (1988)
    L.Lühmann et.al., Phys.Rev. 31C, 828 (1985)
11. R.Bengtsson and S.Frauendorf, Nucl. Phys. A327 (1979) 139; and
    G.Hebbinghaus, Diploma thesis, Jülich 1985, unpublished.
12. J.Dudek, W.Nazarewicz and N.Rowely, Phys.RevC 35, 1489 (1987)
13. W.Nazarewicz, G.A.Leander and S.Tabor, to be published.
14. R.M.Lieder, H.Jäger, A.Neskakis, T.Venkova, C.Michel,
    Nucl.Inst.Meth., 220, 363 (1984)
15. J.Roth, L.Cleemann, J.Eberth, T.Heck, W.Neumann, M.Nolte,
    R.B.Piercey, A.V.Ramayya, J.H.Hamilton,
    Proceedings of 4th Int. Conf. on Nucl. far from Stab., publ. CERN 81-09, 680, (1981)
16. F.Pühlhofer, Nucl. Phys.A280, 267 (1977)
17. A.Gavron, Phys.Rev.C, 21, 230 (1980)
18. J.A.Macdonald et.al., Nucl. Phys A288, 1 (1977)
19. R.Sefzig et. al., Verh. DPG 4/87 (1987); and to be published
20. J.Eberth, L.Cleemann, N.Schmal, Int. Symp. on in-beam Nucl. Spectr.,
    Debrecen, Hungary 1984, p.23, ed. by Z.S. Dombradi and T.Fanyes,
    Akademiai Kiado, Budapest
21. J.Eberth et. al., Verh. DPG 4/87 (1987); and to be published
21. H.Toki and A.Faessler, Nucl. Phys. A253, 231 (1975)
22. W.Nazarewicz, J.Dudek, R.Bengtsson and I.Ragnarsson,
    Nucl. Phys A435, 397 (1985)

Moments and Radii of $^{78-100}$Sr

R. Neugart[1], E. Arnold[1], W. Neu[1], K. Wendt[1], P. Lievens[2],
R.E. Silverans[2], L. Vermeeren[2], F. Buchinger[3], E.B. Ramsay[3], G. Ulm[4],
and The ISOLDE Collaboration[4]

[1]Institut für Physik, Universität Mainz,
 D-6500 Mainz, Fed. Rep. of Germany
[2]Instituut voor Kern- en Stralingsfysika, Leuven University,
 Celestijnenlaan 200 D, B-3030 Leuven, Belgium
[3]Foster Radiation Laboratory, Mc Gill University, Montreal, Canada
[4]CERN, CH-1211 Geneva 23, Switzerland

1. Introduction

The chain of Sr isotopes ranges from the neutron-shell closure at N = 50 into both the N = 38 and N = 60 deformation regions which represent the main topic of this workshop. For a detailed understanding of the nature of these nuclei, laser spectroscopy can provide the ground state spins and moments, as well as the the changes in the mean square charge radii as a function of the neutron number N. Recent experiments at Karlsruhe [1] and at Daresbury [2] essentially cover the neutron-deficient and stable Sr isotopes between N = 40 and N = 50. In order to complement these results, and to extend the measurements into the region of neutron-rich isotopes, we have performed an experiment at the ISOLDE isotope separator facility at CERN.

2. Experimental Remarks

The neutron-deficient isotopes were produced by spallation in a 50 g/cm² niobium powder target, and the neutron-rich ones by fission of uranium (13 g/cm²) which is in the form of UC₂. Both types of reactions are induced by 600 MeV protons. The Sr isotopes, released from the hot (2100 °C) target, are ionized on a tungsten surface and obtained from the on-line mass separator in the presence of the Rb isobars. This beam contamination is only a minor drawback for conventional collinear laser spectroscopy measurements using the selective detection of fluorescence photons. By this technique we have investigated all the isotopes and a few isomers in the sequence $^{78-98}$Sr.

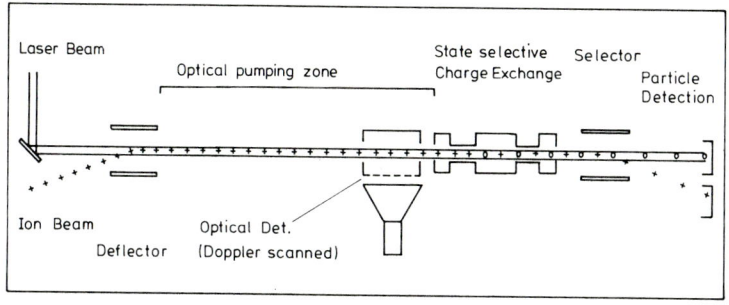

Figure 1. Schematic view of the setup for non-optical detection.

An extension towards the more neutron-rich isotopes in the A = 100 deformation region has become possible by the recently developed non-optical detection scheme using the state selectivity of charge-transfer neutralization to detect optical pumping of an ion [3]. It consists in the optical excitation of the Sr$^+$ ion from the 5s ^{2}S$_{1/2}$ ground state to the excited 5p ^{2}P$_{3/2}$ state and the subsequent decay into the metastable 4d ^{2}D$_{3/2,5/2}$ levels. For these latter the neutralization cross-section in collisions with Na atoms is enhanced, and thus the optical resonance can be detected as an increase of the rate of neutral atoms reaching the secondary-electron detector (Figures 1 and 2). The minimum beam intensity required for such a measurement is well below 10^3 atoms/s. Measurements by the conventional fluorescence technique were performed in the same transition of the alkali-like Sr$^+$ ion which is also the most favourable one for the evaluation of nuclear moments and radii from the hyperfine structures and isotope shifts.

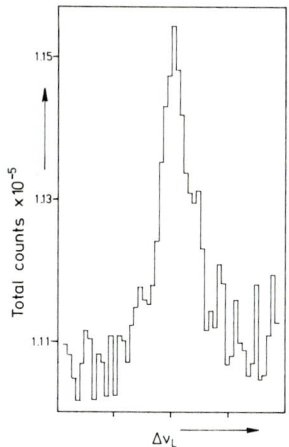

Figure 2. Particle-detected resonance signal 5s ^{2}S$_{1/2}$ - 5p ^{2}P$_{3/2}$ of ^{100}Sr from a 2 s/channel laser frequency scan. The beam intensity was about 10^4 ions per second.

3. Mean Square Charge Radii

From the measured isotope shifts one extracts the changes in the nuclear mean square charge radii, with respect to ^{88}Sr which has a closed N = 50 neutron shell. This evaluation follows the semi-empirical standard procedures [4] using electron densities at the site of the nucleus that are deduced from the contact hyperfine interaction in the $^2S_{1/2}$ ground state or from the atomic energy level sequence (Goudsmit-Fermi-Segre formula). In a medium-mass element like Sr, an important contribution to the isotope shift is due to the change in nuclear recoil energy (mass shift). This contribution is obtained most reliably from a comparison with the isotope shift data for muonic atoms which are available for a few of the stable isotopes [5]. For details see [6] and the previous publications about Sr isotope shifts [1,2].

The results, i.e. the changes in $<r^2>$ relative to ^{88}Sr, are shown in Figure 3. Our $\delta<r^2>$ values largely agree with those of refs. [1] and [2] for the neutron-deficient isotopes. The radii are decreasing as the neutron shell closure at N = 50 is approached from the neutron-deficient side. For N > 50, they increase almost linearly, except for the striking jump between N = 59 and N = 60 which coincides with the onset of strong deformation. In the framework of the droplet model [7] the variation in $<r^2>$ can be interpreted as a combined effect of the change in the volume and the deformation of the charge distribution, including a redistribution correction to account for the Coulomb repulsion. Restricting the deformation part in first order to a change in quadrupole deformation $<\beta^2>$ and assuming a constant diffuseness,

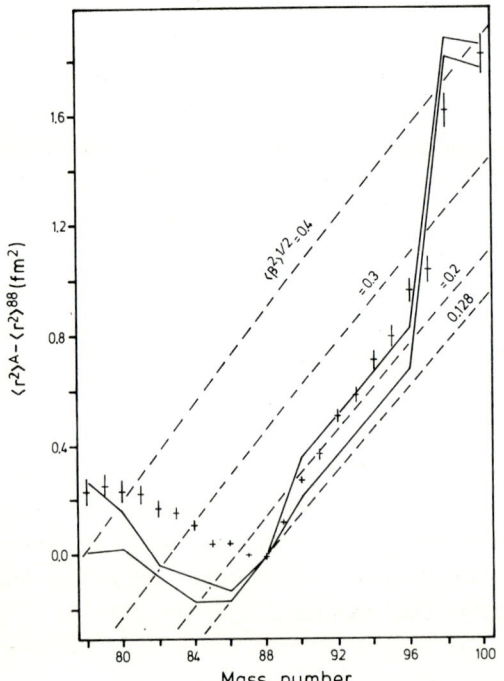

<u>Figure 3.</u> Changes in mean square charge radii for the Sr isotopes. Droplet model isodeformation curves (broken lines) and $\delta<r^2>$ calculated from B(E2) values (error band indicated as full lines) are included.

the droplet values for the changes in $<r^2>$ were calculated using the model parameters of ref. [8]. They are included in Figure 3 as a set of isodeformation lines. From a comparison of the experimental data with these lines, it seems evident that the overall trends of the observed radii are well described by deformation effects within the droplet picture. More quantitatively, the experimental $\delta<r^2>$ values can be compared to calculated ones obtained from the droplet model using $<\beta^2>$ from experimental B(E2) values [9,10]. For a few isotopes (90 ≤ A ≤ 96) the deformation is estimated [11] from the energy of the first excited 2^+ level, because no B(E2) values are available. The calculated curves are also shown in Figure 3. The agreement with the experimental values is rather satisfactory: Deformations of $\beta \approx 0.4$ account well for the radii both at N ≈ 40 and N ≈ 60, and for the neutron-rich isotopes this deformation occurs suddenly between N = 59 and N = 60, while for the light isotopes it develops gradually from N ≈ 46 to N = 40. Still, some unexplained discrepancy remains in this transitional region, which may be related to the observed β and γ softness [12,13].

The trends of our present $\delta<r^2>$ curve for the Sr isotopes agree well with those for the neighbouring odd-Z element Rb [14]. Myers and Rozmej [15] used this case for an investigation of the influence of collective zero-point motion on the development of the mean sqare charge radii. This was mainly to find an explanation for the generally observed pronounced kink at magic neutron numbers which is hardly reproduced by theoretical calculations. They concluded that the calculated collective width $<\beta^2>$ cannot account for the observed pronounced shell effect. Reasons for this discrepancy were sought in higher-order multipole contributions or core polarization effects in the transitional nuclei. However, Figure 3 shows that the droplet model approach including empirical information about $<\beta^2>$ from B(E2) values reproduces the $\delta<r^2>$ data fairly well. Therefore, it should be examined to what extent the collective zero-point motion based on a Nilsson single-particle potential reproduces the experimentally observed E2 strength.

4. Spins and Moments

Hyperfine splittings were measured for the odd-A isotopes $^{79-97}$Sr including the isomers ^{83m,85m,87m}Sr. They yield the ground-state spins and the hyperfine interaction constants $A(^2S_{1/2})$, $A(^2P_{3/2})$ and $B(^2P_{3/2})$ from which the magnetic moments and electric quadrupole moments are deduced. For ^{79}Sr, ^{83m}Sr and all isotopes beyond ^{89}Sr these results constitute the first direct spin measurements. They confirm the assignments from nuclear spectroscopy except for ^{93}Sr which has the spin I = 5/2 and ^{97}Sr which has I = 1/2. The latter case is particularly interesting in connection with shape coexistence in N = 59 nuclei [16]. A discussion of the spin determination is given in ref. [17].

From the A and B factors one deduces the nuclear moments using the precisely known reference values for the stable ^{87}Sr, μ = -1.09282(65) n.m. [18] and Q_s =0.335(20) barn [19]. These moments are plotted in Figure 4 for the nuclei which can be described by the shell-model neutron configurations $p_{1/2}$ and $g_{9/2}$ below N = 50, and $d_{5/2}$ and $s_{1/2}$ above. The magnetic moments show the usual quenching of the single-particle Schmidt moments which is due to core polarization [20], except for those of the $p_{1/2}$ states of ^{81,83,85}Sr. The quenching effect increases with the distance from N = 50, as expected from the increasing admixture of collective states. This trend can be followed in the $g_{9/2}$, $d_{5/2}$ and $s_{1/2}$ ground states which all have μ_{sp} =-1.91 n.m. For the quadrupole moments of the $d_{5/2}$ states in ^{89,91,93}Sr one observes a change from negative to positive values which is similar to the behaviour of the $f_{7/2}$ moments above N = 82 [21]. This is predicted by the shell model and understood also in the collective picture as a change in the projection of a positive Q_0 on the spin axis [22].

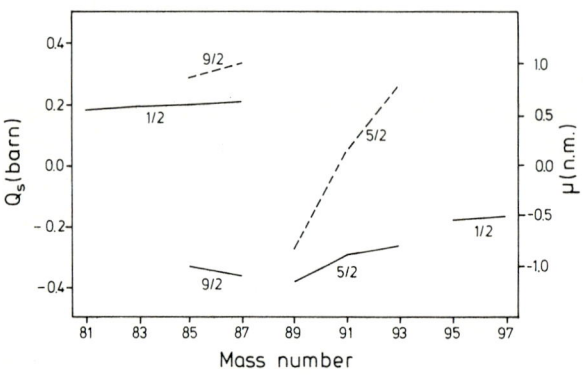

Figure 4. Magnetic moments (full lines) and electric quadrupole moments (dashed lines) of the near-spherical Sr isotopes.

For the I = 3/2 ground state of the deformed ^{79}Sr, the observed moments μ = -0.47 n.m. and Q_s = 0.74 barn are consistent with the assignment of a 3/2-[301] neutron orbital. Calculations of these moments had been performed by Heyde et al. [23] within the model of the odd particle coupled to the deformed even-even core, and the predictions are qualitatively consistent with our experimental results.

The new technique of state-selective charge transfer neutralization gives very good prospects for a continuation of the present collinear fast-beam laser experiments in both deformed regions around A = 100 and A = 78 which should yield further interesting results about the spins and moments of these nuclei.

This work was supported by the Belgian IIKW, the Canadian NSERC, and the German BMFT under the contract number 06MZ458I.

References

1. M. Anselment, K. Bekk, S. Chongkum, S. Göring, A. Hanser, H. Hoeffgen, W. Kälber, G. Meisel, and H. Rebel, Z. Phys. **A 326** (1987) 493
2. D.A. Eastham, P.M. Walker, J.R.H. Smith, D.D. Warner, J.A.R. Griffith, D.E. Evans, S.A. Wells, M.J. Fawcett, and I.S. Grant, Phys. Rev. **C 36** (1987) 1583
3. R.E. Silverans, P. Lievens and L. Vermeeren, Nucl. Instr. Meth. **B 26** (1987) 591
4. K. Heilig and A. Steudel, At. Data Nucl. Data Tables **14** (1974) 613
5. R. Engfer, H. Schneuwly, J.L. Vuilleumier, H.K. Walter, and A. Zehnder, At. Data Nucl. Data Tables **14** (1974) 509
6. R.E. Silverans, P. Lievens, L. Vermeeren, E. Arnold, W. Neu, R. Neugart, K. Wendt, F. Buchinger, E.B. Ramsay, and G. Ulm, submitted to Phys. Rev. Letters
7. W.D. Myers and K.H. Schmidt, Nucl. Phys. **A 410** (1983) 61
8. D. Berdichevsky and F. Tondeur, Z. Phys. **A 322** (1985) 141
9. S. Raman, C.H. Malarkey, W.T. Milner, C.W. Nestor, jr., and P.H. Stelson, At. Data Nucl. Data Tables **36** (1987) 1
10. H. Ohm, G. Lhersonneau, K. Sistemich, B. Pfeiffer, and K.-L. Kratz, Z. Phys. **A 327** (1987) 483
11. L. Grodzins, Phys. Lett. **2** (1962) 88
12. W. Nazarewicz, J. Dudek, R. Bengtsson, T. Bengtsson, and I. Ragnarsson, Nucl. Phys. **A 435** (1985) 397
13. P. Bonche, H. Flocard, P.H. Heenen, S.J. Krieger, and M.S. Weiss, Nucl.Phys. **A 443** (1985) 39
14. C. Thibault, F. Touchard, S. Büttgenbach, R. Klapisch, M. de Saint Simon, H.T. Duong, P. Jacquinot, P. Juncar, S. Liberman, P. Pillet J. Pinard, J.L. Vialle, A. Pesnelle, and G. Huber, Phys. Rev. **C 23** (1981) 2720
15. W.D. Myers and P.Rozmej, Nucl. Phys. **A 470** (1987) 107
16. G. Lhersonneau, B. Pfeiffer, K.-L. Kratz, H. Ohm and K. Sistemich, contribution to this workshop
17. F. Buchinger, E.B. Ramsay, R.E. Silverans, P. Lievens, E. Arnold, W. Neu, R. Neugart, K. Wendt, and G. Ulm, Z.Phys. **A 327** (1987) 361
18. L. Olschewski, Z. Phys. **249** (1972) 205
19. S.M. Heider and G.O. Brink, Phys. Rev. **A 16** (1977) 1371
20. A. Arima and H. Hyuga, in Mesons in Nuclei, ed. by M. Rho and D.H. Wilkinson (North Holland, Amsterdam, 1979), p.683
21. R. Neugart, K. Wendt, S.A. Ahmad, W. Klempt, and C. Ekström, Hyperfine Interactions **15/16** (1983) 181
22. C. Ekström, in Proc. 4th Int. Conf. on Nuclei Far from Stability, Helsingör (1981), CERN 81-09, p.12
23. K. Heyde, J. Moreau, and M. Waroquier, Phys. Rev. **C 29** (1984) 1859

Experimental Evidence for Shape Coexistence in $^{97}Sr_{59}$ and Implications for the Structure of the Odd-Odd Isotone ^{98}Y

G. Lhersonneau[1], K.-L. Kratz[1], H. Ohm[2], B. Pfeiffer[1], and K. Sistemich[2]

[1] Institut für Kernchemie, Universität Mainz,
 D-6500 Mainz, Fed. Rep. of Germany
[2] Institut für Kernphysik, Kernforschungsanlage Jülich,
 Postfach 1913, D-5170 Jülich, Fed. Rep. of Germany

Abstract

Experimental evidence for shape coexistence in the odd-mass N=59 isotone ^{97}Sr is presented. The ground state and the lowest excited levels of ^{97}Sr are shown to be spherical, whereas a rotational band based on a state at 585 keV has been identified. Three further levels of deformed origin are proposed. Nilsson-orbital assignments are supported by RPA shell-model calculations. These findings lead to a better understanding of the structure of the odd-odd N=59 isotone ^{98}Y.

Introduction

The region of neutron-rich nuclei near A=100 is fascinating as a great variety of nuclear structures is encountered within a narrow range of proton and neutron numbers. In particular, for the $_{38}$Sr, $_{39}$Y and $_{40}$Zr nuclei a sudden onset of strong deformation at N=60 has been observed [1], whereas the lighter isotopes up to N=58 keep a stable spherical shape. This raises the question about the nature of the N=59 isotones of Sr, Y and Zr, which are the link between the spherical and deformed regions and for which shape coexistence is expected. In the framework of our systematic study of N=59 isotones, we have reinvestigated the level scheme of ^{97}Sr which was first reported from a β-decay study of ^{97}Rb [2,3]. Recent Laser-spectroscopic measurements have shown that ^{97}Sr has a ground-state (g.s.) spin of I=1/2 which has been interpreted as a $\nu s_{1/2}$ shell-model state [4]. Furthermore, the striking similarities between the lowest-excited levels in the N=59 isotones ^{97}Sr and ^{99}Zr on the one hand, and in the N=57 spherical isotones ^{95}Sr and ^{97}Zr on the other hand [5] strongly suggest that the first and second excited levels in ^{97}Sr are also spherical. The main purpose of the present investigation thus is to search for a possible rotational band as a signature of deformation in ^{97}Sr. From analogy with ^{98}Y [1], such a band might be expected around 500 keV excitation energy. This expectation is further supported by RPA shell-model calculations of the Gamow-Teller (GT) strength in the β-decay of ^{97}Rb to ^{97}Sr [6,7]. These calculations only predict a $\nu s_{1/2}$ g.s. for ^{97}Sr at deformations of β<0.06. On the other hand, the sizeable β-branches to a bunch of levels around 600 keV observed in experiment [2,3] are only reproduced when deformations of β≈0.35 are assumed.

Experimental Procedure

The isotope ^{97}Sr was produced as the β-decay daughter of ^{97}Rb at the on-line mass-separator facility OSTIS at the ILL Grenoble [8]. Singles and γ-γ-t coincidence spectra have been recorded. The main contaminants turned out to be the β-decay activities from ^{97}Sr, ^{97}Y and ^{96}Sr, the latter being produced via β-delayed neutron emission of ^{97}Rb. They provided the standards for internal energy and efficiency calibrations. Conversion coefficients were measured by the fluorescence method. For the lifetime measurements the centroid-shift method was used [9].

Results and Discussion

As mentioned above, the g.s. of ^{97}Sr has recently been measured to have I=1/2, in contrast to the I$^\pi$=3/2$^+$ proposed earlier [10]. The new measurement, together with the M1 multipolarity of the 167.1 keV and the E2 multipolarity of the 141.0 keV transitions leads to 3/2$^+$ and 7/2$^+$ assignments to the levels at 167.1 keV and 308.1 keV, respectively (see Fig.1 and Fig.3 of [3]). Thus, these levels have a character similar to the corresponding ones in the spherical nuclei ^{95}Sr$_{57}$ and ^{97}Zr$_{57}$. It is worth to be mentioned in this context that the same situation is encountered in ^{99}Zr$_{59}$, the isotone of ^{97}Sr.

The I$^\pi$=1/2$^+$ ground state is associated to the $(g_{7/2})^2 s_{1/2}$ neutron configuration, the neutrons being counted from the closure of the $d_{5/2}$ subshell. In our shell-model calculations, a 1/2$^+$ g.s. occurs only at very small deformations.

The first excited state has I$^\pi$=3/2$^+$. It is tempting to assign it to the $d_{3/2}$ single neutron. However, BCS as well as recent IBFFM calculations for the odd-odd nucleus ^{96}Y$_{57}$ [11], indicate that this level is more complex, containing $g_{7/2}$ and $d_{3/2}$ components.

The I$^\pi$=7/2$^+$ state at 308.1 keV can be associated to the $g_{7/2}$ single neutron. Its isomeric character ($t_{1/2}$= 170 ns) results from the E2 transition to the 3/2$^+$ level, with a rate close to the single-particle estimate, and is not associated with the [404]9/2 intruder as proposed in [10].

The level at 522.3 keV is likely to be also spherical, similar to the 3/2$^+$ levels in the N=57 isotones ^{95}Sr (680.7 keV) and ^{97}Zr (1400.0 keV).

In spite of the allowed character of the β-transitions from the deformed I$^\pi$=3/2$^+$ ^{97}Rb parent [12] to the 1/2$^+$ and 3/2$^+$ levels, the logft values are quite high. The reduced transition strengths obviously reflect shape hindrance.

The levels at 585.1, 687.1 and the new one at 822.4 keV (see Fig.1) are proposed to form a rotational band with a K=3/2 band head. Selection rules for β- and γ-transitions allow I$^\pi$=(1/2,3/2)$^+$ for the 585.1 keV level. The analysis of the proposed band in the frame of the rotational model [13] is shown Table 1.

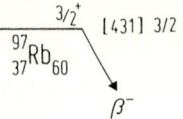

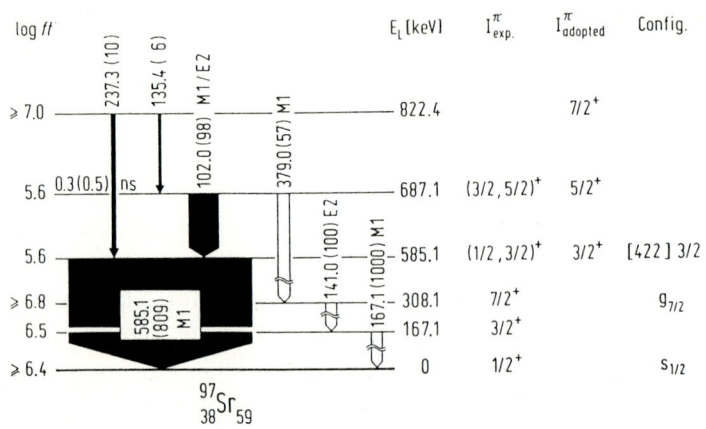

Fig.1: Partial level scheme of ^{97}Sr showing the levels relevant for the discussion of shape coexistence.

For K=3/2, the deduced E2/M1 mixing ratio for the 102.0 keV transition from the 687.1 keV level to the band head, agrees very well with the experimental one from the conversion coefficient [3]. Also the moment of inertia of 80% of the rigid rotor value fits well into the systematics in the A=100 region. The 3/2 assignment for the band head is further supported by the logft ≥7.0 for the 822.4 keV level, which suggests I=7/2 rather than a lower value. The E2 enhancement and the quadrupole deformation can be deduced from the half-life of the first excited level of the band, for which a value of $t_{1/2}$=0.3(5) ns has been measured. After correction for competing γ-rays and conversion, a partial half-life for the 102.0 keV transition of 1.8(30) ns is obtained. Although the accuracy is poor, this value shows clearly the collective character of the 102.0 keV transition (see Fig.2) and gives a lower limit of β≈0.2 for quadrupole deformation. These values, indeed, agree very well with the corresponding ones in the deformed neighbours [9]. From the |(g_K-g_R)/Q_0| band parameter (see Table 1), values for the intrinsic g-factor of g_K=0.58(3) or 0.02(3) are deduced. We used β=0.35 for computing Q_0 and g_R=0.3, a value somewhat lower than Z/A, as usual for odd-neutron nuclei. The error of 0.03 reflects only the uncertainties in the γ-branchings. This yields values of the projection of the spin on the symmetry axis <s_z>, of -0.38 or -0.01 respectively. The first value strongly suggests the [422]3/2 Nilsson orbital to be the band head. The second value implies a mixed configuration. However, the nearest candidate for mixing, namely the [411]3/2 orbital, is too far at the assumed deformation. Therefore, the interpretation of the levels at 585.1, 687.1 and 822.4 keV as forming a rotational

band, with a [422]3/2 band head, seems to be firm. Our RPA calculations predict this
Nilsson orbital around 0.55 MeV with a logft for GT-decay of 5.3.

K	J/J_{rigid}	$(g_K-g_R)/Q_0$	$\delta_{102\ keV}$	$\delta_{exp.}$[3]
1/2	52%	0.315(30)	0.27(3)	
3/2	80%	0.088(10)	0.47(5)	0.42(12)

Table 1: Analysis of the levels at 585.1, 687.1 and 822.4 keV in the frame of the rotational model.

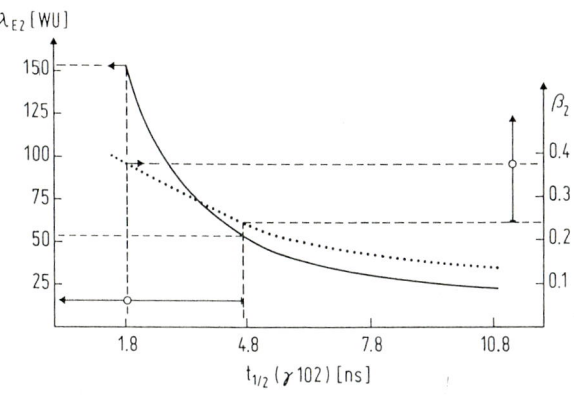

Fig.2: Collectivity of the E2 component in the 102 keV transition (full line) and the corresponding quadrupole deformation parameter β of the proposed band (dotted line) versus the γ-half-life of the 102 keV transition.

The 5/2+ level at 600.5 keV [3], is probably also deformed, as suggested by the quite low logft value of 5.8 to this level. On the one hand, this value is close to the values of 5.6 for the 3/2+ and 5/2+ members of the K=3/2 band identified above, but on the other hand it is significantly lower than the logft values to the low-lying spherical states. The 600.5 keV level could be the [413]5/2 Nilsson orbital which is predicted around 0.67 MeV with a logft(RPA)=5.9.

The level at 644.7 keV [3] has most probably $I^\pi=3/2^-$. Conversion coefficients of the 44.3 and 59.7 keV lines to the levels at 600.5 keV (5/2+) and 585.1 keV (3/2+) respectively (see Table 2) suggest E1 rather than M1 for both transitions. This results in the odd parity for the level at 644.7 keV. Moreover, the multipolarity of the g.s. transition is likely E1 rather than M2. Indeed, the transition rate of 0.6 Weisskopf units, if M2, and the non-observation of this line in the electron spectra of [3] make the M2 alternative unlikely. As for the level at 600.5 keV, the logft value of 5.8 again suggests deformed origin. This level might have the [541]3/2 Nilsson configuration.

The nature of the higher-lying levels is difficult to understand due to the lack of firm spin and parity assignments. Some information can possibly be gained from the

Table 2: Conversion coefficients. Using $t_{1/2}(713.8$ keV$)=1.7(4)$ ns [3], the 69.1 keV transition has an E2 enhancement of 120(80).

E_{keV}	α_K(E1)	α_K(E2)	α_K(M1)	α_K(exp)	α(exp)	Gate	deduced
44.3	1.0	13.8	· 1.3	0.61(33)		600	(E1)
					0.98(50)	69	
59.7	0.41	5.25	0.62	0.24(17)		585	(E1)
					0.43(18)	69	
				0.50(15)		[3]	
69.1	0.27	3.14	0.41	0.53(15)		44	M1/E2
				0.53(12)		59	$\delta=0.18(7)$
				0.48(8)		[3]	

logft values. For example, the level at 985.3 keV [3], which is fed with a logft value of 5.8, might be of deformed origin and can tentatively be assigned as the [420]1/2 Nilsson orbital. This is supported by our RPA calculations which predict this level to lie around 0.86 MeV with a logft≈5.7 for GT-decay.

Structure of the Odd-odd Nucleus $^{98}Y_{59}$

The new information on the odd-neutron nucleus ^{97}Sr allows a better understanding of the structure of its odd-odd isotone ^{98}Y. Recent calculations with the IBFFM model [14] are able to reproduce the lowest levels of the latter isotope in a spherical basis; the g.s. (I^{π}=0-),the levels at 119.3 keV (I^{π}=1-),170.8 keV (I^{π}=2-) and 375.0 keV (proposed I^{π}=4-) result from the coupling of the odd $\pi p_{1/2}$, which is the g.s. of the spherical neighbour ^{97}Y, to the spherical neutron states in ^{97}Sr discussed above, i.e. the g.s. (I^{π}=1/2+), the levels at 167.1 keV (I^{π}=3/2+) and 308.1 keV (I^{π}=7/2+). It is worth to note that the I^{π}=0- assignment to the ^{98}Y g.s. is in agreement with the recently remeasured β-feedings in the A=98 mass chain [14], which indicate first-forbidden character of the g.s. transitions. The 1+ levels in ^{98}Y at 547.8, 600.2 and 986.3 keV [14], which are fed by fast β-transitions with logft values of 4.8, 4.4 and 5.1 respectively, are most probably deformed. The fast decays suggest GT transitions between spin-orbit partner configurations. We propose the following configurations π[431]3/2;ν[413]5/2, π[422]5/2;ν[422]3/2 and π[431]3/2;ν[420]1/2 respectively. Indeed, the [422]5/2 proton orbital is the g.s. of the deformed neighbour ^{99}Y [16,17,18] whereas the [431]3/2 state, although not unambiguously identified, should lie low in energy. Finally, the 2.0 s β-decaying isomer in ^{98}Y, recently assigned $I^{\pi}=5^+$ [19], has been proposed to have the spherical configuration resulting from coupling the $\pi_{g9/2}$ level, at 0.67 MeV in ^{97}Y, to the g.s. of ^{97}Sr.

Conclusions

The present study has shown the existence of a rotational band based on the level at 585.1 keV in ^{97}Sr to which the Nilsson orbital [422]3/2 can be assigned. Further levels are proposed to be of deformed origin at 600.5, 644.7 and 985.3 keV with the tentative assignments [413]5/2, [541]3/2 and [420]1/2, respectively. Together with the recent experimental evidence for sphericity of the low-lying states, this establishes shape coexistence in the N=59 isotone ^{97}Sr. These findings are of crucial importance for the understanding of the sudden shape transition between the spherical N=58 and the deformed N=60 isotopes of $_{38}$Sr, $_{39}$Y and $_{40}$Zr. In particular, the structure of the odd-odd nucleus ^{98}Y$_{59}$ can now be better understood.

Further investigations are planned in order to see if the picture of shape coexistence also persists in the next neighbour of ^{98}Y, the odd-neutron nucleus ^{99}Zr$_{59}$ which has a proton pair more than ^{97}Sr. This should provide information about the influence of the proton-shell closure at Z=40 on the phenomenon of shape coexistence at N=59.

We acknowledge valuable discussions with Drs. L. Peker and R. Neugart. This work was supported by the German BMFT(06 MZ 552).

References

[1] G. Lhersonneau et al., ACS Symp. Ser. 324, 202 (1986)
[2] B. Pfeiffer et al., Report CERN B 81-09, 423 (1981)
[3] K.-L. Kratz et al., Z.Physik A312, 43 (1983)
[4] F. Buchinger et al., Z.Physik A327, 361 (1987)
[5] H. Ohm et al., 15th Int.Conf. on Nuclear Physics, Gaussig GDR (1987)
[6] J. Krumlinde and P.Möller, Nucl.Phys. A417, 419 (1984)
[7] K.-L. Kratz et al., ACS Symp. Ser. 324, 159 (1986)
[8] K.D. Wünsch, Nucl.Instr.Meth. 155, 347 (1978)
[9] H. Ohm et al.,Z.Physik A327, 483 (1987)
[10] R.A. Meyer, Hyperfine Interactions 22, 385 (1985)
[11] S. Brant et al., Z.Physik A329, 301 (1988) (1985)
[12] C. Thibault et al., Phys.Rev. C23, 2720 (1981)
[13] K.E.G. Löbner in The electromagnetic interaction in nuclear spectroscopy
    W.D. Hamilton Ed., 141 . North Holland (1975)
[14] K. Sistemich et al., to be published
[15] H. Mach and R.L.Gill, Phys.Rev. C36, 2721 (1987)
[16] R.A. Meyer et al., Nucl.Phys. A439, 510
[17] F.K. Wohn et al., Phys.Rev. C31, 634 (1985) and AIP Conf.Proc. 164, 407 (1987)
[18] B. Pfeiffer et al., AIP Conf.Proc. 164, 403 (1987)
[19] M.L. Stolzenwald et al., contribution to this workshop, and to be published .

Shape Coexistence and Mixing of Spherical and Deformed Shapes in the N = 60 Isotones

J.C. Hill[1], F.K. Wohn[1], R.F. Petry[2], R.L. Gill[3], H. Mach[3], and M. Moszynski[4]

[1]Ames Laboratory, Iowa State University, Ames, IA 50011, USA
[2]University of Oklahoma, Norman, OK 73019, USA
[3]Brookhaven National Laboratory, Upton, NY 11973, USA
[4]Institute for Nuclear Studies, Warsaw, Poland

Low-lying levels in the $N = 60$ isotones ^{98}Sr and ^{100}Zr have been described in terms of the coexistence of spherical and deformed shapes. A band-mixing calculation using recently determined preliminary values for important excited state lifetimes has been carried out. The results and a discussion of the mixing are presented. Also recent experimental results concerning the level structure of the $N = 62$ isotone ^{102}Zr are given.

INTRODUCTION

Neutron-rich nuclei with mass $A \sim 100$ have attracted considerable interest since they belong to a region of large deformation. Spectroscopic studies have revealed some unusual features of $A \sim 100$ nuclei. One is the rapid onset of deformation that occurs for even-even nuclei at neutron number $N = 60$. This can be seen in Fig. 1 for Sr and Zr nuclei. The 2_1^+ energy drops by a factor of 6 as N increases from 58 to 60. This rapid onset can be attributed to subshell effects that are mutually reinforcing for the combinations $Z = 38,40$ and $N = 56,58$ [1]. Recent considerations of the influence of subshell effects on the onset of deformation show a close analogy between the $A \sim 100$ region and the rare-earth region, although the effect is more than twice as abrupt for the $A \sim 100$ nuclei.

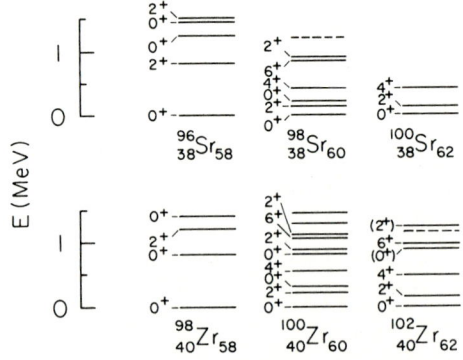

Fig. 1. Systematics for low-lying levels in heavy even-even Sr and Zr nuclei. Levels in ^{102}Zr except for the 6_1^+ are from this work.

Another unusual feature of $A \sim 100$ nuclei is the apparent coexistence of both spherical and highly deformed shapes. All of the $N = 60$ isotones ^{98}Sr, ^{100}Zr, ^{102}Mo, ^{104}Ru, and ^{106}Pd have an "extra" 0^+ state (the 0_2^+ state) that cannot be reproduced in collective nuclear models as a simple collective excitation of the core. The very low-lying 0_2^+ states in ^{98}Sr and ^{100}Zr have been interpreted as evidence for either shape coexistence [2,3] or asymmetric shapes [4]. In contrast, the heavier Mo and Ru isotopes have been interpreted as asymmetric rotors [5] with nonzero values of the geometric asymmetry parameter γ.

The $N = 60$ isotones occupy a special position among the even-even $A \sim 100$ nuclei, as they are the lightest ones to have well-defined rotational bands. The nucleus ^{100}Zr occupies an intermediate position between axially symmetric ^{98}Sr and asymmetric ^{102}Mo or ^{104}Ru. For ^{98}Sr the interpretation [2] of coexisting shapes is supported by a study [3] in which the structure is interpreted as arising from deformed and spherical states that are slightly mixed. The ground-state rotational band (up to spin 6^+) was interpreted as predominately an axially symmetric deformed band and the 0_2^+ state at 215 keV was interpreted as predominately a spherical 0^+ state that has an associated one-phonon 2^+ state at 871 keV [2,3]. Earlier studies of ^{100}Zr by Khan et al. [4] favored an interpretation based on axial asymmetry.

In a recent paper Wohn et al. [6] studied the level structure of ^{100}Zr populated in the decay of the low-spin isomer of ^{100}Y. A total of 64 γ transitions were placed in a level scheme for ^{100}Zr with 20 levels up to 4288 keV. Important information obtained in that study was the definite establishment of the 0_3^+ level at 829 keV by angular correlations and the characterization of the 2_2^+ level at 878 keV. With this new information it was possible to resolve the ambiguity faced by Khan et al. [4] on interpretation of the level structure of ^{100}Zr which can be viewed as shape coexistence between a nearly spherical (S) band consisting of the 0_2^+ and 2_2^+ and an axially symmetric, deformed (D) yrast band (with β and γ vibrations) with appreciable mixing between the 0^+ states of the two bands. Crucial to the analysis of bands in ^{98}Sr and ^{100}Zr are lifetimes for the 2_1^+ and 0_2^+ levels. Recently Mach [7] has remeasured some of these lifetimes. Measurements were made at the TRISTAN separator using a fast BaF γ-ray detector. Preliminary values obtained were sufficiently different from those used by Wohn et al. [6] to justify a reanalysis of the bands in ^{98}Sr and ^{100}Zr.

It is also interesting to study the deformation and coexistence properties of the $N = 62$ isotones of Sr and Zr, namely ^{100}Sr and ^{102}Zr. The $E(4_1^+)/E(2_1^+)$ ratio increases from 2.66 to 3.15 in going from ^{100}Zr to ^{102}Zr and from 3.01 to 3.23 in going from ^{98}Sr to ^{100}Sr. It would be of particular interest to determine the behavior of the spherical band as N increases from 60 to 62. Unfortunately, the 0_2^+ and 2_2^+ states have not been identified in either ^{100}Sr or ^{102}Zr. We present below new information on levels in ^{102}Zr from decay of a low-spin isomer of ^{102}Y.

BAND ANALYSIS FOR THE N = 60 ISOTONES

Using the procedure of Wohn et al. [6] we reanalyzed bands in ^{98}Sr and ^{100}Zr using new level lifetime information obtained by Mach [7] and Ohm [8]. In the upper half of Fig. 2 we present the pertinent experimental data. The lower half gives reduced level energies with respect to the 2_1^+ energy and relative $B(E2)$ transition rates for the 2_2^+ levels. The $B(E2)$ values for the 2_1^+ and 0_2^+ levels are given in Weisskopf units based on the lifetimes given in the upper half of the figure. The similarity between the two deformed isotopes, ^{98}Sr and ^{100}Zr, is very striking. The transition with the strongest $B(E2)$ value from the 2_2^+ level of ^{98}Sr or ^{100}Zr goes to the 0_2^+ level. The 2_2^+ level is observed to have the 2^+ member of the S band as its major component. This is the main reason the 2_2^+ level cannot be the low-lying γ vibration of the Gneuss-Greiner asymmetric rotor model [9].

If we assume mixing of the 0_1^+ and 0_2^+ levels as considered by Khan et al. [4] we obtain

$$|0_1\rangle = \cos\theta|0_D\rangle + \sin\theta|0_S\rangle$$

and $\quad |0_2\rangle = \cos\theta|0_S\rangle) - \sin\theta|0_D\rangle,$

where $\quad \sin^2\theta = E(0_D)/E(0_2).$

Schussler et al. [2] assumed that the D band in ^{98}Sr was that of a rigid symmetric rotor, and hence used the 4_1^+ and 6_1^+ levels to deduce the unmixed 0_D and 2_D of the D band. The Schussler et al. assumption also gave

$$|2_1\rangle = \cos\phi|2_D\rangle + \sin\phi|2_S\rangle$$

and $\quad |2_2\rangle = \cos\phi|2_S\rangle - \sin\phi|2_D\rangle,$

where $\sin^2\phi = [E(2_D) - E(2_1)]/[E(2_2) - E(2_1)]$. The Khan assumption has no 2^+ mixing whereas the Schussler assumption has the maximal creditable 2^+ mixing.

With the mixing determined, only the deformations β_D and β_S of the D and S bands can be freely adjusted to reproduce the three observed transition parameters $B(E2, 0_1 \rightarrow 2_1)$, $B(E2, 0_2 \rightarrow 2_1)$, and $\rho(E0, 0_2 \rightarrow 0_1)$. Assuming only collective intraband $E2$ transitions (hence no "crosstalk" between bands),

$$B(E2, 0_1 \rightarrow 2_1) = (3ZeR^2/4\pi)^2(\beta_D\cos\theta\cos\phi + \beta_S\sin\theta\sin\phi)^2,$$

$$B(E2, 0_2 \rightarrow 2_1) = (3ZeR^2/4\pi)^2(\beta_S\cos\theta\sin\phi - \beta_D\sin\theta\cos\phi)^2,$$

and $|\rho(E0, 0_2 \rightarrow 0_1)| = (3Ze/4\pi)\sin\theta\cos\theta|\beta_D^2 - \beta_S^2|.$

If $\beta_S = 0$, and ϕ is small (so $\cos^2\phi \approx 1$), then β_D is given by the rate $B(E2, 0_1 \rightarrow 2_1)$ and the "mixing angle" θ is overdetermined by the decay rates of the 0_2 level. Thus

$$\tan\theta = 9.9 \times 10^{-6}\, ZA^{4/3}[\rho(E0, 0_2 \rightarrow 0_1)/B(E2, 0_1 \rightarrow 2_1)]$$

and $\tan\theta = [B(E2, 0_2 \rightarrow 2_1)/B(E2, 0_1 \rightarrow 2_1)]^{1/2}.$

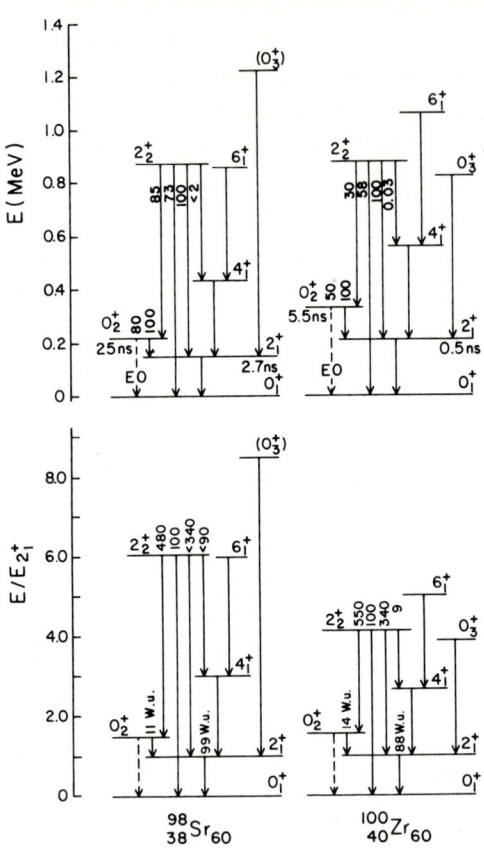

Fig. 2. Systematics for low-lying levels in the $N = 60$ isotones, ^{98}Sr and ^{100}Zr. Level energies, transition intensities, and recent level lifetimes in [7] and [8] are given in the upper half. Reduced level energies with respect to the 2_2^+ energy and $B(E2)$ transition rates for the 2_2^+ levels are shown in the lower half.

Another relationship for $\tan\theta$ can be obtained by considering transition rates from the 2_2^+ level. Unlike the 0^+ levels discussed above, the transition rates from the 2_2^+ are very sensitive to the mixing angle ϕ and the deformation β_S of the spherical S band. Thus, even though the half-lives of the 2_2^+ levels in ^{98}Sr and ^{100}Zr are unknown, ratios of $B(E2)$ transition rates from the level provide a sensitive test of the mixing of the S and D bands. In terms of the two mixing angles θ and ϕ introduced earlier, the $E2$ transition rates from the 2_2^+ level in W.u. (Weisskopf units), are

$$B(E2, 2_2 \to 0_1) = 0.40 Z^2 (\beta_D \cos\theta \sin\phi - \beta_S \sin\theta \cos\phi)^2,$$
$$B(E2, 2_2 \to 2_1) = 0.57 Z^2 (\cos\theta \sin\phi)^2 (\beta_D - \beta_S)^2,$$
$$B(E2, 2_2 \to 4_1) = 1.00 Z^2 (\beta_D \sin\phi)^2,$$
$$B(E2, 2_2 \to 0_2) = 0.40 Z^2 (\beta_D \sin\theta \sin\phi + \beta_S \cos\theta \cos\phi)^2.$$

From the above equations we obtain a third equation for $\tan\theta$ (assuming $\phi = 0$) which is

$$\tan\theta = \left[\frac{B(E2, 2_2^+ \to 0_1^+)}{B(E2, 2_2^+ \to 0_2^+)} \right]^{1/2}$$

For $\beta_S = 0$, the transition with the strongest $B(E2)$ is 2_2^+ to 4_1^+ and the weakest is 2_2^+ to 0_2^+. This is clearly inappropriate for both ^{98}Sr and ^{100}Zr. Another extreme case, $\phi = 0$, would eliminate two transitions, the 2_2^+ to 4_1^+ and the 2_2^+ to 2_1^+, since the 2_1^+ and 4_1^+ levels would then be pure D band. Except for the 2_2^+ to 2_1^+ transition, the observed relative transition strengths (shown in Fig. 2) clearly correspond more closely to the $\phi = 0$ limit than to the $\beta_S = 0$ limit.

One approach to illuminating the nature of the deformation and mixing in the two bands is to assume $\phi = 0$ and determine the corresponding values of the mixing angle θ from the three equations given above for $\tan\theta$. All the expressions for $\tan\theta$ should give the same θ if $\beta_S \approx 0$ and $\phi \approx 0$. For ^{98}Sr the three θ values are 8.5°, 18.8°, and 24.7°. For ^{100}Zr the corresponding values are 13.5°, 21.3°, and 23.3°. In both cases the values are clearly inconsistent with each other when the most recent lifetime values are used indicating that a nonzero value of β_S is necessary. This is different from earlier results of Wohn et al. [6] in which a $\beta_S = 0$ was valid for ^{100}Zr but not ^{98}Sr. Another way of showing the consistency (or lack thereof) of the transition rates of the 0_2^+ and 2_1^+ levels of the $N = 60$ isotones with the assumption of a spherical S band is to use the mean value of the mixing angle θ to predict the value of the deformation β_D of the D band using each of the three experimental rates. Table I presents such an analysis for ^{98}Sr and ^{100}Zr. As can be seen from the results neither can be regarded as consistent with a truly spherical S band.

TABLE I. Values of β_D deduced from transition rates of 0_2^+ and 2_1^+ levels assuming the spherical band has $\beta_S = 0$.

Nucleus	Deformation of β_D deduced from experimental values of:			
	$B(E2, 0_1 \to 2_1)$	$B(E2, 0_2 \to 2_1)$	$\rho(E0, 0_2 \to 0_1)$	Mean value
^{98}Sr	0.44	0.47	0.30	0.40
^{100}Zr	0.40	0.43	0.32	0.38

It is interesting to compare the above θ values with the Khan [4] assumption in which a VMI model is assumed and its parameters determined from the energies of the 2_1^+, 4_1^+, and 6_1^+ levels assuming them to be pure D band. The result gives VMI inertial parameters of 21.57 and 28.86 keV and "stiffness" parameters of 0.0037 and 0.013 respectively for ^{98}Sr and ^{100}Zr. Using the above, the mixing angle θ was determined to be 16.2° and 22.4° respectively assuming $\phi = 0$. The Schussler assumption [2] of maximal 2_1^+ mixing gives $\theta = 25.6°$ and $\phi = 7.9°$ for ^{98}Sr and $\theta = 35.5°$ and $\phi = 13.2°$ for ^{100}Zr.

In conclusion the inconsistency of the θ values and the corresponding β_D values imply that the assumption of $\beta_S = 0$ is not valid for either ^{98}Sr or ^{100}Zr, but θ values calculated by the same method are very similar for the above two nuclei implying that they have similar characteristics. The D band is probably more deformed than originally thought [6] as indicated by new lifetimes for the 2_1^+ states and the β's obtained from $B(E2, 0_1 \rightarrow 2_1)$ as given in Table I. (Of all the β's these are the least sensitive to nonzero values of β_S). Also the corresponding θ's for ^{98}Sr and ^{100}Zr of 18.8° and 21.3° are consistent with the Khan assumption values of 16.2° and 22.4° indicating that the assumption that ϕ is small is reasonable. Also the new value for the lifetime of the 0_2^+ level in ^{100}Zr indicates less mixing of the 0^+ states.

STRUCTURE OF ZR-102

The structure of the $N = 62$ isotones of Sr and Zr are of interest in determining whether the coexisting deformed and spherical bands observed in the $N = 60$ isotones still persist as a neutron pair is added and the deformation increases. Of special interest is the location and properties of the 0_2^+ and 2_2^+ states. No information is available on these states in ^{100}Sr. We present here the results of a recent study of levels in ^{102}Zr populated in the decay of a low-spin isomer of ^{102}Y.

Sources of mass-separated ^{102}Y were produced by the TRISTAN mass separation facility on line to the High Flux Beam Reactor at Brookhaven National Laboratory. Beams of $A = 102$ ions were obtained from a high-temperature thermal ionization source containing a target of 5 g of enriched ^{235}U which was exposed to a neutron flux of 3.0 x 10^{10} n/cm^{2}s. At the temperatures employed the dominant component of the A = 102 beam was Sr. Little primary Y was in the beam. The beams were mass separated and deposited on a movable aluminum-coated Mylar tape.

Levels in ^{102}Zr have been studied earlier by Shizuma et al. [10] at the JOSEF separator. They obtained a $T_{1/2}$ for ^{102}Y of 0.36s and an intensity ratio $I(2_1^+)/I(4_1^+)$ of 2.3. Preliminary values obtained in our study give a $T_{1/2}$ of 0.29s for ^{102}Y and a $I(2_1^+)/I(4_1^+)$ of 10. We view these "inconsistencies" between our results and those of Shizuma et al. [10] as due to the fact that we observe primarily the decay of a low-spin isomer of ^{102}Y populated in the decay of ^{102}Sr while Shizuma et al. [10] were observing primarily the decay of a high-spin ^{102}Y isomer. The "γ strength" in ^{102}Y is much less than in ^{102}Sr implying strong β feeding to the ^{102}Zr ground state.

A preliminary decay scheme for ^{102}Y resulting from our study is shown in Fig. 3. For comparison the level structure of ^{100}Sr obtained from a study of the decay [11] of ^{100}Rb is shown to the right. Of particular interest in the level scheme of ^{102}Zr is a state at 895 keV which is confirmed by coincidences to decay to the 2_1^+ but not the 4_1^+ and 0_1^+ levels. It is a good candidate for the 0_2^+ state. We also observe a state at 1211 keV that decays both to the 0_1^+ and 2_1^+ states. It is a good candidate for the 2_2^+ state.

Assuming the 895 keV level in ^{102}Zr to be the 0_2^+ state, its excitation energy has risen from a value of 331 keV for ^{100}Zr. Similarly the 2_2^+ state's excitation has risen from 878 keV in ^{100}Zr to 1211 keV in ^{102}Zr. There is no strong evidence for a 0_3^+ or 2_3^+ state in ^{102}Zr below 1.3 MeV. One possible explanation of the rise in energy of the 0_2^+ and 2_2^+ states as we go from $N = 60$ to $N = 62$ is that the spherical states have been pushed up to a high energy and that the 0_2^+ and 2_2^+ represent the deformed β and γ bandheads respectively. This is consistent with the lower level density observed around 1 MeV. Also the energy spacing between the 0_2^+ and 2_2^+ states of 316 keV is too low to interpret them as members of a spherical band with small mixing. The above discussion is quite

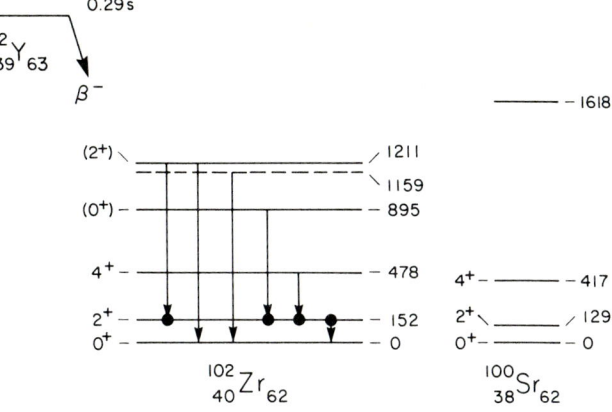

Fig. 3. Decay scheme for ^{102}Y to levels in ^{102}Zr as determined in this work. The levels of ^{100}Sr from Ref. 10 are shown for comparison to the right.

speculative due to the paucity of information on excited states in ^{102}Zr. It would also be useful to carry out a more detailed study of the structure of ^{100}Sr to determine if the 0_2^+ and 2_2^+ states had risen in energy as observed for ^{102}Zr. This work was supported by the U.S. Department of Energy under contracts W-7405-eng-82 and DE-AC02-7600016.

REFERENCES

1. P. Federman and S. Pittel, Phys. Lett. **69B**, 385 (1977); **77B**, 29 (1978); Phys. Rev. C **20**, 820 (1979); R. Bengtsson, P. Möller, J. R. Nix, and J. Zhang, Phys. Scr. **29**, 402 (1984).
2. F. Schussler, J. A. Pinston, E. Monnand, A. Moussa, G. Jung, E. Koglin, B. Pfeiffer, R. V. F. Janssens, and J. Van Klinken, Nucl. Phys. **A339**, 415 (1980).
3. K. Becker, G. Jung, K. H. Kobras, H. Wollnik, and B. Pfeiffer, Z. Phys. A **319**, 193 (1984).
4. T. A. Khan, W. D. Lauppe, K. Sistemich, H. Lawin, G. Sadler, and H. A. Selic, Z. Phys. A **283**, 105 (1977); T. A. Khan, W. D. Lauppe, K. Sistemich, H. Lawin, and H. A. Selic, *ibid*. **284**, 313 (1978); H. A. Selic, W. Borgs, W. D. Lauppe, H. Lawin, and K. Sistemich, *ibid*. **286**, 123 (1978).
5. K. Shizuma, H. Lawin, and K. Sistemich, Z. Phys. A **311**, 71 (1983); K. Sümmerer, N. Kaffrell, E. Stender, N. Trautmann, K. Broden, G. Skarnemark, T. Bjornstad, I. Haldorsen, and J. A. Maruhn, Nucl. Phys. A339, 74 (1980).
6. F. K. Wohn, J. C. Hill, C. B. Howard, K. Sistemich, R. F. Petry, R. L. Gill, H. Mach, and A. Piotrowski, Phys. Rev. C **33**, 677 (1986).
7. H. Mach et al., to be published.
8. H. Ohm, G. Lhersonneau, K. Sistemich, B. Pfeiffer, and K.-L. Kratz, Z. Phys. A **327**, 483 (1987).
9. G. Gneuss and W. Greiner, Nucl. Phys. **A171**, 449 (1971).
10. K. Shizuma, J. C. Hill, H. Lawin, M. Shaanan, H. A. Selic, and K. Sistemich, Phys. Rev. C **27**, 2869 (1983).
11. S. Mattsson, R. E. Azuma, H. A. Gustafsson, P. G. Hansen, B. Jonson, V. Lindfors, G. Nyman, I. Ragnarsson, H. L. Ravn, and D. Schardt, in Proc. 4th Int. Conf. on Nuclei Far From Stability, Helsingor, Denmark, 1981, ed. P. G. Hansen and O. B. Nielsen, CERN Report 81-09, p. 430.

Transition Probabilities and Static Moments in Transitional Nuclei

A. Wolf[1] and R.F. Casten[2]

[1] Nuclear Research Center Negev, Beer-Sheva, Israel
[2] Brookhaven National Laboratory, Upton, NY 11973, USA

1. Introduction

Electromagnetic transition probabilities and static moments of excited nuclear states are known to be good probes of nuclear structure. Therefore, a systematic analysis of the large amount of existing experimental data for these observables is expected to provide valuable information about the respective isotopes. It is the purpose of this talk to show that a combined analysis of static magnetic moments of 2_1^+ states and B(E2) transition probabilities for even-even nuclei can be used to obtain effective numbers of valence nucleons. This kind of information is of particular interest in cases where subshell closures are found. For example, it is well known[1] that for the transitional nuclei in the A=150 region the Z=64 subshell is active when the number of neutrons N<90, but disappears for N≥90. A similar situation exists in the A=100 region, where the Z=38 subshell is active for N≤58.

In the following sections we present the method by which effective numbers of valence protons and neutrons can be deduced from B(E2) and g-factor data, and show applications of this method to the A=150 and A=100 transitional regions. Part of these results were recently published[2].

2. Transition probabilities and magnetic moments in IBA-2.

In order to make a systematic analysis of experimental data for some observable, we need a theoretical model to guide us. In the present case, we use the proton-neutron version of the interacting boson model (IBA-2). There are two main reasons for this. Firstly, we are mainly interested here in properties of even-even collective nuclei and this model is known to provide a good description of these nuclei. Secondly, in IBA-2 we have simple analytical formulas for the B(E2) transition probabilities and the g-factors of 2_1^+ states. For the latter quantity, it was shown[a] that the following linear relation exists:

$$g(2_1^+) = (g_\pi N_\pi + g_\nu N_\nu)/N_t \quad (1)$$

where $N_\pi (N_\nu)$ are the numbers of proton (neutron) bosons, $N_t = N_\pi + N_\nu$, and $g_\pi (g_\nu)$

the g-factors of the single bosons. Equation (1) was derived[3] using only the assumption that F spin is a good quantum number, and is therefore approximately valid for the 2_1^+ states of most even-even nuclei.

For the B(E2) transition probabilities the situation is more complex. There is no single universal B(E2) analytical formula valid for all even-even nuclei. However, simple relations exist for the three dynamical symmetries of the U(6) group[4], namely, U(5), SU(3) and O(6):

$$B(E2: 0_1^+ \rightarrow 2_1^+) = 5/N \, (e_\pi N_\pi + e_\nu N_\nu)^2 \qquad [U(5)] \quad (2)$$
$$= (2N+3)/N \, (e_\pi N_\pi + e_\nu N_\nu)^2 \qquad [SU(3)] \quad (3)$$
$$= (N+4)/N \, (e_\pi N_\pi + e_\nu N_\nu)^2 \qquad [O(6)] \quad (4)$$

Recently, it was shown[5] that for a large number of nuclei throughout the O(6)-SU(3) region a simple approximate analytic expression can be used:

$$B(E2: 0_1^+ \rightarrow 2_1^+) = 0.05(1 - 0.1\chi)^2 (1 + 1/N_t)^2 (e_\pi N_\pi + e_\nu N_\nu)^2 \quad (5)$$

where χ is the parameter of the quadrupole operator.

The method we propose to determine effective numbers of valence particles is basically to use experimental $g(2_1^+)$ and B(E2) values for a specific isotope, and then extract N_π^{eff}, N_ν^{eff} from equation (1) and the appropriate B(E2) expression (i.,e., one of eqs. (2)-(5)) by solving two equations with two unknowns. In this approach we need to know the values of the effective g-factors (g_π, g_ν) and the effective charges (e_π, e_ν) of the (proton, neutron) bosons. These parameters can be deduced by a systematic analysis of $g(2^+)$ and B(E2) data in collective nuclei where we assume that we know the values of N_π, N_ν. We present this analysis in the next section.

3. Effective charges and boson g-factors in collective nuclei.

Equation (1) is a linear relationship between $g(2_1^+)$ and N_π, N_ν. g_π, g_ν can be deduced as the slope and intercept from a linear fit of gN_t/N_ν vs. N_π/N_ν. This was done[6] for 65 nuclei in the range A=70-200. The nuclei were divided into six groups. The resulting g_π, g_ν for these groups are plotted in Fig 1 against the average mass number A of each group. We see that in general, g_π tends to increase while g_ν decreases with increasing A. The sum $g_\pi + g_\nu$ stays constant at about 0.7 in all cases. A notable exception from the smooth dependence of g_π, g_ν with A occurs at A=170. These are the Er, Yb isotopes, and the reasons for their anomalous behaviour were discussed elsewhere[6,7].

The extraction of boson effective charges from experimental B(E2) data is more complicated and has to be done by using the appropriate equation among eqs. (2)-(5). Sala et al.[8] have used this approach and obtained e_π, e_ν for 36 nuclei in the Dy-Os region. They obtained : e_π = 0.15eb, e_ν = 0.11eb. However, they also observed considerable variations of B(E2) values with Z,A which could not be explained by using constant values of the effective charges. In our analysis we need the effective charges in the A=100 and A=150 regions for vibrational and for deformed nuclei. To obtain these values, we made linear fits of $T = k(N_\pi, N_\nu) B(E2)^{1/2}$ vs. N_π/N_ν. The nuclei were classified according to Z and/or A, and structure (i.e., vibrational, rotational). The function $k(N_\pi, N_\nu)$ is obtained from eqs. (2)-(5) after simple algebraic manipulations. We have different $k(N_\pi, N_\nu)$ functions for the different symmetries. The experimental B(E2) values were taken from the compilation of Raman et al.[9] As an example, in fig 2 we present the linear fits for the $^{96-102}$Ru isotopes, and for six neutron-rich Ba, Ce, Gd isotopes (groups 2,5 in Table 1). The latter are known to be deformed nuclei and eq (5) was used with $\chi=-0.52$. A variation of χ of 20% has a negligible effect on the results of the calculation.

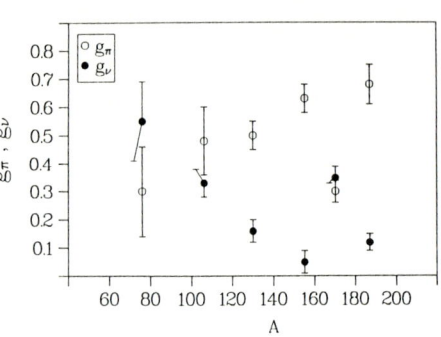

Figure 1 Systematics of g_π, g_ν

The various groups of nuclei for which we extracted effective charges are listed in Table 1. For purposes of comparison, we included the results for heavier nuclei, in the A = 160-190 range. From the table we see that for all deformed nuclei (SU(3), O(6)) the ratio e_ν/e_π is of the order of 0.3-0.6, in agreement with the recent work of Raman et al.[9] For the vibrational (U(5)) nuclei this ratio is considerably larger than 1.00. The fact that neutrons have a larger effective charge than protons was first mentioned by Hamilton et al.[10] for vibrational nuclei in the A=140 region (group 4 in Table 1). Here we report this effect for vibrational nuclei in the A=100 region (groups 1,2). The reason for this effect is at present unclear. It was suggested[10] that it may be related to the different shells occupied by the protons and neutrons, and to the fact that the given values have a (length)^2 factor incorporated in them.

4. Effective numbers of valence nucleons.

We will use now the effective g-factors and charges determined in the previous sections, equations (1)-(5), and experimental B(E2) and $g(2_1^+)$ data to determine the effective numbers of valence protons, neutrons (N_p^{eff}, N_n^{eff}) for transitional nuclei in the A=150 and A=100 regions. In order to be consistent we will follow this procedure only for nuclei which were not included in the linear fits described in the previous section. The results for the A=150 region are given in fig 3. The dissipation of the Z=64 subshell when N increases between $84 \leq N \leq 92$ is clearly seen for the Sm, Nd, and Ce isotopes. The Ba isotopes do not show this effect because Ba is below midshell for both the

Table 1 Effective boson charges for groups of isotopes in the range A=90-200

Group No.	Isotopes	e_π (eb)	e_ν (eb)	e_ν/e_π	Eq. used for fit
1	$^{94-96}$Mo	0.080	0.121	1.51	(2)
2	$^{96-102}$Ru	0.095	0.166	1.75	(2)
3	$^{108-112}$Ru	0.159	0.055	0.35	(4)
4	^{138}Ba-^{146}Nd^a	0.12	0.24	2.00	(2)
5	^{146}Ba-^{160}Gd	0.169	0.095	0.56	(5)
6	$^{156-164}$Dy	0.173	0.064	0.37	(5)
7	$^{180-186}$W	0.234	0.073	0.31	(5)
8	$^{182-192}$Os	0.221	0.083	0.38	(5)

a) From Ref. 10.

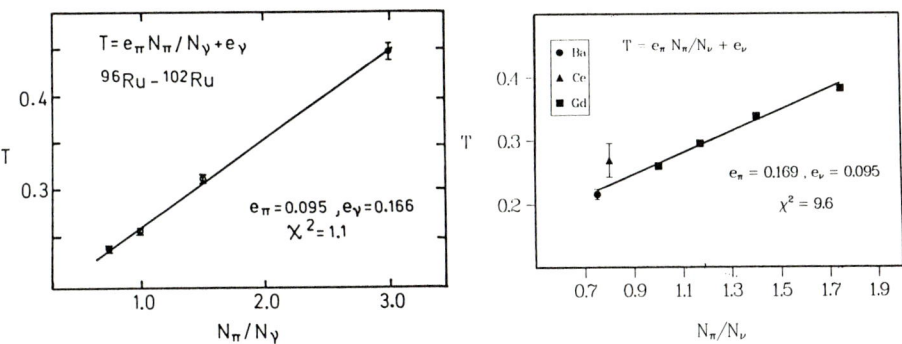

Figure 2 Fits of T vs. N_π/N_ν for vibrational Ru isotopes and deformed neutron-rich Ba, Ce, Gd isotopes.

50-64 and 50-82 shells. It is also remarkable that the deduced numbers of valence neutrons behave as expected from normal counting (solid lines).

In the A=100 region the situation is not as clear. The main problem is the lack of sufficient experimental data, especially magnetic moments. Another difficulty arises because in this region it is not always possible to use analytical formulas which are valid only for the limiting symmetries of IBA. A possible way to circumvent this problem is to perform full numerical calculations. In the following we will make a first estimate of the effective number of valence paricles without using the numerical code. We first consider the ^{84}Kr, ^{86}Sr isotopes. We assume we can use here the U(5) equation for the B(E2) transition probabilities.

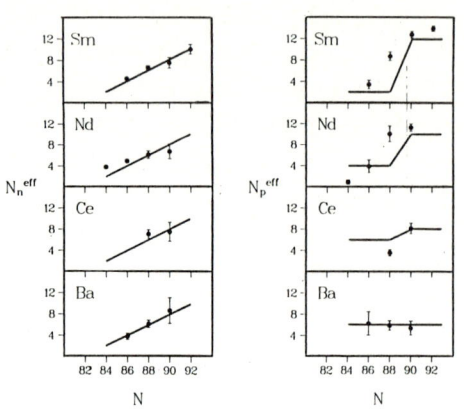

Figure 3 N_p^{eff}, N_n^{eff} around A=150

This assumption is not unreasonable in view of the level structure of these isotopes. We want now to test the hypothesis that for these nuclei the Z=38 subshell is active. Since there is no available data for magnetic moments of 2_1^+ states in this case, we will assume that the numbers of neutrons are those from normal counting. Under this assumption, we can determine the effective numbers of protons from the known experimental B(E2) values. For the effective charges e_π, e_ν we use averages of the respective values of groups 1,2 in Table 1. The resulting numbers of valence protons are given in Table 2. We see that the N_p^{eff} values are much smaller than those from normal counting in the 28-50 shell. We therefore have here direct evidence that a subshell is present, similar to what we have found in the A=150 region.

Table 2 N_p^{eff} in Kr, Sr, Zr, Mo isotopes.

Isotope	N_p (normal counting)	N_p^{eff}
^{84}Kr	8	1.8
^{86}Sr	10	0.5
^{98}Sr	10	4.6
^{100}Zr	10	0.0
^{102}Mo	8	8.5

Additional evidence is provided by $g(2_1^+)$ data for ^{98}Sr, ^{100}Zr, ^{102}Mo.[11-13] Although B(E2) values do exist for these nuclei, we can not use them in the simple way described above because the analytical relations are not valid. Again, in order to obtain a first estimate without using the numerical code we assume that the valence neutrons behave "normally", and calculate N_p^{eff} using eq 1 and g_π, g_ν from the systematics (g_π=0.48, g_ν=0.33). The resulting N_p^{eff} are also given in Table 2. We see that while for ^{100}Zr and ^{98}Sr the Z=38 subshell is still active, for ^{102}Mo it is not.

In conclusion, we have shown that a systematic analysis of B(E2) and $g(2_1^+)$ data can provide direct evidence for the existence of subshell closures. We should emphasize that at this point the evidence in the A=100 region is not as conclusive as in the A=150 region, and more experimental data and theoretical calculations (like, e.g., full numerical IBA computations) are needed to further substantiate the results presented in this work.

<u>References.</u>

1. R.F.Casten et al.: Phys. Rev. Lett. <u>47</u>, 1433 (1981).
2. A.Wolf and R.F.Casten: Phys. Rev. <u>C36</u>, 851 (1987).
3. M.Sambataro and A.E.L.Dieperink: Phys. Lett. <u>107B</u>, 249 (1981).
4. J.N.Ginocchio and P. Van Isacker: Phys. Rev. <u>C33</u>, 365 (1986).
5. R.F.Casten and A.Wolf, Phys. Rev.: <u>C35</u>, 1156 (1987).
6. A.Wolf, R.F.Casten and D.D.Warner: Phys. Lett. <u>190B</u>, 19 (1987).
7. R.F.Casten, K.Heyde and A.Wolf: to be published.
8. P.Sala, A.Gelberg and P. von Brentano: Z. Phys. <u>A32</u>, 281 (1986).
9. S.Raman et al.: At. Data Nucl. Data Tables <u>36</u>, 1 (1987);
    S.Raman et al.: Phys. Rev. <u>C37</u>, 805 (1988).
10. W.D.Hamilton, A.Irbäck and J.P.Elliott: Phys. Rev. Lett. <u>53</u>, 2469 (1984).
11. A.Wolf, K.Sistemich et al.: to be published.
12. A.Wolf et al.: Phys. Lett. <u>97B</u>, 195 (1980).
13. G.Menzen et al.: Z. Physik <u>A321</u>, 593, (1985).

Evidence for Shape Coexistence in Neutron-Rich Rh and Ag Nuclei

J. Rogowski[1], N. Kaffrell[1], H. Tetzlaff[1], N. Trautmann[1], D. De Frenne[2],
K. Heyde[2], E. Jacobs[2], G. Skarnemark[3], J. Alstad[4], M.N. Harakeh[5],
J.M. Schippers[5], S.Y. van der Werf[5], W.R. Daniels[6], and K. Wolfsberg[6]

[1]Institut für Kernchemie, Universität Mainz,
 D-6500 Mainz, Fed. Rep. of Germany
[2]Laboratorium voor Kernfysica Proeftuinstraat 86, B-9000 Gent, Belgium
[3]Department of Nuclear Chemistry,
 Chalmers University of Technology, S-41296 Göteborg, Sweden
[4]Department of Chemistry, University of Oslo, N-0315 Oslo, Norway
[5]Kernfysisch Versneller Instituut, NL-9747 Groningen, The Netherlands
[6]Los Alamos National Laboratory, Los Alamos, NM 87545, USA

Abstract

Neutron-rich nuclei of Rh and Ag have been investigated by γ-ray spectroscopic measurements as well as single-particle transfer reactions like (d,^{3}He) and (^{3}He,d). The obtained level schemes of ^{105,107,109,111}Rh and ^{113,115}Ag show two different types of excitation: (I) proton-hole states related to a spherical shape of the nucleus and (II) proton-particle states with a rotational band-like structure (intruder band) pointing towards deformation of the nucleus. The "fingerprints" for such a shape coexistence and resulting systematics in these nuclei and the possible interpretation of the intruder band as a rotational band built upon the 1/2+[431] Nilsson configuration are discussed.

Introduction

The phenomenon of shape coexistence in medium-heavy and heavy odd-mass nuclei has recently been reviewed in detail by Heyde et al. [1]. For the odd-mass In (Z=49) and Ag (Z=47) nuclei, it was shown that the $2d_{5/2}$ and/or $1g_{7/2}$ shell-model states intrude across the Z=50 shell closure giving rise to a rotational-like positive-parity band with J$^\pi$=1/2+,3/2+,5/2+,... (intruder band) coexisting with spherical hole states ($1g_{9/2}, 2p_{1/2}, 2p_{3/2}, 1f_{5/2}$) and $1g_{9/2}$ and $2p_{1/2}$ core coupled configurations. In the In nuclei, were the most extensive information is available, an interpretation of these intruder bands as decoupled bands built on the $2d_{5/2}$ and $1g_{7/2}$ orbitals with some degree of mixing is favoured [1] over a description as a single deformed band built on the $1/2^+[431]$ Nilsson orbital.

For the Ag isotopes, the data are less complete and especially nothing is known on the more neutron-rich nuclei including the mid-shell region at neutron number $N \approx 66$, where a maximum quadrupole deformation, i.e. lowest energy for these intruder bands, is expected on the basis of the residual proton-neutron interaction. In odd mass Rh (Z=45) isotopes, five nucleons outside the Z=50 shell closure, these intruder bands are not known yet at all. Here, the more deformed underlying cores may probably favour an identification and interpretation as rotational states built on a single Nilsson configuration.

Therefore, we have started a systematic investigation of neutron-rich Rh and Ag isotopes by post β^--decay γ-ray spectroscopic measurements and by single-particle transfer reactions like (d,³He) and (³He,d) as far as stable target material has been available. For experimental details see [2-5].

The results of these experiments enabled us to check the "fingerprints" for shape coexistence [1] in the investigated Rh and Ag nuclei: (a) a rotational-like band structure of the levels which are assumed to be intruder states having their origin above the Z=50 shell closure, (b) enhanced E2 transitions within the intruder band, (c) retarded electromagnetic transitions between members of the intruder band and states related to a spherical shape, and (d) a strong excitation of the intruder states as particle states in stripping reactions and no or only weak excitation of these states in pick-up reactions.

Results and discussion

The complete level schemes of ^{105,107,109}Rh and ^{113,115}Ag can be found in [2-5]. Spin and parity assignments have been made from the ℓ-transfer results of the (d,³He) and (³He,d) reactions together with the analyzing power results of Flynn et al. [6], where available. As example angular distributions of deuterons from the ^{104}Ru(³He,d)^{105}Rh reaction together with one-step DWBA calculations are shown in Fig. 1 for five states being candidates for members of an intruder band and for four "normal" states. In addition, results from the decay studies, like selection rules for β- and γ-decay and a few conversion coefficients as well as systematic trends of the levels in well-known neighbouring nuclei have been used. For ^{111}Rh spin and parity assignments can only be based on informations from systematics. In Fig. 2 the low-energy negative-parity levels for ^{103,105,107,109,111}Rh are shown. Here, and for the corresponding states in ^{109,111,113,115}Ag only a smooth change in energy with increasing neutron number is observed. An interpretation of the first 3/2- and 5/2- states as core coupled configurations (coupling of the $2p_{1/2}$ to the first 2+ state in the even-even core) is favoured by the fact that in the pick-up reactions most of the $2p_{3/2}$ and $1f_{5/2}$ strength is taken by the second 3/2⁻ and 5/2⁻ states. For the positive-parity states presented in Fig. 3, the first 7/2⁺ and 9/2⁺ states show a similar smooth change in energy with increasing neutron number for both, Ag and Rh isotopes.

The other positive-parity states, drawn as thick lines in Fig. 3, are candidates for being the 1/2⁺, 3/2⁺, 5/2⁺ and 7/2⁺ members of an intruder band. They show a decrease in energy with increasing neutron number and in the Ag isotopes they reach a minimum at N=66, i.e. exactly in the middle of the N=50 and N=82 shell closures. For the Rh isotopes this minimum is already reached at N=64. This is also expected from the systematics of the excitation energies of the lowest intruder states in this mass region shown in Fig. 4. In this comparison not only neighbouring In and Ag isotopes but also the corresponding Sb, I and Cs isotones from above the Z=50 shell closure have been considered; i.e. isotones

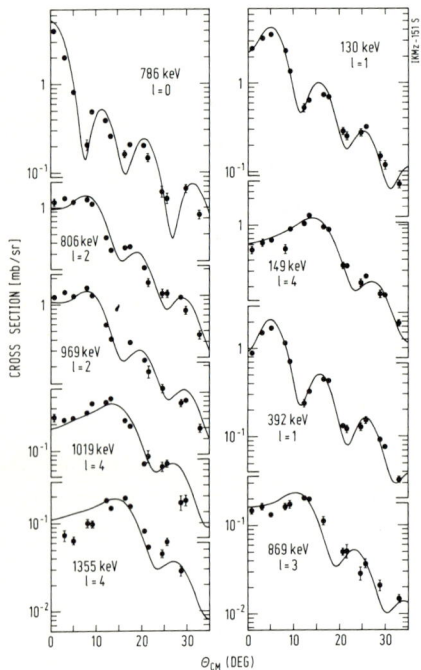

Fig. 1. Angular distributions of deuterons from the ¹⁰⁴Ru(³He,d)¹⁰⁵Rh reaction at $E_{^3He}$=50 MeV for the candidates of the intruder band (left part) and for four "normal" states. The curves are one-step DWBA predictions for the indicated ℓ-transfers.

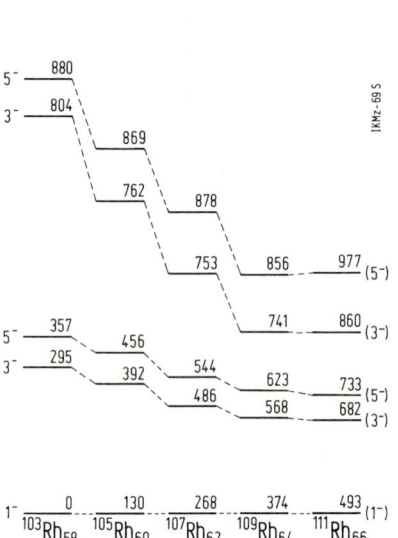

Fig. 2. Low-lying negative-parity states in odd-mass neutron-rich Rh isotopes. Level energies are given in keV; the indicated spin values are twice the actual values. Data are taken from [1-6, 9-12].

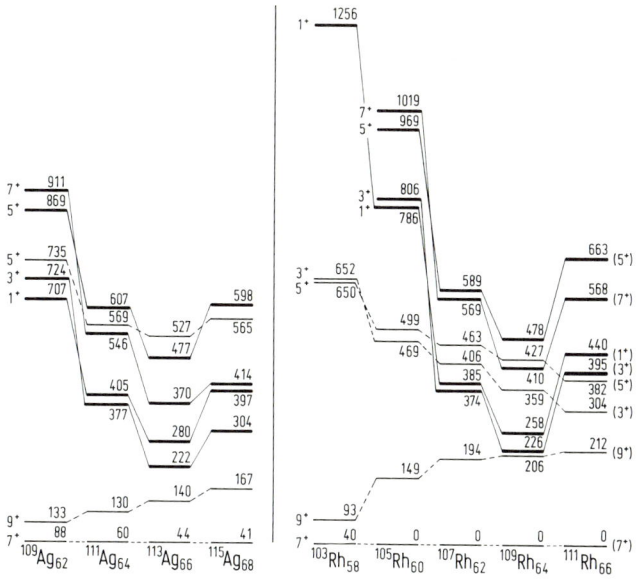

Fig. 3. Same as Fig. 2 for positive-parity states in Ag and Rh isotopes.

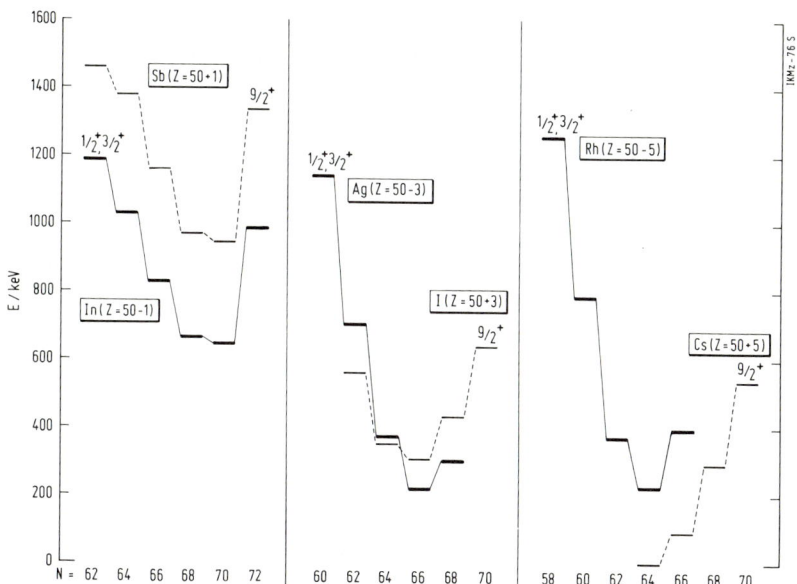

Fig. 4. Comparison of low-energy intruder states in In, Ag and Rh nuclei ($J^\pi=1/2^+$ or $3/2^+$) with those in Sb, I and Cs nuclei ($J^\pi=9/2^+$). The level energies are taken from [1-5,13].

with the same number of proton holes and proton particles, respectively, are compared, and a quite obvious similarity in the energy pattern is observed. For the Cs nuclei (Z=50+5) an increase in energy for the intruder state occurs already at neutron number N=66. Hence, if systematics hold, in the corresponding Rh nuclei (Z=50-5) this increase should occur already in ^{111}Rh (N=66), which is in good agreement with the experimental result. According to the "fingerprints" for shape coexistence described in [1], the intruder states should form a rotational-like band. A fit of the rotational formula for K=1/2 bands to the level energies of the four band members leads to an agreement of the experimental and calculated level energies within 3 keV for the Rh and 4 keV for the Ag isotopes. Under the assumption of a rotational band, strong E2 intraband transitions should be observed. This is not in conflict with the decay pattern of these levels and can clearly be proved for the 31.8 keV transition (73%M1+27±4%E2) connecting the 1/2+ and 3/2+ band members in ^{109}Rh. From the measured lifetime of 28.7 ns for the 258 keV level in ^{109}Rh, an experimental B(E2) value of 5400 e^{2}fm^4 could be deduced, which corresponds to an enhancement of 174 Weisskopf units for the 31.8 keV E2 transition. Similar results for ^{113,115}Ag are reported by Fogelberg et al. [7]. From this experimental B(E2) value of the 31.8 keV transition in ^{109}Rh the intrinsic quadrupole moment Q_o can be obtained and then a deformation parameter ϵ_2=0.27±0.03 results.

For ^{105}Rh band mixing calculations taking into account all Nilsson orbitals from the N=4 harmonic oscillator shell [8] have been performed for the intruder band. The spectroscopic factors resulting from the (^{3}He,d) reaction are compared with calculated values for the different members of the lowest K$^\pi$=1/2+ band (mainly the 1/2+[431] Nilsson orbital) for deformations ϵ_2=0.25, 0.30 and 0.35. The agreement is quite reasonable for the higher ϵ_2 values and is improved in an important way over a pure 1/2+[431] band description, especially for the 5/2+ and 7/2+ levels.

Another hint for the onset of deformation in ^{105}Rh is the strong splitting of the ℓ=0 strength in the ^{104}Ru(^{3}He,d)^{105}Rh reaction. The number of excited 1/2+ states within the measured energy range of 3.3 MeV increases from two in ^{103}Rh (N=58) to eight in ^{105}Rh (N=60).

In contrast to the enhanced transitions between members of the intruder band, a hindrance of transitions between intruder states and states connected with a spherical shape of the nucleus should be observed [1]. For example, the lifetime of the 226 keV level in ^{109}Rh has been determined to 1660 ns and multipolarity E2 for the transition to the 7/2+ ground state has been obtained via the α_K value. A comparison with the Weisskopf estimate shows a hindrance of this transition by a factor of F_W=58. In ^{111}Rh a lifetime of 86 ns for the corresponding level at 395 keV leads to a hindrance factor $F_W \approx$100 for the respective transition.

As expected, the intruder states are not or only weakly excited in pick-up reactions and show a strong excitation in stripping reactions as demonstrated for ^{105}Rh (Fig. 1). Unfortunately, this information from stripping reactions is restricted to lighter Rh and Ag nuclei due to the lack of stable target material.

Acknowledgements

The authors are grateful to H. Folger, GSI Darmstadt, for preparation of the targets used in the nuclear reaction studies and to K.H. Gläsel for his help during the experiments at the Mainz reactor. We would like to thank the US Transplutonium Research Committee for making ^{249}Cf available. This research was supported by the Bundesministerium für Forschung und Technologie (Germany), the Interuniversity Institute for Nuclear Sciences (Belgium), the Natural Science Research Council (Sweden), the Research Council for Science and Humanities (Norway), the Stichting voor Fundamenteel Onderzoek der Materie (FOM) with financial support from the Nederlandse Organisatie voor Zuiver Wetenschappelijk Onderzoek (ZWO) and the Department of Energy (USA).

References

[1] K. Heyde, P. Van Isacker, M. Waroquier, J.L. Wood and R.A. Meyer, Phys. Reports 102, 291 (1983), and references given therein.
[2] N. Kaffrell, P. Hill, J. Rogowski, H. Tetzlaff, N. Trautmann, E. Jacobs, P. De Gelder, D. De Frenne, K. Heyde, G. Skarnemark, J. Alstad, N. Blasi, M.N. Harakeh, W.A. Sterrenburg and K. Wolfsberg, Nucl. Phys. A460, 437 (1986), and references given therein.
[3] N. Kaffrell, P. Hill, J. Rogowski, H. Tetzlaff, N. Trautmann, E. Jacobs, P. De Gelder, D. De Frenne, K. Heyde, S. Börjesson, G. Skarnemark, J. Alstad, N. Blasi, M.N. Harakeh, W.A. Sterrenburg and K. Wolfsberg, Nucl. Phys. A470, 141 (1987).
[4] J. Rogowski, Diploma thesis, Universität Mainz (1985).
[5] J. Rogowski, N. Kaffrell, D. De Frenne, K. Heyde, E. Jacobs, M.N. Harakeh, J.M. Schippers and S.Y. van der Werf, Phys. Lett. B, in print.
[6] E.R. Flynn, F. Ajzenberg-Selove, R.E. Brown, J.A. Cizewski and J.W. Sunier, Phys. Rev. C24, 902 (1981); C25, 2851 (1982); C27, 2587 (1983).
[7] B. Fogelberg, E. Lund, Y. Zongyuan and B. Ekström, Proc. 5th Int. Conf. on nuclei far from stability, Rosseau Lake, Canada 1987, p. 296.
[8] K. Heyde, M. Waroquier, P. Van Isacker and H. Vincx, Nucl. Phys. A292, 237 (1977).
[9] R.E. Anderson, J.J Kraushaar, I.C. Oelrich, R.M. DelVecchio, R.A. Naumann, E.R. Flynn and C.E. Moss, Phys. Rev. C15, 123 (1977).
[10] J. Blachot, Nucl. Data Sheets 41, 111 (1984).
[11] D. De Frenne, E. Jacobs and M. Verboven, Nucl. Data Sheets 45, 386 (1985).
[12] N.K. Aras and W.B. Walters, Phys. Rev. C11, 927 (1975).
[13] W.F. Piel, P. Chowdhury, U. Garg, M.A. Quader, P.M. Stwertka, S. Vajda and D.B. Fossan, Phys. Rev. C31, 456 (1985).

Shape Coexistence and the Structure of Even-Even Se, Pd, and Ru Nuclei

R.A. Meyer[1,2*], D.F. Kusnezov[1,3+], M.A. Stoyer[1,4**], and R.P. Yaffe[1,5++]

[1]Lawrence Livermore National Laboratory, Livermore, CA 94550, USA
[2]Institute of Theoretical Physics, Universität Frankfurt am Main,
 D-6000 Frankfurt, Fed. Rep. of Germany
[3]Physics Department, Princeton University, Princeton, NJ 08540, USA
[4]Chemistry Department, University of California, Berkeley, CA 94720, USA
[5]Chemistry Department, San Jose State University, San Jose, CA 95192, USA

Abstract New multiparameter (t, p-gamma-gamma-time) data are used to construct a ^{112}Pd level scheme, which is combined with ISOL data and previous studies to establish the systematics of even-even Pd nuclei. The level structure of the even-even Pd nuclei is accounted for by mixed configuration IBM-II calculations, which include coexisting intruder state bands. The structure of ^{70}Se and ^{102}Ru are suggested to be modified by coexisting deformed and spherical bands, respectively.

1. Introduction

There has been considerable interest in the properties of the even-even Pd nuclei.[1-12] Recently, proton gated gamma-ray and conversion-electron spectra following the ^{110}Pd(t,p)^{112}Pd reaction using enriched targets have been reported.[8] Here, we give an interim report on the reanalysis of the in-beam data. These results, when combined with past experiments and recent on-line mass separator spectroscopy studies,[12] give a detailed spectra for the levels of ^{112}Pd. We then compare our mixed-configuration IBA-II calculations with these data to account for the level structure of the experimentally-observed heavier Pd nuclei within the context of intruder state modification of O(6) nuclei. The lighter Pd nuclei, such as ^{102}Pd, are suggested to possess proton-$d_{3/2}$, neutron-$d_{5/2}$ intruder states. Also, we discuss the oversight of the spin dependant nature of B(E2) values in the Zr region nuclei.

2. Experimentally known levels

The details of the ^{110}Pd(t,p)^{112}Pd in-beam multiparameter experiment are given elsewhere.[8,10] In Table I we give the present results of our reanalysis, our assignments, and a comparison to the mass separator (ISOL) results of Äystö et al.[12]. Our new data consisted of more detailed coincidence spectra, new analysis to rid of non-proton gated events, and comparison with parallel experiments that emphasized possible contaminants.

The 1097-keV level has been proposed[12] as the $3^+(1)$ level in ^{112}Pd, hence its branching ratios are of some importance in testing the IBA-II calculations. Our branching ratios to the lower energy 2^+ levels agree with the ISOL results. The ISOL study places a transition between the 1097-keV level the and 883-keV {$4^+(1)$}level with a branching ratio of 1.9%. However, detailed analysis of our spectra can place a limit of $\leq$0.18% for this branching ratio.

Table I. Energies and Intensities of Gamma Rays of ^{112}Pd in (t,p) and ISOL Experiments

Energy[a] (t,p)	ISOL	I(Δ)	Assgnmt From	To	Energy[a] (t,p)	ISOL	I(Δ)	Assgnmt From	To
193.4	---[b]	10(1)	1952	1759	1053.8	---[b]	17(1)[e]		
---	212.9	≤0.4	1097	883	1061.6	---[b]	6(1)	2612	
235.7	---[b,c]	e⁻0.15	1126	890[c]	1069.2	---[b]	3.4(1)	1952	883
348.7	348.7	1000(50)	349	GS	1074.7	1074.7	4.9(5)	1423	349
359.6	359.7	124(6)	1097	737	1099.1	1098.6	21(2)[e]	2195	1097
362.8	---[b,d]	0.8(5)	1423	1096	1115.8	---[b]	3.8(8)	1999	883[k]
388.1	388.2	73(15)	737	349	1125.6	---[c]	0.0015	1126	GS[l]
433.7	---[b]	4.4(8)			1140	---[b]	1.5(8)	1140	GS[m]
479.2	479.7	22(2)	1363	883	1153.7	---[b]	10(1)	2037	883
534.2	534.6	354(18)	883	349	1161.1	---[b]	9(1)		
541.3	541.6	6(1)[d]	890	349	1215.4	---[b]	9.8(8)	1952	737
574.9	---[b]	20(1)	1312	737	1262.1	---[b]	3.4(8)	1999	737[k]
625.3	626.1	93(5)	1363	737	1266.2	---[b]	8(1)	2628	1362[k]
635.7	---[b]	3.8(8)[e]			1300.1	---[b]	12(1)	2037	737
640.4	640.8	31(2)	1952	1312[f]	1311.9		3.0(8)	2195	883[n]
650.4	---[b]	11(1)			1312.6	1312.8	4.2(8)	1312	GS[n]
662.7	663.1	46(3)	1760	1097	1317.9	---[b]	4.5(8)	2201	883[k]
667.4	667.9	103(5)	1551	883	1376.1	---[b]	3.8(8)		
685.6	686.2	9(1)	1423	737[h]	1386.9	1386.9	31(2)	2270	880[o]
736.8	737.2	33(1)	737	GS[i]	1399.4	---[b]	8(1)	2496	1097[k]
747.6	748.1	94(5)	1097	349	1423.1	---[b]	6.4(8)	1423	GS[p]
768.1	---[b]	13(1)	2318	1551	1471.9	1471.7	3.8(8)	2355	883
776.9	777.5	13(1)	1126	349	1493.1	p--[b]	9(1)		
791.2	791.6	17(1)	1140	349	1511.7	p--[b]	12(1)		
810.6	---[b]	20(1)[j]			1552.9	p--[b]	4(1)		
832.2	---[b]	6.4(8)	1715	883	1580.2	p--[b]	5(1)		
834.5	---[b]	4.9(8)[e]			1603.5	p--[b]	0.5(3)	1952	349
855.6	---[b]	7.5(8)	1952	1097	1688.2	p--[b]	18(1)	2037	349
876.9	---[b]	4(1)	1759	883	1729.1	p--[b]	5(1)	2612	883
924.3	---[b]	2.6(8)			1758.7	p--[b]	6(1)	2496	737[k]
963.0	---[b]	0.8(3)	1312	348	1891.9	p--[b]	5(1)	2628	737[k]
978.6	979.2	8.7(8)	1716	737	1897.3	p--[b]	2.6(8)		
1014.0	---[b]	4.9(8)	1362	349	1990.6	p--[b]	4(1)		

a) Äystö et al. (Ref. 12) appear to have a systematic error in their energies.
b) Not reported in the ISOL investigation of Äystö et al.
c) E0 transition. Observed in the electron spectra only.
d) This gamma ray was not reported before.

e) This gamma ray was originally attributed to the $(t,2n)^{111}$Ag reaction product.

f) This gamma ray was originally assigned as the sole deexcitation of the 2002-keV level.

g) The energy given in this work is the calculated level energy difference.

h) There is a large contribution from a ^{111}Cd photopeak of the same energy.

i) Our reanalysis agrees with that given by Äystö et al.

j) Originally attributed to the beta decay of ^{111}Ag$^{m\&g}$.

k) This level has been established by Reitz principle.

l) The 1126-keV E0 intensity has been reported earlier by us.[8]

m) Determined from gamma-gamma coincidence information.

n) Intensity partition done using gamma-gamma coincidence information.

o) Gamma ray placed from the 2756-keV level by Äystö et al. (see text).

p) High energy lines were recalibrated.

The ISOL work has identified the $0^+(2)$ intruder band head at 890-keV and assigned a 541-keV transition as its sole gamma-ray deexcitation path. We can identify the presence of the 541-keV gamma ray in our spectra with a relative intensity of 6 units with respect to the 349-keV transition having a 1000 units of intensity. The low intensity for the population of this level in the (t,p) experiment is consistent with our (t,p) results for the ^{110}Pd nucleus and the low intensity observed for the population of its known $0^+(2)$ level. We propose a new level at 1312 keV of excitation and identify the 574.9-, 963.0-, and 1312.8-keV gamma rays as deexciting this level. The 574.9-keV transition was originally placed as deexciting a level at 924.0 keV which we exclude. However, we find evidence suggesting that the 574.9-keV gamma ray is in coincidence with the transitions that deexcite the 737-keV level. This and our identification of the doublet nature of the 1313-keV photopeak.

The ISOL investigation placed a 1386.9-keV gamma ray as a transition from the 2756-keV level to a 1369-keV level, which, in turn, deexcited by a sole 485.7-keV transition. Such a placement in the ISOL results yielded an unrealistic negative beta-intensity for the population of the level. Further, such a placement requires the presence of an intense 560-keV gamma ray, which we do not observe. The 1386.9-keV photopeak does appear in both our 348- ($2^+(1)$ to $0^+(1)$) and 534-keV ($4^+(1)$ to $2^+(1)$) gates. Thus we propose the reverse of the order of deexcitation proposed by Äystö et al. and the existence of a new level at 2269.8 keV, which eliminates the existence of a level at 1369 keV (nota bene: Äystö et al. have the 1386.9-keV transition populating the wrong level in their Fig. 6 decay scheme but do have the correct assignment in their Table 3). A photopeak was originally given an energy of 1781.6-keV. However, upon comparison of ^{104}Ru(t,p)^{106}Ru and ^{108}Pd(t,p)^{110}Pd spectra, taken under the same conditions, this photopeak was identified with the 1778.895-keV photopeak of ^{28}Al and used to recalibrate the higher energy portion of the spectra.

When we compare our in-beam ^{110}Pd and ^{112}Pd results with the ISOL studies, the systematics of the heavy even-even Pd nuclei results. These systematics show that the low-energy structure of the mid-shell Pd nuclei remain remarkably similar. As in the case of the Cd nuclei,[4,6] an additional set of levels with a 0^+ band head obfuscates a simple collective character to the excitation spectra. Further, this extra set of levels, again as in the Cd case, occurs at its lowest energy at mid-shell, an effect that has been discussed in detail elsewhere.[6] However, what has not been recognized is that the intruder configuration drops in energy in the same way as the second quadrupole phonon 0^+ state. This leads to the question: Does this apparent "accidental" decrease indicate a close structural tie between the particle-hole nature of the intruder and the seniority four quasiparticle nature of the two-phonon state? Next we discuss a mixed-configuration IBM-II description of these states as intruder states.

3. Coexistence in Pd and Related Nuclei

3.1 Shape Coexistence in the Heavy Even-Even Pd Nuclei.

Until now the low energy structure of heavier Pd nuclei has not been accounted for by any of the collective models. Stachel and coworkers,[5] using an early version of the Interacting Boson Model (IBM-1), were not able to account for all the features of these nuclei. A particular problem was the number and nature of the low energy (E $\leq$ 2.3 MeV) 0^+ levels. However, as we discuss herein, we can account for all the levels observed in ^{110}Pd and ^{112}Pd. Our using mixed configuration IBM-II calculations take recognition of the existence of proton cross-shell intruder-state configurations and are an extension of our earlier work on the neighboring Cd nuclei.[4] Our mixed configuration IBM-II calculations take two separate collective bands separated by an energy $\triangle$ (less the binding energy of the configurations) and mixes them together with a simple mixing Hamiltonian. In calculating the collective bands, we have used the same Hamiltonian as in our Cd calculations and have kept fixed the mixing strength (0.08) while we have put the energy gap between configurations at $\triangle$ = 5.0 MeV and the Majorana strength at 0.06. Further, we have assumed x_v to be a strict function of neutron number and estimated x_p by assuming its variation is linear within a shell starting at a value of $-(7)^{1/2}/2$ at the empty shell and going to a value of $+(7)^{1/2}/2$ at full shell. We have also used the same neutron-neutron boson interaction strength (C_{0n}, C_{2n}) from previous Pd calculations.[7] The remaining 4 parameters (e and K in each configuration) were set by a fit to ^{110}Pd. The ^{112}Pd was then predicted by modifying these parameters in the same wa[3] as ^{114}Cd was predicted from ^{112}Cd. These levels, compared with our experimental levels (Experimental/Calculated) are (spin-parity values are given in brackets []): 0/0 [$0^+(1)$], 348/387 [$2^+(1)$], 737/858 [$2^+(2)$], 883/908 [$4^+(1)$], 890/916 [$0^+(2)$], 1096/1488 [$3^+(1)$], 1126/1238 [$0^+(3)$], 1140/1216 [$2^+(3)$], 1362/1458 [$4^+(2)$], 1423/1638 [$2^+(4)$], 1551/1541 [$6^+(1)$], 1716/1667 [$4^+(3)$], 1759/- [$5^+(1)$], and 2318/2271 [$8^+(1)$]. Elsewhere,[10] we compare our results for ^{110}Pd as well as the branching ratios for the level deexcitations.

3.2 Shape Coexistence in the Light Pd Nuclei and Spin Dependance of B(E2) Values.

Recent studies of the low energy excitations of ^{102}Pd and ^{104}Pd have revealed several uncommon features.[2] First, an excited 0^+ state was traced from its lowest energy in ^{102}Pd to its highest known energy in ^{106}Pd. Such a mass dependence of increasing energy with increasing mass is the opposite of the trend for collective states in the heavy-mass region of the experimentally known Pd isotopes up to mid-shell ^{112}Pd. For ^{102}Pd, the 0^+ level possesses a 14.3 ns half life and is almost degenerate with a second 0^+ level. The deexcitation properties of the intruder state bandhead were reported to be abnormal.^{2} First, the predominant deexcitation was by a 59-keV E2 transition which possesses a B(E2) value of 96 SPU. This is in contrast to the limit of $<4 \times 10^{-4}$ for the E2 deexcitation path to the first 2^+ vibrational state. A 2^+ level 350 keV higher in energy is observed to deexcite to the 0^+ band head with a relative B(E2) value 240 times greater than that observed for the transition to the ground state. However, we suggest that these features can be accounted for within the framework of the strong interaction between cross-shell $d_{3/2}$-protons and neutrons which are filling the $d_{5/2}$ orbital in this mass region of the Pd isotopes. That is properties of ^{102}Pd are consistent with the occurrence of a $pd_{3/2}$ intruder configuration similar in nature to the $pg_{9/2}$-$ng_{7/2}$ intruder structure that coexists in double closed subshell ^{98}Zr.^{9}

In general, comparison is made for the spin-loss deexcitation of levels and, in general, to a common final state. Thus any spin dependance drops out. However, for the special case where we wish to compare the **SPIN-GAIN** E2 deexcitation of one level, with the **SPIN-LOSS** E2 deexcitation of a common connecting level (e.g. a $0^+(2)$--(E2)-->$2^+(1)$--(E2)-->$0^+(1)$ sequence), a significant difference arises for the relative B(E2) values. Here, we must take 1/5 of the B(E2) value for the $0^+(2)$--(E2)-->$2^+(1)$ transition and compare it with the full B(E2) value for the $2^+(1)$--(E2)-->$0^+(1)$ transition, or $\{1/5\}\{B(E2;0^+(2)$-->$2^+(1))\} = B(E2;0^+(2)$<<==$2^+(1))$. In the recent studies of ^{102}Pd have concluded that "the B(E2; $0^+(2)$-->$2^+(1)$) value reveals strong collectivity" because of its 96 Wu B(E2) value.^{2} As we have discussed, in order to correctly compare the intruder state bandhead deexcitation with the relative transition strength of the $2^+(1)$-->$0^+(1)$ transition, we must take 1/5 of the $0^+(2)$-->$2^+(1)$ value or $B(E2;0^+(2)$<<==$2^+(1)) = \{1/5\}B(E2;0^+(2)$-->$2^+(1)) = (1/5)(96) = 19$ Wu, which is less than the 33 Wu for the $2^+(1)$-->$0^+(1)$ B(E2) value. We have shown^4 that configuration mixing can be the cause of such a relatively "slow" intraband $2^+(i)$-->$0^+(i)$ B(E2) value of 17 Wu (cancellation) in comparison with the 19 Wu for the $0^+(i)$<<==2^+(gs) transition (additivity).

3.3 Shape Coexistence in ^{70}Se and ^{102}Ru and Spin Dependent B(E2) Values

Two other cases where incorrect spin dependant B(E2) comparisons led to misconceptions are: ^{70}Se and ^{102}Ru. In ^{70}Se a "strong" collectivity of 53 Wu for the $0^+(2)$--(E2)-->$2^+(1)$ B(E2) value was reported. Instead, if we compare the $0^+(2)$<<==(E2)==$2^+(1)$ value of (1/5)(53)~10 Wu with 21 Wu for the $2^+(1)$--(E2)-->$0^+(1)$ transition, we find this in the realm of understanding within the context of the mixing of coexisting systems. A recent report on ^{102}Ru could not account for its level structure within a simple symmetry model.^{11} However, we can account for the ^{102}Ru structure within the framework coexisting states where we identify the 984- and 1581-keV levels as the 0^+ and 2^+ states of a coexisting vibrational system while the gs, 475-, 1103-, 1106-, and 1521-keV levels are associated with the $0^+(1)$, $2^+(1)$, $2^+(2)$, $4^+(1)$, and $3^+(1)$ members of the O(6)-like coexisting system reported for ^{98}Zr, which is the N=58 isotone of ^{102}Ru. That the 1581-keV 2^+ level is associated with the 0^+ fp-d "spherical" system is supported by the B(E2) ratio of ~400 for the intra- versus inter-band transition to the $0^+(2)$ versus the $0^+(1)$ level. Also, if we take the B(E2;0^+<<==2^+) value rather than the B(E2;0^+-->2^+) value we find a comparitively slow transition. It is important to note that this may be one of the <u>first identifications</u> of **an excited spherical structure** coexisitng with a more deformed ground state structure.

4. Summary

The low-energy spectra of ^{112}Pd has been established by combining a reanalysis of in-beam (t,p) spectroscopy studies with ISOL beta-decay spectroscopy studies. We showed that mixed-configuration IBA-II calculations can be used to account for the observed excitation spectra of this nucleus. Occurrence of a proton-$d_{3/2}$, neutron-$d_{5/2}$ intruder configuration in the even-even Pd nuclei with A~102 was proposed. We compared spin-increase B(E2) values with spin-decrease B(E2) values for the case of $0^+(2)$-->$2^+(1)$-->$0^+(1)$ cascades and suggested that a convienent and correct method of comparision is to compare the $2^+(1)$-->$0^+(1)$ B(E2) value with the $0^+(2)$<<==$2^+(1)$ B(E2) value which is equivalent to 1/5 of the $0^+(2)$-->$2^+(1)$ B(E2) value. Evidence is given for coexisting structures in ^{70}Se, ^{102}Pd, and ^{102}Ru.

This work was supported, in part, by the Deutsche Akademische Austauchdienst, performed, in part, under the auspices of the U.S. Department of Energy at Lawrence Livermore National Laboratory under contract Nr. W-7405-Eng-48, in part, by NATO grant Nr. NATO-RG.86/0452, and in part, by Associated Western Universities, Inc.

References

* Deutsche Akademische Austauschdienst visiting Professor 1987, Permanent address: LLNL.

\+ Participating guest at LLNL. Permanent address: Princeton University.

** LLNL participating guest. Permanent address: University of California, Berkeley.

++ LLNL-AWU participating guest. Permanent address: San Jose State University.

1. A. Passoja, et al., Nucl. Phys. <u>A441</u>, 261 (1985).
2. M. Luontama, et al., Zeit. fur Physik <u>A324</u>, 317 (1986).
3. K. Heyde, et al., Phys. Rev. C <u>25</u>, 3160 (1982).
4. D. Kusnezov, et al., Helv. Phys. Acta <u>60</u>, 456 (1987).
5. J. Stachel, et. al., Phys. Rev. C <u>25</u>, 650 (1982).
6. A. Aprahamian, et al., Phys. Lett. <u>140B</u>, 22 (1984).
7. P. Van Isacker and G. Puddu, Nucl. Phys. <u>A348</u>, 125 (1980).
8. R.J. Estep, et. al., Phys. Rev. C <u>35</u>, 1485 (1987) and sub. to Phys. Rev. C (1988).
9. R.A. Meyer, et al., Phys. Lett. B <u>177</u>, 271 (1986).
10. R.A. Meyer, et al., UCRL-85261, sub. to Phys. Rev. C (1988).
11. A. Giannatiempo, et al., Phys. Rev. C <u>33</u>, 1024 (1986).
12. J. Äystö, et al., Nucl. Phys. <u>A480</u>, 104 (1988).

Monopole Degrees of Freedom in the Odd-Mass Neighbours of Even-Even Nuclei with Low-Lying 0^+ Levels

E.F. Zganjar[1] and J.L. Wood[2]

[1]Department of Physics and Astronomy,
 Lousiana State University, Baton Rouge, LA 70803, USA
[2]School of Physics, Georgia Institute of Technology,
 Atlanta, GA 30332, USA

Abstract

We show that when an even-even nucleus in a region has a low-lying excited 0^+ level, one often observes a sizeable E0 contribution to several of the $I^\pi \to I^\pi$ transitions in the odd-mass neighbours. Based on our work in neutron deficient, odd-mass nuclei near $Z = 80$ and $N = 104$, we propose that this E0 enhancement is associated with transitions between nuclei with very different mean-square radii and that these bands arise from particle/hole couplings to the corresponding shape coexisting structures in the even-even neighbours. Under this assumption, the E0 enhancement of transitions in the odd-mass neighbours of even-even nuclei may serve as indicators of shape coexistence. The application of these ideas to the Zr region, where some of the lowest lying excited 0^+ states are observed, is discussed.

Introduction

Shape coexistence, intruder orbitals, and dramatic shape changes are quite common phenomena in the $Z \approx 40$ region of nuclei. The neutron-rich Sr (Z=38) and Zr (Z=40) nuclei, for example, exhibit a sudden onset of deformation between neutron number N=58, with nearly spherical shapes, and N=60, where the ground states are deformed. All of the N=60 isotones, from ^{98}Sr to ^{106}Pd, have 0^+_2 states that are not accounted for by any collective model as a simple collective excitation of the core [1]. For the low-Z, N=60 isotones ^{98}Sr and ^{100}Zr, shape coexistence is supported by extremely low-lying 0^+_2 states at 215 and 331 keV respectively [2,3]. These values are shown in table I along with all known 0^+_2 states which lie below 636 keV. We have used this table in conjunction with table II to show that where one observes low-lying 0^+_2 states and shape coexisting bands interconnected by transitions with large B(E0) values in even-even nuclei, one also observes transitions with large B(E0) values in the neighbouring odd-mass nuclei [15]. We have further shown [15] that the enhanced monopole strength can be used to identify the shape coexisting structures in the odd-mass nuclei which arise out of the coupling of an odd nucleon to the even-even core. The odd-particle can then be used to probe the even-even core and to indicate the quasiparticle configurations which give rise to the shape coexistence itself. Note that most of the odd-mass nuclei listed in Table II are neighbours of the even-even

Table I. The lowest first excited 0^+ states known in doubly-even nuclei. All known cases with $E_x \leq 635$ keV are shown.

Isotope	E_x	ref.	comments
^{98}Sr	215	2	
^{100}Zr	331	3	
^{182}Hg	(350)	4,5	est. from $I \geq 2^+$ members (in-beam)
^{184}Hg	375	6,7	
^{178}Pt	(422)	8	0^+ spin inferred
^{176}Pt	(433)	8	0^+ spin inferred
^{180}Hg	(450)	5	est. from $I \geq 2^+$ members (in-beam)
^{186}Pt	472	9	
^{184}Pt	492	10	
^{182}Pt	500	11	
^{180}Pt	(500)	–	est. from systematics
^{186}Hg	522	12	
^{152}Gd	615	13	
^{230}Th	635	14	

nuclei listed in table I. This is graphically displayed in fig. 1 where we have added, for illustrative purposes, an example from the rare-earth region and one from the actinide region, both of which exceed the limits set in tables I and II.

Shape coexistence is also observed on the neutron-deficient side of the Zr region, although there the 0_2^+ states are much higher in energy. For example, a 0_2^+ state in ^{72}Se is observed at 978 keV [29] and predictions [30] indicate considerable shape coexistence in the region. This may, in fact, be the region where the much sought-after strongly deformed oblate structures will be found.

Table II. The lowest-energy transitions with E0 components known between low-lying levels in odd-mass nuclei. All known cases with $E_\gamma \leq 429$ keV are shown.

Isotope	E_γ (keV)	E_i (keV)	I_i^π	ref.
^{185}Au	205	497	$5/2^+$	16,17
^{187}Pt	260	260	$3/2^-$	18,19
^{187}Pt	262	288	$5/2^-$	18,19
^{185}Au	281	515	$5/2^+$	16
^{183}Au	284	----not certain----		20
^{185}Au	289	289	$5/2^-$	16
^{183}Au	297	----not certain----		20
^{185}Au	313	322	$9/2^-$	16,21,22
^{195}Pb	318	1126	$11/2^+$	23
^{187}Au	323	444	$9/2^-$	24,25
^{185}Au	330	371	$3/2^+$	16,17
^{185}Pt	340	521	$3/2^-$	26
^{187}Au	388	742	$13/2^-$	24,25
^{195}Pb	401	1177	$9/2^+$	23
^{185}Au	427	648	$13/2^-$	16,21,22
^{231}Th	429	634	$7/2^-$	27

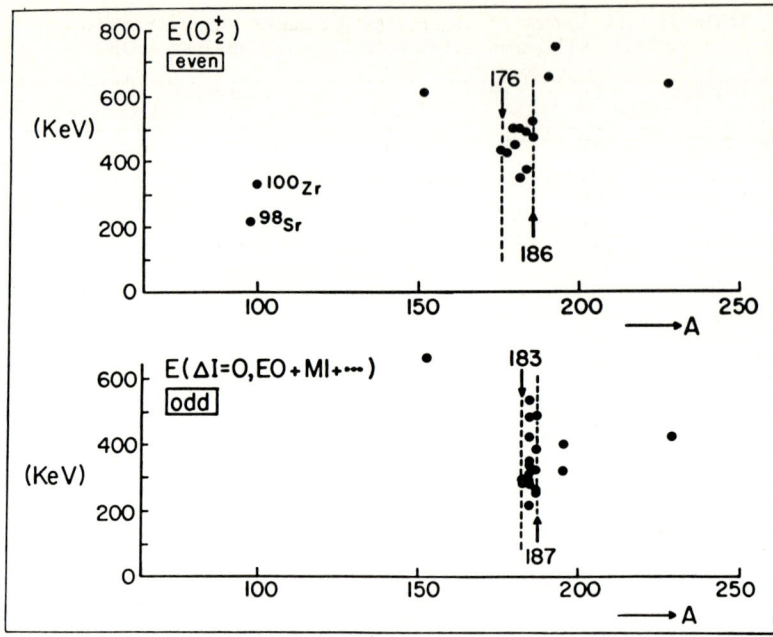

Fig. 1. Graph of the data presented in tables I and II showing the energy of the 0_2^+ states in even-even nuclei (upper part) and the energy of transitions in odd-mass nuclei which have large E0 components (lower part). Points for ^{192,194}Pb [28] as well as examples from the rare-earth (A ~ 150) and actinide (A ~ 230) regions, all of which exceed the limits set in the tables, have been added for illustrative purposes.

Heyde et al. [31,32,33] attribute the low-lying 0_2^+ states to the proton-neutron interaction arising from particle-hole (p-h) excitations across major shells. In this picture the major configuration of the 0_2^+ states involves the excitation of pairs of particles across major shells into intruder orbitals, and the interaction of those valence nucleons with nucleons of the other type through the proton-neutron force. This interaction is maximal then when the "other" nucleons are midway between closed shells. For the Z=82 region, discussed below, the major intruder configuration involves $(\pi h_{9/2})^{+2}$ with the maximum interaction (minimum 0_2^+ energy) occurring between the closed neutron shells of 82 and 126, that is, at N=104. Heyde [33] has put the occurrence of low-lying 0_2^+ intruder states near major closed shells and the onset of stable deformation in the 0^+ ground states near subshells on an equal footing. Thus, it is instructive to use the experimental information gleaned in a region where only a major shell is operative (Z=82) and where extensive (and smooth) systematics are available to aid in the study of nuclei in the Zr region. This region is crossed by a major shell (N=50) and several subshells (N=56, Z=38,40), and the systematics are cut short and fraught with sudden jumps. With that premise, we present some information from our study of monopole degrees of freedom in the Z=82 region.

The Z = 82 Region

Nuclei in the far from stability region near Z = 82 and N = 104 comprise the most extensive region known exhibiting coexisting nuclear shapes [31,34]. For the even-even nuclei in the region, the excited 0^+ states, upon which the structure coexisting with the ground state is based, lie quite low in energy. An example of this is presented in fig. 2 and the upper part of fig. 3 for the even Hg isotopes. In fig. 2 the energy of the excited 0^+ is plotted as a function of mass. The coexisting band structures for three of these isotopes, $^{186-190}$Hg, are shown in the upper part of fig. 3. These are selected for display because of their proximity to ^{185,187}Au which provide the focus of this report. Note, from fig. 2, that the minimum energy of the excited 0^+ occurs quite close to the mid-neutron shell at N = 104, as one would expect on the basis of the picture of Heyde et al. [32,33]. While these even-even Hg isotopes provide the classic example of the coexistence of levels built on different shapes (2h ground state and 2p-4h intruder configurations), it is the even-even Pt isotopes in this region which are more interesting in that it can be demonstrated (see discussion below) that the intruder configuration, which forms the excited 0_2^+ band in ^{188}Pt and all heavier Pt isotopes, becomes the ground state in ^{186,184}Pt.

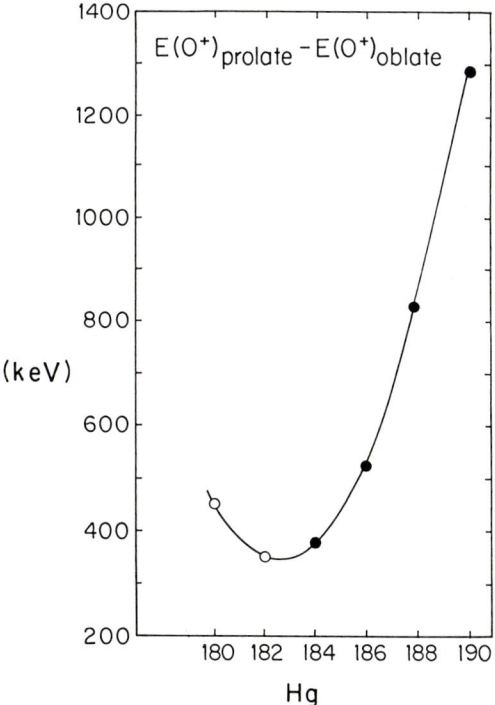

Fig. 2. Plot of the energy of the 0_2^+ state in the even-even Hg isotopes. Data on A = 182-188 are from refs. [4,6,12,35], that for A = 180 from ref. [5] and A = 190 from ref. [36]. The open circles represent rotational extrapolations to the 0_2^+ energy from the I ≥ 2^+ members of the band seen in-beam. The black dots represent directly measured energies (by internal conversion).

In table III and table IV we show the experimental conversion coefficients for the even-even cases of fig. 3 and the complete list of E0 enhanced transitions (E > 430 keV included) for the odd-mass nuclei listed in table II. Obvious missing entries include Hg and Tℓ isotopes. Data for $^{189-197}$Hg [38] ^{195}Hg [39] and ^{187}Hg [40] are either preliminary or the transitions are much higher in energy. Studies on ^{193,195}Tℓ are underway [41].

Table III. K-conversion coefficients for the $I_i^\pi \to I_f^\pi$ (I≠0) transitions in the Hg and Pt isotopes shown in fig. 3.

Isotope	E_γ (keV)	E_i (keV)	I_i^π	Experiment α_K	ref.	Theory [37] α_K(M1)	α_K(E2)
^{186}Hg	216	621	2^+	3.1 (7)	12	0.76	0.14
^{188}Hg	468	881	2^+	0.087 (6)	35	0.093	0.022
	203	1208	4^+	0.99 (11)	35	0.89	0.16
^{190}Hg	1155	1571	2^+	0.035 (4)	36	0.0093	0.0037
	933	1975	4^+	0.047 (5)	36	0.015	0.0055
^{184}Pt	682	845	2^+	0.28 (6)	10	0.031	0.010
	799	1235	4^+	0.049 (10)	10	0.020	0.0074
^{186}Pt	606	798	2^+	>0.10*	9	0.041	0.013
	733	1223	4^+	0.057 (22)	9	0.025	0.0091
^{188}Pt	849	1115	2^+	0.22 (1)	9	0.016	0.0067

*Doublet. Lower limit computed from I_K/I_γ given in ref. 9. The other component is probably E2.

Table IV. K-conversion coefficients for $I_i^\pi \to I_f^\pi$ transitions with E0 components for those cases shown in table II.

Isotope	E_γ (keV)	E_i (keV)	I_i^π	Experiment α_K	ref.	Theory [37] α_K(M1)	α_K(E2)
^{183}Au	284	--not certain--		1.6	20	0.33	0.069
	297	--not certain--		3.9	20	0.28	0.064
^{185}Au	205	497	$5/2^+$	2.5 (5)	16,17	0.82	0.16
	281	515	$5/2^+$	0.77 (41)	16	0.34	0.073
	289	289	$5/2^-$	0.52 (20)	16	0.32	0.068
	313	322	$9/2^-$	0.58 (15)	16,21,22	0.26	0.056
	330	371	$3/2^+$	2.8 (8)	16,17	0.22	0.049
	427	648	$13/2^-$	0.33 (3)	16,21,22	0.11	0.027
	492	712	$11/2^-$	0.21 (2)	16,21,22	0.077	0.020
^{187}Au	323	444	$9/2^-$	0.66 (17)	24,25	0.12	0.035
	388	742	$13/2^-$	0.64 (9)	24,25	0.23	0.054
	657	881	$11/2^-$	0.10 (2)	25	0.036	0.011
^{185}Pt	340	521	$3/2^-$	0.35	26	0.19	0.042
	542	723	$3/2^-$	0.19	26	0.055	0.015
^{187}Pt	260	260	$3/2^-$	3.2 (6)	18,19	0.39	0.086
	262	288	$5/2^-$	5.7 (14)	18,19	0.38	0.084
	499	508	$3/2^-$	0.8 (3)	18,19	0.068	0.018
^{195}Pb	318	1126	$11/2^+$	4.9 (32)	23	0.34	0.056
	401	1177	$9/2^+$	5.4 (27)	23	0.17	0.033
^{231}Th	429*	634	$7/2^-$	0.66 (4)	27	0.28	0.037

*Additional cases for ^{231}Th are not shown.

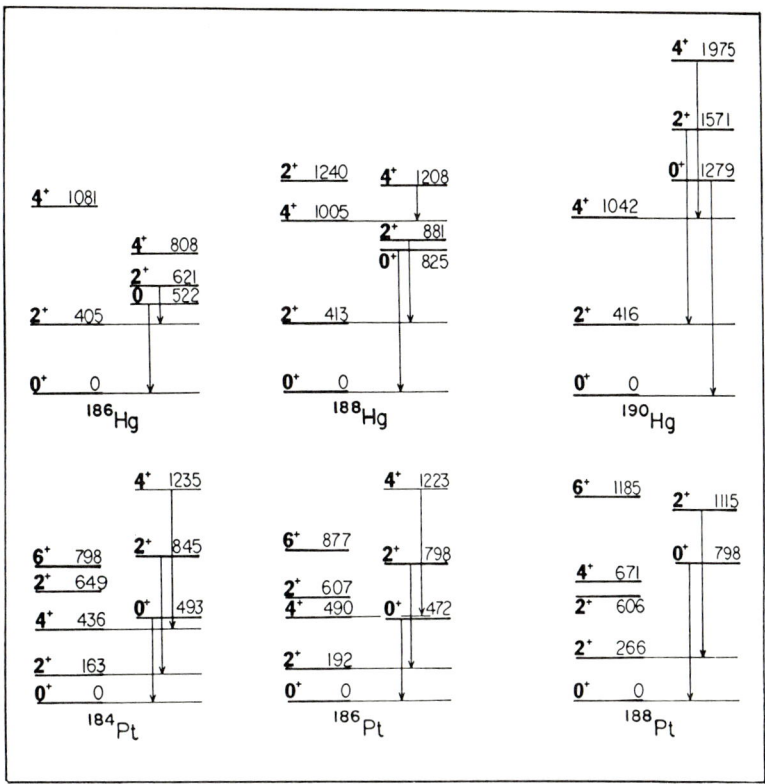

Fig. 3. Systematics of the low-lying levels in ^{186,188,190}Hg, ^{184,186,188}Pt. Transitions which are pure E0 or with E0 admixtures are indicated. Data are from refs. ^{186}Hg [12], ^{188}Hg [35], ^{190}Hg [36], ^{184}Pt [10], and ^{186,188}Pt [9].

We have shown that the very converted transitions which we observed in our study of the odd-mass Au isotopes [15,42] are entirely consistent with an E0 multipole component and that they closely match low-energy E0 transitions in the neighbouring doubly even Hg and Pt isotopes. Examples of this are shown in figs. 4 and 5 for ^{185,187}Au. The systematic pattern of transitions with E0 components in ^{185,187}Au bears a close resemblance to E0 and E0 admixed transitions in the neighbouring doubly-even Pt and Hg isotopes, shown in fig. 3, which can be considered as cores for certain configurations in the odd-mass Au isotopes. The numbers in boxes represent the ratio α_K(expt)/α_K(M1 theory). Since it can be demonstrated by lifetime measurements that multipoles greater than E2 are not involved in these transitions, a ratio greater than unity indicates an E0 contribution. The 11/2$^-$ state at 220 keV (^{185}Au) and 224 keV (^{187}Au), shown in fig. 4, is the proton $h_{11/2}$ hole state where the appropriate cores for proton-hole configurations in ^{185}Au and ^{187}Au are ^{186}Hg and ^{188}Hg respectively. The 9/2$^-$ state at 9 keV (^{185}Au) and 121 keV (^{187}Au), shown in fig. 5, is the proton $h_{9/2}$ particle (intruder) state where the appropriate cores for proton particle configurations in ^{185}Au and ^{187}Au are ^{184}Pt and ^{186}Pt respectively. Pairs of bands interconnected by transitions with intense E0 admixtures are known in strongly-

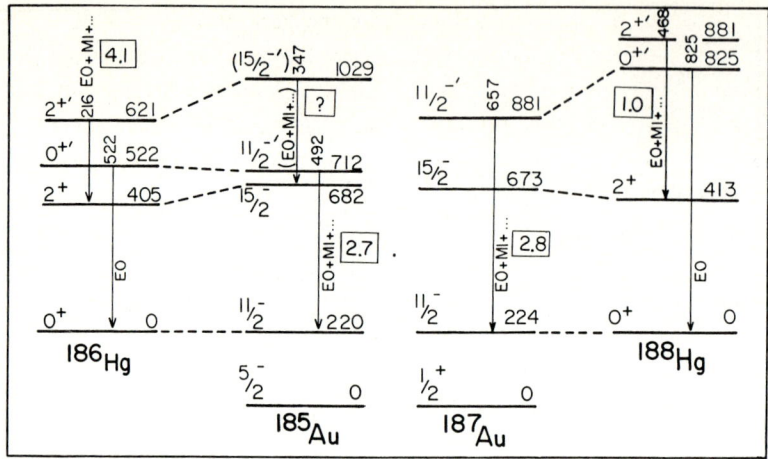

Fig. 4. Portions of the $h_{11/2}$ and $h'_{11/2}$ bands in ^{185,187}Au compared to the 0^+ and $0^{+'}$ bands in ^{186,188}Hg. Only transitions with $\alpha_K > \alpha_K(M1)$ are shown. The numbers in boxes represent the ratio $\alpha_K(\text{expt})/\alpha_K(M1\ \text{theory})$. The ^{185m,g}Hg decay contains three 347 keV transitions and thus a multipolarity for the 1029 → 682, 347 keV transition could not be determined. The data for the even-even isotopes are from refs. [12,35].

deformed odd-mass nuclei. Some examples of note are ^{231}Th [27], ^{155}Eu [43], and ^{173}Lu [44]. The systematic occurrence of such pairs has also been reported [45] in the actinides. To our knowledge, however, the present work reports the first systematic occurrence of pairs of bands interconnected by transitions with intense E0 admixtures outside of the traditional strongly-deformed regions, and it is significantly lower in energy.

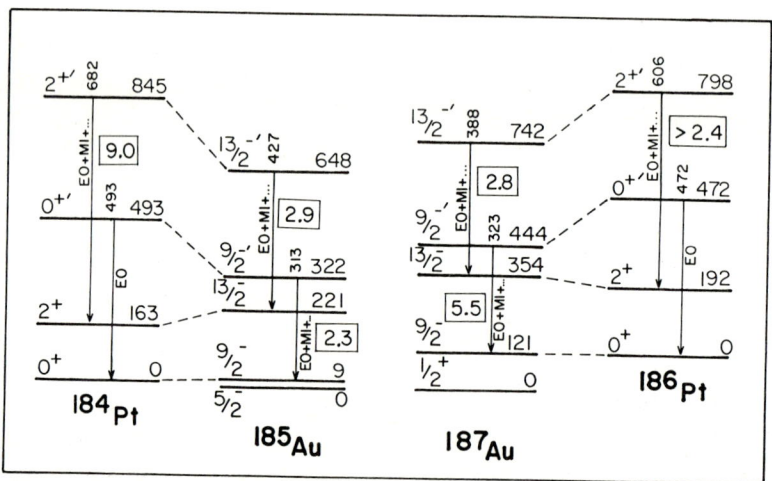

Fig. 5. Portions of the $h'_{9/2}$ and $h_{9/2}$ bands in ^{185,187}Au compared to the 0^+ and $0^{+'}$ bands in ^{184,186}Pt. Only transitions with $\alpha_K > \alpha_K(M1)$ are shown. The numbers in boxes represent the ratio $\alpha_K(\text{expt})/\alpha_K(M1\ \text{theory})$. The data for the even-even isotopes are from refs. [9,10].

The Pt isotopes are particularly interesting because they undergo a pronounced structure change between ^{188}Pt and ^{186}Pt. This is interpreted as being due to the 2p-6h intruder configuration which is observed as an excited band for A ≥ 188, and which becomes the ground state for A=186,184. This proposal was first presented in 1981 by Wood [46] on the basis of systematics of the N = 106 isotones and on our work [24] on ^{187}Au which used a blocking argument to deduce that the ground-state configuration is 2p-6h. Our recent work [47] on ^{189}Au clearly shows that the blocking argument is valid. That is, it indicates that the ground state of ^{188}Pt is 4h. The transition of the 2p-6h configuration from excited to ground state as one moves from ^{188}Pt to ^{186}Pt has also been confirmed by the in-beam data of Dracoulis et al. [48].

Conclusions

The most outstanding example [49] of an electric monopole transition between states of very different shapes is probably the decay of the fission isomer to the ground state in ^{238}U. The B(E0) for the decay of ^{238m}U is the smallest ever measured [49]. While E0 and E0 admixed transitions appear to be a promising fingerprint for shape coexistence, shape change alone would not seem to be the solution to understanding the strong, low-energy, E0 transitions in the neutron-deficient Pb, Au and Hg isotopes, even though these shape changes appear to be large in some cases. E0 transitions need to be quantified if they are to be used as a fingerprint for coexisting bands. This will, of course, require lifetime measurements in the 50 ps range.

It should also be noted that E0 transitions and isotope shifts are dynamic and static aspects of the same operator -- the square of the electric charge radius [33]. A quantitative determination of the E0 components in the $I^{\pi} \to I^{\pi}$ transitions is also needed. This is impossible by conversion-electron measurements alone, but can be accomplished in combination with γ anisotropies from nuclei oriented at low temperature. This is vitally important when an E0+E2 admixture masquerades as M1 in the internal-conversion process. The studies described here can be applied to any region of the nuclear chart where excitations across shells or subshells are driven by the proton-neutron interaction. The region of nuclei near Zr, with a major shell (N=50) and several subshells (N=50, Z=38,40) should be especially rich in information on these interactions.

Acknowledgements

We wish to acknowledge the assistance and advice of our many colleagues at UNISOR and HHIRF. The work was supported in part by the US DOE through grant/contract DE-FG05-84ER40159 (LSU), DE-FG05-87ER40330 (Ga. Tech.), and DE-AC05-76OR00033 (UNISOR).

References

1. F. K. Wohn et al., Phys. Rev. C33, 677 (1986).
2. F. Schussler et al., Nucl. Phys. A339, 415 (1980).
3. T. A. Khan et al., Nucl. Phys. A283, 105 (1977).
4. W. C. Ma et al., Phys. Lett. B139, 276 (1984).
5. G. Dracoulis et. al., Phys. Lett. B, preprint (1988).
6. J. D. Cole et al., Nucl. Phys. A283, 105 (1977).
7. J. D. Cole, Ph.D. thesis, Vanderbilt, 1978.
8. E. Hagberg et al., Nucl. Phys. A318, 29 (1979).
9. M. Finger et al., Nucl. Phys. A188, 369 (1972).
10. M. Cailliau et al., J. de Phys. 35, 469 (1974).
11. J. P. Husson et al., in Proc. Third Int. Conf. on Nuclei Far From Stability, Cargese, 1976, ed. R. Klapisch, CERN 76-13, p 460.
12. J. D. Cole et al., Phys. Rev. C16, 2010 (1977).
13. C. M. Baglin, Nucl. Data Sheets 30, 1 (1980).
14. Y. A. Ellis-Akovali, Nucl. Data Sheets 40, 385 (1983).
15. E. F. Zganjar et al., in Proc. of the Fifth Intl. Conf. on Nuclei Far From Stability, Rosseau Lake, AIP Conf. Proc. 164, I. S. Towner ed., American Institute of Physics, New York (1988), p. 313.
16. C. D. Papanicolopulos, Ph.D. thesis, Ga. Tech., 1987.
17. C. D. Papanicolopulos, private communication.
18. A. Ben Braham et al., Nucl. Phys. A332, 397 (1979).
19. B. E. Gnade et al., Nucl. Phys. A406, 29 (1983).
20. M. I. Macias-Marques et al., Nucl. Phys. A427, 205 (1984).
21. E. F. Zganjar et al., in Nuclei Off the Line of Stability, ed. R. A. Meyer and D. S. Brenner, ACS Symp. Series 324, 1986, p. 245.
22. E. F. Zganjar et al., in Nuclear Structure Reactions and Symmetries, Vol. 2, ed. R. A. Meyer and V. Paar (World Scientific Publ. Co., Singapore, 1986), p. 716.
23. J. C. Griffin, Ph.D. thesis, Ga. Tech., 1987.
24. E. F. Zganjar et al., in proc. Fourth Intl. Conf. on Nuclei Far From Stability, Helsingør, Denmark, 1981, ed. P. G. Hansen and O. B. Nielsen, CERN Report 81-09, p. 630.
25. M. A. Grimm, Ph.D. thesis, Ga. Tech., 1978.
26. B. Roussiere et al., Nucl. Phys. A438, 93 (1985).
27. D. H. White et al., Phys. Rev. C35, 81 (1987).
28. P. Van Duppen, et al., Phys. Rev. C35 1861 (1987).
29. J. H. Hamilton et al., Phys. Rev. Lett. 32, 239 (1974).
30. W. Nazarewicz et al., Nucl. Phys. A435, 397 (1985).
31. K. Heyde et al., Phys. Rept. 102, 291 (1983).
32. K. Heyde et al., Nucl. Phys. A466, 189 (1987).
33. K. Heyde, in proc. of the Fifth Intl. Conf. on Nuclei Far From Stability, Rosseau Lake, AIP Conf. Proc. 164, I. S. Towner ed., American Institute of Physics, New York (1988), p. 255.

34. J. H. Hamilton et al., Rep. Prog. Phys. 48, 631 (1985).
35. J. D. Cole et al., Phys. Rev. C30, 1267 (1984).
36. M. O. Kortelahti et al., Phys. Lett. B, preprint (1988).
37. F. Rosel et al., At. Data Nucl. Data Tables 21, 291 (1978).
38. G. M. Gowdy, Ph.D. thesis, Ga. Tech., 1976.
39. G. M. Gowdy et al., Nucl. Phys. A312, 56 (1978).
40. R. A. Braga et al., Phys. Rev. C19, 2305 (1979).
41. C. R. Bingham, private communication.
42. M. O. Kortelahti, et al., Z. Phys., preprint (1988).
43. P. T. Prokofjev et al., Nucl. Phys. A455, 1 (1986).
44. E. G. Funk et al., Phys. Rev. C10, 2015 (1974).
45. T. von Egidy et al., Phys. Lett. 81B, 281 (1979).
46. J. L. Wood, in proc. Fourth Intl. Conf. on Nuclei Far From Stability, Helsingør, Denmark, 1981, ed. P. G. Hansen and O. B. Nielsen, CERN Report 81-09, p. 612.
47. M. O. Kortelahti et al., to be published.
48. G. D. Dracoulis et al., J. Phys. G12, L97 (1986).
49. J. Kantele et al., Phys. Rev. Lett. 51, 91 (1983).

Part III

High Spin Effects

Techniques for the Study of the A ~ 80 and A ~ 100 Nuclei Far from Stability

W. Gelletly[1], Y. Abdelrahman[2], A.A. Chishti[2], J.L. Durell[2], J. Fitzgerald[2], C.J. Lister[2], J.H. McNeill[2], W.R. Phillips[2], and B.J. Varley[2]

[1]SERC Daresbury Laboratory, Daresbury, Warrington, WA44AD, UK
[2]Department of Physics, Schuster Laboratory, University of Manchester, Manchester, M139PL, UK

Techniques used at Daresbury Laboratory for the study of nuclei far from stability in the A~80 and A~100 regions are reviewed. Developments of both the Recoil Mass Separator and the neutron detectors used in modular, gamma ray arrays are discussed and examples given of their performance. The use of heavy ion induced fission for the study of the neutron-rich nuclei with A~100 is described. The results of studies of the prompt gamma rays from heavy ion induced fission are presented.

1. Introduction

At first sight the subject of this conference, namely the 'Nuclear Structure of the Zirconium Region', may seem to be a very restricted one. As it transpires, however, the atomic nuclei with Z~40 on both sides of the Valley of Nuclear Stability exhibit a very wide variety of behaviour, and, in microcosm, exhibit the properties we can observe in nuclei throughout the Segré chart. We find nuclei with unusually large deformations, co-existing states of different shape, rapid changes in shape with Z, N, and rotational frequency, intruder orbitals with large deformations, and pronounced shell effects. The occurrence of large shell gaps in deformed nuclei both at low and high spin is now a well established phenomenon in many regions of the Periodic Table. The richness of these phenomena and the abruptness of the observed changes makes this region an ideal one for the testing of our theoretical understanding of Nuclear Structure.

At Daresbury we have had a long-standing interest in the Sr-Zr-Mo nuclei on both sides of the Valley of Stability. Ample testimony to this interest is given in contributions (1-4,13) to the present proceedings. In this contribution I will confine myself to our in-beam studies in the A=80 region and our studies of prompt gamma rays from neutron rich fission fragments produced in heavy ion induced fission. Radioactive decay studies at Daresbury centre on the on-line Isotope Separator [5] and involve both laser resonance fluorescence measurements of mean square charge radii [6] and nuclear orientation. The former is not represented here but Stone [4] will present the results of measurements carried out at Daresbury of the moments of some Br isotopes by Nuclear Orientation.

On the neutron-deficient side of stability we have concentrated on technically demanding experiments designed to produce and study nuclei on or near the N=Z line. The present status of these measurements will be briefly reviewed, and the factors which limit them will be discussed. The results are presented in greater detail in refs. 1 and 7. The technical improvements required to allow spectroscopic studies of heavy ion fusion evaporation reactions with cross sections of $\sim 1\mu b$ will then be discussed. Turning to the nuclei with A$\sim$100 the use of heavy ion induced fission to study states of intermediate spin in the nuclei with A$\sim$100 will be discussed. Some representative results of recent experiments [8,9] on the prompt gamma rays emitted from nascent fission fragments will be presented.

2. Studies of A$\sim$80 Nuclei near the N=Z line

It has been pointed out elsewhere that the neutron-deficient nuclei with A=70-80 are the heaviest nuclei we can study where neutrons and protons occupy the same orbitals. They are thus of particular interest because of the effects of n-p pairing. Such nuclei are difficult to produce and study. The main means of production is the heavy ion induced, fusion-evaporation reaction.

Fig. 1 shows in simplified form how such reactions are thought to proceed, taking as an example the ^{92}Mo(^{40}Ca, α 2p)^{126}Ce reaction. If the system fuses it forms a highly excited (hot) compound nucleus rotating at high frequency. Initially this highly unstable system evaporates energetic particles, with neutron emission favoured near stability but with increasing proton and alpha particle emission as we move away from stability and the relative binding energies of neutrons and charged particles change. Particle evaporation continues until we are approximately one

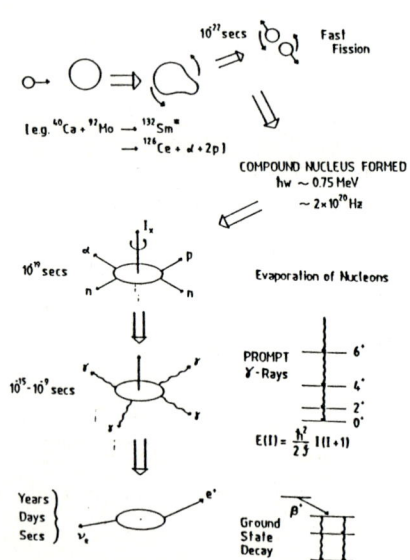

Fig. 1 *A simple picture of a typical heavy ion, fusion-evaporation reaction*

neutron separation energy from the yrast line. The large amount of angular momentum which remains is then dissipated in the emission of a cascade of 20-30 γ-rays. Initially these are statistical in nature but eventually they lead to states on or near the yrast line and de-excite by a series of stretched E2 transitions leading down into the ground state band in an even-even nucleus. In the absence of long-lived isomers this de-excitation process is over in 10^{-9} secs or so. The ground state then lives for a relatively long time before beta decay occurs.

For our purposes a number of features of this reaction stand out. Firstly if we are to form and study the nuclei farthest from stability we must devise some means of selecting the reaction channels involving neutron emission. Secondly the reaction proceeds via states of high spin and we can only expect to learn about high spin states in cold nuclei on or near the yrast line. In contrast the γ-rays emitted following the beta decay of the ground state may provide information about states of low spin.

There are a variety of ways in which we can select the reaction channel we wish to study. At Daresbury we have adopted two main methods, namely the Neutron Wall [10] and the Recoil Separator [11]. In the former method γ-rays are detected in coincidence with neutrons and charged particles and a measurement of the multiplicities of the associated particles allows one to determine the final nucleus from which the γ-rays are emitted. In the latter the A and Z of the recoiling nuclei are determined directly.

2.1 Neutron Detection

In its final form the Manchester Neutron Wall [10] consisted of an array of BGO suppressed Ge detectors at backward angles, a silicon detector charged particle telescope behind the target and an array of 37 neutron detectors forming a 1m^2 wall of liquid scintillator.

This apparatus was very effective but had two main limitations. Firstly this method does not distinguish between γ-rays from contaminants in the target and γ-rays from target nuclei. Ultimately the cleanliness of the target sets a limit to the sensitivity. Secondly the close-packed geometry of the neutron detectors leads to a high fraction of false, two-neutron coincidences due to scattering. This effect is illustrated in fig.2. The two-neutron gated spectrum is then only obtained by eliminating all coincidences between adjacent detectors.

To overcome the latter problem we have built a new series of modular neutron detectors, which are shaped to mimic a BGO suppression shield. This means that they fit into any of our standard γ-ray arrays [12]. The detectors again use NE213 liquid scintillator of volume ~1ℓ. Fast rise time photomultiplier tubes, of the type XP2041, are used to give good timing. They are used with a 1"Pb absorber in front of them to reduce the gamma flash, and pulse shape discrimination is used to

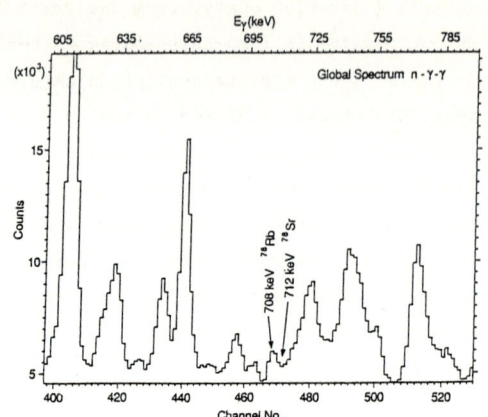

Fig. 3 Part of the global n-γ-γ spectrum from the ^{58}Ni + ^{24}Mg reaction. The position of the 712 keV gamma ray from ^{78}Sr is indicated.

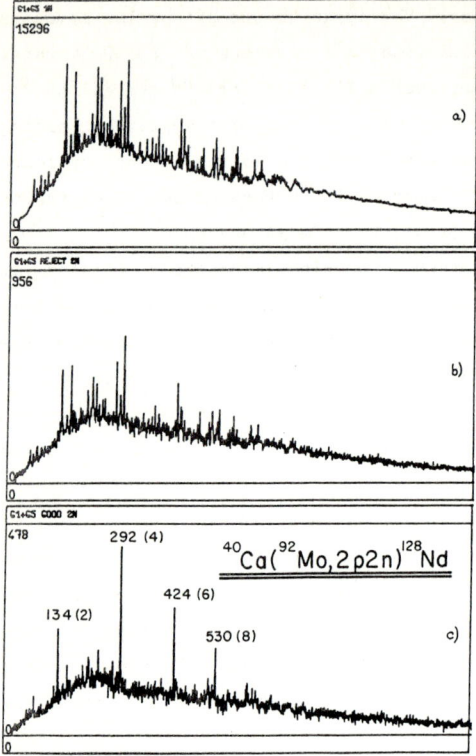

Fig. 2 The effects of the close-packed geometry in the Neutron Wall. From top to bottom we see the γ-ray spectrum (0-1250keV) gated by a single neutron, gated by adjacent detectors only and gated by non-adjacent neutron detectors

separate neutrons from γ-rays. Fifteen of these detectors were recently used[13] in a study of the ^{58}Ni(^{24}Mg, 2p2n)^{78}Sr reaction. The neutron detectors were placed in the forward hemisphere with fifteen, Compton-suppressed, Ge detectors in the backward hemisphere. Fig. 3 shows part of the global n-γ-γ spectrum from this reaction. The position in the spectrum of the 712 keV γ-ray from ^{78}Sr is indicated. It is almost lost in the noise.

Fig. 4 shows the same spectrum gated by two neutrons and the 712 keV gamma ray from ^{78}Sr. The γ-rays de-exciting the yrast line stand out clearly.

A level scheme for ^{78}Sr [13] has been constructed from these data. Clearly the physical separation of the neutron detectors has greatly reduced the scattering problem. The overall efficiency of the fifteen neutron detectors was determined to be 20-30%, from a comparison of the photopeak intensities in the n-γ-γ and γ-γ spectra.

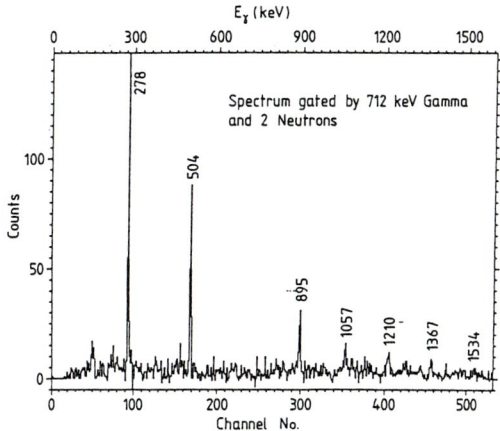

Fig. 4 Gamma ray spectrum gated by 2 neutrons and the 712 keV γ-ray in ^{78}Sr

The single neutron channel is also seen strongly in this experiment, and leads to the much improved level scheme for ^{78}Rb shown in fig. 5.

2.2 Recoil Separator

Fig. 6 shows the optical layout of the Daresbury Recoil Separator (14) in schematic form. The first Wien filter removes the beam and the second provides an energy dispersed beam of evaporation residues at the velocity slit. Residues with an energy spread of ±3% then pass through a dipole magnet which separates them by A/q, where q is the ionic charge. A position spectrum is obtained at the focal plane by a proximity-focussed carbon foil detector and an ionization chamber is used to measure both ΔE and E for the recoils. Taken together these measurements allow us to determine A and Z for the recoils in coincidence with an array of up to 20 BGO suppressed Ge detectors round the target. Since the system is modular any number of the suppressed Ge detectors may be replaced by neutron detectors.

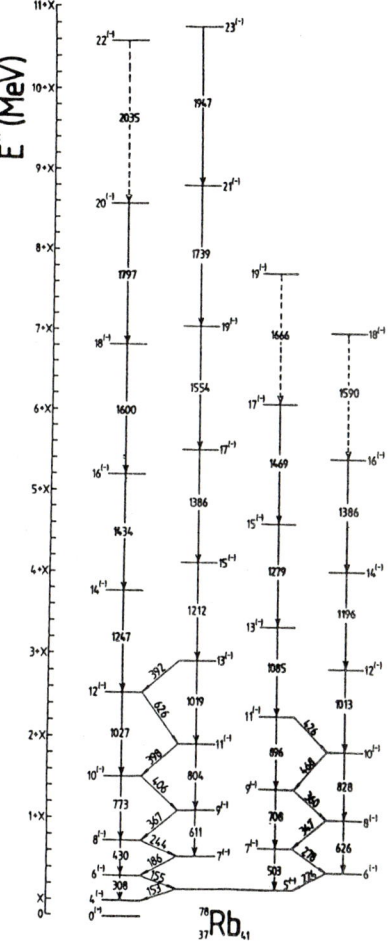

Fig. 5 Level scheme for ^{78}Rb

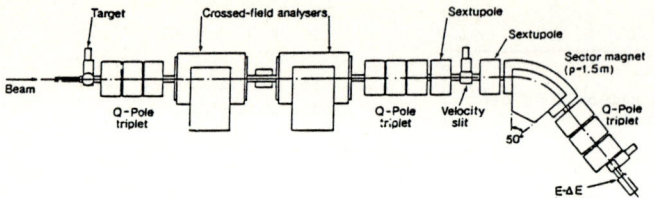

Fig. 6 The optical layout of the Daresbury Recoil Separator

Fig. 7 shows an example of the A/q spectrum from the $^{12}C + ^{58}Ni$ reaction. Adjacent masses are clearly resolved and a peak due to an A/q ambiguity is seen.

Fig. 8 shows γ-rays from the N=Z nuclei ^{80}Zr and ^{72}Kr measured with the Recoil Separator. They were measured in reactions with cross-sections of 10μb and 75μb and represent our best efforts so far in terms of sensitivity.

Fig. 9 shows the photopeak intensities of three of the γ-rays seen in the $^{16}O + ^{58}Ni$ reaction as a function of ΔE in the Ionization Chamber. They are from ^{72}Se, ^{72}Br and ^{72}Kr. They are clearly separated in Z but our ability to detect the one farthest from stability, namely ^{72}Kr, is

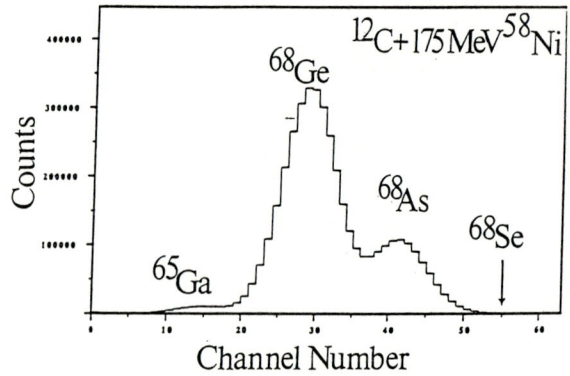

Fig. 7 A/q spectrum from the $^{12}C + 175$ MeV ^{58}Ni reaction

Fig. 8 γ-rays from ^{80}Zr and ^{72}Kr evaporation residues identified with the Recoil Separator

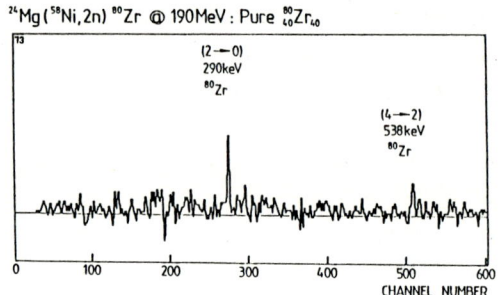

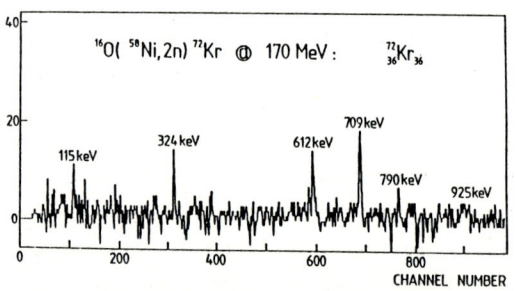

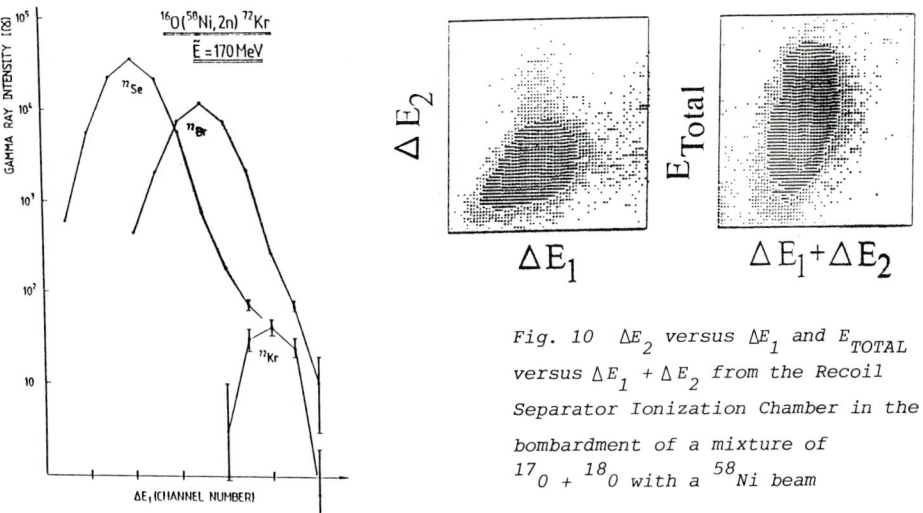

Fig. 10 ΔE_2 versus ΔE_1 and E_{TOTAL} versus $\Delta E_1 + \Delta E_2$ from the Recoil Separator Ionization Chamber in the bombardment of a mixture of $^{17}O + ^{18}O$ with a ^{58}Ni beam

Fig. 9 Intensities of the photopeaks of γ-rays from ^{72}Se, ^{72}Br and ^{72}Kr as a function of the energy loss signal in the Ionization Chamber of the Recoil Separator

restricted by the tail on each curve. This effect is thought (15) to be due to scattering of the recoils from the carbon in the isobutane gas. Support for this idea comes from the results of a short run with a modified ionization chamber with three electrodes, involving a tungsten oxide target (50% ^{17}O and 50% ^{18}O) bombarded by a beam of ^{58}Ni ions.

Fig. 10 shows a plot of the energy loss signals ΔE_2 versus ΔE_1 from this experiment. In addition to the expected diagonal pattern of events we clearly see events which are larger in <u>either</u> ΔE_2 or ΔE_1. They constitute 1/500th of the total number of events which is consistent with the size of the tails seen in fig. 9. Although this is consistent with the idea that the effect is due to scattering, analysis is continuing to confirm that this is the correct explanation.

By summing the ΔE_2 and ΔE_1 signals we can reconstitute our normal E-ΔE map, which is shown on the right in fig. 10. The various Z values are not clearly separated but by selecting slices in ΔE and making careful subtractions we can produce γ-ray spectra associated with a single Z value. The upper part of fig. 11 shows the intensities of γ-ray photopeaks from various nuclei produced in these reactions as a function of the ΔE signal shown in fig. 10. In this example the Z-separation is clear but it is nearing its limits. An alternative is to measure recoil-neutron-gamma coincidences, and use the neutron gating to select Z. This was done in the same experiment, and the results are shown in the lower part of fig. 11, where we can see that ^{73}Se is suppressed relative to ^{73}Br by a factor of two. It does not disappear completely because ^{73}Se is produced in a neutron-emitting channel in a reaction on the ^{18}O in the target. In a suitable case this technique could be

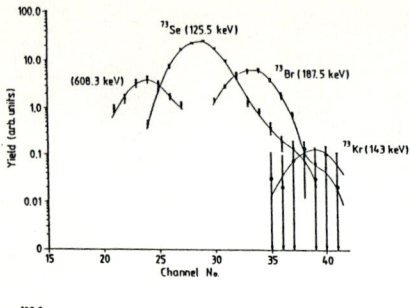

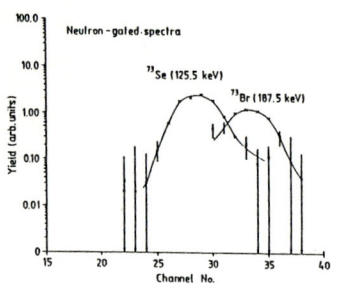

Fig. 11 The intensities of γ-ray photopeaks from ^{73}Se, ^{73}Br and ^{73}Kr as a function of the ΔE signal in the Recoil Separator Ionization Chamber. The upper half is ungated and the lower half is gated by a neutron (see text).

used to completely suppress channels which do not involve neutron emission.

How can we improve on present performance as shown in fig. 9? Roughly speaking we want to be able to study reactions with 1μb cross-sections or better, rather than 10μb as at present. Clearly removing the tail on ΔE will improve the sensitivity, and neutron detection will improve the cleanliness as well. In the way we operate the Recoil Separator currently a single charge state is used. Plans are in hand to use ray-tracing techniques to utilise three charge states simultaneously. These developments together with improved γ-ray detection efficiency, which will be achieved by a combination of better geometry and larger detectors, should allow us to measure at the 1μb level in the near future.

3. Prompt γ-rays from Fission

Until recently studies of Nuclei far from stability were dominated by radioactive decay studies, with spallation and fission being the favoured means of production. The development of in-beam γ-ray spectroscopy to take advantage of heavy ion induced, fusion-evaporation reactions has meant that we have learned a great deal about neutron-deficient nuclei at high spin. The neutron-rich nuclei remain neglected in this respect.

One avenue we have now explored [16] is to study the prompt gamma rays from nascent fission fragments. The fission is induced by heavy ions, and by varying the target-projectile combination we gain some control of the nuclear species we produce.

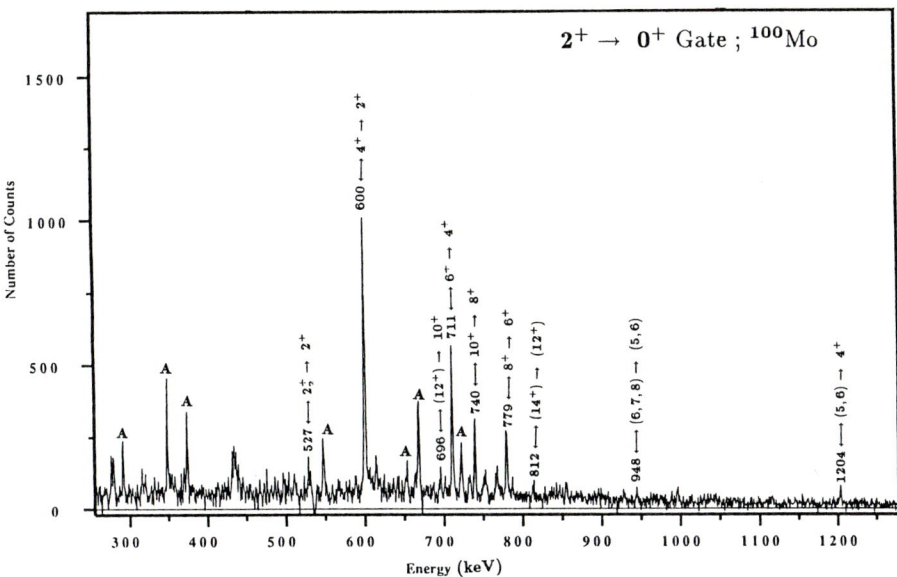

Fig. 12 Spectrum of γ-rays in coincidence with the 2^+-0^+ transition in ^{100}Mo produced in the fission of the ^{19}F + ^{197}Au system. γ-rays from complementary fragments are indicated by an A

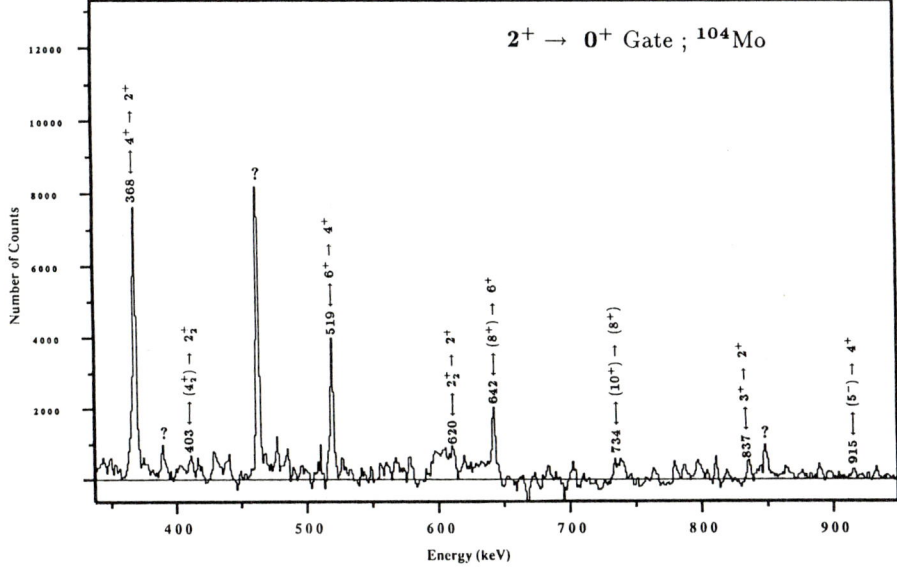

Fig. 13 Spectrum of γ-rays in coincidence with the 2^+-0^+ transition in ^{104}Mo produced in the fission of the ^{18}O +^{232}Th system

In simple terms Fission is a process in which a heavy nucleus distorts because of the long-range Coulomb force and then breaks up into two large fragments. The two fragments share the large amount of energy released. Thus they are born with high kinetic energy and internal excitation energy, and share a large amount of angular momentum. In many ways the nascent fragment is in a similar state to the compound nucleus formed in an (α,xn) reaction. The de-excitation of the fragments will proceed in a fashion similar to that described earlier in relation to fig. 1, with neutron emission being followed by a cascade of γ-rays. The γ-ray multiplicities have been measured [16,17] in various fissioning systems and the results suggest that one may expect to observe states in the neutron-rich fragments up to say $20\hbar$ if the techniques used are sufficiently sensitive.

Earlier experiments [18,19] on the γ-rays from the spontaneous fission of ^{252}Cf have shown that this route to studying neutron-rich nuclei is promising. We have now studied the γ-rays from both the ^{19}F + ^{197}Au reaction at Argonne National Laboratory and the ^{18}O + ^{232}Th reaction at Daresbury. In both cases a thick target was used. In the case of the ^{19}F + ^{197}Au reaction eight BGO suppressed Ge detectors were used with an array of BGO detectors for multiplicity measurements. At Daresbury the ESSA30 array of thirty BGO suppressed Ge detectors was used. In both experiments there was no channel selection and the nuclei formed were identified on the basis of γ-rays known from radioactive decay studies. To date we have concentrated on the even-even fission products and the γ-rays from some 50-60 nuclides have been observed. Examples of the γ-γ-coincidence spectra from the two reactions are shown in figs. 12 and 13. The γ-ray gates are set on the 2^+-0^+ transitions in ^{100}Mo and ^{104}Mo respectively. Two types of coincidence are seen. Firstly coincidences between two γ-rays from a single fragment and secondly coincidences between γ-rays from complementary fragments. The latter allow one to identify γ-rays in an unknown nucleus from the observation of coincidences with γ-rays known to be in a particular complementary fragment.

Fig. 14 shows the level scheme for ^{104}Mo based on the ^{18}O + ^{232}Th study. It also shows the percentage population of the levels as a function of spin. As suggested earlier the pattern of feeding is similar to that observed in (α, xn) reactions.

Fig. 15 shows the same level scheme as deduced from the measurements on ^{252}Cf spontaneous fission. The fall-off in intensity associated with the reduced $<M_\gamma>$ is clearly seen. However, this is partly compensated by the improved sensitivity associated with bench top measurements.

Fig. 16 shows the level scheme deduced for ^{108}Mo, the heaviest even-even Mo isotope we have seen so far.

Fig. 17 shows the most intensely populated fragments for each of the fissioning systems studied as a function of Z and A. The lines connecting the peak yields are shown, and they indicate clearly how one can tailor the experiment to maximise the yield of a particular nuclide or region of nuclides with the choice of beam and target.

111

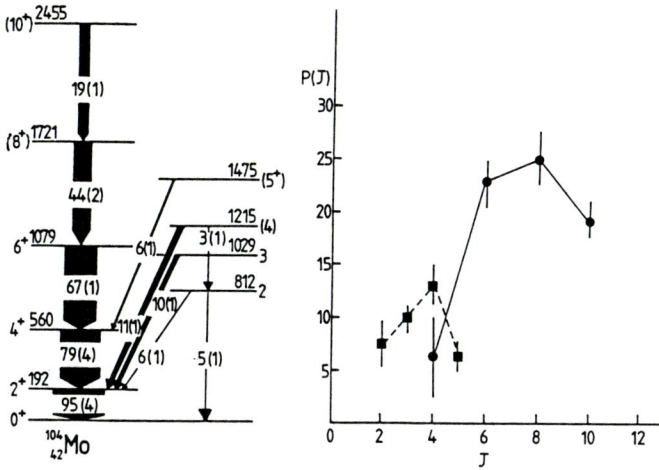

Fig. 14 Level scheme for ^{104}Mo deduced from studies of γ-rays from the ^{18}O+^{232}Th fissioning system. The percentage populations of the levels are shown as a function of spin on the right

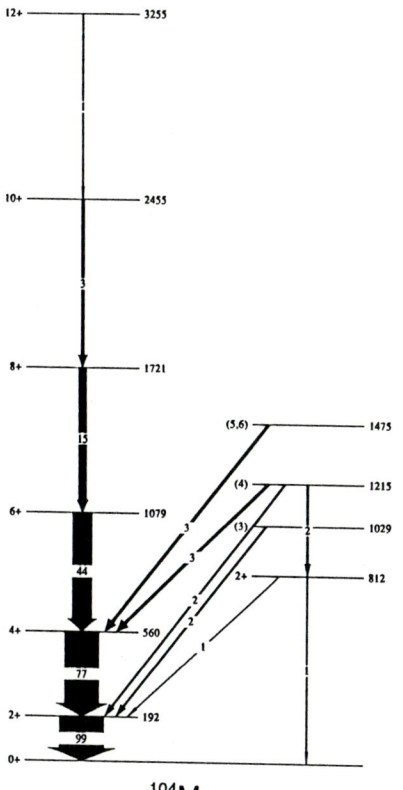

Fig. 15 Level scheme for ^{104}Mo deduced from ^{252}Cf spontaneous fission studies

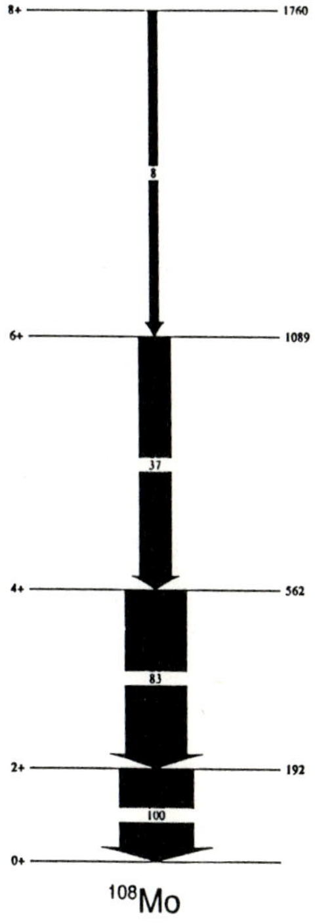

Fig. 16 Level scheme for ^{108}Mo from ^{252}Cf studies

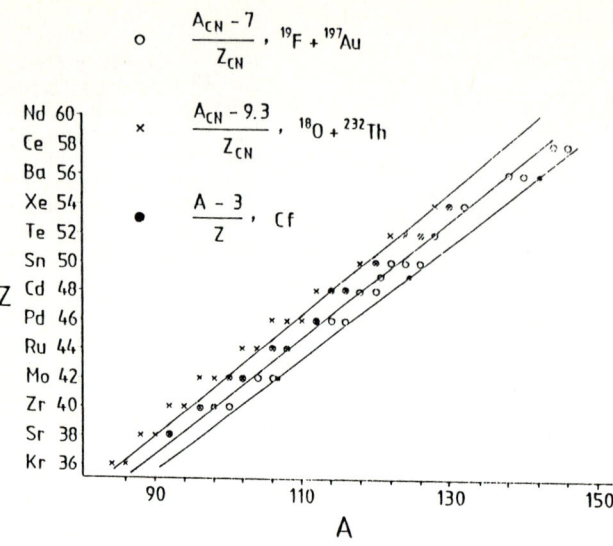

Fig. 17 The even-even isotopes of largest yield observed in three fissioning systems are shown as a function of Z and A. The straight lines join the points of maximum yield in each case. The algebraic expressions for the straight lines are given on the figure.

Fig. 18 shows $2\mathcal{J}/\hbar^2$ versus $(\hbar\omega)^2$ for the even-even Mo isotopes we have studied. At low spin we see clearly the effects of shape co-existence in the lighter isotopes. In ^{100}Mo we observe a back-bend at $\hbar\omega=0.4$MeV. Our data do not tell us whether this is due to protons or neutrons, but one may well speculate that it is due to the alignment of a pair of neutrons from the h11/2 intruder orbital. This would be consistent with the role of such intruder orbitals in other parts of the Periodic table.

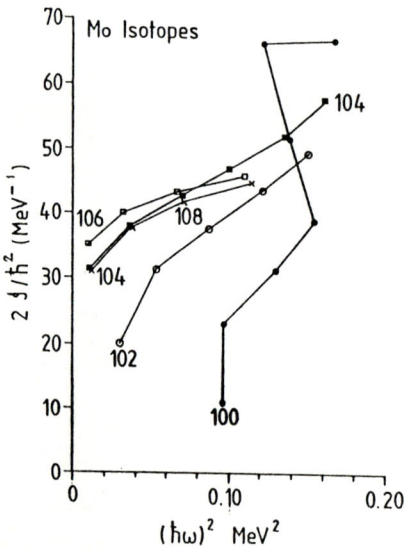

Fig. 18 $2\mathcal{J}/\hbar^2$ versus $(\hbar\omega)^2$ for the yrast bands in the even Mo isotopes

4. Conclusions

In many ways the studies described here are in their infancy. Improvements in technique and an increase in the efficiency of the Recoil Separator will lead to greater sensitivity and allow us to do proper spectroscopy on the N=Z nuclei with A ~80 and allow us to tackle the remaining N=Z nuclei which can be reached up to ^{100}Sn. We are already attempting to study the Sn nuclei with A=101-104.

Our fission experiments are still primitive but are promising. A method of mass selection based on time-of-flight is being developed at Manchester and this should allow us to look at fragments produced in lower abundance and study the odd-A nuclei. At the same time we are developing a spectrometer which will allow us to tag γ-rays from the products of deep inelastic reactions, a type of reaction which will be more readily available to us on completion of the superconducting linear accelerator we are adding to the NSF at Daresbury.

This work is supported by the United Kingdom Science and Engineering Research Council. The authors acknowledge the help of the staff of Daresbury Laboratory in carrying out some of the experiments described here.

References

1. C.J. Lister, A.A. Chishti, B.J. Varley and W. Gelletly, Proceedings of the conference.
2. A.A. Chishti et al., Proceedings of this conference.
3. A. Kucharska, J. Billowes and C.J. Lister, Proceedings of this conference.
4. N.J. Stone, Proceedings of this conference.
5. I.S. Grant et al., Nucl. Inst. and Methods B26 (1987) 95.
6. D.A. Eastham et al., Phys. Rev. C36 (1987) 1583.
7. C.J. Lister, B.J. Varley, W. Gelletly, A.A. Chishti, A.N. James, T. Morrison, H.G. Price, J. Simpson and O. Skeppstedt, Proceedings of 5th International Conference on Nuclei Far From Stability, Lake Rosseau, Ontario, Canada, ed. I.S. Towner (1987) p.354.
8. C.J. Lister et al., Phys. Rev. Letters 59 (1987) 1270.
9. B.J. Varley et al., Phys. Letters B194 (1987) 463.
10. L. Goettig et al., Nucl. Phys. A464 (1987) 159.
11. A.N. James et al., Nucl. Inst. and Methods A267 (1988) 144.

12. P.J. Twin et al., Nucl. Phys. A409 (1983) 343C.

13. C.J. Gross et al., Proceedings of this conference. To be published.

14. A.N. James, Daresbury Laboratory Study Weekend DL/NUC/R20 (1979) 84.

15. Udo Lenz (Munich), private communication.

16. Y. Abdelrahman et al., Phys. Letters B199 (1987) 504.

17. J.R. Leigh et al., Phys. Letters 159B (1988) 9.

18. E. Cheifetz et al., Phys. Rev. Letters 25 (1970) 38.

19. W.R. Phillips et al., Phys. Rev. Letters 57 (1986) 3527.

M1 and E2 Transition Probabilities in Kr Nuclei

L. Funke and G. Winter

Zentralinstitut für Kernforschung Rossendorf, Postfach 19,
DDR-8051 Dresden, G.D.R.

INTRODUCTION

In the mass region A=70-90 a richness of nuclear structure effects is caused by energy gaps in the single-particle scheme at different nuclear shapes (oblate, prolate, spherical). These gaps reflect themselves in several minima in the β_γ-plane of the potential energy surfaces /1/. Due to the competition of these minima shape coexistence is a general feature in the mass region considered. The nuclei in the central part of that mass region exhibit a very large deformation of $\varepsilon_2 \approx 0.4$ already near the ground state /2-4/. Furthermore, very recently the predicted oblate shapes have been proved /5/ in ^{69,71}Se.
The series of the Kr isotopes between the N=Z nucleus ^{72}Kr and the N=50 nucleus ^{86}Kr covers the entire range from well deformed via transitional to spherical nuclei. The most recent and/or most complete papers regarding the nuclear structure in the vicinity of the yrast line in the Kr nuclei are compiled in the list of references. There the denotations E and T indicate whether new data on excitation energies or lifetimes, respectively, are involved /6-25/.
While the lighter-mass Kr nuclei have been investigated to a large extent in heavy-ion experiments using arrays of Compton-suppressed germanium detectors, the heavier-mass Kr isotopes were mainly investigated in α and ^{7}Li induced reactions. In this respect the Rossendorf group in collaboration with groups from the AFI Stockholm and from the FTI Leningrad have contributed not only by establishing new excited states but also by determining many mean lifetimes. For this purpose different experimental methods have been applied: the Dopplershift methods (DSAM, RDDS), the γ-γ- and particle-γ-timing as well as rf.-γ-timing methods. Thus, a great number of absolute transition probabilities could be determined, from which two classes, namely the enhanced M1 transitions and the stretched E2 transitions, will be discussed in some detail.

ENHANCED M1 TRANSITIONS

Fast $\Delta I=0$ M1 transitions appear between mixed states, e.g. in the band crossing region, if the level mixing and the difference of the magnetic moments of the mixed configurations is large /26/.

A new kind of enhanced M1 transitions with $\Delta I=1$ has been observed within certain 3-quasiparticle (qp) configurations of the odd-mass nuclei ^{79,81,83}Kr /20-22,29/. In the positive-parity yrast sequence of these nuclei the nuclear structure changes drastically at spin 21/2, where in addition to the odd $g_{9/2}$ neutron a pair of $g_{9/2}$ protons is broken and the two spin vectors are aligned along the rotational axis. This band crossing is associated with a shape change from near oblate at lower angular momenta to a prolate shape in the 3qp configuration. The most striking feature of the crossing is the change of the M1 transition probabilities which jump just at spin 21/2 by a factor of 10 in ^{81}Kr and a factor of 3 in ^{79}Kr. The origin of the enhancement and the difference between ^{79}Kr and ^{81}Kr is well understood on the basis of a semiclassical coupling scheme developed by Dönau and Frauendorf /27,28/ to explain our experimental findings in ^{81}Kr /29/.

In this scheme the M1 radiation strength within a collective many-qp band is determined by the components of the magnetic moments perpendicular to the total angular momentum. Since in the 3qp configuration $\nu(g_{9/2})\pi(g_{9/2})^2$ the odd neutron is deformation-aligned and the two protons rotation-aligned and, furthermore, the g-factors of protons and neutrons have opposite signs, the proton and neutron components add up and lead to the M1 enhancement. The smaller B(M1) values in ^{79}Kr (compared to ^{81}Kr) are also quantitatively understood /20/ on the basis of the smaller values of both the aligned angular momentum i_{2p} (extracted from the excitation energies) and the value of K_n. The change of these two quantities is caused by the larger deformation and the lower position of the Fermi surface in the neutron system, respectively. In this way our new results on M1 transitions in ^{79}Kr may be considered as an additional experimental evidence for the validity of this coupling scheme.

Strong M1 transitions have also been observed in negative-parity 3qp bands in the Kr isotopes /20-23/. However, inspite of rather complete data in some cases no straightforward explanation of the negative-parity bands has been found. As an example the sequence from $13/2^-$ to $27/2^-$ excited in ^{81}Kr and ^{83}Kr above 2.5 MeV should be mentioned. The B(M1) values are as large as about 1 W.u. at the bottom of this structure but decrease rapidly with increasing spin. The excitation energies follow roughly an I(I+1) dependence and are similar in ^{81}Kr and ^{83}Kr, but the B(E2) values are only in the order of 5-10 W.u.. A similar sequence of negative-parity states seems to exist in ^{79}Kr above 2.5 MeV /20/.

Table 1: Compilation of B(E2) values (in W.u.) in the positive-parity sequence of even-mass Kr nuclei. The data are taken from /7-15/ and references therein.

$I_i \rightarrow I_f$	^{74}Kr	^{76}Kr	^{78}Kr	^{80}Kr	^{82}Kr	^{84}Kr	^{86}Kr
$2_1 \rightarrow 0$	78(16)	89(8)	65(4)	37(3)	21(1)	12(1)	11(1)
$4_1 \rightarrow 2$	62(6)	105(10)	87(7)	44^{+11}_{-7}	30(15)	22(4)	0.05(1)
$4_2 \rightarrow 2$						0.16(3)	
$6_1 \rightarrow 4$		100(30)	87(17)	50^{+31}_{-13}	7(3)	7(2)	
$6_2 \rightarrow 4$					8(2)		
$8_1 \rightarrow 6$		120(15)	100(15)	88^{+80}_{-40}	5(1)	2.3(3)	
$8_2 \rightarrow 6$					15(8)	≈0.2	
$10_1 \rightarrow 8_1$		100(25)	65(12)	46^{+29}_{-17}	11(4)	6(2)	
$10_2 \rightarrow 8_1$			30(6)	12(7)	<1		
$10_2 \rightarrow 8_2$					66(10)	30(14)	
$12_1 \rightarrow 10_1$			95^{+95}_{-35}	84^{+90}_{-30}	20(5)	3.6(4)	
$12_2 \rightarrow 10_2$					30(6)		
$14_1 \rightarrow 12_1$			58^{+30}_{-15}				

E2 TRANSITIONS AS A PROBE OF COLLECTIVITY

The B(E2) values of the positive-parity sequence observed in even-mass Kr nuclei are listed in table 1. Several trends can be seen from these numbers. As expected the E2 transition rates decrease when approaching the N=50 nucleus ^{86}Kr. In ^{84,86}Kr even extremely small values have been measured. A decrease of the B(E2) values appears also in almost all Kr nuclei at about spin 6-10 due to the band crossing and mixing effects. It should, however, be mentioned that e.g. in ^{78}Kr the sum of the two $10^+ \rightarrow 8^+$ transitions corresponds to the full collective strength. Whereas in ^{78,80}Kr a broken and rotation-aligned pair of $g_{9/2}$ protons is responsible for the band crossing /11,12/, in ^{82}Kr in addition to the protons the $g_{9/2}$ neutrons come into play and lead to a complicated 3-band mixing situation /13/. The B(E2) values indicate a higher collectivity in the $g_{9/2}$ proton band than in the $g_{9/2}$ neutron band. Such a statement is supported by the observation of a level sequence in ^{84}Kr that is likely related to the broken $g_{9/2}$ proton pair and shows also some collectivity /14/.

Furthermore, we found in ^{84}Kr the first 4qp isomer in the whole mass region. On the basis of a g-factor measurement a 12^+ state at 5.3 MeV with a half-life of 45 ± 5 ns could be uniquely identified as the stretched configuration $\nu(g_{9/2}^{-2})_{8^+}\pi(f_{5/2}^{-1}p_{3/2}^{-1})_{4^+}$, where the two 2qp partners involved have also been observed at lower energies. The configuration $\pi(f_{5/2}^{-1}p_{3/2}^{-1})_{4^+}$ has turned out to be the lowest 2qp excitation in the heavier Kr nuclei and its deexcitation is remarkably slow. The B(E2) values of 0.16 and 0.05 W.u. for the $4\to2_1$ transitions in ^{84}Kr and ^{86}Kr, respectively, are extremely small and not yet fully understood. However, a close resemblance to the N=82 nuclei ^{138}Ba, ^{140}Ce and ^{142}Nd /30/ should be mentioned, where similarly small B(E2) values have been found for the transitions deexciting the 6^+ and 4^+ members of the proton configuration $\pi(g_{7/2}d_{5/2})^{-2}$. The origin of the strong hindrance of such E2 transitions is being investigated by shell model calculations.

It should also be noticed that in ^{85}Kr, where we found a $17/2^+$ isomer /24/ and assigned it to the $\nu(g_{9/2}^{-1})\pi(f_{5/2}^{-1}p_{3/2}^{-1})$ configuration, the deexciting E2 transition to the $13/2^+$ level is almost 2 orders of magnitude faster (B(E2)=4 W.u.) than the $4\to2$ transition in the core nucleus ^{86}Kr. The cancellation of the hindrance in that case may be explained by the assumption that the $13/2^+$ level is not a simple coupling of the neutron hole to the 2^+ core state but rather contains both the 2^+ and 4^+ core states.

We like to express our gratitude to all members of the 3 nuclear spectroscopy groups in Rossendorf, Stockholm and Leningrad for a nice collaboration during the systematic study of the Kr nuclei.

REFERENCES

/1/ R. Bengtsson
    W. Nazarewicz } Contributions to this Conference
    I. Ragnarsson
/2/ R.B. Piercey et al., Phys. Rev. Lett. 47(1981)1514
/3/ C.J. Lister et al., Phys. Rev. Lett. 49(1982)308
/4/ C.J. Lister et al., Contribution to this Conference
/5/ M. Wiosna et al., Phys. Lett. 200B(1988)255
/6/ B.J. Varley et al., Phys. Lett. 194B(1987)463 ^{72}Kr, E
/7/ J. Roth et al., J. Phys. G10(1984)L25 ^{74}Kr, E, T
/8/ F.J. Bergmeister et al., Contribution to the Conference
    on Nuclear Structure, Reactions and Symmetries, ^{76}Kr, E
    Dubrovnik, 1986

/9/ G. Winter et al., ZfK-Annual Report 1981, p. 40 ^{76}Kr, T
and Contribution to the INS Int. Symposium on Dynamics
of Nuclear Collective Motion, Mt. Fuji, Japan, 1982
/10/ B. Wörmann et al., Nucl. Phys. A431(1984)170 ^{76,77}Kr, T
/11/ G. Winter et al., J. Phys. G11(1985)277 ^{78}Kr, E, T
/12/ L. Funke et al., Nucl. Phys. A355(1981)228 ^{80}Kr, E, T
/13/ P. Kemnitz et al., Nucl. Phys. A425(1984)493 ^{82}Kr, E, T
/14/ H. Rotter et al., Phys. Lett. 163B(1985)323 ^{84}Kr, E, T
and to be published
/15/ G. Winter et al., ZfK-Annual Report 1987 ^{86}Kr, T
and to be published
/16/ W. Gelletly et al., Contribution to this Conference ^{73}Kr, E
/17/ A.A. Chishti et al., Contribution to this Conference ^{75}Kr, E
/18/ S. Skoda et al., Contribution to this Conference ^{75}Kr, E, T
/19/ C.J. Gross et al., Phys. Rev. C36(1987)2601 ^{77}Kr, E
/20/ G. Winter et al., J. Phys. G14(1988)L13 ^{79}Kr, E, T
and to be published
/21/ L. Funke et al., Nucl. Phys. A455(1986)206 ^{81}Kr, E, T
/22/ P. Kemnitz et al., Nucl. Phys. A456(1986)89 ^{83}Kr, E, T
/23/ M.F. Kudojarov et al., Izv. AN SSSR 48(1984)1887 ^{83}Kr, T
/24/ G. Winter et al., ZfK-Annual Report 1987 ^{85}Kr, E, T
and to be published
/25/ A.E. Sobov et al., Contribution to the 36. Conference ^{85}Kr, E, T
on Nuclear Spectroscopy and Theory, Kharkov, USSR, 1986
/26/ P. Kemnitz et al., Phys. Lett. 125B(1983)119
/27/ F. Dönau and S. Frauendorf, Proc. Int. Summer School,
Poiana Brasov, Romania, 1982
/28/ F. Dönau, Nucl. Phys. A471(1987)469
/29/ L. Funke et al., Phys. Lett. 120B(1983)301
/30/ C.M. Lederer and V.S. Shirley (ed.), Table of Isotopes,
7th ed. (Wiley, New York, 1978)

Lifetimes and Sidefeeding Times in A = 70–83 Nuclei

K.P. Lieb, F. Cristancho, W. Fieber, C. Gross, J. Heese, T. Osipowicz,
S. Ulbig, and B. Wörmann

II. Physikalisches Institut der Universität Göttingen,
D-3400 Göttingen, Fed. Rep. of Germany

Lifetimes of high spin states in ^{72,73}Br, ^{78,79}Sr, ^{83}Y and ^{83}Zr measured with the recoil distance and Doppler shift attenuation techniques will be reviewed with the emphasis on optimized detector arrangements. We will also discuss Monte-Carlo calculations of the γ-ray flux and sidefeeding times. Finally, first DSA lifetime measurements in ^{54}Cr following thermal neutron capture and employing the NBS/ILL double flat crystal spectrometer GAMS4 will be reported on.

Heavy ion fusion reactions are ideally suited for recoil distance (RD) and Doppler shift attenuation (DSA) measurements as they produce ensembles of rapidly moving and well collimated recoil nuclei. In recent years hundreds of lifetimes have been measured in the A $\approx$ 80 region, some 150 of them by our group. Detailed nuclear structure information on the development of shapes with spin and nucleon number and on the alignment effects of $g_{9/2}$ particles has evolved [1]. With the advent of heavier projectiles and sophisticated detector arrays like OSIRIS and ESSA30, it has become possible to identify γ-ray transitions in neutron deficient N=Z nuclei, in combination with a recoil separator [2] and to perform full yrast spectroscopy in the N = Z+2 isotopes (e.g. in ^{70}Se [3], ^{72}Br [4], ^{74}Kr [5], ^{78}Sr [6,7]).

1. Detector arrays in lifetime measurements

Rather than discussing nuclear structure effects, we would like to emphasize here the various detector arrangements employed in the lifetime measurements. The compound nuclei produced by fusing N$\approx$Z targets and projectiles mainly decay via evaporation of 2-4 charged particles. Single neutron emission is comparably weak ($\sigma_{1n} \approx 30$ mb) and double neutron evaporation is a very rare process ($\sigma_{2n} \leq 10$ mb). Table I summarizes the recent lifetime measurements performed by us.

On channels: R(D) measurements in the strong On channels are straightforward: The only gate required is that from an additional γ-detector which enhances the high γ-multiplicity events in the evaporation channels over activity transitions and removes the strong Coulomb excitation lines from the spectra. A typical set-up used in the ^{83}Y experiment [8] is shown in Fig. 1: R(D) data are taken in the 0°

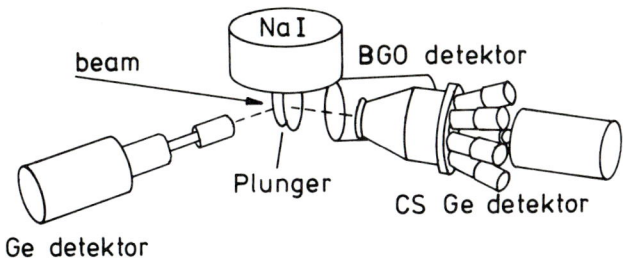

Fig. 1: Set-up for RD measurements in the reaction $^{58}Ni(^{28}Si,3p)^{83}Y$ using 1-fold gamma gating [8].

Compton suppressed Ge detector and gated by 1-fold events in the NaI(Tl) or BGO detectors.- Clean Doppler broadened lineshapes can be best measured in the $\gamma\gamma$ event-by-event mode, with gates set onto the discrete transitions in these bands. This requires a CS Ge detector at 0° and several Ge detectors around 90°. As an example, Fig. 2 illustrates a set-up used at VICKSI for measuring lifetimes in the $^{40}Ca(^{36}Ar,xpyn)^{72,73}Br$ reactions. DSA lineshapes were taken at 0° and 145° in coincidence with six 90° CS Ge detectors from the OSIRIS array.

Table I: Lifetime measurements in ^{70}Se, $^{72,73}Br$, $^{78,79}Sr$, ^{83}Y and ^{83}Zr

Reaction, E_{lab} (MeV)			Evaporation Channel	Technique, detectors[a]	Ref.
$^{40}Ca+^{36}Ar$	125	A	$^{73}Br+3p$	DSA: 1 CS/0°; 6 CS/90°;	[9]
			$^{72}Br+3pn$	2 CS/145°; $\gamma\gamma$-mode	[4]
	115		$^{70}Se+\alpha p$	RD: 3 Ge/0°,70°,142°	[3]
			$^{73}Br+3p$	RD: like ^{70}Se	[9]
$^{58}Ni+^{16}O$	39	B	$^{72}Br+pn$	RD: 3 Ge/0°,53°,138°; 1 NE213 n-gate	[4]
$^{58}Ni+^{24}Mg$	110	C	$^{78}Sr+2p2n$	DSA: 15 CS; 15 NE213; 2n-gate	[7]
			$^{78}Rb+3pn$	like ^{78}Sr; n$\gamma\gamma$-mode	
	90	D	$^{79}Sr+2pn$	RD: 1 CS/0°; 1 Ge/140° NaI and BGO γ-gates	[10]
				DSA: 1 CS/0°; 4 NE213 n-gates	
$^{58}Ni+^{28}Si$	90	D	$^{83}Y+3p$	RD: like ^{79}Sr	[8]
$^{54}Fe+^{32}S$	103	B	$^{83}Y+3p$	DSA: 1 CS/0°; 2 Ge/90°; $\gamma\gamma$-mode	
			$^{83}Zr+2pn$	RD: 4 Ge/0°,58°,140°; NE213 n-gate	

a) 1 CS/0° = Compton suppressed Ge detector at $\Theta = 0°$
A = VICKSI HMI Berlin; B = Cologne FN tandem;
C = NSF Daresbury; D = Oxford folded tandem

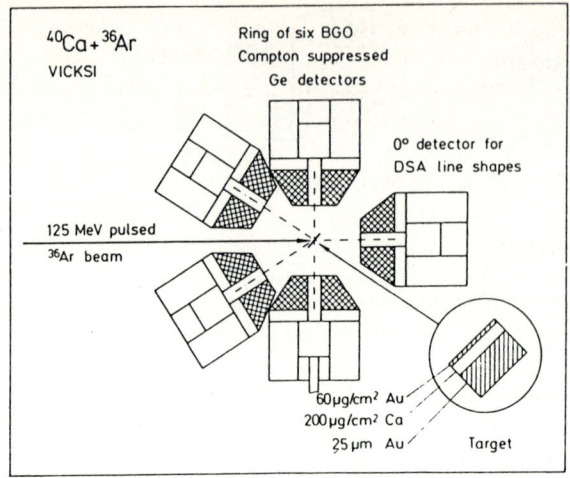

Fig. 2: Set-up for measuring DSA lineshapes in ^{72,73}Br in the reaction ^{40}Ca + ^{36}Ar [4,9]

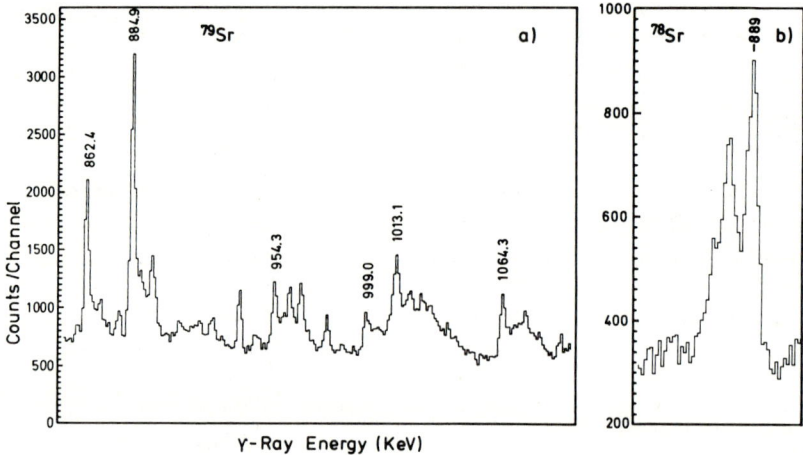

Fig. 3: Doppler broadened line shapes of transitions in ^{79}Sr (a: with neutron gate) and ^{78}Sr (b: with nn-gate)

1n channels: Reasonably good RD and DSA data in the weaker 1n- channels can be obtained by gating the 0° spectra with events from several NE213 neutron detectors. Due to the low neutron multiplicity, a large solid angle has to be covered by these detectors. We have recently used this technique for RD and DSA measurements in ^{72}Br, ^{79}Sr and ^{83}Zr [4,10]. Fig. 3a displays Doppler broadened lineshapes of transitions in ^{79}Sr accumulated via the reaction ^{58}Ni(^{24}Mg,2pn). Evidently, small contaminants on the Doppler tails might introduce uncertainties in the analysis, but the yrast cascades can be studied in this way.

2n channels: While the previous detector arrangements are still rather simple, spectroscopy of 2n channels becomes much more difficult. We have so far performed a single experiment of this type in which we looked for yrast transitions and lifetimes in the ^{78}Sr ground band by means of the reaction ^{58}Ni(^{24}Mg,2p2n) at 110 MeV beam energy [7]. Signals from 15 CS Ge detectors of the ESSA30 array in the backward hemisphere and 15 NE213 neutron detectors at forward angles were routed into nnγγ coincidence events. Fig. 3b shows a Doppler broadened line shape of the 889 keV $8^+ \to 6^+$ transition observed with double neutron gating; here the 280 µg/cm² ^{58}Ni target was backed by a Ta foil.

2. Feeding time calculations of heavy ion reactions

In most rotational nuclei studied we have encountered lifetimes as short as 0.1 ps pointing to large quadrupole moments (Q ≈ 3 b) and deformations (β_2 ≈ 0.3-0.4). In view of such short lifetimes, the problem of delayed feeding had to be considered very carefully, in particular for the reactions induced by heavier projectiles (i.e. ^{28}Si-^{40}Ca). The time development of the fusion-evaporation process and subsequent γ-ray cascades was Monte-Carlo simulated [11] yielding predictions of the evaporation cross sections and entry state, γ-ray multiplicity and side feeding time distributions. Care was taken to adjust the discrete yrast line in the final nucleus considered to the Fermi gas level density. Collective E2 transitions as fast as for the discrete yrast transitions (30-130 Wu) were employed in a 3 MeV wide region parallel to the yrast line; in the statistical decay, E1 and M1 strengths of 5 10⁻⁴ resp. 10⁻² Wu were used. Fig. 4 illustrates the calculated side feeding times of the reactions ^{40}Ca(^{36}Ar,3p)^{73}Br and ^{62}Ni(^{16}O,p2n)^{75}Br in comparison with measured values of $\langle \tau_F \rangle$ and state lifetimes [9,12]. Note the dramatic increase of $\langle \tau_F \rangle$ at low spin. Further feeding time measurements are urgently needed to verify this approach.

3. DSA measurements after thermal neutron capture reactions

Although this workshop mainly deals with high spin spectroscopy, we would like to mention a new instrument for measuring picosecond lifetimes, the double flat crystal spectrometer GAMS4 installed at the ILL high flux reactor [13]. The idea of the method is rather simple: A nucleus formed after thermal neutron capture emits a 5-7 MeV primary γ-ray from which it gets a small initial recoil of E_r ≈0.4 keV. If the secondary transition is measured with an energy resolution of ≤20 ppm, its Doppler broadened line shape pictures the slowing down process of the recoil nucleus in the solid target. The crystal spectrometer GAMS 4 combines the features of extremely good energy resolution and sufficient luminosity for MeV γ-radiation.

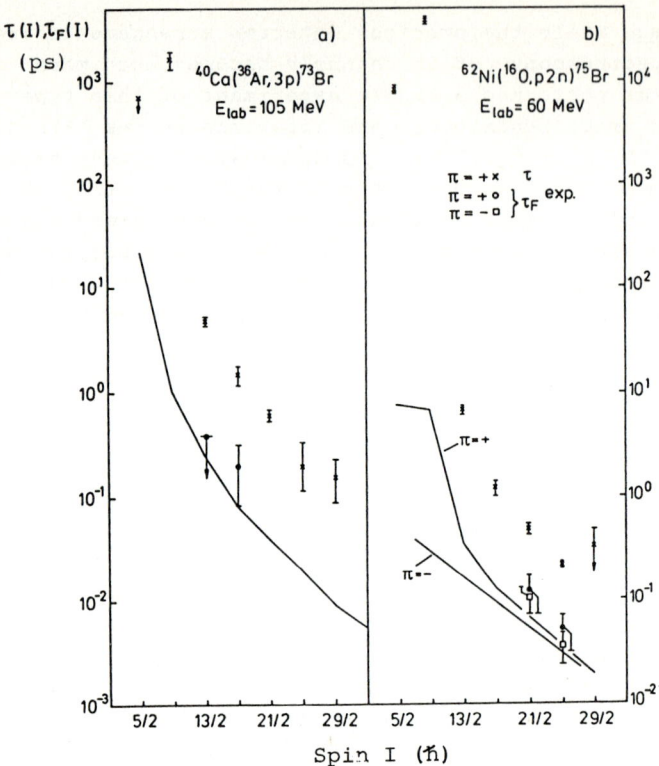

Fig. 4: Measured and calculated average side feeding times $\langle\tau_F\rangle$ and measured state lifetimes τ in ^{73}Br and ^{75}Br [3,11,12]

We have applied this technique to the measurement of lifetimes in ^{54}Cr and ^{57}Fe [14]. These cases are ideally suited as they combine large capture cross sections ($\sigma \approx 3$-10 b) with strong primary decay branches ($\approx 10\%$), thus ensuring prompt feeding and a well-defined initial recoil velocity. Fig. 5 illustrates the Doppler broadened lineshape of the 2239 keV $2_3^+ \to 2_1^+$ transition in ^{54}Cr in comparison with the spectral resolution; the lifetime deduced is 12 fs $\leq \tau \leq$ 25 fs. This level appears to be most interesting from the point of view of the Interacting Boson Model IBM-2 [15] which predicts the 2_3^+ state to be 70 % of "mixed symmetry" character. In Fig. 6 the experimental level energies and lifetimes are compared with the results of shell model calculations [16] and an IBM-2 fit. In IBM-2, lifetimes were calculated with the correct transition energies, effective boson charges $e_\pi = e_\nu$ =0.066 eb derived from $B(E2, 2_1^+ \to 0_1^+)$, and effective boson g-factors $|g_\nu - g_\pi| = 1.0$ μ_N [17]. The IBM-2 analysis also predicts a 1^+ mixed symmetry state at E = 3770 keV, with $\tau \approx$ 8 fs and $B(M1, 1^+ \to 0^+) \approx$ 0.1 μ_N^2. We have just performed a GAMS4 DSA measurement for the 3720 keV $(1,2)^+$ state which appears to have the characteristics of such a mixed symmetry 1^+ state; data analysis is in progress.

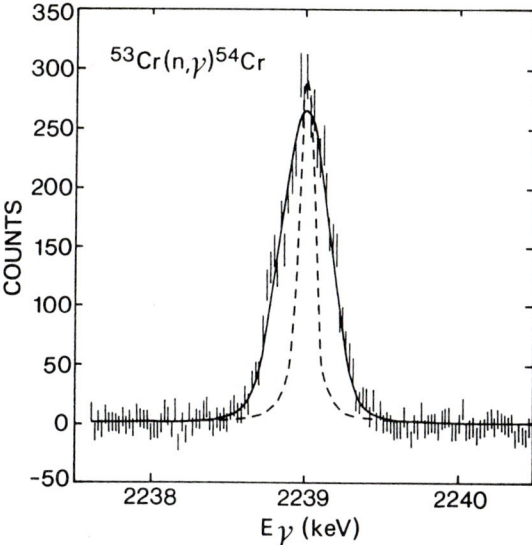

Fig. 5: Doppler broadened lineshape of the 2239 keV transition in ^{54}Cr measured with GAMS4 in the reaction $^{53}Cr(n,\gamma)$ [14]

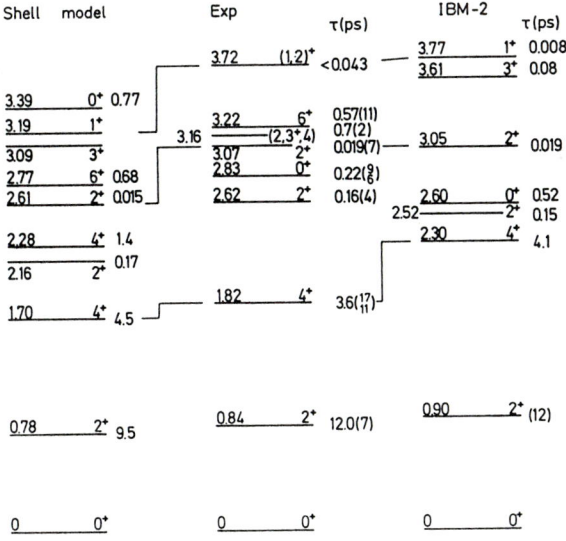

Fig. 6: Experimental level energies and lifetimes in ^{54}Cr compared with shell model [16] and IBM-2 calculations

We would like to express our gratitude to our colleagues in the ^{78,79}Sr heavy ion and the ^{54}Cr GAMS4 experiments, mainly to J. Billowes, W. Gelletly and C.J. Lister (University of Manchester) and H. Börner, M.S. Dewey, J. Jolie and S. Robinson (Institut Laue-Langevin, Grenoble) for the permission to quote unpublished results. This work has been funded by the German BMFT under the contract number 06 Gö 456.

References

[1] K.P. Lieb, in "Reactor Physics and Nuclear Spectroscopy Research", World Scientific, Singapore 1986, p. 619; in "Weak and Electromagnetic Interactions in Nuclei", (Springer, Berlin, 1986) p. 106
[2] C.J. Lister et al., these Proceedings
[3] J. Heese et al., Z. Phys. A325, 45 (1986)
[4] S. Ulbig et al., Z. Phys. A329, 51 (1988)
[5] R.B. Piercey et al., Phys. Rev. Lett. 47, 1514 (1981)
[6] C.J. Lister et al., Phys. Rev. Lett. 40, 308 (1982)
[7] C.J. Gross, J. Heese, K.P. Lieb, C.J. Lister, A.A. Chishty, W. Gelletly, B.J. Varley, these Proceedings
[8] C.J. Lister, B.J. Varley, W. Fieber, J. Heese, K.P. Lieb, E.K. Warburton, J.W. Olness, Z. Phys. A329 (1988) in press
[9] J. Heese et al., Phys. Rev. C36, 2409 (1987)
[10] J. Heese, K.P. Lieb, S. Ulbig, B. Wörmann, J. Billowes, A.A. Chishty, W. Gelletly, C.J. Lister, to be published
[11] F. Cristancho, K.P. Lieb, submitted for publication
[12] L. Lühmann et al., Phys. Rev. C31, 828 (1985)
[13] H.G. Börner, et al., 6th Symposium on "Capture Gamma Ray Spectroscopy", Leuven 1987, P. van Assche, ed., in press
[14] K.P. Lieb, H.G. Börner, M.S. Dewey, J. Jolie, S. Robinson, S. Ulbig, C. Winter, submitted for publication
[15] T. Otsuka, A. Arima, F. Iachello, I. Talmi, Phys. Lett. 76B, 139 (1978)
[16] A.E. Stuchbery et al., Nucl. Phys. A337, 1 (1980)
[17] S.A.A. Eid, W.D. Hamilton, J.P. Elliott, Phys. Lett. 166B, 267 (1986)

The Moment of Inertia of ^{78}Sr *

C.J. Gross[1], J. Heese[1], K.P. Lieb[1], C.J. Lister[2,+], B.J. Varley[2],
A.A. Chishti[2], and W. Gelletly[3]

[1]II. Physikalisches Institut der Universität Göttingen,
 D-3400 Göttingen, Fed. Rep. of Germany
[2]Department of Physics, Schuster Laboratory,
 Manchester University, Manchester M13 9PL, UK
[3]SERC Daresbury Laboratory, Daresbury, Warrington WA 4AD, UK
[+]present address: Yale University, P.O. Box 6666,
 272 Whitney Avenue, New Haven, CT 06511, USA

Introduction and experimental procedures

Extreme prolate deformations ($\beta_2 \geq 0.35$) have been observed[1-4] in the neutron deficient strontium isotopes. These nuclei are among the most deformed in the mass 80 region. Recently, detailed spectroscopic work has been completed on $^{79-81}$Sr (ref. 2-4). Theoretical work predicting changes in deformations and collectivity as a function of neutron number has been carried out in ref. 5. The work discussed in this contribution was initiated to study ^{78}Sr and to compare the observed rotational structure with other Sr isotopes as one moves closer to the N=Z line.

High spin states in ^{78}Sr were populated in the reaction ^{58}Ni(^{24}Mg,2p2n)^{78}Sr at 110 MeV at the NSF tandem facility at Daresbury Laboratory. A stack of two enriched ^{58}Ni self-supporting targets of 500 μg/cm^2 was placed perpendicular to the direction of the beam. Due to the low production cross section of ^{78}Sr (approximately 10 mb), 15 neutron detectors were placed at forward angles and 15 BGO Compton suppressed Ge detectors were placed at back angles with respect to the beam. Three types of coincidence events were recorded and sorted on to three 2048 x 2048 channel matrices: γγ, nγγ, and nnγγ. For additional details on the experimental procedure, refer to the contribution to these proceedings by Gelletly, et al.

Spectra and level scheme

The nnγγ gating allowed clear identification of transitions in ^{78}Sr. Fig. 1 is the summed coincidence spectrum of the 278.5, 504, 712, 895, and 1057 keV gates. The spectrum is fairly free of contaminants except for some transitions attributed to the 3pn channel ^{78}Rb. The 278, 503 and 894 keV lines from ^{78}Rb are the sources of these contaminants.

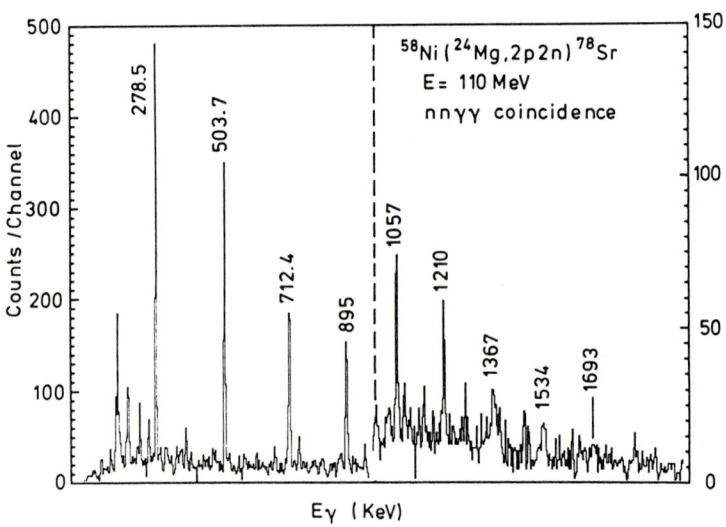

Fig. 1. Summed coincidence spectrum for ^{78}Sr consisting of gates set on the 278, 504, 712, 895, and 1057 keV transitions.

The level scheme shown in fig. 2 is based on the previous work[1] and the present data. The level scheme is supported by the observed intensities of the transitions in each gate and the expected rotational behavior in this mass region. All lines are visible in gates set on the individual transitions in the nnγγ data with the exception of the upper three transitions in the 1367 and 1534 keV gates. The nnγγ gated spectra, however, allowed the determination of narrow gates to be placed in the nγγ matrix. The higher statistics of this data provided the confirmation for the placement of the 6024 and 7558 keV energy levels. The highest energy level (indicated by dashed lines) is tentatively placed in ^{78}Sr. In the nγγ gated data, ^{78}Sr transitions are visible in the 1696 keV gate. In addition, the 1696 keV transition can be observed in the nnγγ data summed over the lower five transitions of the ground band (fig. 1). The placement of the (18$^+$) state is consistent with the energy spacing between the levels of the lower spin states. So far, no side bands have been observed in ^{78}Sr. The tentative spins assigned to the levels are consistent with the stretched E2 transitions expected for rotational bands.

Analysis of the moment of inertia

The kinematical and dynamical moments of inertia $J^{(1)}/\hbar^2$ and $J^{(2)}/\hbar^2$ in ^{78}Sr and ^{80}Sr have been plotted in fig. 3. One visible characteristic is that the $J^{(2)}$ values are consistently higher than the corresponding $J^{(1)}$ values. As the frequency increases, the $J^{(1)}$ curve for ^{78}Sr approaches $J_{RIG}/\hbar^2=22.1$ MeV^{-1} evaluated at the predicted[5] deformation parameters $\beta_2=0.38$, $\gamma=0°$ for the ^{78}Sr ground state. At I=20 $\hbar$ this calculation[5] predicts $\beta_2\approx0.30$, $\gamma\approx+15°$ (Lund convention). The $J^{(1)}$ moments of inertia indicate that ^{78}Sr is more strongly deformed than ^{80}Sr at low frequencies. However, the two curves intersect at $\hbar\omega\approx0.55$ MeV which has been reported[2,5] to be a band crossing involving the alignment of two $\pi g_{9/2}$ quasiparticle spins to the rotational axis. This effect is only evident in the $J^{(2)}$ plots as broad bumps. From this one may conclude that the interaction between the two bands is quite strong. Since the bump for ^{78}Sr is

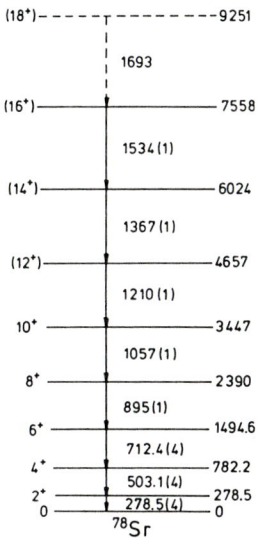

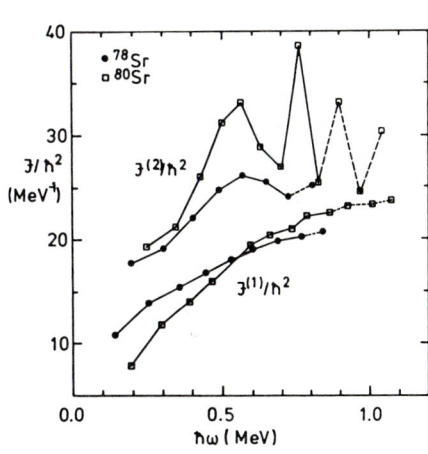

Fig. 2. Proposed level scheme.   Fig. 3. Moment of inertia plots.

considerably lower than that for ^{80}Sr, the strength of this interaction is larger in ^{78}Sr. This coincides with the prediction of ref. 5 that γ softness increases as N increases in the strontium isotopes.

In a perturbative approach, if the interaction between two bands is large, then the overlap between the nuclear wavefunctions must be large, too. Hence, the deformations of the two bands are very similar. Because the interaction is stronger in ^{78}Sr than in the case of ^{80}Sr, the resulting minimum in the total energy surface should be better defined for ^{78}Sr, as is predicted[5] for the ground states. This leads to a nucleus that is less susceptible to γ softness as a function of spin. Thus, one would expect the moment of inertia for ^{78}Sr to be little affected by the increasing rotational frequency. This trend can be seen in the $J^{(1)}$ curve in fig. 3.

Conclusions

Spectroscopic results on residual nuclei with very small cross-sections may be obtained using an array of many neutron and Compton suppressed Ge detectors. The use of these neutron detectors as a neutron multiplicity filter, produced nn$\gamma\gamma$ gated data with very little contamination from strong competing reactions. The yrast rotational states of ^{78}Sr have been extended up to atleast 16 $\hbar$. This nucleus displays good rotational behavior with little obvious effects from particle alignment. Cranked shell model analysis[5] has inferred that two strongly interacting bands based on similar deformations are present in ^{78}Sr. As a result, the moment of inertia is a smoothly varying function of the spin. --- Extensions of the yrast bands in ^{75}Br, ^{76,78}Kr, and ^{78}Rb by some 8 $\hbar$ have been established; further analysis is continuing.

[1] C.J. Lister, et al., Phys. Rev. Lett., **49**, 308, (1982).
[2] R.F. Davie, et al., Nucl. Phys. **A463**, 683, (1987).
[3] J. Heese, K.P. Lieb, S. Ulbig, J. Billowes, A.A. Chishti, C.J. Lister, and B.J. Varley, in preparation.
[4] E.F. Moore, P.D. Cottle, C.J. Gross, D.M. Headly, U.J. Hüttmeier, S.L. Tabor, and W. Nazarewicz, submitted to Phys. Rev. C.
[5] W. Nazarewicz, et al., Nucl. Phys., **A435**, 397, (1985).

g-Factors Near the First Backbend in ^{82}Sr and ^{84}Sr

A.I. Kucharska[1], J. Billowes[2], and C.J. Lister[2]

[1] Nuclear Physics Laboratory, Oxford University,
Oxford, OX1 3RH, UK
[2] Department of Physics, Schuster Laboratory,
Manchester University, Manchester M13 9PL, UK

1. Introduction

The gyromagnetic ratio is an important quantity in spectroscopic studies of deformed nuclei because its value is sensitive to proton or neutron quasiparticle admixtures in the nuclear wavefunction. When the lifetimes of these states are only a few picoseconds the transient field (TF) technique seems to provide the best means of measuring their g-factors. In this method the magnetic field is experienced by the nucleus of an ion only as it moves with high velocity ($\geq 0.01c$) through a ferromagnet such as iron or gadolinium. Although the very short stopping times of ions in solids ($\sim$ 1 ps) limits the duration of the field, measurable nuclear precessions of typically $\Delta\phi \simeq 2°$ are produced which are directly proportional to the nuclear g-factor:

$$\Delta\phi = -\frac{g\mu_N}{\hbar} \int_{t_i}^{t_o} B_{TF}(v) e^{-t/\tau} dt \qquad (1)$$

The field strength B_{TF} depends on the instantaneous velocity of the ion and is therefore a function of time. The limits t_i and t_o are the entry and exit times of the ferromagnetic foil and τ is the mean lifetime of the nuclear state. The application of the TF technique to states near the backbend of deformed nuclei has been hampered by the relatively long times taken to populate these levels by the γ-ray cascade following heavy-ion reactions; in the rare-earth and mass-130 regions the side-feeding times to discrete yrast states are comparable to the stopping time of the ions and at which time the transient field disappears. Such difficulties seem to be greatly reduced in the mass-80 region; a number of experiments using recoil-distance and Doppler-shift attenuation methods [1,2] find rather fast side-feeding times of the order of 0.1 ps to states near the first backbend ($E_x \approx 4$ MeV), particularly when the excited nucleus is formed at low spin. It should therefore be possible to measure g-factors of these states in a TF experiment. In the present work we have applied the TF technique to states in the transitional nuclei ^{82}Sr and ^{84}Sr which exhibit backbending around spin 8^+ due to alignment of either proton or neutron $g_{9/2}$ quasiparticles.

2. Experimental description

The experimental method is very similar to that of Ward et al [3] who measured average g-factors for different spin regions in ^{78}Kr. States of interest in the ^{82,84}Sr isotopes were populated in the inverted reactions ^{12}C(^{72,74}Ge,2n) at beam energies between 210 MeV and 230 MeV supplied by the 20 MV tandem accelerator at Daresbury Laboratory. The nuclei were formed at low spin, $7-8\hbar$, with a high initial recoil velocity. The target foil was a four-layer sandwich; the 300μg/cm^2 ^{12}C layer was separated from a 6μm iron foil by a layer of tantalum sputtered on the iron to a thickness of 2μm. The purpose of the tantalum layer was to allow time (about 0.2 ps) for fast feeding to take place before the nuclei experienced the transient field. The foil was backed with a sputtered layer of 5mg/cm^2 copper. The velocity range of the Sr ions in the iron foil was between 0.054c and 0.011c. They emerged from the iron layer approximately 0.8 ps after formation and came to rest in the copper where they completed their γ-ray cascade to the ground state.

The foil was clamped between the poles of an electromagnet and polarised in a vertical field of 0.17 T. The γ-radiation was observed with four Ge(Li) detectors set in the horizontal plane at $\pm 60°$ and $\pm 120°$ to the beam axis. A 12.5cm $\times$ 15cm NE213 scintillation counter was fixed at 0° and pulse-shape discrimination was used to identify neutrons from the reaction. Singles and neutron-gated γ-ray spectra were collected for both directions of the magnetic field. The cascades observed are shown in figs.1(a) and 1(b) and the details of the principal transitions are listed in tables 1 and 2.

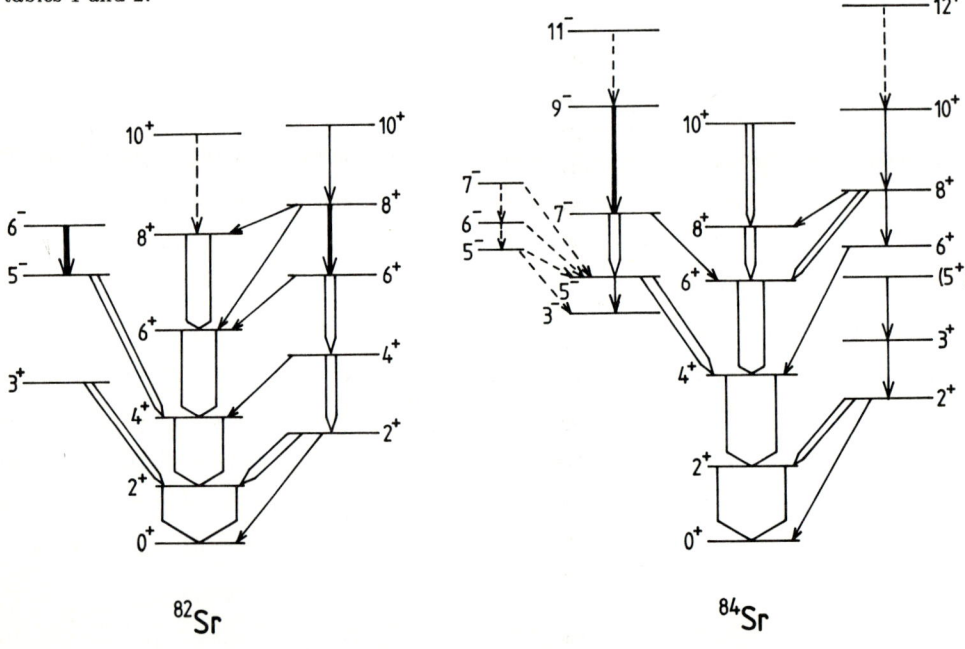

Fig.1(a) Transitions observed in ^{82}Sr   Fig.1(b) Transitions observed in ^{84}Sr

The nuclear precessions were deduced from the rotations of each γ-ray angular distribution in the conventional way using the double ratio of counting rates for a pair of detectors, defined by

$$\rho = \frac{N_1(\uparrow)}{N_1(\downarrow)} \times \frac{N_2(\downarrow)}{N_2(\uparrow)} \qquad (2)$$

where the arrow indicates the field direction. The rotation $\Delta\phi$ of the γ-ray distribution about the magnetic field axis is then

$$\Delta\phi = \epsilon/S(\theta) \qquad (3)$$

where $\epsilon = (\sqrt{\rho} - 1)/(\sqrt{\rho} + 1)$. The logarithmic slopes, $S(\theta) = (1/W)(dW/d\theta)$, of the γ-ray distributions were deduced during the course of the experiment by measuring the double ratios ρ when the γ-ray detectors were displaced by $\pm 4°$ from their mean position to mimic a rotation of the γ-ray distribution. The results were corroborated by a measurement of the full angular distribution between 0° and 90° using the reactions ^{72,74}Ge(^{12}C,2n) at the same centre of mass energies.

3. Analysis and results

The feeding pattern of the γ-ray cascades was established from the intensities in the singles spectra (unless a line was contaminated, in which case the n-coincidence spectra were used). The experimental values for $S(\theta)$ of the γ-ray distributions must be consistent with the nuclear feeding pattern; this requirement was met by finding the set of slopes $S(\theta)_{model}$ based on the observed cascade that best fitted the experimental values. This was done by initially assuming that all stretched E2 transitions had the same values of the A_k angular distribution coefficients before allowing for the effect of the finite detector size. Additional loss of alignment due to discrete feeding transitions of mixed multipolarity were calculated explicitly. The best fit of the model values to the experimental data was found by varying A_2 and the mixing ratios. The A_4 coefficient was related to A_2 by the alignment-yrast curve described by Zobel et al [4]. This procedure gave a normalised χ^2 of 1.5 for the ^{84}Sr cascade and 1.3 for ^{82}Sr and agreement between model and experiment was better than 2% for the well determined values.

The double ratios ρ defined in equation (2) were obtained from the singles spectra for all but the weakest transitions, except in the case of the $8_1^+ \to 6_1^+$ (1013 keV) transition in ^{82}Sr where the n-coincidence spectra were used because of contamination in singles. Only the counts in the stopped part of the lineshapes were used to form the double ratios (the Doppler tail was excluded). As a consequence, the precessions $\Delta\phi$, shown in tables 1 and 2, are deduced for those nuclei that have experienced the full transient field and emerged from the iron foil.

The Doppler-broadened profile of the $8_1^+ \to 6_1^+$ transition in ^{82}Sr observed in n-coincidence

Table 1. Principal results for the ^{84}Sr cascade.

$J_i^\pi \to$	J_f^π	E_γ(keV)	$W(60°)$	$S(60°)_{model}$	$\epsilon(\times 10^{-3})$	$-\Delta\phi$ (mrad)	g
$2_1^+ \to$	0^+	793	100	−0.527(7)	+12.1(5)	+22.9(10)	[+0.419(47)]^a
$4_1^+ \to$	2_1^+	974	68(4)	−0.528(8)	+14.7(6)	+27.9(12)	
$6_1^+ \to$	4_1^+	1040	40(8)	−0.623(11)	+15.3(9)	+24.5(15)	
$8_1^+ \to$	6_1^+	524	13(1)	−0.563(36)	+7.9(20)	+14.0(37)	−0.1(2)
$10_1^+ \to$	8_1^+	1116	8(3)	−0.692(51)	+9.6(41)	+14(7)	+0.2(1)
$8_2^+ \to$	6_1^+	872	7(1)	−0.64(10)	+29.0(38)	+45(9)	+0.9(1)
$10_2^+ \to$	8_2^+	854	2.3(2)	−0.64(10)	+27.7(69)	+43(13)	+0.8(2)
$5^- \to$	4_1^+	1001	21(1)	+0.218(25)	−11.6(14)	+53(9)	+1.6(2)^b
$7^- \to$	5^-	719	11(1)	+0.581(75)	+10.5(26)	+18(5)	+0.6(2)
$9^- \to$	7^-	1148	7(2)	−0.581(75)	+1(1)	+2(2)	+0.00(4)

^{a} From reference [8] using same transient field parametrization.
^{b} Mean value for unobserved feeding states.

Table 2. Principal results for the ^{82}Sr cascade.

$J_i^\pi \to$	J_f^π	E_γ(keV)	$W(60°)$	$S(60°)_{model}$	$\epsilon(\times 10^{-3})$	$-\Delta\phi$ (mrad)	g
$2_1^+ \to$	0^+	574	100	−0.441(6)	+9.8(4)	+22.1(8)	
$4_1^+ \to$	2_1^+	755	67(5)	−0.529(8)	+16.2(5)	+30.6(10)	
$6_1^+ \to$	4_1^+	901	45(4)	−0.538(9)	+16.6(11)	+30.9(20)	
$8_1^+ \to$	6_1^+	1013	29(4)	−0.542(10)	+21(7)	+38(13)	+0.7(1)
$8_2^+ \to$	6_2^+	786	5.7(6)	−0.542(10)	+23.4(43)	+43(8)	+0.7(1)
$10_2^+ \to$	8_2^+	801	3(1)	−0.542(10)	+27(12)	+50(22)	+1.1(5)
$5^- \to$	4_1^+	1489	11.0(6)	+0.265(5)	−8.4(21)	+32(8)	+0.3(4)
$6^- \to$	5^-	522	5.7(4)	+1.05(6)	−46.2(33)	+44(4)	+0.9(1)

provided the best measurement of the side-feeding time. Analysis of the lineshape yielded a lifetime of $\tau(8_1^+) = 1.1(2)$ ps and a feeding time of 0.0(2) ps.

Since the TF precession of one level is passed down to the next level by the γ-decay, the observed precessions must be consistent with the cascade pattern. A computer code was written that followed the time-evolution of the nucleus in femtosecond steps from its formation until its exit from the iron foil where it was allowed to complete its decay to the ground state. The program fitted the experimental rotations by varying the individual level g-factors and initial populations to obtain the lowest χ^2 value. The lifetimes of the various levels were allowed to vary within their experimental limits except where they were greater than 4 ps when they were fixed at their values recorded in the literature [5,6]. The program was run for three different values of the side-feeding time: 0.0, 0.15 and 0.3 ps. The choice of value for the continuum g-factor was based on the rotations obtained by Ward et al [3] giving $g_c = +0.6(1)$; since the side-feeding time was short, the value of g_c had very little effect on the results. The transient field parametrization of Shu et al [7] was used for B_{TF} in equation (1) since this has already been applied to the 2_1^+ states of the stable Sr isotopes [8] and direct comparisons of g-values are then possible. The best determined g-factors from this procedure are shown in the last columns of tables 1 and 2.

4. Conclusions

As expected, the states with the best determined values are those near the top of the cascade and near the entry spin of the reaction where the feeding is fastest and the slow (discrete) feeding corrections are small. Although the errors are quite large — $\delta g \approx 0.2$ — it is easy to see the effect of the $g_{9/2}$ neutron or proton alignments; the 8_1^+ and 10_1^+ states in ^{84}Sr contain a strong neutron component (a pure configuration for the 8^+ state would have $g = -0.24$) as was first reported by Broude et al [9]. The 9^- state in the same nucleus must also have a stretched neutron configuration and therefore does not belong in a band build on the 3^- or 5^- states. Proton alignments are seen in the 8_2^+ and 10_2^+ states in both nuclei (a pure 8^+ $\pi g_{9/2}^2$ configuration would have $g = +1.37$). The largest g-factor was assigned to the 5^- state in ^{84}Sr which is probably more representitive of the negative parity states feeding into it (which we could not unravel). These states are therefore *proton* particle-hole states with configurations such as $g_{9/2}p_{3/2}^{-1}$ and $g_{9/2}f_{5/2}^{-1}$. Similar negative parity states are seen in the heavier Sr isotopes at about the same excitation energy.

The transient field technique has wide applicability in this mass region for states with similar spin and excitation energy. Since these states are only weakly populated, the results have large statistical errors but are sufficient to distinguish the large g-factor variations expected with nucleon alignments.

References

[1] L Lühmann, M Debray, K P Lieb, W Nazarewicz, B Wörmann, J Eberth and T Heck, *Phys.Rev.* C31 (1985) 828.

[2] J Heese, K P Lieb, L Lühmann, F Raether, B Wörmann, D Abler, H Grawe, J Eberth and T Mylaeus, *Z. Phys.* A325 (1986) 45.

[3] D Ward, H R Andrews, A J Ferguson, O Häusser, N Rud, P Skensved, J Keinonen and P Taras, *Nucl. Phys.* A365 (1981) 173.

[4] V Zobel, L Cleemann, J Eberth, H P Hellmeister, W Neumann and N Wiehl, *Nucl. Instrum. Methods* 171 (1980) 223.

[5] A Dewald, U Kaup, W Gast, A Gelberg, K O Zell and von Brentano, *Proc. 4th Int. Conf. Nucl. Far from Stability, Helsingør*, ed. L O Skolen (CERN 81-09) p.418.

[6] A Dewald, U Kaup, W Gast, A Gelberg, H-W Schuh, K O Zell and von Brentano, *Phys. Rev.* C25 (1982) 226.

[7] N K B Shu, D Melnik, J M Brennan, W Semmler and N Benczer-Koller, *Phys. Rev.* C21 (1980) 1828.

[8] A I Kucharska, J Billowes and M A Grace, *J. Phys. G: Nucl. Phys.* 14 (1988) 65.

[9] C Broude, E Dafni, A Gelberg, M B Goldberg, G Goldring, M Hass, O C Kistner and A Zemel, *Phys. Lett.* 105B (1981) 119.

Core-Related Effects on Electromagnetic Transition Strengths in Medium-Mass Nuclei

W. Andrejtscheff, L.K. Kostov, P. Petkov, and Y. Sy Savane

Bulgarian Academy of Sciences,
Institute for Nuclear Research and Nuclear Energy,
1784 Sofia, Bulgaria

1. Introduction

The investigation of γ-ray transition probabilities allows quantitative conclusions on the important question to what extent the coupling of an odd particle is influenced by the underlying core motion. The transition rates considered here in nearly spherical nuclei are deduced from in-beam delayed-coincidence measurements analyzed according to the generalized centroid-shift method [1] which facilitates lifetime determinations with germanium detectors down to ≈ 0.3 ns.

Low-lying excitations in even $Z=50$ nuclei (sec.2) strongly affect the transitions in odd-A $Z=49$ isotopes (sec.3). Similar phenomena are observed in the $A \approx 90$ region (sec.4). In sec.5, l-forbidden M1 transitions in $Z=46,48$ nuclei are considered.

2. Non-Collective E2 Transition Strengths in $^{106-112}$Sn Involving Two-Phonon State Admixtures

The delicate interplay of colectivity and single-particle degrees of freedom in the light even semimagic tin isotopes [2-4] is manifested by a characteristic feature: the low E2 transition strength between the 6^+ and 4^+ levels [5]. In a recent experiment performed on the Cologne tandem we determined the lifetime of the 6^+ level in ^{106}Sn and remeasured the corresponding values in ^{110,112}Sn. Together with the measurement in ^{108}Sn [6], a full set of data on this transition is now available in $^{106-112}$Sn (fig.1). The corresponding $B(E2, 6^+ \to 4^+)$ values are in the order of 1 W.u. The transition strengths $B(E2, 6^+ \to 4^+ \to 2^+)$ in ^{112}Sn are about one order of magnitude higher.

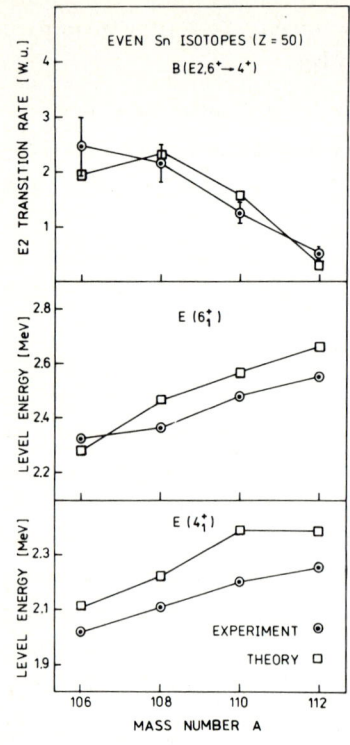

Fig.1. Experimental and theoretical level energies and B(E2) values. The latter were calculated with $e_{eff}=0.1$.

Microscopic calculations were performed within the quasiparticle-phonon model [7], the parameters being determined by the usual prescription (fig.1). The (calculated) structure of the 6^+ state consists mainly of the non-collective $[6^+_1]$ phonon predominantly determined by the two-quasineutron configuration $[2d_{5/2}1g_{7/2}]$. In the 4^+ states, beside one-phonon terms $[4^+_1]$, $[4^+_2]$, $[4^+_3]$ a significant contribution is given by the two-phonon component $[2^+_1 \times 2^+_1]$. Further components are very small.

It is of particular interest that just very small components in the wave functions turn out to be of decisive importance for the $6^+ \rightarrow 4^+$ transition rates. In the formation of the matrix elements $\langle 4^+ \| E2 \| 6^+ \rangle$, the largest contributions arise from terms of the type $\langle [2^+_1 \times 6^+_1]_{4^+} \| E2 \| [6^+_1] \rangle$ and $\langle [4^+_1] \| E2 \| [2^+_1 \times 4^+_1]_{6^+} \rangle$. The corresponding two-phonon components involved $[2^+_1 \times 6^+_1]_{4^+}$ and $[2^+_1 \times 4^+_1]_{6^+}$ are only weakly ($\approx 1\%$) participating in the wave functions.

3. Transition Probabilities and Hole-Core Coupled Configurations in Odd-Indium Isotopes

Let us consider the systematics of some relevant isomers in ^{107,109,111}In (table 1) including thereby recent results in ^{107}In [8]. For comparison, the transition strengths $B(E2, 6^+ \to 4^+)$ in the Sn core nuclei are also given.

From the relative positions of the levels in ^{107}In with respect to the 2^+ level in ^{108}Sn and comparison with heavier In isotopes [9]

Table 1. Some isomers and E2 transitions in ^{107,109,111}In and ^{108,110,112}Sn

Nucleus	Level [keV]	E_γ [keV]	J_i^π	J_f^π	B(E2) [W.u.]	$B(E2, 6^+ \to 4^+)_{core}$ [W.u.]
^{107}In	1853.4	438.8	$17/2^+$	$13/2^+$	0.7	2.1
^{109}In	2101.8	(M3)	$19/2^+$			1.2
^{111}In	2716.9	255.3	21/2	17/2	1.1	0.5

it may appear that the $13/2^+(17/2^+)$ state in ^{107}In arises from a coupling of the type $\pi g_{9/2}^{-1} \times 2^+ (\pi g_{9/2}^{-1} \times 4^+)$. However, this would provide difficulties in the explanation of the $17/2^+ \to 13/2^+$ E2 transition strength. We suggest appreciable admixtures from the configuration $\pi g_{9/2}^{-1} \times 6^+ (\pi g_{9/2}^{-1} \times 4^+)$ in the $17/2^+(13/2)^+$ state. Under assumption of weak coupling:

$$B(E2, 17/2^+ \to 13/2^+, ^{107}\text{In}) = 0.68 \cdot B(E2, 6^+ \to 4^+, ^{108}\text{Sn}) \qquad (1)$$

is expected. The experimental values (table 1) provide an agreement of eq.1 within a factor of two. We note that within a hole-core coupling model including configuration mixings, cancellations (opposite phases) should occur in the formation of the total transition matrix element in order to reproduce the low $B(E2, 17/2^+ \to 13/2^+)$ value.

4. Multiplet Structure and Core-Related Effects on Transition Rates in Some A≈90 Nuclei

Subnanosecond lifetimes in ^{87}Y were investigated [10] on the Rossendorf cyclotron (fig.2). Let us consider transitions between $21/2^+$ and $17/2^+$ states in some odd-A nuclei as well as between 8^+ and 6^+ states in the presumed core nuclei (table 2).

The 8^+ and 6^+ levels in each even (core) nucleus are considered as members of the $\nu g_{9/2}^{-2}$ (or $\pi g_{9/2}^2$) multiplet. The $21/2^+$ and $17/2^+$ states in the odd-A nuclei arise by the inclusion of an additional

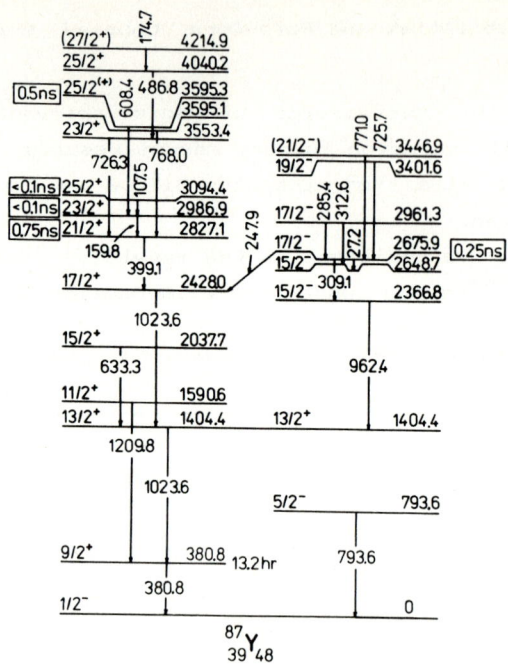

Fig.2. Partial level scheme of ^{87}Y according to [13]. The framed lifetimes are measured in [10]

Table 2. E2 transitions in some odd-A and even core nuclei. $R_{exp} = B(E2, 21/2^+ \rightarrow 17/2^+)/B(E2, 8^+ \rightarrow 6^+)$

Nucleus	J_i^π	J_f^π	Configuration	B(E2)(W.u.)	R_{exp}
$^{87}_{39}Y_{48}$	$21/2^+$	$17/2^+$	$(\nu g_{9/2}^{-2})(\pi g_{9/2})$	3.2±0.4	
$^{86}_{38}Sr_{48}$	8^+	6^+	$(\nu g_{9/2}^{-2})$	2.8±0.2	1.1±0.2
$^{89}_{41}Nb_{48}$	$21/2^+$	$17/2^+$	$(\nu g_{9/2}^{-2})(\pi g_{9/2})$	1.5±1.0	
$^{88}_{40}Zr_{48}$	8^+	6^+	$(\nu g_{9/2}^{-2})$	1.3±0.2	1.2±0.8
$^{89}_{40}Zr_{49}$	$21/2^+$	$17/2^+$	$(\pi g_{9/2}^2)(\nu g_{9/2}^{-1})$	3.2±0.1	
$^{90}_{40}Zr_{50}$	8^+	6^+	$(\pi g_{9/2}^2)$	2.6±0.1	1.15±0.10
$^{91}_{42}Mo_{49}$	$21/2^+$	$17/2^+$	$(\pi g_{9/2}^2)(\nu g_{9/2}^{-1})$	1.7±0.2	
$^{92}_{42}Mo_{50}$	8^+	6^+	$(\pi g_{9/2}^2)$	1.3±0.1	1.3±0.2

$\pi g_{9/2}(\nu g_{9/2}^{-1})$ particle (hole). In this case, the weak coupling model predicts:

$$R_{th} = B(E2, 21/2^+ \to 17/2^+)/B(E2, 6^+ \to 4^+) = 0.8 \tag{2}$$

The experiment provides, however, $R_{exp} > 1.1$ (table 2). At this point, we again consider mixings from core configurations other than 8^+ and 6^+, respectively. Using mixed wave functions obtained [11] for $^{89}Zr/^{90}Zr$, we derive:

$$B(E2, 21/2^+ \to 17/2^+) = |0.136 \langle (\pi g_{9/2}^2)_6^+ \| E2 \| (\pi g_{9/2}^2)_8^+ \rangle + 0.081 \langle (\pi g_{9/2}^2)_4^+ \| E2 \| (\pi g_{9/2}^2)_6^+ \rangle |^2 \tag{3}$$

and consenquently

$$R_{th}(^{89}Zr/^{90}Zr) = 1.04 \tag{4}$$

what is already closer to the experiment compared to (2). It is worth mentioning that both r.h.s. terms in (3) provide almost equal contributions.

5. The l-Forbidden M1 Transitions $\nu(g_{7/2} \leftrightarrow d_{5/2})$ in Pd and Cd isotopes

In a recent experiment on the Cologne tandem, lifetimes of $7/2^+$ states in ^{99}Pd (N=53) and ^{109}Cd (N=55) were determined [5]. These data help to illumunate the systematic behaviour of the above transition in nuclei approaching N=50 (fig.3). It is of importance that in the isotopes considered, both configurations form low-lying states and

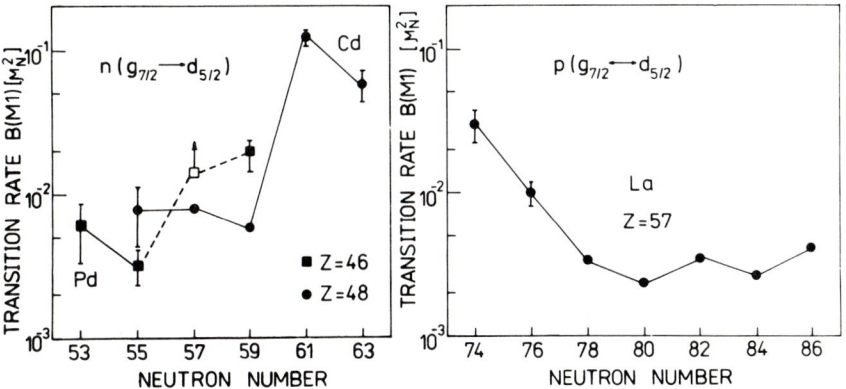

Fig.3. Some l-forbidden M1 transitions around closed shells

even one of them is mostly the ground state, i.e. little fragmentation is expected. Approacing N=50, l-forbidden M1 strengths of this type generally decrease (fig.3). In La isotopes around N=82, the increasing M1 strength away from N=82 (fig.3) was associated with increasing deformation of the core producing rotational M1 admixtures [12].

Fruitful collaboration, kind hospitality and financial support by the nuclear spectroscopy groups at ZfK Rossendorf and Institut für Kernphysik der Universität zu Köln as well as Committee for Science, Bulgaria (contract No258) are greatly acknowledged.

References

1. W. Andrejtscheff et al., Nucl. Instrum. Methods 204(1982)123
2. G. Bonsignori et al., Nucl. Phys. A432(1985)389
3. A. van Poelgeest et al., Nucl. Phys. A432(1985)389
4. M. Waroquier et al., Phys. Rep. 148, No.5(1987)
5. W. Andrejtscheff et al., to be published
6. W. Andrejtscheff et al., Bull. Amer. Phys. Soc. 27,1(1982)27
7. V. G. Soloviev et al., Nucl. Phys. A288(1977)376
8. W. Andrejtscheff et al., Z. Phys. A328(1987)23
9. W. H. A. Hesselink et al., Nucl. Phys. A299(1978)60
10. L. K. Kostov et al., to appear in Z. Phys. A (1988)
11. A. M. Bizzeti-Sona et al., Z. Phys. A324(1986)311
12. W. Andrejtscheff et al., Nucl. Phys. A368(1981)95
13. C. A. Fields et. al., Z. Phys. A295(1980)365

Microscopic Description of the Ground-State and High-Spin Properties of the Light Strontiums

H.C. Flocard

Div. Phys. Théo., I.P.N. Orsay, F-91406 Orsay Cedex, France

Introduction

Today it seems fair to say that the nuclear Hartree-Fock (HF) theory has come out of age. By this we mean that it is no longer just a set of equations providing a conceptual connection between the basic ingredients of a microscopic theory of nuclear structure: the participating nucleons and their two-body interaction, and the macroscopic collective properties of nuclei. HF has become a tool that one can put to test on any nucleus for almost any kind of collective motion. In some cases it even appears that by renouncing to any sort of parametrization, however clever and physically motivated it may seem, HF has freed himself from constraints which are restricting the predictive power of more phenomenological models. The region of light strontiums which provides a remarkable variety of collective nuclear phenomenon appeared to us as one of the natural places where we should put to work the recently acquired potentialities of the HF method

Constrained Hartree-Fock with the Skyrme interaction

2.a) The Hartree-Fock equations

The derivation of the HF equations is based on a variational principle for the energy E which states that given a variational space, the optimal choice for a wave-function Ψ within this space is determined by a minimization of the energy defined as the expectation value of the Hamiltonian $\hat{H}$. The HF equations are obtained when the variational space of Ψ is restricted to the set of Slater determinants[1]. Let us define the total Hamiltonian $\hat{H}$ as:

$$\hat{H} = \sum_{i,j}\langle i|t|j\rangle\, a_i^+ a_j \; + \frac{1}{4}\sum_{i,j,k,l}\langle i,k|v|j,l\rangle\, a_i^+ a_k^+ a_l\, a_j \quad , \tag{1}$$

where $\langle i|t|j\rangle$ and $\langle i,k|v|j,l\rangle$ denote the matrix elements of respectively the kinetic energy operator and the Skyrme effective two-body interaction contained in $\hat{H}$. The minimisation of E with respect to Ψ (or equivalently the single particle wave-functions Φ_λ's) yields the HF eigenvalue equations:

$$\sum_j \langle i|h|j\rangle\, \Phi_\lambda(j) = e_\lambda\, \Phi_\lambda(i) \quad ; \quad \langle i|h|j\rangle = \langle i|t|j\rangle + \sum_{k,l}\langle i,k|v|j,l\rangle\, \rho_{lk} \quad . \tag{2}$$

This equation combined with the definition of the one body density matrix ρ in terms of Ψ defines self-consistently the single particle Hamiltonian introduced in (2), which in turn generates the single particle spectrum. These equations are therefore non-linear. As a result their solution may break the symmetries of the Hamiltonian like for exemple the rotational invariance.

A realistic description of nuclei requires that one takes into account the residual pairing interaction. We have chosen to avoid the technical difficulties associated with HFB which is the natural extension of HF. Instead we introduce pairing effects by means of BCS with a seniority interaction whose intensity G is parametrized with a simple formula adjusted to reproduce the quasi-particle energies of a selected set of nuclei. The corresponding HF+BCS equations are solved self-consistently. Technical details on the method can be found in [BFH85].

---

[1] To derive the HFB equations, the variational space for the Ψ's is enlarged and includes the set of all BCS wave-function.

2.b) The constrained Hartree-Fock equations

Solving HF equations (eq. 2) gives an approximation for the binding energy and for the wave-function of the ground state. It gives also a description of the excited states built on few uncorrelated single-particle excitations. On the other hand, experimental evidence demonstrates that most of the nuclear dynamics is governed by slow (mostly quadrupole) collective motions. One of the attractive feature of the HF theory is its ability to incorporate the collective dynamics into a model of microscopically interacting particles. This is achieved by the introduction of external constraints and leads to the constrained Hartree-Fock (CHF) equations. The idea at the core of the CHF approach is that collective nuclear dynamics can be described in terms of a restricted number of degrees of freedom which can be described by relatively simple (generally one-body) operators. The addition of these operators with appropriate Lagrange multipliers to the total binding energy, leads to a new variational principle in which the energy is minimized, subject to constraints which determine the values of the collective parameters. The knowledge of the variation of the binding energy in terms of the collective degrees of freedom provides then a crucial information on the characteristics of spectroscopy of the nucleus. To be more specific, let us assume that the collective degree of freedom we want to study is associated with the measure of the operator $\hat{Q}$. We look then for a Slater determinant Ψ which minimizes the energy and such that the expectation value of the collective operator is equal to some chosen number q. This problem is solved by a minimization of the constrained energy $\mathcal{E}$

$$\mathcal{E} = \langle \Psi | \hat{H} - \mu \hat{Q} | \Psi \rangle \quad , \tag{3}$$

where μ is the Lagrange multiplier which must be adjusted to ensure that $\langle Q \rangle = q$. The mathematical structure of the resulting CHF equations is the same as that obtained with HF (eq. 2), the only modification being the addition of the matrix elements $-\mu \langle i | \hat{Q} | j \rangle$ to the one-body HF Hamitonian. Solving the CHF equations as a function of q provides the deformation energy curve

$$E(q) = \mathcal{E} + \mu q \quad , \tag{4}$$

associated with the collective path determined by $\hat{Q}$.

It should be mentionned that the relevant constraints are of a quite different nature depending on the physics one wants to describe. The above discussion implies that we are considering local operators acting on the shape of the nuclear density (quadrupole, octupole, ...). On the other hand to study the evolution of the nucleus with increasing spin, the natural constraining operator [RSC82] is the projection of the total angular momentum on one of the inertial axes of the density; this is the so-called self-consistent cranking method. With the symmetries required in all our calculations, we can take these inertial axes along those of the intrinsic frame: Ox, Oy and Oz and select a cranking constraint proportional to $-\omega \hat{J}_z$. It may be noticed that while the multipole operators lead only to a breaking of the space symmetries (spherical symmetry or parity), the cranking constraint will additionally break the time-reversal invariance and will lead to structurally more complex CHF equations.

2.c) Extension of constrained Hartree-Fock by the generator coordinate method

Extensions of CHF are also desirable when the collective energy surface does not exhibit a well defined minimum (on a typical energy scale of about 1MeV). In such a case, the sole HF minimum cannot take into account all the implications of the collective dynamics. In fact a more physical wave-function can be constructed from the set of all CHF states $\Psi(q)$. This is achieved by the generator coordinate method (GCM).

In the GCM method one looks for an improved approximation of the wave-function with the

following form:
$$\Phi = \int dq\, f(q)\, \Psi(q) \quad . \tag{5}$$

It is then possible to write the minimization problem for $\hat{H}$ in terms of the unknown mixing function $f(q)$. The calculation is however more complex than HF because it involves the evaluation of matrix elements of $\hat{H}$ between states of the non-orthogonal set $\{\Psi(q)\}$.

2.d) **Method of solution of the HF equations**

The HF equations are a set of non-linear coupled integro-differential equations. The number of coupled equations can be large; a typical order of magnitude being the number of nucleons. This number can sometimes be significantly reduced by an appropriate use of the symmetries of the solution (rotational or time-reversal invariance for instance). In our method the HF equations are solved directly on a cartesian mesh. The number of variational parameters per orbital becomes of the order of the number of points used to discretize the problem. A typical dimension is three to four thousand. This may seem a large number. It is however the minimum price to pay if one wants a completely unbiased solution of the CHF equations. In particular all multipole moments of the density and of the current will be treated with equal accuracy. The variation will be complete with respect to all degrees of freedom. The original point in our method is that we avoid the almost untractable diagonalization of the Hamiltonian h by noticing that nuclear properties are entirely determined by occupied orbitals [DFK80].

Deformation Properties of the Light Strontiums

3.a) **Deformation Energy Surfaces**

The light isotopes of Sr are among the best exemples of the diversity of deformation properties (see figure 3.1) one can expect in this area of the mass table. Due to the magic number $N = 50$, the surface energy of the nucleus ^{88}Sr is that of a stiff spherical nucleus. For decreasing values of N the nucleus remains first spherical, while its surface softens gradually. The triaxial shell effect which for $N = 44$ moves the HF ground state to $\gamma \approx 30°$ shows up already in the surface of ^{84}Sr. At $N = 42$, the energy surface exhibits one spherical and three other minima; one oblate one triaxial and one prolate. The latter corresponds to a shell effect whose intensity culminates at $N = Z = 38$; the nucleus ^{76}Sr is a well deformed rotor with only one minimum. This rapid variation of the intrinsic deformation (spherical then triaxial then oblate) with a change of few units of the neutron number is confirmed by the available spectroscopy [HAM82].

3.b) **Variation of the root mean square radii**

Recently the variation of the root mean square radii with decreasing N has been experimentally determined [EWR87]. Its essential features are a steady increase of the charge radius till ^{80}Sr followed by a small decrease for ^{78}Sr. The HF results display the same qualitative behavior but exaggerate the amplitude of the variation. In addition since the increase of radius is generated within HF by the onset of deformation, HF does not explain why the radius of the spherical ^{86}Sr is larger than that of the heavier ^{88}Sr. The evolution of the deformation energy surfaces points to a possible explanation for this phenomenon in terms of collective quadrupole vibrations. The collective wave-function of the softer ^{86}Sr will certainly contain a larger admixture of deformed states than that of the rigid ^{88}Sr. We have therefore begun a complete GCM calculation based on the CHF+BCS energy surface. Our first results obtained by a replacement of the integral in formula (5) by a sum over only three states showed a trend in the right direction. We expect to finish soon the complete calculation. As an illustration we give on figure 3.3 an exemple of the overlap energy surface of a fixed deformed prolate CHF ($Q = 1$b) state of ^{82}Sr with all the other CHF states. From the slow decrease of the overlap function one can expect rather significant collective effects.

Rotation Properties of $_{80}$Sr

4.a) Collective Rotations; $0 \leq J \leq 60\hbar$

The deformation energy surface of ^{80}Sr displays two local minima (one oblate, one prolate) in addition to the triaxial ($\gamma \approx 10°$) HF+BCS ground state. At low spins the two cranking solutions associated with the ground-state (one for positive and one for negative values of γ) compete with the band constructed on the (more deformed) prolate minimum. On the left side of figure 4.1 we display as a function of spin the variation of the total energy of the lowest positive signature bands. Below $16\hbar$ the band h associated with the HF ground-state $\gamma = -10°$ remains Yrast. Its trajectory in the Q-γ plane (see left side of figure 4.2) which remains in the second sextant ($-60° \leq \gamma \leq 0$), indicates a steady decrease of the quadrupole deformation as a function of spin.

Between $17\hbar$ and $33\hbar$ the Yrast solution has a negative signature (band l in lower part of figure 4.1). Its trajectory corresponds to even more negative values of γ and to smaller deformations. With the exception of the band A build on the prolate minimum and c on the $\gamma = 10°$ ground-state the lowest excited bands (p, b, e for $S = +$; I, q, Q, u for $S = -$) correspond to single particle excitations built on either band h or band l. It can be seen on figure 4.2 that the self-consistency often leads to a significant displacement of the trajectory in the Q-γ plane.

Four bands have been identified in the experimental work of Davie et al.[DSO87] In particular a positive parity and a negative signature have been assigned to two side bands. We propose to describe these two bands by the cranking solutions q and Q which result from the promotion of a proton either within the pf or the sdg shell (attempts in terms of the promotion of a neutron) led to more excited solutions. On figure 4.3 we compare our findings with experiment. The too compressed CHF spectrum can be explained by the neglect of pairing correlations in our calculation. Our results agree generally well with those of cranked-Nilsson-Strutinsky [NDB85]. Some differences can however be noted. We find that all our bands tend to move more as a function of J in the Q-γ plane than theirs do. The fact that [NDB85] does not propose solutions similar to our bands q and Q is most probably only a proof that it is difficult to foresee all of the potentially interesting single particle excitations (contrarily to us, the authors of [NDB85] did not have the incentive provided by experiment!). As a general rule we tend to see more low excited solutions than they do. In some cases we find that several CHF bands can be assigned to one of their solutions. For instance this is the case for the intermediate values of spin $36\hbar \leq J \leq 56\hbar$: to the band 3 of ref [NDB85] correspond three of our bands (f,F,C). On the other hand, these three solutions of CHF, already characterized by the same set of single particle quantum numbers, have somewhat similar values of Q and γ so that they may well describe the same physical band. An other difference with the Nilsson-Strutinsky approach is the slightly larger value of the moment of inertia for the average Yrast line ($\hbar^2/2\mathcal{I} = 0.018$ MeV for CHF compared with 0.2 MeV).

For values of J between $36\hbar$ and $60\hbar$ the Yrast bands are located in the sextant $0 \leq \gamma \leq 60°$. With the exception of the negative signature band r the bands forming or near the Yrast line have deformations larger (typical value 9bn) than that of the ground state (7bn) by no more than 30%. Such bands can therefore not be considered to be superdeformed.

4.b) Non Collective Rotations; $0 \leq J \leq 60\hbar$

One of the zero-spin HF+BCS local minima is oblate. We checked that the possible collective rotation associated with this minimum (band a) is always very excited (the same is true for the band build on the $\gamma = -110°$ solution). On the other hand the oblate minimum at $\gamma = 60°$ provides a basis on which one can build non-collective configurations to be used as starting points of CHF calculations. Our results are shown on figure 4.4. On this figure we have also drawn the collective Yrast line to single out the more interesting non-collective states. Only three such states are significantly (1 MeV or more) lower than the collective line. Two of them

($J = 22\hbar, 43\hbar$) are also predicted in [NBD85]. We propose in addition a state at $J = 16\hbar$ based on a $p_6(p3/2)^3(g9/2); n_6(g9/2)^4$ configuration. As compared to [NBD85] we note that we can connect **continuously** these three interesting non-collective states to other **oblate** states (dashed lines on figure 4.4). This phenomenon had already been observed in an earlier CHF calculation for ^{24}Mg [BFH87] and had been explained by hexadecapole deformations of the Y_4^4 type (quantization axis along the rotation axis).

4.c) **Superdeformed bands; $60\hbar \leq J$**

The Yrast line and a few excited bands above $60\hbar$ are shown on figure 4.5 (note that the average moment of inertia has become much larger: $\hbar^2/2\mathcal{I} = 0.0135$ MeV). The right side of the figure displays the trajectory associated with these bands. The band j which is Yrast between 60 and $66\hbar$ has been followed down to $30\hbar$ (where it is excited by more than 20MeV) and to its extreme limit of stability at $80\hbar$. Between 30 and $70\hbar$ the quadrupole moment of the nucleus stays remarkably constant at a value of 14bn. Above $70\hbar$ there is a gradual stretching of the nucleus which eventually leads to fission.

Above $70\hbar$ two bands (d and z) make the Yrast line. All these bands are slightly triaxial (γ between $+5°$ and $-5°$). These values of γ may appear small; they lead however to a significant (several MeV's) improvement of the Yrast energies as compared to solutions constrained to $\gamma = 0$. The four superdeformed bands studied here have similar proton configurations and differ either by their neutron filling or because they correspond to different solutions of the same nonlinear CHF equations.

In order to study the stability of the superdeformed states against fission we have introduced a constraint on the magnitude of the quadrupole moment. The fission barriers of band j for several values of spin are plotted in the left part of figure 4.6. One understands the process leading to stretching and ultimately to fission: as deformation increases the shell effects beyond superdeformation become more and more competitive and the nucleus moves in a gradually tilting well.

It can be noticed that the shell effect at $Q = 25$bn is never "used" by band j. Such is not the case for the band z (right part of figure 4.6) for which there exists a rather well defined "hyperdeformed" minimum. This minimum corresponds to a slightly negative ($-2°$) value of γ. Thanks to this increased stability the band z survives to much higher spins (around $92\hbar$).

To conclude this section we note that the band g which has the quantum numbers of a configuration of two colliding nuclei of ^{40}K is always more excited than the Yrast line.

4.c) **The shapes of the rotating ^{80}Sr**

Within the CHF formalism one keeps track of the cranking and deformation properties of all the nucleons. To compute all the multipole moments of the density one does not have to estimate separately the bulk (liquid drop) contribution and the shell contribution and resort to models connecting the deformation parameters of the mean-field to those of the liquid drop. The electric moments are defined by the deformations of the density. On figures 4.7 and 4.8 we show for several values of the total spin the different shapes taken by the nucleus from the ground state configuration h to the unstable configuration (bande z at $J = 94\hbar$) which marks the end of the Yrast line. In particular there are drawings of the super and hyperdeformed configurations.

Figure captions

Figure 3.1 Deformation energy surfaces of the light strontiums. Deformations are given in fm^2. The separation between contour lines is .5 Mev for ^{76}Sr and ^{78}Sr, .25 MeV otherwise.

Figure 3.2 Experimental (1) and HF (2) variation of the rms radii (in fm^2) of strontium isotopes. ^{88}Sr is used as reference.

Figure 3.3 Equicontour lines in the $Q - \gamma$ plane of the overlap between CHF solutions for ^{82}Sr and the prolate CHF solution at $Q = 1b$. The interval between contour lines is 0.1.

Figure 4.1 Energy of the collective bands of ^{80}Sr as a function of angular momentum. The energies are displayed relative to an average Yrast line. The left (resp. right) part of the figure corresponds to positive (resp. negative) signature bands. Positive parity bands are drawn with solid lines and negative parity bands with dashed lines.

Figure 4.2 Trajectory of the collective bands of ^{80}Sr as functions of angular momentum. The left (resp. right) part of the figure corresponds to positive (resp. negative) signature bands. To facilitate drawing, the Q-γ plane has been rotated by $-60°$ so that the oblate non collective axis has become the positive horizontal axis.

Figure 4.3 Comparison between the experimental and CHF spectra of ^{80}Sr . The CHF and experimental energies of the 12^+ state have been adjusted

Figure 4.4 Energies of the non collective configurations of ^{80}Sr . The energies are displayed relative to an average Yrast line. The solid line indicates the position of the collective Yrast line. The dashed lines join the non-collective states which can be connected in a continuous manner by the CHF equations.

Figure 4.5 Left: energy of the superdeformed bands of ^{80}Sr as a function of angular momentum. The energies are displayed relative to an average superdeformed Yrast line. Right: trajectory of the superdeformed bands of ^{80}Sr as functions of angular momentum.

Figure 4.6 Left: fission barriers of the band j for several values of the angular momentum. The triaxial degree of freedom is not constrained. Right: comparison of the fission barriers of the superdeformed bands j (solid line) and z (dashed line) at $J = 80\hbar$.

Figure 4.7 Contour energy surfaces $\rho = 0.06 \text{fm}^{-3}$ for several configurations of the rotating nucleus ^{80}Sr . Rotation takes place around the vertical axis. a) ground-state band (h) (collective triaxial $\gamma = -10°$) $J = 10\hbar$. b) oblate band (A) $J = 10\hbar$. c) oblate collective band (a) ($\gamma \approx -55°$) $J = 10\hbar$. d) non-collective state $J = 16\hbar$. e) triaxial band (l) ($\gamma \approx -30°$) $J = 35\hbar$. f) triaxial band (C) ($\gamma \approx 15°$) $J = 44\hbar$.

Figure 4.8 Contour energy surfaces $\rho = 0.06 \text{fm}^{-3}$ for several configurations of the rotating nucleus ^{80}Sr . Rotation takes place around the vertical axis. g) axial band (s) $J = 51\hbar$. h) superdeformed band (j) $J = 80\hbar$. i) superdeformed band (z) $J = 80\hbar$. j) hyperdeformed band (z) $J = 80\hbar$. k) superdeformed band (z) $J = 92\hbar$ last point of the Yrast line. l) fissioning superdeformed (z) $J = 94\hbar$. (Note that the scale is not the same for all drawings)

References

[BFH85] P. Bonche, H. Flocard, P.H. Heenen et al., Nucl. Phys **A443** (1985) 39

[BFH87] P. Bonche, H. Flocard and P.H. Heenen, Nucl. Phys. **A467** (1987) 115

[DFK80] K.Davies, H. Flocard, S. Krieger et al., Nucl. Phys **A342** (1980) 111

[DSO87] R. Davie, D. Sinclair, S. Ooi et al., Nucl. Phys. **A463** (1987) 687

[EWR87] D.Eastham, P.Walker, J. Smith et al., Phys. Rev. **C36** (1987) 1583

[HAM82] J. Hamilton, Proc. Int. Summer School on HI Collisions, (1982) 287, E. Madurga and J. Lozano editors, Springer Verlag (Berlin)

[NDB85] W. Nazarewicz, J. Dudek, R. Bengtsson, et al., Nucl. Phys. **A435** (1985) 397

[RSC82] P. Ring and P. Schuck, The Nuclear Many-Body Problem, (1982) Springer Verlag (Berlin)

Aknowledgements

Collaborators to this work are P. Bonche (DPhT Saclay) and P.H. Heenen (ULB Bruxelles). We thank the CCVR for extended computing facilities. This research was partly supported by the NATO grant RG 85-0195. P.H. Heenen thanks also the FNRS for a computational grant.

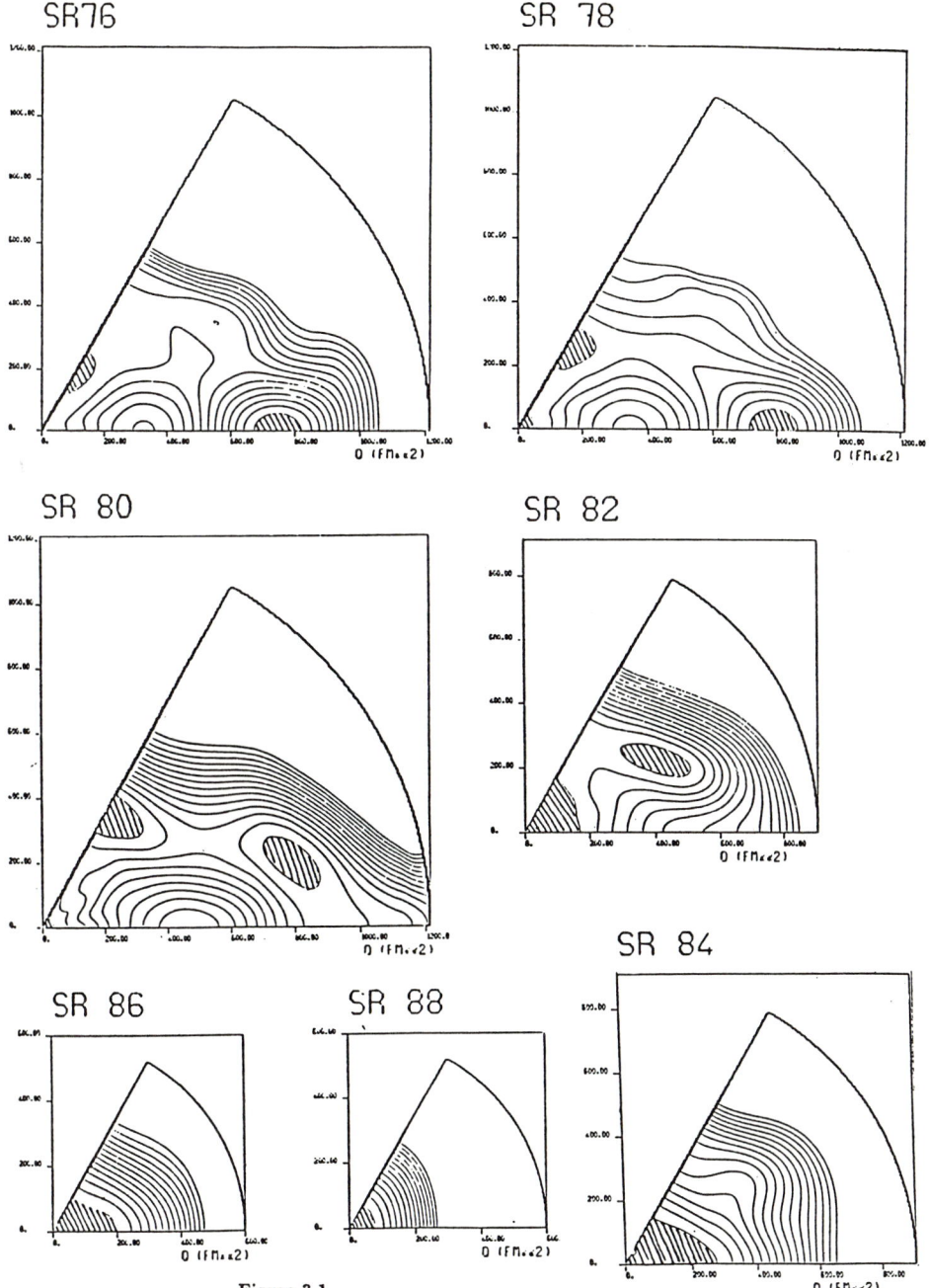

Figure 3.1

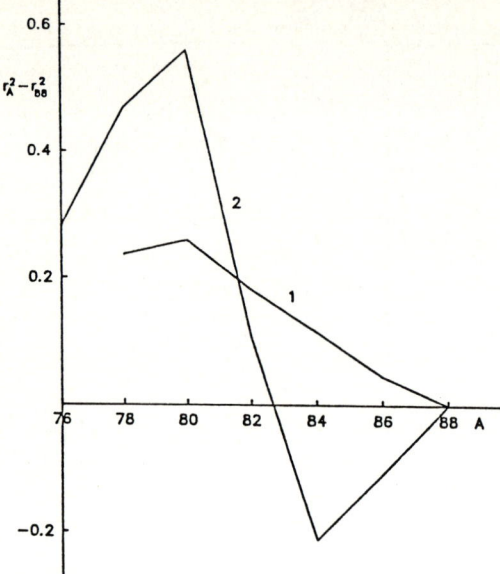

Figure 3.2

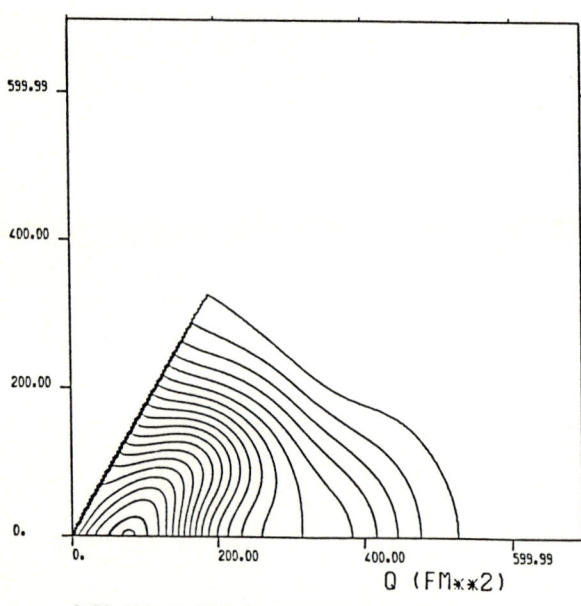

Figure 3.3

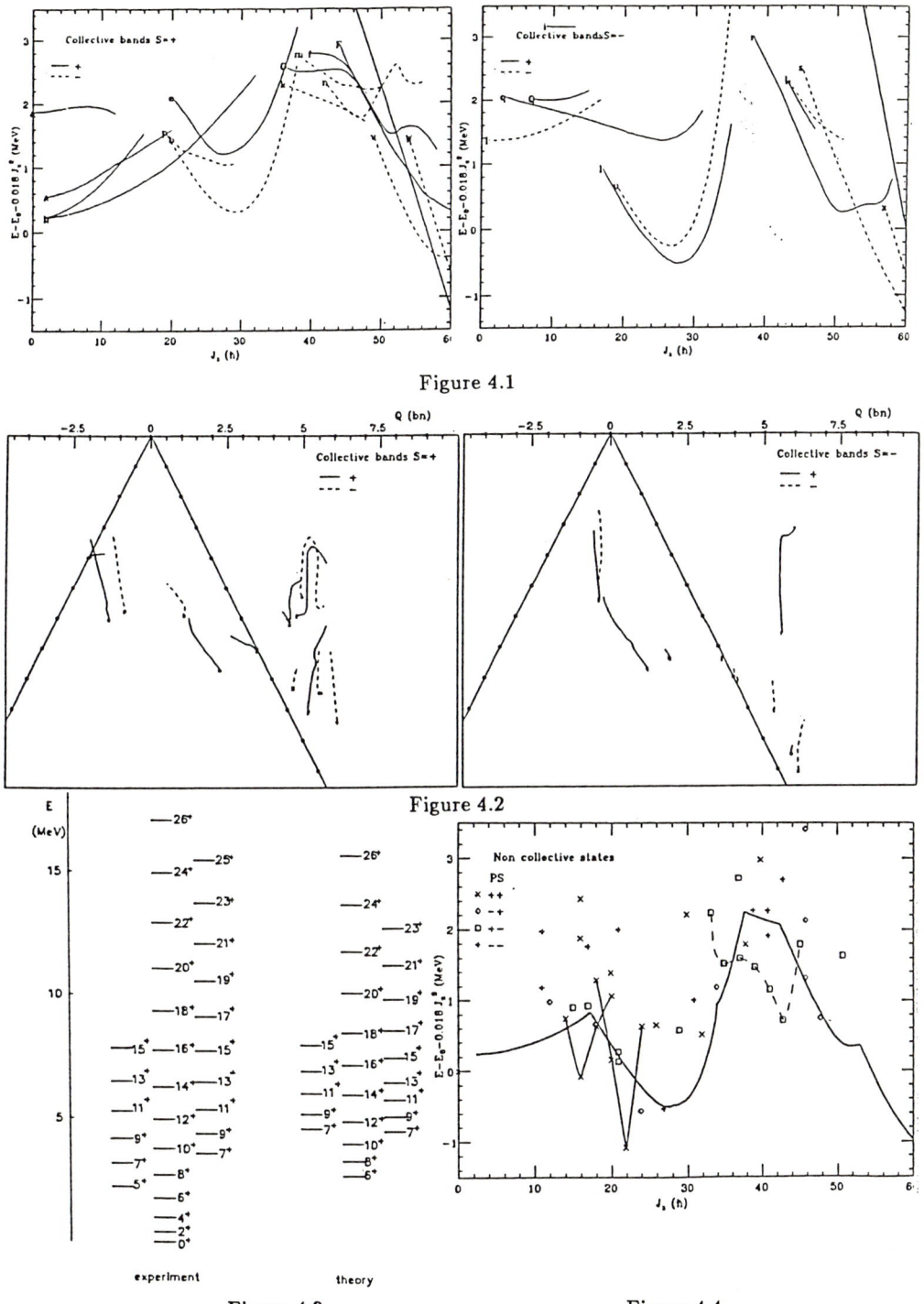

Figure 4.1

Figure 4.2

Figure 4.3

Figure 4.4

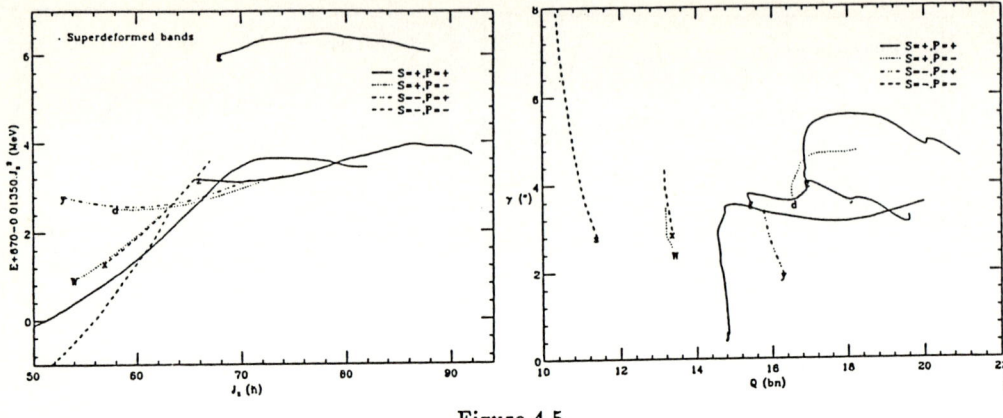

Figure 4.5

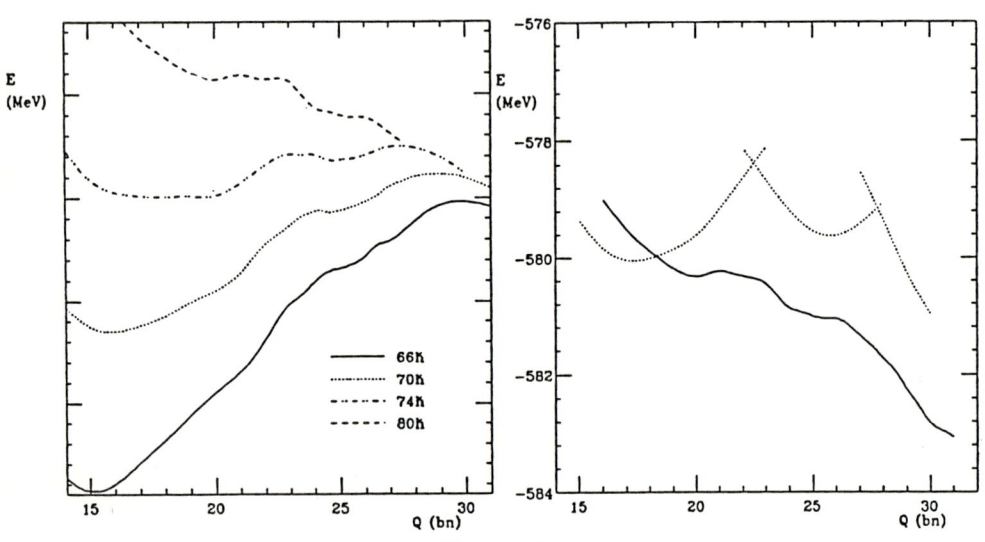

Figure 4.6

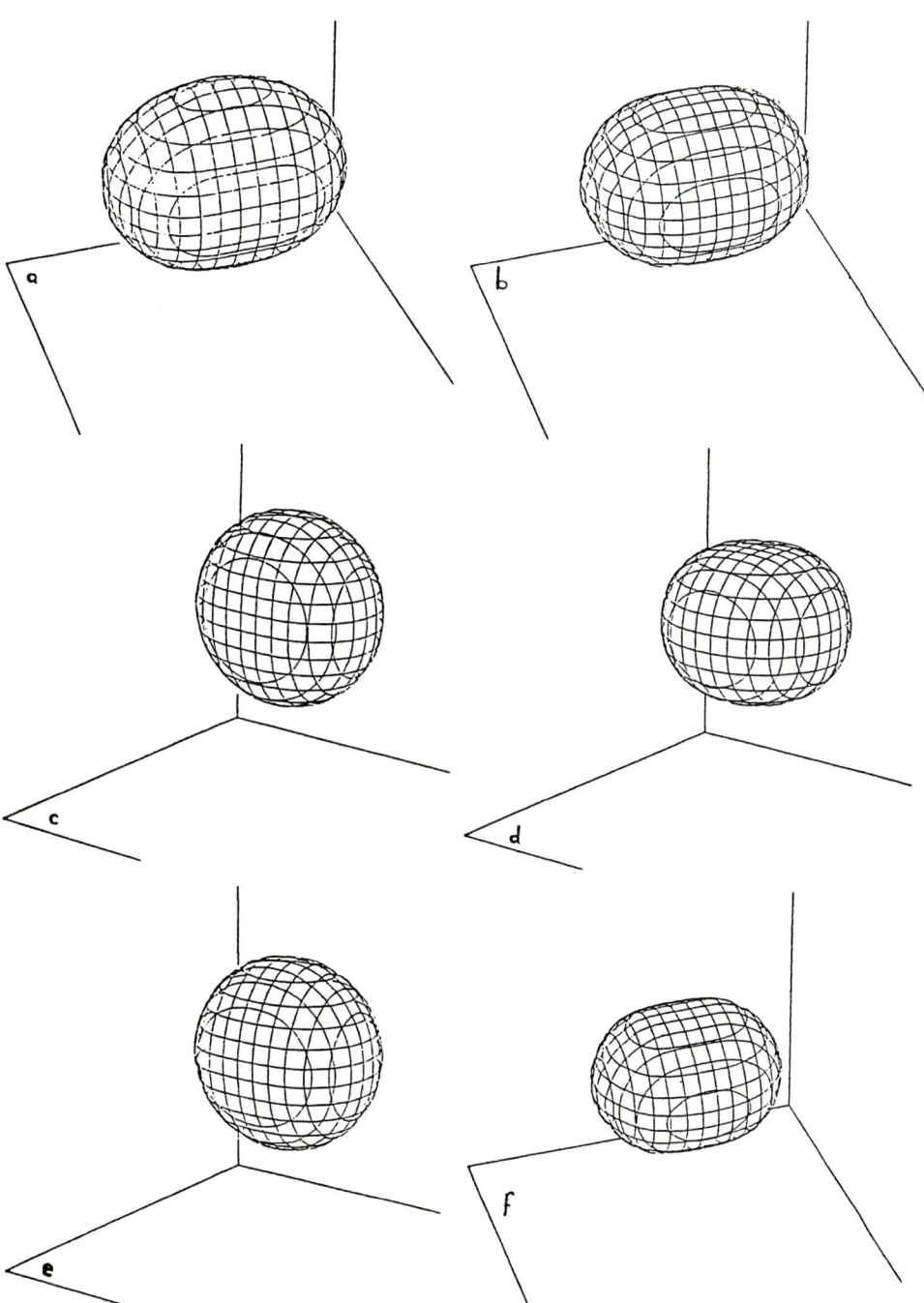

Figure 4.7

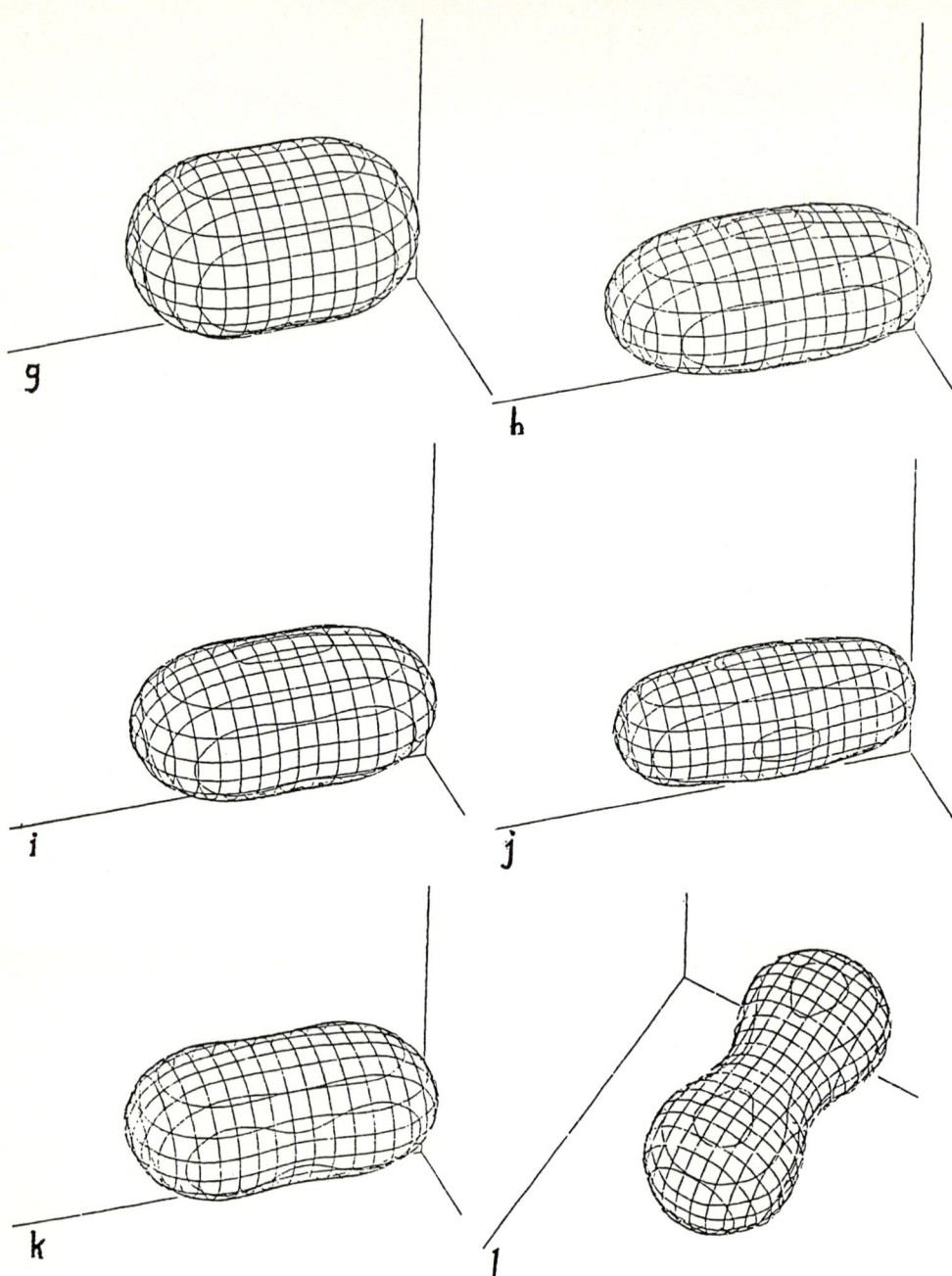

Figure 4.8

Part IV

Complete Spectroscopy

Complete Spectroscopy

P. von Brentano[1], A. Dewald[1], W. Lieberz[1], R. Reinhardt[1], K.O. Zell[1], and V. Zipper[2]

[1]Institut für Kernphysik der Universität zu Köln,
 Zülpicher Str. 77, D-5000 Köln 41, Fed.Rep.of Germany
[2]Silesian University, Institute of Physica,
 40–007 Katowice, Poland

Introduction:
In the field of nuclear structure physics with in-beam gamma-ray spectroscopy one can identify three main lines of investigation: High spin states, states at a high temperature and states in nuclei far from stability. In future we hope to reach all three together. With H.I. fusion reactions it has been possible to achieve two of them together, namely the study of high spin states in nuclei far from stability. In this talk we report on the study of states at higher temperature. Before discussing this matter, we will briefly asses the reasons why we want to study states at higher temperatures in nuclei, that are states which lie by 1-2 MeV or even higher above the yrast-line. One reason is that we have very little knowledge about such states. There are only a few nuclei in which even the first three 0^+, 2^+ or 4^+ levels are known. Yet significant tests of nuclear models such as e.g. the Interacting Boson Model [1,2], the Shell-Model or the Excited VAMPIR-Model [3] need states above the yrast line. A crucial test of nuclear models should not only consist in finding some of the predicted levels of this model but it must also confirm the absence of levels which are not predicted by the model. This brings the concept of complete spectroscopy into the discussion. A further incentive to study "all" nuclear levels in a given spin and energy window is that we can learn from the distribution of nuclear levels very much about nuclear chaos, a field which has become quite exciting in recent years as is discussed in the paper by Wladimir Paar [4]. In the following we will first discuss the most useful non-selective nuclear reactions. Then the notion of completeness is discussed and some examples of complete spectroscopy are given.

Non-selective reactions:

There are 3 groups of useful non-selective reactions.

1. Neutron induced reactions as the thermal (n,γ) reaction studied e.g. at ILL Grenoble, the average neutron resonance capture reaction (ARC) studied e.g. at BNL and the (n,n'γ) reaction. Of these the ARC method is the best example of a complete reaction. The thermal (n,γ) and the (n,n'γ) reactions may approach completeness. As an example for the (n,γ) work we give the level scheme of ^{178}Hf from the work of the Köln-BNL-ILL group [5]. This level scheme is indeed very rich and it contains 17 bands. With this investigation ^{178}Hf is now about as well known as ^{168}Er [6] which has become the darling of nuclear theorists.

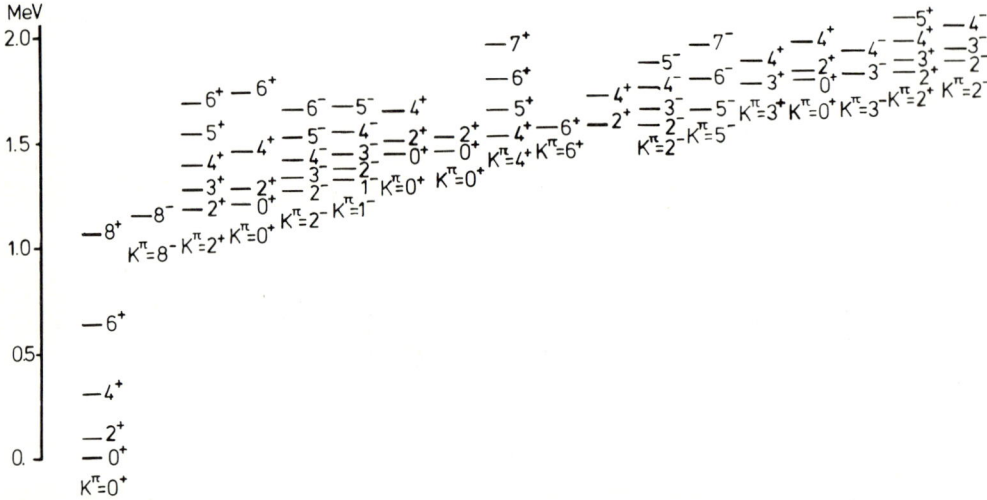

Fig. 1: Level scheme of ^{178}Hf measured with thermal neutron and average resonance neutron reactions by Haque et al. [5].

2. A second group of non-selective reactions are "direct" reactions with low projectile energies, very good energy resolutions of about $2*10^{-4}$ and low backgrounds obtained with magnetic spectrometers at Tandem Van de Graaff accelerators. Such work is carried out e.g. at the TU Munich by von Egidy [7]. It is clear that with a combination of (d,p), (d,t) and (p,p') reactions one obtains a large number of states and one may even approach completeness. Both the neutron reactions and the "direct" reactions populate, however, only a small spin window and need stable targets.

3. We will now focus the discusion on the third method namely in-beam

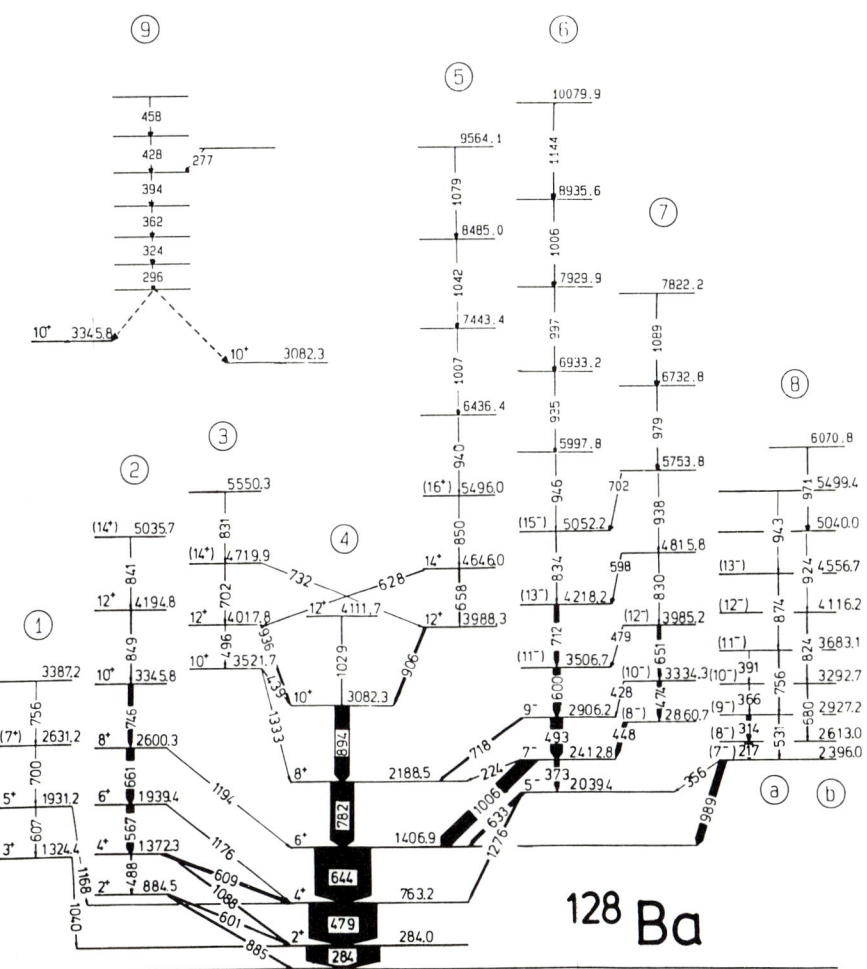

Fig. 2: Level scheme of ^{128}Ba as observed in the (^{22}Ne,4n) and in the (^{13}C,3n) reaction at 96 MeV and 53 MeV, respectively [9]. Note, that 6 levels were observed with spin 12, but only two levels with spin 2, 4 and 6, respectively. thus the (^{13}C,3n) as well as the (^{22}Ne,4n) compound nucleus reaction is incomplete in the window 0<I<7, E<3 MeV.

gamma-ray spectroscopy with fusion reactions. Since the discovery of this method by Morinaga and Gugelot [8] there has been a tremendous progress and this method has become one of the most powerful tools in nuclear structure physics. In particular the use of escape suppressed Germanium detector arrays has led to a vast increase in the amount of data obtained by it. A typical example is given in fig. 2 which shows the decay scheme of ^{128}Ba obtained with the (^{13}C,3n) and (^{22}Ne,4n) reactions [9] using the accelerators at Köln and at Berlin with the OSIRIS-spectrometer [10]. This level scheme of ^{128}Ba contains 13 bands which comes close to the number of bands observed by (n,γ) in ^{178}Hf. It is interesting to note that in these reactions we observe 6 spin I=12 states whereas only 2 spin I=2 states are seen. Clearly the fusion reaction is not complete for I=2 states. On the other hand we observe a great number of I=12 states in a very tiny energy window of 250 keV above the yrast-band. It is remarkable that although this reaction was reinvestigated with the OSIRIS 12-detector coincidence array still no more than two I=2 states were observed. Thus the question arises whether we can find fusion reactions which populate more low spin states. In order to reach this goal it is crucial to produce the compound nucleus with low spin. This can be done by taking light projectiles and projectile energies near the fusion barrier. Ideal probes are the (p,n) and the (α,n) reactions [11,12]. A summary of some nuclei which were recently studied by us at Köln with the (α,n) reaction is given in table 1. Particular impressive is the ^{123}Te (α,n)^{126}Xe reaction [12] which was studied in Köln using the OSIRIS spectrometer

Table 1: Nuclei investigated in Cologne with fusion reactions at the barrier.

nucleus	reference	reactions	number of observed levels with		
			J^π	J	all
^{73}Se	[13]	(α,n)	22	26	56
^{126}Xe*	[14]	(α,n)	25	40	150
^{126}Xe*	[15]	(^{13}C,3n)	19	24	60
^{128}Xe	[16]	(α,n)	20	35	75
^{143}Nd	[18]	(α,n)	30	40	64

* measured with the OSIRIS spectrometer

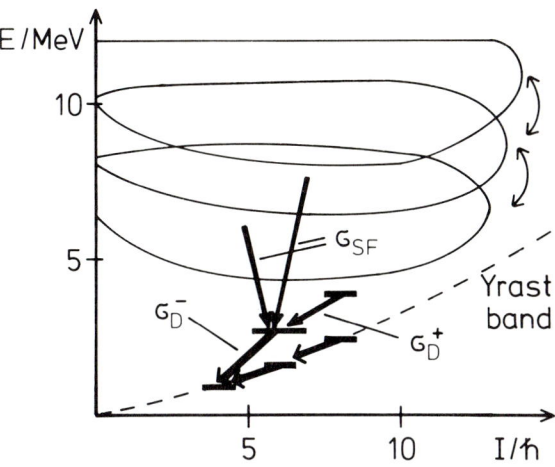

Fig. 3: This figure shows in a schematic way the decay of an excited nucleus built up by a (HI,xn) reaction. After neutron evaporation the nucleus decays mainly with a few fast E1-transitions to continuum states forming new population clouds or to discrete levels. The average multiplicity of these E1-transitions correspond to the average number of clouds. The population of a discrete level consists of two parts:
   a) Population coming from other discrete levels (σ_D^+)
   b) Side feeding population (σ_{SF}) from continuum states.
The side feeding cross section of a certain level is the difference of the cross section for depopulation (σ_D^-) and that one for discrete population (σ_D^+) of this level:
$\sigma_{SF} = \sigma_D^- - \sigma_D^+$

and where 150 levels were observed with the (α,n) reaction and 60 levels with the (^{13}C,3n) reaction [15] of which 28 were also observed with the (α,n) reaction. So clearly the (α,n) reaction gives a most rich level scheme. By using in addition more heavy projectiles one can extend the level scheme to much higher spins.
One problem with the (α,n) reaction is that one is able to give unique spin and parity assignments to only a part of the observed levels as is shown in table 1. This has three reasons:
a) It is difficult to assign unique spin and parity values by any in-beam γ-spectroscopy method, if there are no bands.
b) It is difficult to group states at high excitation energy above the yrast-line into bands because the inband γ-transitions are strongly

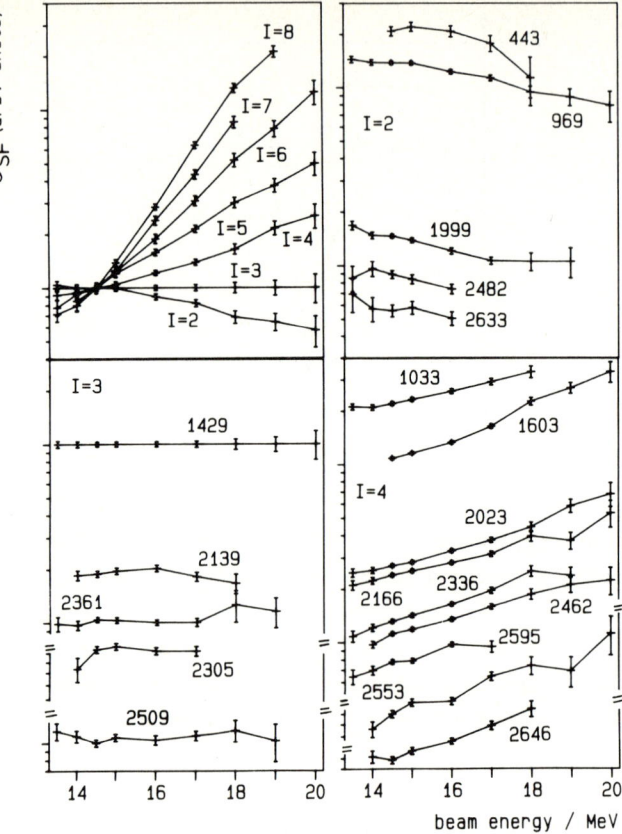

Fig. 4a: Energy dependence of side feeding cross section $\sigma_{SF}(I, E_x, E_p)$ versus projectile energy E_p for levels with different spins and comparison of side feeding excitation functions of levels for several spin 2, 3 and 4 states. One observes the following properties:

The slope of the side feeding excitation function depends on spin I and in general the side feeding intensity decreases with increasing level energy.

All data shown in figure 4a and 4b are compiled from ^{128}Xe [16].

disfavored by the energy factor.
c) many levels observe in the γ-spectrum of the $(\alpha,n\gamma)$ reaction are dublets or even triplets.
One of the most powerful method, which we have used for spin

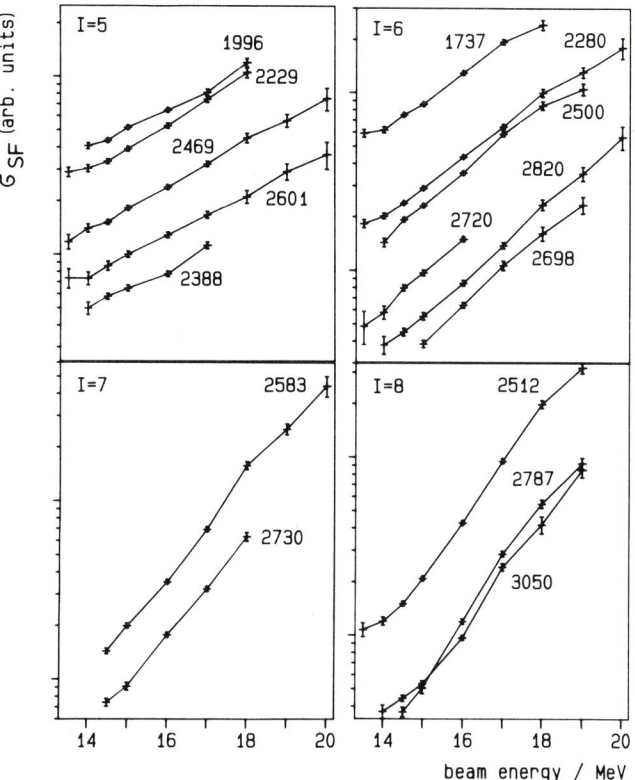

Fig. 4b: Side feeding excitation functions of levels with spin 5, 6, 7 and 8. See also fig. caption 4a.

determination of a level k has been the investigation of the side feeding cross section which comes directly from statistical gamma rays as is shown in fig. 3. We obtain σ_{kSF} from the experimental data by subtracting the cross section $\sigma_{kD}{}^+$ for discrete line population of the level (k) from the cross section $\sigma_k{}^-$ for (discrete) depopulation of the level (k): $\sigma_{kSF} = \sigma_{kD}{}^- - \sigma_{kD}{}^+$. The fact that σ_{SF} has a statistical origin leads to the assumption that σ_{SF} is independent of the configuration of the state and depends only on a few quantum numbers as e.g. E, I, π, K and the projectile energy E_P. As an example we give σ_{SF} from the ^{125}Te $(\alpha,n)^{128}$Xe reaction in fig. 4a,b [16]. We note that

$$R_k = \log \sigma_{kSF}(E_k, I, E_P)$$

has the following properties A, B and C.

A: States with the same spin have the same slope (d R_k/d E_P)

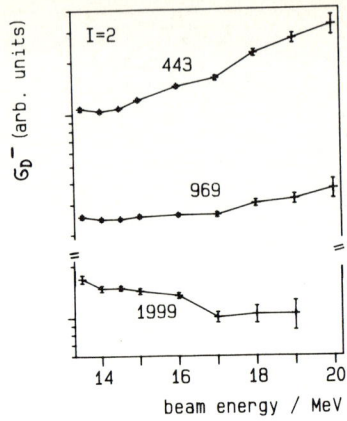

Fig. 5: Excitation functions of total cross section $\sigma_D^-(2^+, E_x, E_p)$ of gamma ray transitions depopulating 2^+ states of ^{128}Xe [16] with 443 keV, 969 keV and 1999 keV excitation energy, respectively. Clearly, the slope of all three excitation functions is vastly different, although all levels have the same spin. The levels were populated via the (α,n) reaction.

B: States with different spins have different slopes
C: R_k decreases (about) linearly with the excitation energy E_k.
These properties allow the determination of spins from $\sigma_{SF}(E_p)$. We call this method "spin meter". The property C is the basis of completeness properties of the reaction. The fact that the total fusion cross section σ_k^- to a level k depends strongly on E_p and I is widely used for spin determination. It is, however, a dangerous tool because σ_k^- depends on the level scheme and various spin 2 levels can have a very different energy dependence. An example is given in fig. 5. In order to see whether our experimental results on the properties of the side feeding cross section are of a general nature we have performed calculations with the computercode "CASCADE" [17]. The results are given in fig. 6, which shows that one can reproduce the spin dependence of σ_{SF} very well by this code and we confirm also the other properties A, B and C which we have listed above from such calculations.

Completeness:
When we compare level energies to those given by a theoretical model the notion of completeness gets important if the number of levels with

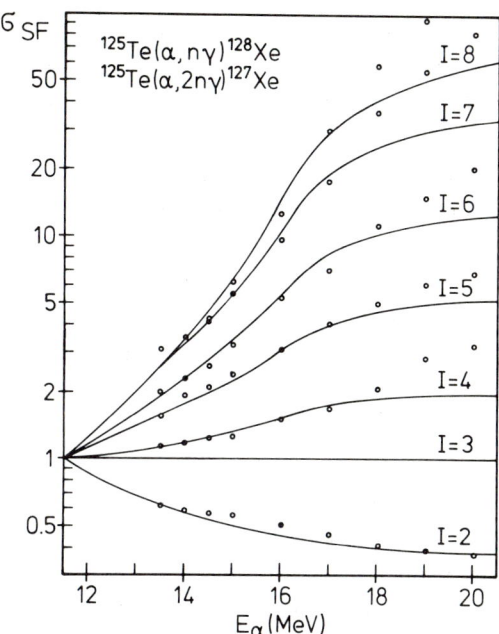

Fig. 6: A comparison of calculated side feeding excitation functions (solid lines) and the experimental ones (circles) for states with different spins. All excitations functions were normalized to the same beam energy E_α. The calculations were performed with the code "CASCADE" [17].

the same spin and parity is large in a certain energy interval. Then it is not possible to compare the levels according to their energies and one does not know which level corresponds to which theoretical one. Only the spectroscopy of all states with the same spin and parity in the considered energy interval allows a meaningful one by one comparison of experimental to calculated levels. Yet a discussion of this concept is often avoided. Experimentalists make often no comparison with theory so they avoid the issue and the theorists consider this discussion to be the job of the experimentalist. Both groups tend to avoid the word "complete" because one will be immediately confronted with the question: - How complete is complete and are there exceptions?- So it seems safer to avoid the issue, but if we avoid it too much, we make the comparison between experiments and nuclear models in cases meaningless. Thus there seems to be no way to avoid a discussion of completeness and we have to face it. Completeness is an experimental concept and there are experimental errors involved in the statement of completeness. Clearly one should state completeness with qualifications and one might give it a confidence level. The most simple notion of completeness is to claim that one knows "all" levels in a given spin window $I_1 \leq I \leq I_2$ and energy window $E \leq E_0$.

The most reliable method in complete spectroscopy is the average resonance capture reaction (ARC) in which we observe the spectrum of the primary transitions from a group of capture states to various final states. Any state in the given window should be populated and from the absence of a line in the spectrum one can infer the absence of a level in the window.

The question of possible "completeness" with direct reactions is more complicated. In this case we have to rely on the fact that nuclear levels are mixing to a certain extent and each level will contain small admixtures of single particle levels. Clearly direct reactions are bound to be somewhat incomplete, but if a combination of reactions is studied with low background and high statistics they may become very powerful tools.

The fusion reaction at the barrier has promising completeness properties. Namely the side feeding cross section σ_{sF} decreases monotonically with increasing excitation energy E. Thus if we have observed a particular state (0) with energy E_0 spin I_0, which can be depopulated by N_0 transitions, and with a sidefeeding σ_{sF} sufficiently large so that σ_{sF}/N_0 is larger than the intensity of the weakest transition, which could be assigned in the experiment, we should observe all states (i) with the same spin, which lie below this state ($E_i < E_0$). This statement has to be qualified, however. Often it is possible to assign lines only in the coincidence spectra. In this case we might have to allow for additional levels which directly decay to the ground state or to a long lived isomer.

This point should be made explicit in a suitable formulation of completeness in this case. On the experimental side such an exception can be lifted if by the use of additional channel selective devices, such as a neutron detectors, all lines in the single spectra are identified. An additional problem is given by the occurence of dublets. Again an appropriate qualifying statement should be made in the statement of completeness. With these qualifications we still can obtain very important completeness information from fusion reactions at the barrier and this is the only way by which one can obtain such information for nuclei off stability.

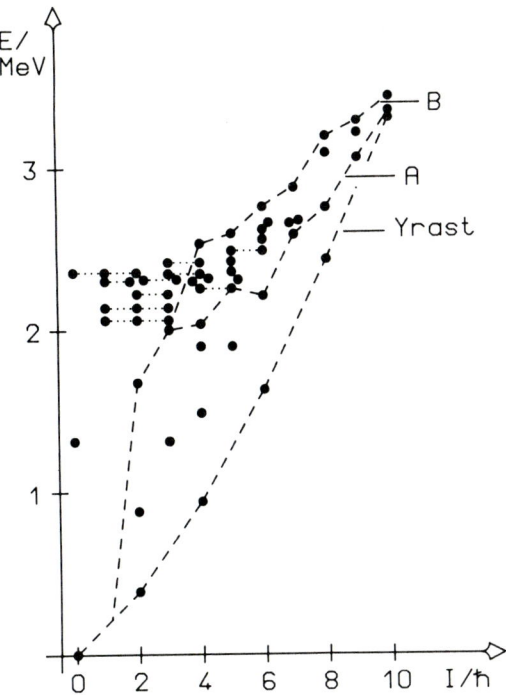

Fig. 7: This energy-spin diagram shows the levels of ^{126}Xe as obtained in the (α,n) reaction [14]. Only levels, which have unique spin assignment (single point) or a spin estimation (two or three points connected with a dotted line), are included. Line B defines the energy interval as a function of spin in which the spectroscopy should be complete. If experimental qualifications are taken into account a reduced energy window is obtained, which is indicated by line A.

Applications:

We want to mention finally a few applications of complete spectroscopy. The first is the investigation of particle phonon interaction by the study of complete particle phonon multiplets. In this respect the Köln-Krakow-Bucharest-collaboration have investigated the 3^-x$f7/2$ multiplets in the N=83 isotones [18]. These investigations have not only established the complete octupole particle multiplets in these nuclei but they have also given us a full understanding of the splitting of these multiplets which can be described rather well by the quadrupole-quadrupole interaction. A second type of application is the study of the dynamical symmetries of the Interacting Boson Model

[1,2]. As an example of such work we show in fig. 8 a comparison of the low lying positive parity levels of ^{126}Xe with the predictions of the O(5) dynamical symmetry [14]. Clearly we find all levels predicted by the O(5) dynamical symmetry and there are no additional levels in the window: 1<I<9; E<2 MeV . Of particular interest in the establishment of the O(5) symmetry for the nucleus ^{126}Xe are the finding of a new K=4 band and of a new quasi-beta band which was observed up to a spin 6. For the test of the O(5) symmetry it is very important to obtain the transition ratios of the γ-decay from highly excited states. These transition ratios are given in table 2. We find that all experimental transition ratios which are forbidden by the O(5) symmetry have less than 10% of the strongest transition. We remark that even these high lying states have a rather interesting pattern of the branched electromagnetic decay. A similar observation has been recently made concerning the γ-decay of the $2^+{}_5$ and $2^+{}_4$ levels in ^{196}Pt by Casten et al [19]. This transition ratios are a striking proof of the validity of the O(5)

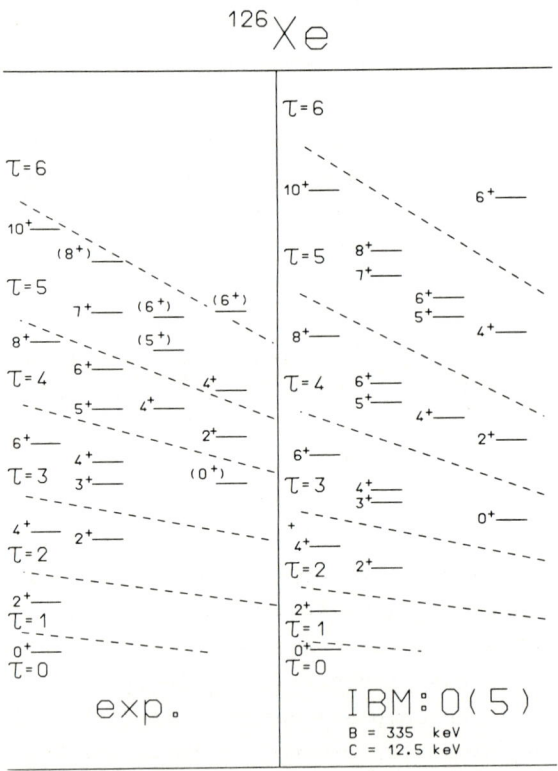

Fig. 8: Energy levels of ^{126}Xe compared to the predicted IBM-1 O(5) symmetry levels. The theoretical levels were calculated with the expression: $E(N=\sigma,\tau,J) = B\tau(\tau+3) + CJ(J+1)$

Table 2: Experimental and O(5) transition ratios of ^{126}Xe. The experimental ratios are obtained from the coincidence measurement.

Level	Trans.	$B(E2)^{Exp*}$	$B(E2)^{(O5)}$
2_2^+	$\to 2_1^+$	100.	100
	$\to 0_1^+$	1.4	0
$(0^+)_2$	$\to 2_2^+$	100.	100
	$\to 2_1^+$	8.0	0
3_1^+	$\to 2_2^+$	100.	100
	$\to 4_1^+$	33.4	40
	$\to 2_1^+$	2.4	0
4_2^+	$\to 2_2^+$	100.	100
	$\to 4_1^+$	83.0	91
	$\to 2_1^+$	0.4	0
2_3^+	$\to 0_2^+$	100.	100
	$\to 3_1^+$	52.7	125
	$\to 4_1^+$	1.5	0
	$\to 2_2^+$	2.8	0
	$\to 2_1^+$	0.1	0
	$\to 0_1^+$	0.1	0
4_3^+	$\to 3_1^+$	40.2	115
	$\to 4_2^+$	100.	100
	$\to 4_1^+$	3.9	0
	$\to 2_2^+$	2.7	0
5_1^+	$\to 3_1^+$	100.	100
	$\to 4_2^+$	50.4	46
	$\to 6_1^+$	72.7	45
	$\to 4_1^+$	2.8	0
$4_4^{(+)}$	$\to 2_3^+$	100.	100
	$\to 4_1^+$	10.2	0
	$\to 2_1^+$	1.2	0
6_2^+	$\to 4_2^+$	100.	100
	$\to 6_1^+$	72.2	47
	$\to 4_1^+$	0.4	0

* M1-portions are neglected

symmetry for these nuclei.

Conclusion:

In this paper we have discussed various non-selective reactions with

which one can study many states with the same spin and parity. We have
shown that fusion reactions at the barrier are a very powerful tool in
this respect and they are the only methods which can be used for
nuclei far from stability. Finally we have given a few applications
which show that levels at higher excitation energy can have a very
definite and regular electromagnetic decay pattern, which is not at
all "chaotic". Thus the study of these levels seems to open an
exciting field for the future.

Acknowledgements:
We thank R.F. Casten, Prof. A. Gelberg, A. Granderath, S. Freund and
J. Wrzesinski for discussions. This work was supported by the BMFT
under Contract No. 06OK272.

References

[1] F. Iachello, A. Arima, The interacting boson model,
Cambridge University Press 1987

[2] R.F. Casten, D.D. Warner, the interacting boson approximation,
submitted to reviews of modern physics

[3] K.W. Schmid, F. Grümmer, M. Kyotoku, A. Faessler, Nucl. Phys.
A452 (1986) 493

[4] W. Paar, in these proceedings

[5] A.M.I. Haque, R.F. Casten, I. Förster, A. Gelberg, R. Rascher,
R. Richter, P. von Brentano, G. Berreau, H.G. Börner, S.A. Kerr,
K. Schreckenbach and D.D. Warner, Nucl. Phys. A455 (1986) 231

[6] W.F. Davidson, D.D. Warner, R.F. Casten, K. Schreckenbach,
H.G. Börner, J. Simić, M. Stojanović, M. Bogdanović, S. Koički,
W. Gelletly, G.B. Orr, M.L. Stelts, J. Phys. G 7 (1981) 455,
J. Phys. G 7 (1981) 843

[7] T. von Egidy et al., in: Jahresbericht der TU München 1982 und
1983 (unpublished)

[8] H. Morinaga, P. C. Gugelot, Nucl. Phys. 46 (1963) 210

[9] H. Wolters, K. Schiffer, A. Gelberg, J. Eberth, R. Reinhardt
     K.O. Zell, P. von Brentano, D. Alber, H. Grawe,
     Z. Phys. A328 (1987) 15

[10] R.M. Lieder, H. Jäger, A. Neskakis, T. Venkova, C. Michel,
     Nucl. Instr. Meth. 220 (1984) 363

[11] A. Dewald, R. Reinhardt, J. Panqueva, K.O. Zell,
     P. von Brentano, Z. Phys. A 315 (1984) 77

[12] P. von Brentano, A. Dewald, A. Gelberg, W. Lieberz,
     R. Reinhardt, J. Panqueva, K.O. Zell, Int. Symp. 1984, p. 189
     In-beam nuclear spectroscopy, Debrecen, Hungary
     edited by Z. Dombrádi, T. Fényes

[13] F. Seiffert, R. Wrzal, K.O. Zell, K.P. Schmittgen,
     R. Reinhardt, W. Lieberz, A. Dewald, A. Gelberg,
     P. von Brentano, in these proceedings

[14] W. Lieberz et al., to be published

[15] W. Lieberz, S. Freund, A. Granderath, A. Gelberg, A. Dewald,
     R. Reinhardt, R. Wirowski, K.O. Zell, P. von Brentano,
     Z. Phys. (1988)

[16] R. Reinhardt, A. Dewald, A. Gelberg, W. Lieberz, K. Schiffer,
     K.P. Schmittgen, K.O. Zell, P. von Brentano, Z. Phys. A 329
     (1988) 507 and to be published

[17] F. Pühlhofer, Nucl. Phys. A 280 (1977) 267

[18] L. Trache, J. Wrzesinski, C. Wesselborg, D. Bazzacco,
     R. Reinhardt, C.F. Moore, P. von Brentano, G.P.A. Berg,
     W. Hürlimann, I. Katayama, J. Meissburger, J.G.M. Römer,
     J.L. Tain, Phys. Lett. 131B (1983) 285 and to be published

[19] R.F. Casten, J.A. Cizewski, Phys. Lett. B 185 (1987) 293

^{73}Se Investigated by the $(\alpha,n\gamma)$ Reaction

F. Seiffert, R. Wrzal, K.O. Zell, K.P. Schmittgen, R. Reinhardt, W. Lieberz,
A. Dewald, A. Gelberg, and P. von Bretano

Institut für Kernphysik der Universität zu Köln,
Zülpicher Str. 77, D-5000 Köln 41, Fed. Rep. of Germany

In order to extend our experimental investigations on the odd Se isotopes we used the ^{70}Ge$(\alpha,n)^{73}$Se reaction to populate medium spin states of the ^{73}Se nucleus. The α-beam was provided by the Cologne FN-Tandem accelerator.

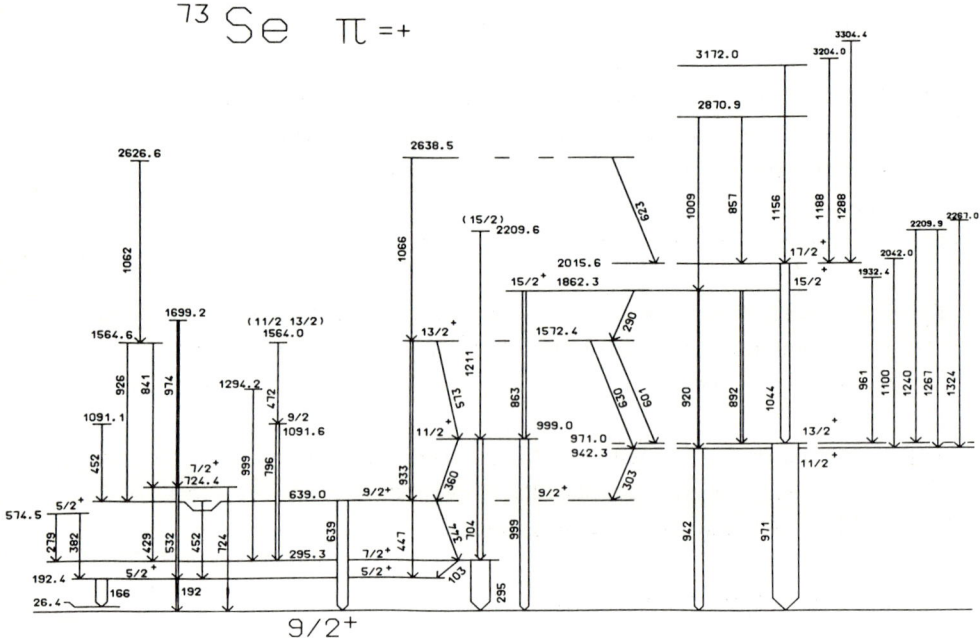

fig.1: partial levelscheme of ^{73}Se with positive parity states

The beam energy was 20 MeV except for the excitation function measurements, for which γ-singles spectra were taken at 14, 16, 18 and 20 MeV. Enriched (98.45%) ^{70}Ge backed targets and germanium detectors with an active volume of about 45 cm^3 were used. $2.5 \ast 10^6$ $\gamma\gamma$-coincidences were stored on magnetic tapes and sorted off line into a $4k \times 4k$ matrix. The beam energy was 20 MeV. In a γ angular distribution measurement Gamma-singles spectra were taken at 17°, 30°, 40°, 55°, 65° and 90° relative to the beam axis.

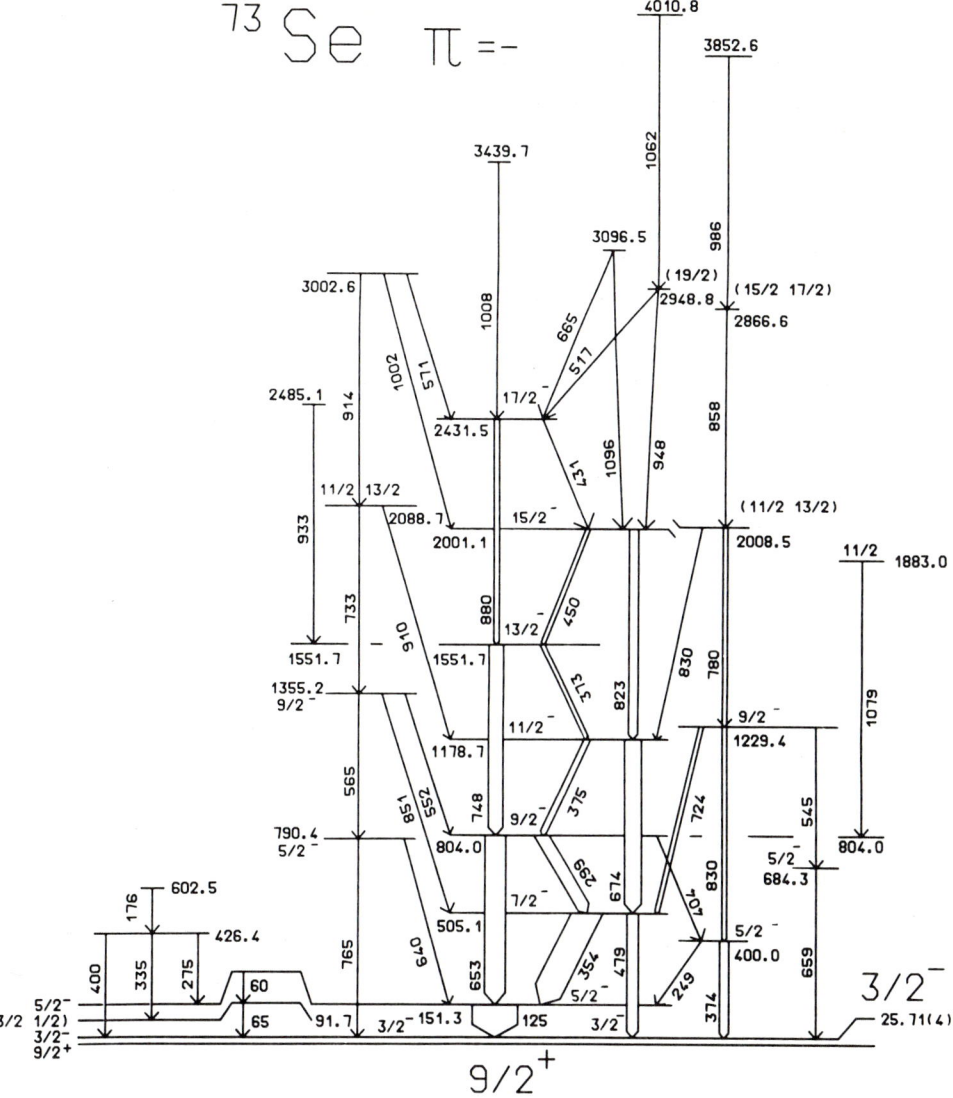

fig.2: partial levelscheme of ^{73}Se with negative parity states

The function $W_{exp}(\Theta) = A_0 + A_2 P_2(\cos\Theta) + A_4 P_4(\cos\Theta)$ was fitted to the data points. The resulting A_2/A_0 and A_4/A_0 coefficients are given in Table 1. For a linear polarisation measurement the γ-rays scattered in a germanium detector were detected by means of two other germanium detectors. The latter were placed perpendicular and parallel to the beam axis. To suppress background we set an off line gate to suppress events, where the angle between the first and the *Compton* scattered photon was not within the range close to 90°. Using this gate we were able to suppress all those events produced by photons which were not scattered within the first detector (scatterer) and analysed in one of the other detectors. To determine the multipolarity of the transitions we used a grid-search method to minimize the following χ^2 :

$$\chi^2 = \sum_i \left(\frac{W_{exp}(\Theta_i) - W_{th}(\Theta_i)}{\Delta W_{exp}(\Theta_i)}\right)^2 + \left(\frac{Pol_{exp} - Pol_{th}}{\Delta Pol_{exp}}\right)^2$$

From W_{th} we obtained the multipolarity and the mixing ratios δ (Table 1.). We propose the level scheme which is splitted in two parts (fig.1 ,fig.2). It extends the level scheme of ^{73}Se reported previously by Zell et al. [1] to 96 transitions and 56 levels. Due to the excitation function data, the multipolarities and the polarisation data we were able to assign unique spin for 26 levels and for 22 of them also unique parity. All parities given in the first part of the levelscheme are positive and all parities given in the second part are negative. We could not find any connection between both parts. Two of the negative parity bands are *strongly coupled*. Calculations made by Dewald et al. [2] describe these bands in the framework of a rotor plus particle model. These calculations were able to reproduce the tendencies of the level energies and of the $B(E2)$ values using $\beta \approx 0.3$. Due to the new mixing ratios of the intraband transitions we could derive new or better $B(E2)$ values for these transitions (Table 1.) Due to a change of $\beta-$ and $\gamma-$ deformation the coupling behavior of positive parity bands in Se isotopes changes (fig.4). We used the *AROVM2 Asymmetric Rotor with Variable Moment of inertia* model [3] to describe states of positive parity in ^{73}Se. The best agreement we obtained so far is shown in fig.3. The deformation parameters β and γ were found to be 0.3 and 32°, respectively. At least the ordering of the levels is given correctly though the spacings are not so well reproduced. Dewald et al. [2] tried to describe the positive parity states using a symmetric rotor model. They found $\beta = 0.2$ and $\gamma = 0°$. But so far no calculation is available which gives good agreement between experiment and theory for these states. To throw more light on the deformation of positive- and

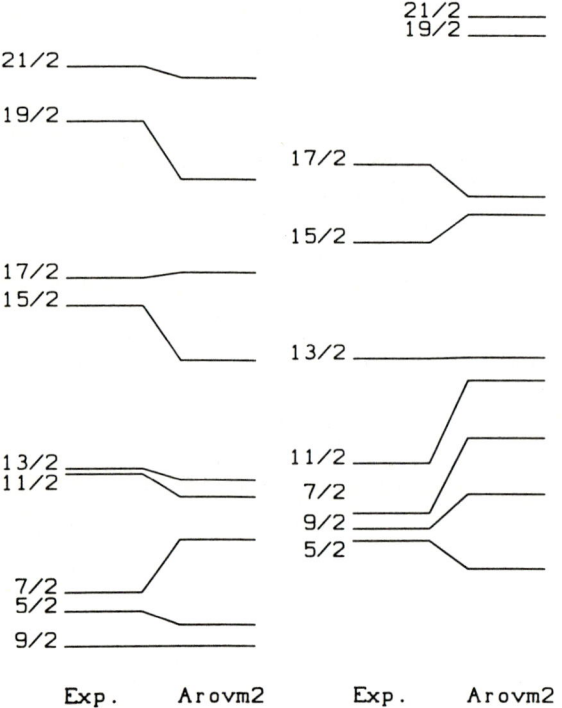

fig.3: calculation for positive parity states in ^{73}Se

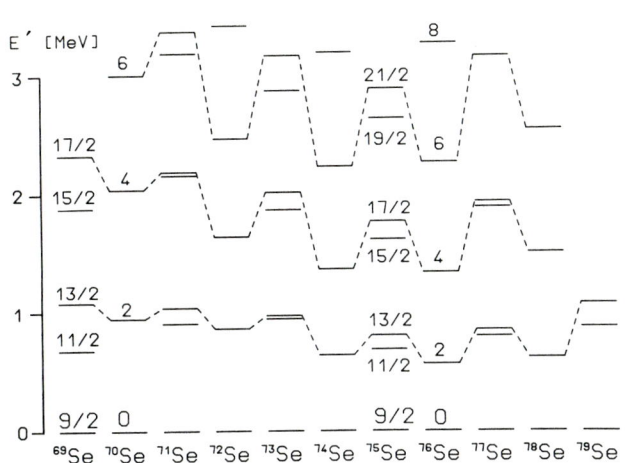

negative parity states in ^{73}Se new calculations using AROV-M2 with different core energies and the model of Larsson, Leander and Ragnarsson [4] are in process.

fig.4: Comparison of positive parity bands in the Selenium isotopes

E_x [keV]	E_γ [keV]	δ_{exp}	B(E2) $e^2 fm^4$	A_2/A_0	A_4/A_0
151.3	125			-0.431(1)	0.015(1)
505.1	479	$-0.00^{+0.02}_{-0.02}$	1260^{+253}_{-240}	0.230(4)	-0.084(5)
	354	$-0.430^{+0.25}_{-0.67}$	2450^{+2580}_{-2500}	-0.622(1)	0.025(2)
804.0	653	$-0.00^{+0.02}_{-0.02}$	1080^{+120}_{-119}	0.282(3)	-0.103(3)
	299	$-0.240^{+0.38}_{-1.00}$	2080^{+8640}_{-2100}	-0.554(3)	0.024(3)
1178.7	674	$-0.00^{+0.02}_{-0.02}$	1410^{+379}_{-375}	0.293(3)	-0.105(4)
	375			-0.583(2)	0.039(3)
1551.7	748	$-0.059^{+0.05}_{-0.05}$	1540^{+288}_{-261}	0.255(3)	-0.100(4)
	373	$-0.3^{+0.91}_{-1.26}$	1380^{+8670}_{-1400}	-0.491(7)	0.033(9)
971.0	971	$-0.00^{+0.02}_{-0.02}$	726^{+55}_{-55}	0.268(2)	-0.097(2)
2015.6	1044	$-0.15^{+0.05}_{-0.05}$	> 940	0.092(5)	-0.027(6)

Table 1.

Mixing ratios δ and B(E2) values derived from them in ^{73}Se

References
[1] K. O. Zell, B. Heits, W. Gast, D. Hippe, W. Schuh, and P. von Brentano, Z. Phys. **A 279** (1976) 373
[2] A. Dewald, A. Gelberg, U. Kaup, R. Richter, K. O. Zell, and P. von Brentano, Z. Phys. **A 326** (1987) 509
[3] H. Toki and A. Faessler, Nucl. Phys. **A 253** (1975) 231.
[4] S. E. Larsson, G. Leander and I. Ragnarsson, Nucl. Phys. **A 307** (1978) 189.

Complete Spectroscopy of ^{87,88,89}Sr with (n,γ) and (d,p) Reactions? *

C. Winter[1], B. Krusche[1], K.P. Lieb[1], G. Hlawatsch[2], T. v. Egidy[2], F. Hoyler[3], and R.F. Casten[4]

[1]II. Physikalisches Institut der Universität Göttingen, D-3400 Göttingen, Fed. Rep. of Germany
[2]Physik Department TU München, D-8046 Garching, Fed. Rep. of Germany
[3]Institut Laue-Langevin, F-38046 Grenoble, France
[4]Brookhaven National Laboratory, Upton, NY 11973, USA

I. Introduction

Over the recent years the nuclear structure around the N=50 shell closure, which is very pronounced in the strontium and zirconium isotopes, has been the subject of extensive experimental and theoretical work[1-5]. On the proton side Z=38 and Z=40 provide fairly closed sub-shells[1]. In the strontium isotopes the $1g_{9/2}$ neutron shell is closed at ^{88}Sr, supplying relatively pure neutron-hole and neutron-particle states with large spectroscopic factors in ^{87}Sr and ^{89}Sr, as well as core-coupled states. The mass region is thus ideally suited to examine the transition from a correlated to an uncorrelated (chaotic?) excitational behavior[6,7]. These two types are characterized e.g. by the density of excited states, the transition strengths, and the spectroscopic factors observed in transfer reactions.

We conducted (n,γ) and (d,p) reactions[8-10] leading to ^{87,88,89}Sr in addition to ^{88}Sr(d,t)^{87}Sr and 24 keV neutron capture in ^{88}Sr. The vast amounts of data were used to analyze the level densities and transition strengths in the statistical model[11]. Correlations were found between the selective population of states dominating the (d,p) process, and the statistical decay of the compound state formed after thermal neutron capture. In the course of the analyses we established essentially complete level schemes (within limited spin windows centered around the spin of the capture state) up to around half the neutron binding energy. The E1 partial decay widths were found to agree with the results of giant resonance calculations[12]. We intend to give an overview of the experiments and outline the results of the model calculations.

II. Experiments

(n,γ): The 96-99.99% enriched targets were exposed to a thermal neutron flux of $5.5*10^{14}$ n cm^{-2} s^{-1}, which is available close to the core of the high flux reactor of the Institut Laue-Langevin. The γ-spectra following neutron capture were measured with Ge- and pair spectrometers over periods of between 3 and 6 days. The excellent energy resolution of 1.9 keV at 1.33 MeV and 4.2 keV at 6 MeV and the good statistics enabled us to identify all transitions contributing significantly to the γ-flux in the final nuclei. The spectra were energy and intensity calibrated with respect to the reaction ^{35}Cl(n,γ)^{36}Cl[13]. - To circumvent the low thermal capture cross section of ^{88}Sr (σ=5.8 mb) we measured the γ-decay following 24 keV neutron capture at the Brookhaven National Laboratory filtered beam facility[14]. A spectrum with a resolution comparable to the one given above was recorded with the available pair spectrometer.

(d,p),(d,t): The reactions were studied at the Munich Tandem Laboratory with the Q3D spectrometer. Spectra were recorded under 20 and 30 degrees with respect to the beam axis. The multiwire proportio-

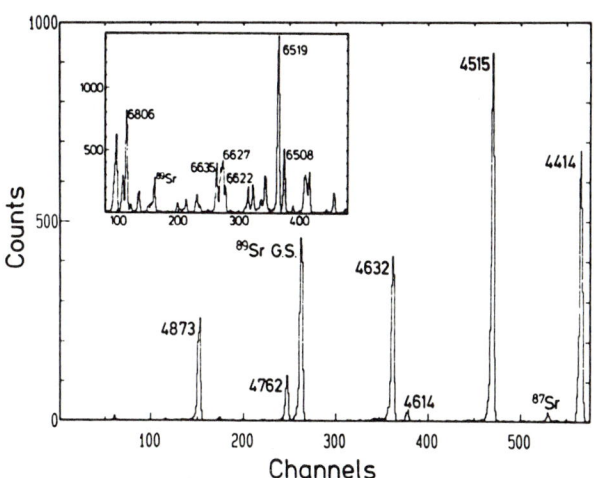

Fig. 1: Proton spectrum of the reaction ^{87}Sr(d,p)^{88}Sr. The energies are given in keV. The resolution is around 3.8 keV FWHM.

nal counter chamber with particle identification[15] yielded spectra with very low background from other than (d,p) (or (d,t)) reactions, and a wide dynamic range of observable intensities was obtained. Two spectra of the reaction ^{87}Sr(d,p)^{88}Sr illustrating the quality of the background suppression and the detector resolution are depicted in fig.1. The minimum cross section that was still detected was estimated as 10 μb/sr. The details of the (n,γ) and (d,p) data reduction are described in ref.[8]. The following table gives a brief summary of the results of the analyses:

final nucleus	σ [b]	number γ's obs.	% of flux assigned	number of states (n,γ)	(d,p)	E_b [keV]
^{87}Sr	5.8	759	90	80	94	8428.2(2)
^{88}Sr	16	661	88	47	85	11112.7(2)
^{89}Sr	0.0058	221	50	19	55	6358.6(2)

III. Statistical model calculations

In the statistical model the density of excited states $\Phi(E_x)$ can be parametrized with three parameters: the nuclear temperature T, the energy of the fictitious ground state (or backshift) E_0 and the spin cut-off parameter σ, which is a measure for the "most probable spin" in the nucleus. The parametrization of the level density $\Phi(E_x)$ in the Constant Temperature Fermi Gas Model[16] (CTFG) can then be written as

$$\Phi(E_x) = 1/T \sum_{I_1 \leq I \leq I_2} f(I,\sigma) \exp\{(E_x - E_0)/T\}$$

with

$$f(I,\sigma) = \exp(-I^2/2\sigma^2) - \exp(-(I+1)^2/2\sigma^2).$$

The interval $[I_1, I_2]$ defines the range of observable spins for the final states. Taking the values for σ from ref. [16] the parameters T and E_0 were calculated from a fit[11] to the experimental density in the complete part of the level scheme and to the density of neutron resonances D at the binding energy[17-19]. The spin window $[I_1, I_2]$ was restricted to an interval which could be populated by E1 and M1 radiation from the capture state. The following table summarizes the results:

final nucleus	I^π cap.state	I^π pop.	D [keV]	fit region [MeV]	σ	T [MeV]	E_0 [MeV]
^{87}Sr	$1/2^+$	1/2,3/2	0.50(7)	0 - 4	3.68	1.10(10)	-1.6(6)
^{88}Sr	4^+ *	3-5	0.12(1)	0 - 4.5	4.04	1.06(10)	0.6(5)
^{89}Sr	$1/2^+$	1/2,3/2	3.2 (2)	0 - 3	3.56	0.94(10)	-0.8(6)

*) The dominance of only one capture spin is reported in ref. [17].

For ^{87}Sr the model seems to give a good description of the data[8]. Fig. 2 suggests that the integrated level density below 4 MeV is essentially that of the model and that the level scheme is thus "complete" below this energy. At higher excitation energies the

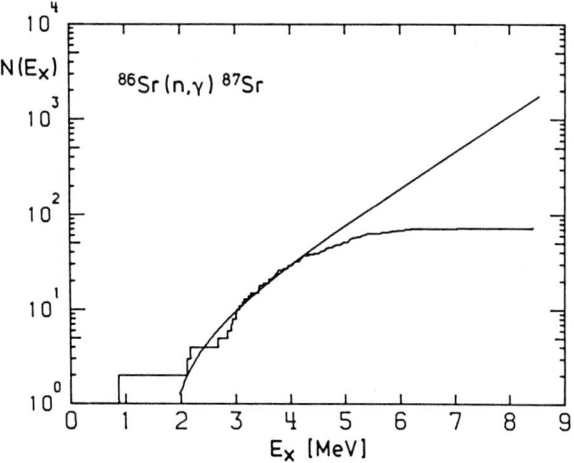

Fig. 2: Total number of states observed in the reaction ^{86}Sr(n,γ) ^{87}Sr up to the excitation energy E_x. The solid line indicates the integrated CTFG level density. The level scheme is essentially complete up to 4 MeV ($I^\pi \in$ [1/2,3/2]).

available phase space for primary transitions becomes considerably smaller and weak transitions can no longer be detected or assigned. The good agreement with the statistical model may be attributed to the fact that, even at low excitation energies, a large number of possible configurations exist, which involve several neutron and proton shells. The effect of the ^{88}Sr shell gap is thus greatly reduced[20].

The nucleus ^{88}Sr reveals interesting consequences of the N=50 shell closure: a) There are no states within the E1/M1 spin window below 2.7 MeV. This too leads to the high positive backshift E_0. b) Above 4 MeV excitation energy the $2d_{5/2}$, $3s_{1/2}$ and $2d_{3/2}$ neutron orbits become available. The single-particle states show little fragmentation, which leads to irregularities in the level density. These show a perfect correlation with the expected single particle structure obtained from the (d,p) analysis[2-4].

In ^{89}Sr the statistical approach is poor in describing the level density as the low-lying spectrum is dominated by mostly unfragmented single-particle states (g.s., 1.03, 2.01, 2.45 MeV)[5]. The deviations show the same pattern as in ^{88}Sr justifying the interpretation given above. It is interesting to note that the parameters T and E_0 still follow the expected tendency of a lower, possibly negative backshift and a temperature similar to that of the other Sr nuclei. A reason may be that local correlations are smoothed out by the averaging process and that the values for T and E_0 depend strongly on the value of D, the resonance spacing at the binding energy.

The fast neutron capture in ^{88}Sr (populating the 23.6 keV $3/2^-$ resonance) is dominated by the valence process. In the A≈90 region the 3p single particle orbit is known to be located around the neutron binding energy[22]. We found a correlation of ϱ=0.7 with a significance of 82% between the E1 partial radiative widths and the spectroscopic factors[4] of the first few excited states[10]. The interpretation within the valence model is in agreement with the observed strong correlation (ϱ=0.96, P(ϱ)>99%) between the reduced neutron and radiative widths of $p_{3/2}$ resonances in ^{88}Sr[19].

In all three nuclei the (d,p) reaction populates states over a wider spin range than (n,γ) but, since it selectively feeds only states containing significant single-particle strength, only subset of CTFG-predicted states is observed. Fig. 3 illustrates this fact. In all three nuclei a separate fit to the (d,p) density yields a similar value for T (within the error) but a between 0.4 and 1.1 MeV higher value for the backshift E_0. This leads to the conclusion, that a fixed fraction of *all* states in the each of the nuclei contains single-particle strength, *independent* of the excitation energy[9,10]. This fraction is as low as 36% in ^{88}Sr and rises to 66% in ^{87}Sr and ^{89}Sr. The low percentage of observed states in ^{88}Sr is partly explained by

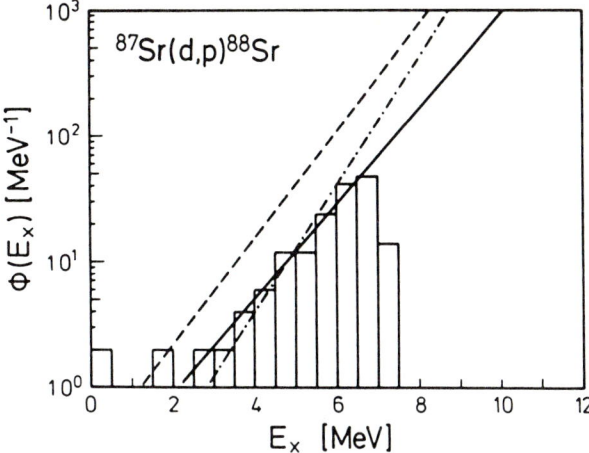

Fig. 3: The density of excited states observed in the (d,p) reaction. Histogram: experimental values; dashed: (n,γ) extrapolated denity; dot-dashed: fit to (d,p) data with resonance spacing value; solid: fit to only the (d,p) data.

the 4 MeV distance between the $1g_{9/2}$ and $2d_{5/2}$ neutron orbits. Below this energy the spectrum is dominated by collective and proton excitations. Neutron excitations become important only above this energy. The fractions of the strengths of the $d_{5/2}$, $s_{1/2}$ and $d_{3/2}$ orbits are then distributed over few states[4] between 4 and 6 MeV giving rise to the irregularities in the level density.

IV. Giant resonance calculations

A simple approximation of the influence of the E1 giant resonance predicts an $I_\gamma \propto E_\gamma^5$ energy scaling for the primary γ-intensities[23]. This differs from the $I_\gamma \propto E_\gamma^3$ single particle model[24]. If the structural effects in the level density are properly accounted for[8], ^{87}Sr and ^{88}Sr definitely favor the E_γ^5 scaling. Fig. 4 shows the comparison of the summed primary E1/M1 intensity in ^{87}Sr with the n=5 and n=3 exponential scaling predictions. In ^{89}Sr the level density is not sufficient for such a comparison. A more detailed analysis of the primary E1, M1 and E2 strengths shows[12] that the giant dipole and giant quadrupole models account for all of the observed γ-strengths.

The fluctuations around the average energy scaling are described by Porter-Thomas fluctuations of one degree of freedom. This is in agreement with the assumption that the capture state is dominated by a single resonance of defined spin and parity.

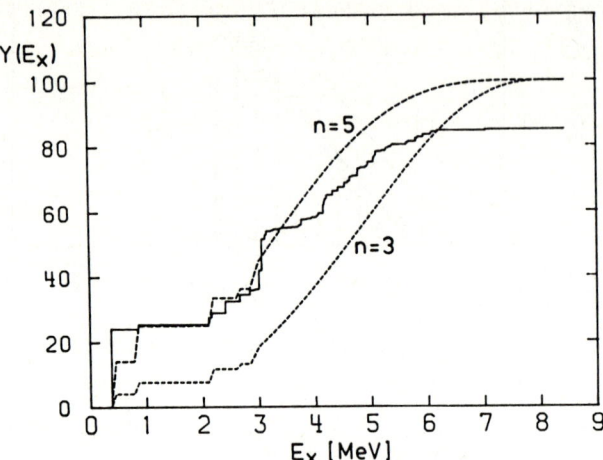

Fig. 4: The integrated primary intensity in the reaction $^{86}Sr(n,\gamma)^{87}Sr$. The step function indicates the experimental values. The two dashed lines give the predictions of the CTFG model for $I_\gamma \approx E_\gamma^n$, n=3,5 scalings. Below 3 MeV the experimental density of states was used to calculate the theoretical values.

*) Work supported by the Deutsches Bundesministerium für Forschung und Technologie, Bonn, under contract 06GOE456.

References

[1] A.I. Kucharska et al., J.Phys. G14 (1988) 65
[2] P.C. Li and W.W. Daehnik, Nucl. Phys. A462 (1987) 26
[3] B.L. Burks et al., Nucl. Phys. A457 (1986) 337
[4] A. Saganek et al., J. Phys. G10 (1984) 549
[5] S.M. Abekasis, J. Davidson and M. Davidson
     Phys. Rev. C22 (1980) 2237
[6] H. Weidenmüller, Comm. Nucl. Part. Phys. 16 (1986) 199
[7] T. v.Egidy, H.H. Schmidt and A.N. Behkami
     Proc. 6th Int. Symp. on Capt. Gamma-Ray Spectr., Leuven 1987
[8] Ch. Winter et al., Nucl. Phys. A460 (1986) 501
[9] Ch. Winter et al., Nucl. Phys. A473 (1987) 129
[10] Ch. Winter et al., in preparation
[11] B. Krusche and K.P. Lieb, Phys. Rev. C34 (1986) 2103
[12] Ch. Winter and K.P. Lieb, in preparation
[13] B. Krusche et al., Nucl. Phys. A386 (1982) 245
[14] R.C. Greenwood and R.E. Chrien, Nucl. Inst. Meth. 138 (1976) 125
[15] A. Chalupka et al., Nucl. Inst. Meth. 217 (1983) 113
[16] A. Gilbert and A.G.W. Cameron, Can J. Phys. 43 (1965) 1446
[17] S.F. Mughabghab, M. Divadeenam and E. Holden
     Neutron Cross Sections Vol.1A, Acad. Press, N.Y.,1981
[18] G.C. Hicks et al., Aust. J. Phys. 35 (1982) 267
[19] J.W. Boldeman et al., Nucl. Phys. A269 (1976) 397
[20] J.E. Kitching, Z. Physik 258 (1973) 22
[21] Ch. Winter et al., Proc. 6th Int. Symp. on Capt. Gamma-Ray Spectr., Leuven 1987
[22] S.K. Rathi and H.M. Argawal, J. Phys. G7 (1981) 53; J.E. Lynn, The theory of neutron resonance reactions, Clarendon Press, 1968
[23] P. Axel, Phys. Rev. 126 (1962) 671
[24] J.M. Blatt and V.F. Weisskopf, Theoretical Nuclear Physics, Wiley, New York, 1952

Quantum Chaos and the Boson-Fermion Approach to A ~ 100 Region

V. Paar[1,*], D. Vorkapić[1,*], S. Brant[2,**], H. Seyfarth[2], and V. Lopac[3,*]

[1]Prirodoslovno-matematički fakultet, University of Zagreb,
41000 Zagreb, Yugoslavia, and
Institut für Kernphysik, Kernforschungsanlage Jülich,
Postfach 1913, D-5170 Jülich, Fed. Rep. of Germany

[2]Institut für Kernphysik, Kernforschungsanlage Jülich,
Postfach 1913, D-5170 Jülich, Fed. Rep. of Germany

[3]Tehnološki fakultet, University of Zagreb, 41000 Zagreb, Yugoslavia, and
Institut für Kernphysik, Kernforschungsanlage Jülich,
Postfach 1913, D-5170 Jülich, Fed. Rep. of Germany

Abstract: Chaoticity of nuclei in the A~100 region is investigated in the frame of boson-fermion approach. Nearest neighbour spacing distribution, Δ_3 statistic of Dyson and Metha and χ^2 comulant are calculated for the IBM, IBFM and IBFFM spectra of nuclei in the A ~ 100 region. Sizeable deviations from chaoticity have been found, in accordance with the phenomenological Abul-Magd-Weidenmüller observation. The relation between symmetry and chaoticity is investigated and it was shown that the "supraregular" behaviour is generated by superposition of near-chaotic partial distributions.

The interplay between regular and chaotic dynamics is an interesting aspect of classical and quantum systems, which is receiving much attention in recent years [1]. The nuclear system provides a particularly convenient case for the study of fluctuation properties. On one hand, a combined set of nuclear resonance energy data exhibits a close agreement with the Gaussian orthogonal ensemble (GOE) predictions, i.e. reveals a chaotic behaviour [2]. On the other

---

* Assisted by Internationales Büro, KFA Jülich
** Alexander von Humboldt fellow, on leave of absence from University of Zagreb

hand, recent analyses [3,4] of a new compilation of the low-energy level spacings have revealed sizeable deviations from chaoticity in direction of regularity, in contrast to the previous indications [5,6]. Previous chaoticity analysis of theoretical nuclear spectra obtained by shell-model calculations gave the results consistent with complete chaoticity [6].

Recently, the chaoticity investigations were extended to the framework of the interacting boson model (IBM) [7], which accounts for both the collective and the dynamical symmetry aspects of the nuclear system. It was shown for the O(6) and SU(3) limits that the 2^+, 4^+ levels exhibit tendency towards regularity and the 0^+, 3^+ levels towards chaoticity [8]. This theoretical result is consistent with the phenomenological result obtained by Abul-Magd and Weidenmüller [3], which was obtained by the analysis of the available systematics of experimental level spacings. It was shown [9] that the origin of this anomalous behaviour lies in the K-degeneracy associated with the SU(3) limit of IBM: All the J=0 states have K=0, while the J=2 states have K=0 or K=2. Thus, there appear two subsets of J=2 levels: J=2, K=0 and J=2, K=2. For each of these two subsets the Δ_3-pattern turns out to be similar to the Δ_3-pattern for the J=0 levels. The regular or "supraregular" Δ_3-pattern (in which case the results for Δ_3 lie above the line corresponding to regularity) corresponds to the total set of J=2 levels, i.e. to the superposition of the J=2, K=0 and J=2, K=2 classes. Consequently, in the SU(3) limit the strong deviation from chaoticity is associated with the appearance of K-quantum number.

In a related investigation the Δ_3 statistic and number variance have been calculated for the energy spectra of ^{147}Gd and ^{115}In [10], which have been previously calculated in a particle-core coupling model, eventually including the intruder states (1 particle - 1 hole excitations across the major closed shell). It was shown that the inclusion of intruder states leads to an increase of the degree of chaoticity.

Sizeable degree of chaoticity associated with IBM has been also studied with approaching the classical limit in the boson space [11]. It was shown that the degree of chaoticity diminishes with the increase of the total boson number N: the boson system approaches the regularity in the classical limit of

IBM. Thereby, there appears an interesting effect of oscillations around the line of regularity, as illustrated in fig.1.

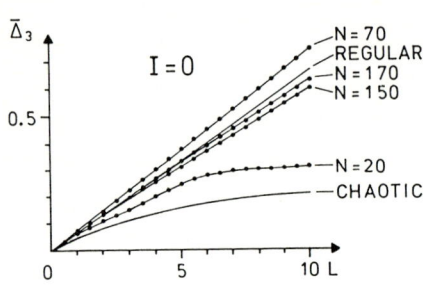

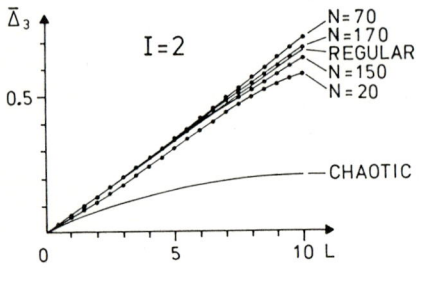

Fig. 1: $\bar{\Delta}_3$ statistic for the IBM spectra in the O(6) limit for angular momenta I=0 and I=2. The O(6) parameters are A=0.17, B=0.333, C=0.01. The $\bar{\Delta}_3$ statistic is displayed for the boson numbers N=20, 70, 150 and 170.

The preliminary chaoticity analysis (using $\bar{\Delta}_3$ statistic and/or nearest-neighbour spacing distribution) has been performed for several theoretical spectra calculated in the boson-fermion approach: The ^{40}K-energy spectrum calculated in IBFFM [12], the N=79-energy spectra calculated in IBFM [13] and the ^{198}Au-energy spectrum calculated in IBFFM [14].

Systematic investitgation of chaoticity is under way for the nuclei in A~100 region, which is characterized by the most sudden transition of nuclear shape. In particular, these nuclei have been calculated in the framework of IBM/IBFM/IBFFM [15]. In figs. 2 and 3 we present the preliminary results for $\bar{\Delta}_3$ statistic and number variance $\Sigma^2(r)$ calculated for the IBFFM energy spectrum of ^{98}Y from ref. [15]. In order to reduce the size of computations we have reduced the boson number and renormalized the SU(5) parameters so that the same type of boson core is reproduced.

The two solid curves in each drawing correspond to the Poisson ensemble (PE) and Gaussian overlap ensemble (GOE), which are attributed to complete regularity and chaoticity, respectively. The results are not presented for angular moments J=0,1 because in these cases the number of levels is smaller

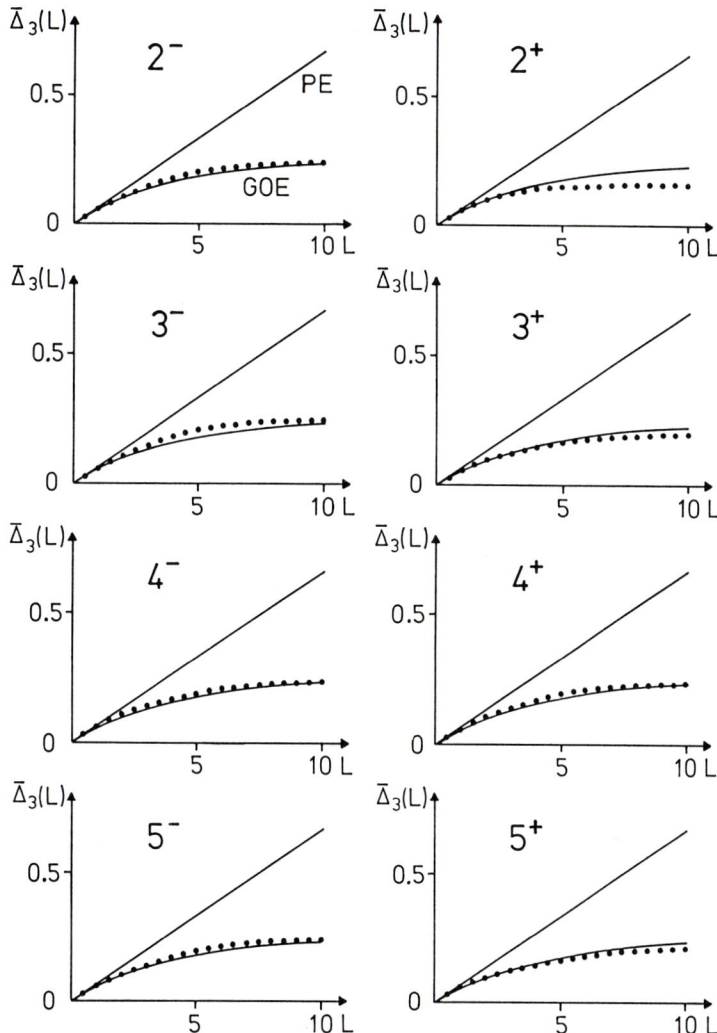

Fig. 2: $\bar{\Delta}_3$-statistic for positive- and negative-parity levels of IBFFM energy spectrum for ^{98}Y.

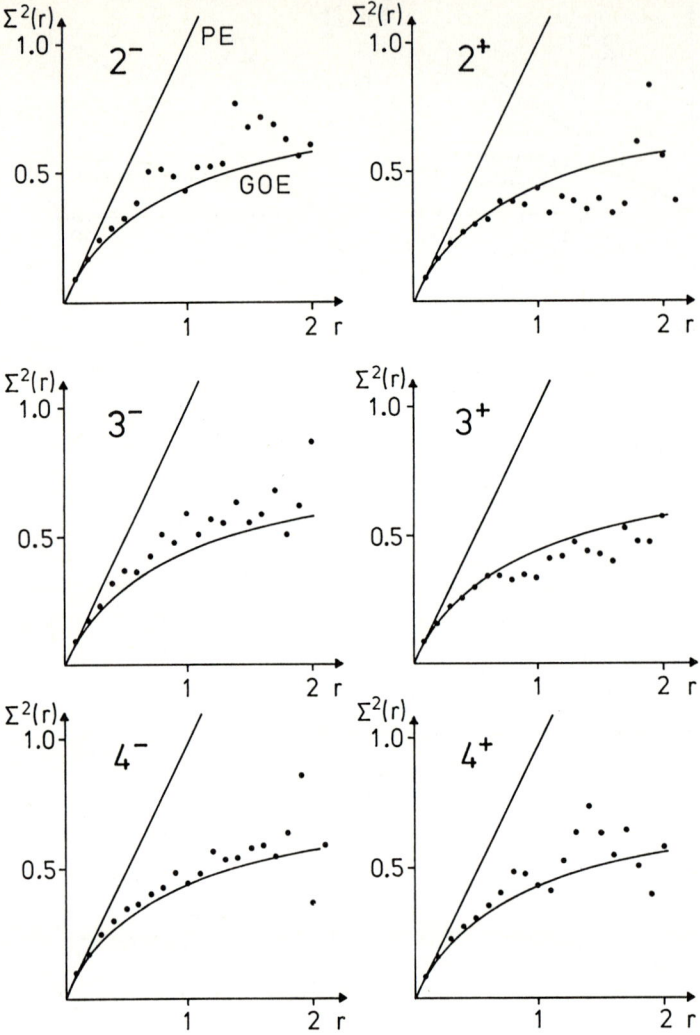

Fig. 3: Σ^2 (r) number variance for positive- and negative-parity levels of IBFFM energy spectrum for ^{98}Y.

than 50, which makes the statistical analysis less reliable. The number of levels for each J^π-value in figures lies in the range between 66 (for 2^+ states) and 186 (for 3^- states). As seen from figs. 2 and 3, both the $\bar{\Delta}_3$ (L) and the $\tilde{\chi}^2$ (r) pattern exhibit somewhat stronger tendency towards chaoticity for the positive-parity (i.e. unnatural parity) levels than for the negative-parity (i.e. natural parity) levels. This is in accordance with the phenomenological effect pointed out by Abul-Magd and Weidenmüller [3].

References

[1] O. Bohigas and M.J. Giannoni, in Lecture Notes on Physics (Springer, Berlin, 1984), Vol 209, p.1 and references therein.

[2] R.U. Haq, A. Pandey and O. Bohigas, Phys. Rev. Lett. 48 (1982) 1086; O. Bohigas, R.U. Haq and A. Pandey, Phys. Rev. Lett. 54 (1985) 1645.

[3] A.Y. Abul-Magd and H.A. Weidenmüller, Phys. Lett. 162B (1985) 223

[4] T. von Egidy, A.N. Behkami and H.H. Schmidt, Nucl. Phys. A454 (1986) 109

[5] E. Cota, J. Flores, P.A. Mello and E. Yepez, Phys. Lett. 53B (1974) 32

[6] T.A. Brody, J. Flores, J.B. French, P.A. Mello, A. Pandey and S.S.M. Wong, Rev. Mod. Phys. 53 (1981) 385

[7] A. Arima and F. Iachello, Adv. Nucl. Phys. 13 (1984) 139

[8] V. Paar and D. Vorkapić, Phys. Lett. 205 B (1988) 5

[9] V. Paar and D. Vorkapić, submitted for publication to Phys. Rev. C.

[10] V. Paar, D. Vorkapić and K. Heyde, Phys. Lett. B (1988), in print.

[11] V. Paar, D. Vorkapić and A.E.L. Dieperink, to be published.

[12] H. Seyfarth, S. Brant, P. Göttel, V. Paar, D. Vorkapić and D. Vretenar Z. Phys. A (1988), in print.

[13] R.A. Meyer, B.K.S. Koene, R. Chrien, S. Brant, V. Paar and V. Lopac, submitted to Phys. Rev. C.

[14] S. Brant, V. Paar, D. Vorkapić, V. Lopac, R.A. Meyer and T. von Egidy, to be published.

[15] S. Brant, K. Sistemich, H. Seyfarth, O.W.B. Schult, H. Ohm, M.L. Stolzenwald, V. Paar, D. Vretenar, D. Vorkapić, V. Lopac, R.A. Meyer, G. Lhersonneau, K.-L. Kratz and B. Pfeiffer, contribution to these Proceedings.

Part V

Shell Effects

Part V

The Approach to 2:1 Deformations Induced by Particle Numbers Around 40 and 60

I. Ragnarsson and T. Bengtsson

Department of Mathematical Physics, Lund Institute of Technology,
P.O. Box 118, S-221 00 Lund, Sweden

Abstract: The 2:1 shell effects of the harmonic oscillator and their possible manifestation at low spin for nuclei around ^{80}Zr and ^{102}Zr are briefly reviewed. Attempts to describe the spectrum of ^{84}Zr observed to a tentative 34^+ state are discussed. It is concluded that the high-spin branch of this band does not fit into any conventional theory. The importance of more and even better high-spin data in this region is pointed out.

It is now well established that the shell gaps of harmonic oscillator potential corresponding to 2:1 elongated shapes ($\varepsilon=0.6$) manifest themselves also in realistic nuclear potentials. However the corresponding magic numbers, N = ..., 16, 28, 40, 60, 80, 110, 140, ... are shifted to somewhat higher particle numbers in a similar way as the magic spherical nuclei are found for example for nucleon numbers 82 and 126 corresponding to the oscillator gaps 70 and 110. In the case of 2:1 shapes, fission isomers have been detected for nuclei with N = 140-150 (e.g. [Me79]) while superdeformed high-spin states have been found for $^{152}_{66}$Dy$_{86}$ [Tw86], $^{149}_{64}$Gd$_{85}$ [Ha87] and neighbouring nuclei. If the single-particle orbitals are drawn as functions of quadrupole deformation (fig. 1), the 2:1 shell gaps are seen as regions of low level density around $\varepsilon=0.6$ intersected by some strongly down-sloping (polar) orbitals but with no up-sloping (equatorial) orbitals. This can be seen as a manifestation of the level density in the pure oscillator with degenerate eqatorial orbitals as discussed in ref. [Ra78] where it was pointed out that all orbitals except those of high-j intruder shells behave essentially as in the pure oscillator justifying the term pseudo-oscillator symmetry [Du87b]. Indeed, it seems that the general features discussed there can be seen in the superdeformed high-spin spectra [Du87b,Be88].

The neutron shell energy is plotted as a function of quadrupole deformation and neutron number in fig. 2. It is immediately obvious that the particle numbers 64-66 and 84-86 are situated in a shell energy valley at $\varepsilon=0.5-0.6$. The associated low shell energies explain why superdeformed bands are observed in e.g. $^{152}_{66}$Dy$_{86}$ [Tw86] and $^{149}_{64}$Gd$_{85}$ [Ha87]. Similarly, the strongly deformed bands in the Nd/Ce region (e.g. [No85,Be87]) appear to be associated with the low shell energies at $\varepsilon \simeq 0.4$ for particle numbers ~60 and ~70.

In this contribution, we are especially interested in the low shell energies for Z,N$\simeq$40-44 and N$\simeq$60-64 at $\varepsilon=0.4-0.6$. After the first evidence [Ch70] for large

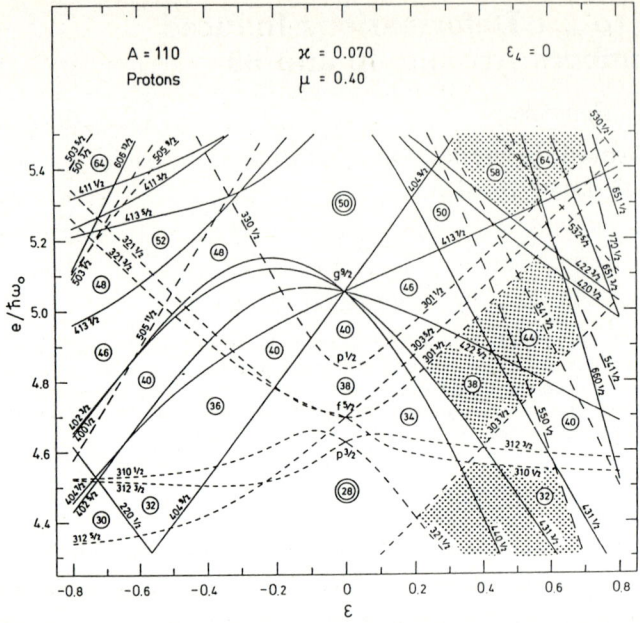

Fig. 1. Nilsson diagram calculated with parameters used for protons in the A~110 region. The general features of the diagram should be valid for protons as well as neutrons in the Z=40 region. Regions of low level density for ε≈0.4-0.6 are indicated.

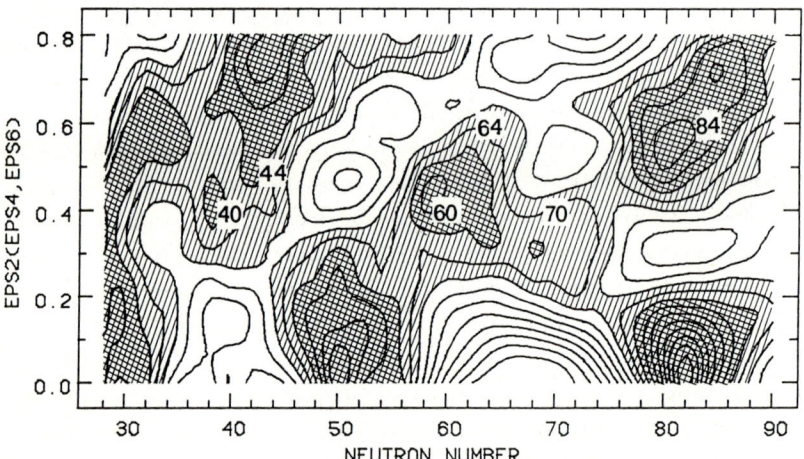

Fig. 2. The shell energy landscape for neutrons [Ra84] as a function of neutron number and elongation. In shaded regions, the shell energy is between -1 and 1 MeV while it is below -1 MeV in cross-hatched regions. Some favoured particle numbers of experimental interest are indicated. The qualitative features of the plot remains unchanged for protons or, at large deformation, if rotation is added.

deformations around $^{100}_{40}\text{Zr}_{60}$, it was speculated [Sh72] that these nuclei might have ground state deformations with ε close to 0.6. Special importance was given to the 2^+ energies in this region as exhibited in fig. 3. The very low energies which seemed to be present for Z=40 and N=40 and 62-64 ("the decisive "chameleon" role played by Z=40", [Sh72]) made it very natural to associate with the 2:1 shell gaps suggesting very large deformations around ε=0.6. However, as confirmed by more recent life-time measurements (e.g. [Az79]), detailed calculations could not allow deformations larger than ε=0.3-0.4. On the other hand, also these deformations can to some extent be considered as a manifestation of the 2:1 shell gaps. Furthermore, the special character of the Z=40 particle number being magic for both spherical and very deformed shapes suggested the possibility of shape coexistence in these nuclei [Sh72]. For example the very low-lying 0^+ states which have been found in more recent experiments [Kh78,Sc80] give strong support for such ideas.

The recent possibilities to study rotational bands up to very high spins give much improved possibilities to study strongly deformed nuclei. This is so because the centrifugal forces help to deform the nucleus. At present it is essentially only for neutron-deficient nuclei where very high-spin rotational bands can be detected.

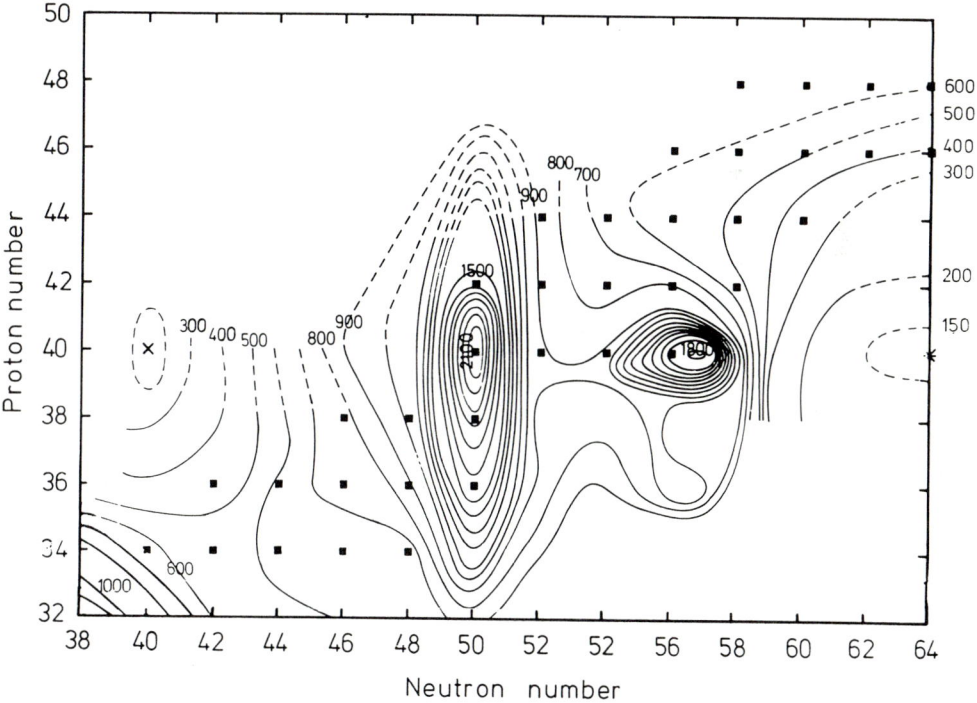

Fig. 3. Contour map of the experimental excitation energies of the first 2^+ state for even-even nuclei in the intermediate mass region 70<A<110 as presented in ref. [Ra78].

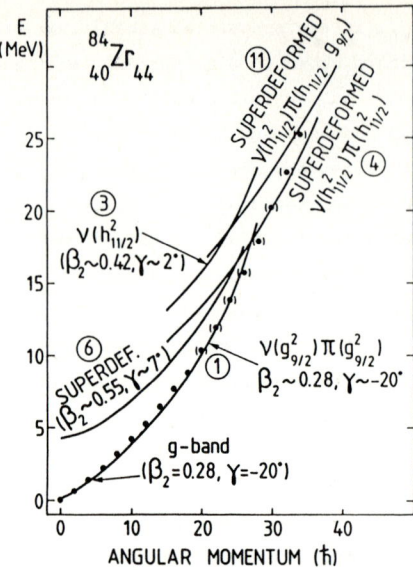

Fig. 4. Total energy calculated using the Woods-Saxon potential and the particle-number-projection (RFBCS) method vs angular momentum for characteristic rotational bands in ^{84}Zr: calculated bands are labeled by means of the aligned quasiparticles and by deformation parameters. The bandhead energies were taken from Strutinsky calculations with pairing (from ref. [Du87a].

Thus, the nuclei around ^{102}Zr are of less interest than those around ^{80}Zr. In this context, the band in ^{84}Zr which has been identified [Pr83] to a tentative 34^+ state is of special interest. The band shows many of the features expected [Pr83] for a strongly deformed band but it has also been argued [Du87a] that this band, or at least the major parts of it, can be understood as a small-deformation triaxial band. The observed band is shown in fig. 4 together with some bands calculated in the Woods-Saxon model [Du87a] and in fig. 5 together with present calculations carried out in the modified oscillator. It seems that both calculations suggest that the observed band up to spin I≈26-30 is explained by a triaxial small deformation band with no excitations out of the valence space. Detailed calculations in [Du87a] considering for example different band crossings support this assumption. Relative to an ^{80}Zr core, the band in fig. 5 has the configuration $\pi(N=3)^{-4}(g_{9/2})^4 \nu(N=3)^{-2}(g_{9/2})^6$ where "(N=3)" denotes the orbitals $p_{3/2}$, $f_{5/2}$ and $p_{1/2}$. The maximum spin of this configuration is I=34 and as seen in fig. 5, the termination occurs in an energetically rather unfavoured state. Other bands within the valence space show similar terminations and all states within this space become energetically unfavoured above I=30. Therefore it seems essentially excluded that a very regular band, as the observed one, could be formed all the way up to I=34.

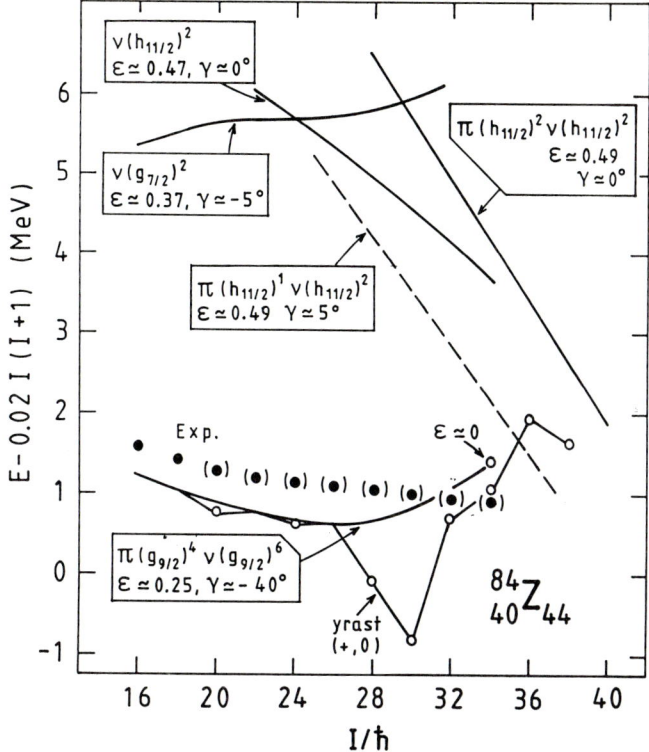

Fig. 5. *The rotational band observed experimentally in ^{84}Zr is shown relative to a rigid rotation reference together with the calculated even spin yrast states of positive parity $((\pi,\alpha)=(+,0))$ and some selected bands with a fixed configuration. The calculations were carried out using the modified oscillator potential and ignoring pairing within the formalism of ref. [Be85]. The bands are labelled by the <u>total number</u> of particles in specific j-shells and by approximate deformations. Aligned states at $\gamma=60°$ are drawn by open rings.*

As discussed above, favoured configurations are formed at large deformations. As suggested from the single-particle diagram in fig. 1 and as seen in fig. 5, the configuration with 2 $h_{11/2}$ neutrons and no $h_{11/2}$ protons is relatively low-lying in the modified oscillator calculations while its moment of inertia come out essentially as in experiment in the Woods-Saxon calculations [Du87a]. The negative parity configuration $\pi(h_{11/2})^1\nu(h_{11/2})^2$ is however more low-lying in both calculations. Another problem with the configurations with particles in $h_{11/2}$ is how they could enter into the observed band. At low spins, it is essentially excluded that the yrast states have any appreciable amplitudes of $h_{11/2}$ particles. One would then expect band with two or more $h_{11/2}$ particles to be separate from the observed low-spin states just as no or very weak connections have been found between strongly deformed bands and smaller deformation bands in heavier nuclei [No85, Be87, Tw86, Ha87]. In our opinion, this is the strongest argument against a suggestion that the observed band should be superdeformed involving $h_{11/2}$ particles.

A third possibility is then that particles are excited to the [431 1/2] orbital which emerges from the N=4 subshells ($d_{5/2}$ and $g_{7/2}$) above the Z=N=50 gap. With two neutrons in this orbital we calculate a rotational band as shown in fig. 5. The general properties of this band appear to be in reasonable agreement with experiment but its high excitation energy makes it difficult to believe that it is responsible for the observed band.

In summary, we have discussed the shell effects favouring large deformations up to $\varepsilon \gtrsim 0.6$ for particle numbers around 40 and 60. Especially, we pointed out the very nice experiment identifying a rotational band to a tentative 34^+ state in ^{84}Zr. In agreement with ref. [Du87a], we found it somewhat difficult to explain this band from the large deformation shell effects. One very important factor in this context was that such a band should involve 2 $h_{11/2}$ neutrons and is then not expected to form one regular band when combined with the low-spin yrast band which should have no $h_{11/2}$ neutrons. However, we found it equally difficult to explain the observed band as formed within the valence space at low deformations. This is provided the tentative 32^+ and 34^+ states are confirmed because these spins are very close to the highest possible spins which can be formed in the valence space. They appear to be energetically rather unfavoured and therefore, it is difficult to see how a very smooth band could be formed up to I=34. Thus, the nucleus ^{84}Zr stands as a challenge to both experimentalists and theorists. Especially, it seems very important to find out if the tentative spectrum to I=34^+ can be confirmed and maybe expanded and also to extract even better and more complete transition probabilities not discussed here.

References

[Az79]   R.E. Azuma et al., Phys. Lett. 86B (1979) 5.
[Be87]   E.M. Beck et al., Phys. Lett. 195B (1987) 531.
[Be85]   T. Bengtsson and I. Ragnarsson, Nucl. Phys. A436 (1985) 14.
[Be88]   T. Bengtsson, I. Ragnarsson and S. Åberg, subm. to Phys. Lett.
[Ch70]   E. Cheifetz et al., Phys. Rev. Lett. 25 (1970) 38.
[Du87a] J. Dudek, W. Nazarewicz and N. Rowley. Phys. Rev. C35 (1987) 1489.
[Du87b] J. Dudek et al., Phys. Rev. Lett. 59 (1987) 405.
[Ha87]   B. Haas et al., Phys. Rev. Lett. 60 (1988) 503.
[Kh78]   T.A. Khan et al., Z. Phys. A284 (1978) 313.
[Me79]   V. Metag, Proc. Fourth IAEA Symp. on Physics and Chemistry of Fission, Jülich 1979, (IAEA, Vienna, 1980) vol I, p. 153.
[No85]   P.J. Nolan et al., J. Phys. G11 (1985) L17;
[Pr83]   H.G. Price et al., Phys. Rev. Lett. 51 (1983) 1842.
[Ra78]   I. Ragnarsson, S.G. Nilsson and R.K. Sheline, Phys. Rep. 45 (1978) 1.
[Ra84]   I. Ragnarsson and R.K. Sheline, Physics Scripta 29 (1984) 385.
[Sc80]   F. Schussler et al., Nucl. Phys. A339 (1980) 415.
[Sh72]   R.K. Sheline, I. Ragnarsson and S.G. Nilsson, Phys. Lett. 41B (1972) 115.
[Tw86]   P.J. Twin et al., Phys. Rev. Lett. 57 (1986) 811.

IBMF and IBFFM Approach to Nuclei in the A ≃ 100 Region

S. Brant[1,*], K. Sistemich[1], H. Seyfarth[1], H. Ohm[1], M.L. Stolzenwald[1],
V. Paar[2], D. Vretenar[2], D. Vorkapić[2], V. Lopac[3], R.A. Meyer[4],
G. Lhersonneau[5], K.-L. Kratz[5], and B. Pfeiffer[5]

[1]Institut für Kernphysik, Kernforschungsanlage Jülich,
  Postfach 1913, D-5170 Jülich, Fed. Rep. of Germany
[2]Prirodoslovno-matematički fakultet, University of Zagreb,
  41000 Zagreb, Yugoslavia
[3]Tehnološki fakultet, University of Zagreb, 41000 Zagreb, Yugoslavia
[4]Lawrence Livermore National Laboratory, Livermore, CA 94550, USA
[5]Institut für Kernchemie, Universität Mainz,
  D-6500 Mainz, Fed. Rep. of Germany

Abstract:
An overview of the calculations in IBFM and IBFFM for nuclei in the A ≃ 100 region is presented. The application to these nuclei with a complex structure including the rapid transition from spherical to deformed nuclear shapes provides a stringent test for the capacities of this theoretical approach. From the result of the studies of the heavy Yttrium isotopes and the N=59 isotones it is concluded that it can account for the basic structure and the phase transition in these nuclei.

1. Introduction

In the last decade the interacting boson model (IBM) [1-3] has been applied with considerable success to the description of even-even nuclei. Its extension to boson-fermion systems, referred to as the interacting boson-fermion model (IBFM) [4-6], was applied to the description of odd-even nuclei. Recently, this approach was further extended to the systems with bosons, odd proton and odd neutron fermions [7]. This model, referred to as the interacting boson-fermion-fermion model (IBFFM) was applied to the interpretation of odd-odd nuclei [8-13].

* Alexander von Humboldt fellow, on leave of absence from University of Zagreb

In recent years the IBFM/IBFFM approach has been applied to nuclei in the region of A ≃ 100 nuclei. Particularly interesting are the calculations for the sequence of Yttrium isotopes ^{96,97,98,99}Y because of their strongly different character and for the sequence of N=59 isotones of ^{98}Y, ^{101}Mo, ^{103}Ru, ^{105}Pd.

Neutron-rich nuclei in the region around A=100 exhibit a rapid change from the shell-model type of structure to a pronounced collective rotational pattern. Thus, the high-spin states in the immediate neighbours of doubly submagic ^{96}Zr exhibit rather pure shell-model configurations [14], while already ^{99}Y has the properties of a symmetric rotor [15] and ^{98}Y contains a rotational band based on the 495 keV level [16].

Such a rapid change in the nuclear structure presents a challenging test for the IBM/IBFM/IBFFM. With the exception of Zr isotopes, that are pure shell-model type nuclei (particularly ^{96}Zr), all other nuclei in this region are candidates for the interacting boson description. This is due to a pronounced quadrupole collectivity that is present in the neighbourhood of Zr isotopes. It can be expected that the IBM/IBFM/IBFFM results will be the better the more the proton number differs from Z=40 (due to greater collectivity). The N=59 nuclei establish a special problem since the coexistence of spherical and deformed shapes is expected, which will introduce configuration mixing. We concentrated our attention on the most challenging Yttrium chain and on N=59 isotones. If a rather good description of these complex nuclei is obtained than one gains confidence that the interaction boson description of other nuclei in the A ≃ 100 region can be even better.

A pre requisite of calculations in any nuclear model is sufficient knowledge on the nuclear levels of the investigated nuclei as well as of the neighbours. This is a weak point of all studies at A ≃ 100 since most of these isotopes are shortlived and can only reached through fission. Nevertheless, some basic information on these nuclei has become available in the recent years, which justified the present theoretical studies.

2. The Yttrium Isotopes

2.1. Odd-mass isotopes

2.1.1. ^{97}Y

A prominent example for the IBFM calculations is the case of ^{97}Y. The nucleus ^{97}Y, with one proton hole and two neutron particles beyond ^{96}Zr, has been recently investigated experimentally [17]. A particularly interesting feature of this nucleus is a family of unique-parity levels, associated with $g_{9/2}$ proton configuration. No direct information on the spins of these levels is available. But the energies and the deexcitation properties suggest that they are based on the coupling of a $g_{9/2}$ proton on core configurations determined by the two neutrons. Employing the boson-fermion rules for the unique-parity band pattern [18,19], it was predicted that the low-lying part of this positive-parity band in ^{97}Y is of decoupled type, i.e. the 1336.3 keV and 1657.6 keV levels are assigned as $13/2^+$ and $11/2^+$, respectively [20]. Adopting this spin-assignment, we have performed the IBFM calculation for the positive-parity levels in ^{97}Y (For the negative parity states see section 2.2.2.). The even-even core nucleus ^{96}Sr has been described by an U(5) core with the parameters $h_1=0.815$, $h_{40}=0$, $h_{42}=-0.37$, $h_{44}=0.22$, $N=4$. (For definition of parameters see ref. [21].) The corresponding core spectrum is presented in fig. 1a.

In the IBFM calculation [20] for the positive-parity states in ^{97}Y we have included the proton quasiparticle state $\tilde{g}_{9/2}$ with the occupation probability $v^2=0.044$. The fermion-boson interaction strengths employed in the calculation are $\Gamma_o=0.4$, $\Lambda_o=4$, $\chi = -\frac{\sqrt{7}}{2}$. (Parameters are defined according to ref. [21].) These values were adjusted to the low-lying positive-parity states of ^{97}Y. We note that the strength of the exchange force, $\Lambda_o=4$, is similar to the value used for the calculation of the positive-parity levels in ^{91}Rb, where $\Lambda_o=5$ was used [22]. It was argued that such a large value is consistent with a sizeable blocking effect for the unique-parity state due to its presence in the internal structure of the boson.

The energy spectrum obtained by this IBFM calculation is presented in fig. 1b. The experimental levels are well reproduced. We note that the yrast line band pattern changes its character at the angular momentum 19/2: the levels 9/2, 13/2, 11/2, 17/2, 15/2 exhibit a decoupled-type band pattern, which is followed by the 19/2, 21/2 levels which appear in the normal (strongly-coupled band type) ordering. Such a change of band-type is associated with the anharmonicities in the SU(5) boson core.

Employing the wave functions obtained by diagonalization, the E2 and M1 transitions have been also calculated in ref. [20], in rather good agreement with experiment.

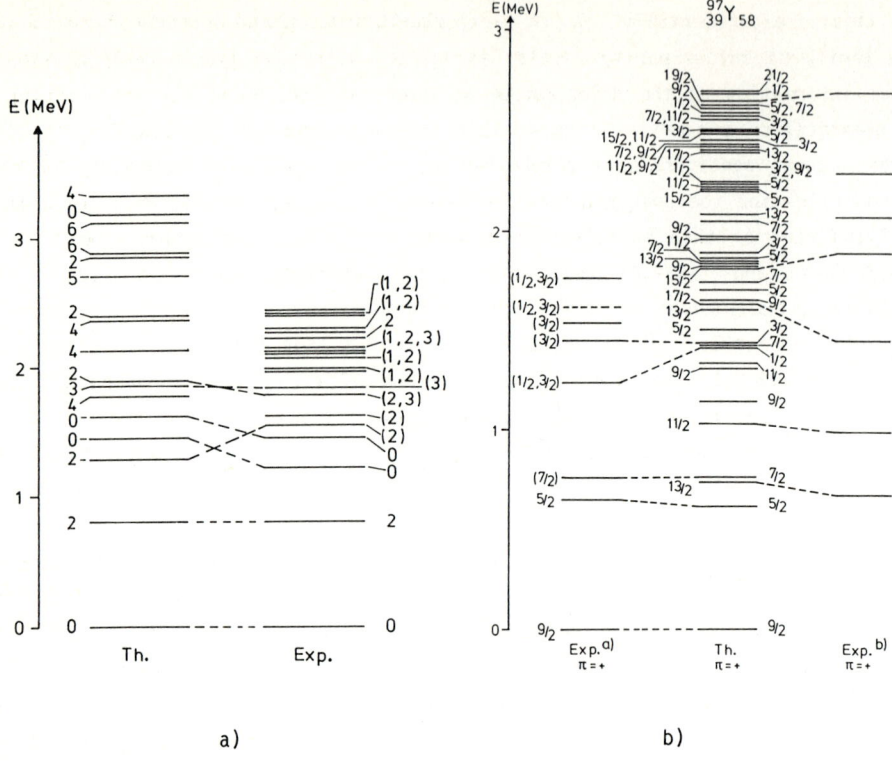

Fig. 1: IBFM calculation for the positive parity states of ^{97}Y: (1a) IBM core spectrum in comparison to the low-lying spectrum of ^{96}Sr, (1b) IBFM positive parity spectrum of ^{97}Y in comparison to the experimental data.

2.1.2. ^{99}Y

The next heavier odd-A isotope ^{99}Y has a complete different character. There appears the rotational-like band based on the $5/2^+$ ground state (g.s.), exhibiting very closely the $I(I+1)$ energy rule, with members up to $19/2^+$ [23]. In ref. [24,25] evidence for additional bands has been reported, in particular, two bands based on heads at 1009 and 1119 keV. The ground-state band and these two bands have been tentatively associated with the [422 5/2], [431 1/2] and [431 3/2] Nilsson configurations, respectively. In ref. [25] the 0.536 MeV band-head was identified as $3/2^-$. However, on the basis of log ft and γ-multipolarity arguments in ref. [26] it was argued that the 0.536 MeV level has the $3/2^+$ assignment.

Details of the IBFM calculation for ^{99}Y are given in [26]. The boson core has been determined in order to reproduce the low-lying positive-parity spectrum of the ^{98}Sr core, with the parametrization rather close to the SU(3) limit of IBM. The $g_{9/2}$ quasi-particle was coupled to this core, with the boson-fermion interaction strengths determined by fitting to the g.s. band in ^{99}Y: Γ_0=0.477, Λ_0=4.77, $\chi = -\frac{\sqrt{7}}{2}$, with $v^2_{g_{9/2}}$ = 0.2. The resulting positive-parity energy spectrum of ^{99}Y [27] is presented in fig. 2. It is seen that the 3/2$^+$-band head appears close to the observed 0.536 MeV level. At about 1 MeV there appear two calculated band heads with 1/2$^+$ and 7/2$^+$ spin/parity assignments.

We note that the IBFM bands associated with the SU(3) limit of the boson core can be classified according to the quantum number K_{IBFM}, associated with the boson-fermion structure. There are two ways to introduce this quantum number algebraically: one method is associated with truncated strong-coupling limit from ref. [6] and the other is associated with approximate super-symmetry limit from ref. [28].

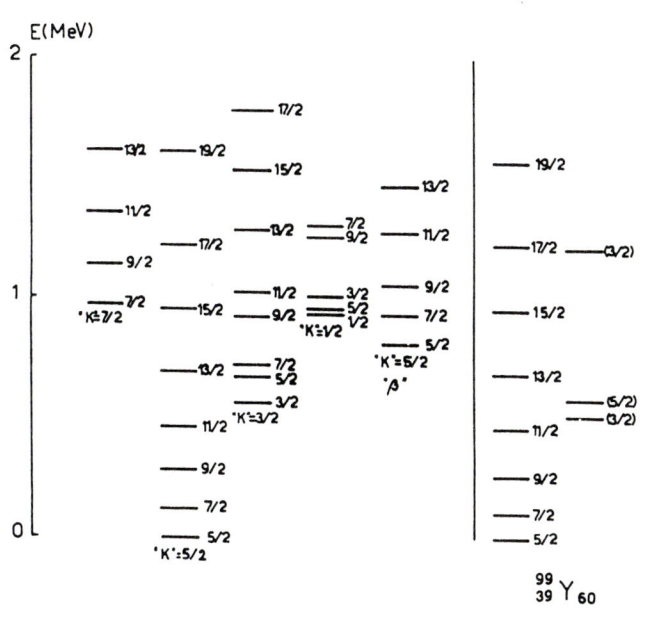

Fig. 2: IBFM calculation for the low-lying positive-parity bands in ^{99}Y.

It is seen that the phase transition from ^{97}Y to ^{99}Y is obtained in IBFM as a consequence of changing the boson core from the SU(5) to the SU(3) type.

On the other hand, we note that the IBFM in the SU(3) limit has much lower flexibility than the Nilsson-plus-BCS model. Thus, in the latter model, the position of the 3/2$^+$ band head is very sensitive to the parameters, and can be fitted to any desired position in a wide range [29,30].

Only for a particular range of the parameters gives the Nilsson-plus-BCS model calculation the results similar to the IBFM calculated spectrum. Thus, the flexibility of IBFM in description of rotational-like nuclei is much smaller. This is due to the rather simple form of boson-boson interaction included in the IBM Hamiltonian. Detailed analysis seems to indicate that, while giving good description of the $5/2^+$ g.s. band, the IBFM calculation does not contain enough flexibility to describe side bands [29,30].

2.2. Odd-odd nuclei

2.2.1. ^{96}Y

^{96}Y is a very fortunate case, since the knowledge about the properties of the levels is relatively extended. Thus an especially good basis for the comparison with the results of the calculations [31] in IBFFM is given. The $1/2^-$ g.s. in ^{95}Y and the $1/2^+$ g.s. in ^{95}Sr are associated with $\tilde{\pi p}_{1/2}$ and $\tilde{\nu s}_{1/2}$ quasiparticles, respectively. Therefrom the lowest-lying states in ^{96}Y are predicted to arise from the ($\tilde{\pi p}_{1/2}$ $\tilde{\nu s}_{1/2}$) 0_1^-, 1_1^- doublet. The corresponding Nordheim number is $N=0$, and due to the parabolic rule [32] it is predicted $E(0_1^-) < E(1_1^-)$, which is in accordance with experiment. (We note that the parabolic rule can be interpreted as being associated with the SU(5) limit of IBFFM in the leading order.)

A pronounced pattern of the ^{96}Y experimental spectrum is that no low-spin level was observed in the energy interval between 1_1^- (122 keV) and 2_1^- (652 keV) levels. Accordingly, in our discussion we consider that there is no 1^- level in that energy range. On the other hand, there is a $(3/2^+)$ level in ^{95}Sr at 352 keV; if we associate this level with the $\tilde{\pi d}_{3/2}$ quasiparticle, we would predict the 1^-, 2^- doublet, arising from ($\tilde{\pi p}_{1/2}$ $\tilde{\nu d}_{3/2}$) 1^-, 2^- configurations, in the energy gap between 122 keV and 652 keV. Since no 1^- level is observed in that energy range, we conclude that the 352-keV level in ^{95}Sr is not based on $\tilde{d}_{3/2}$ configuration, but is of different character. If we associate the 686-keV level in ^{95}Y with $\tilde{\pi p}_{3/2}$ quasiparticle, we obtain the corresponding doublet in ^{96}Y, based on ($\tilde{\pi p}_{3/2}$, $\tilde{\nu s}_{1/2}$) 1^-, 2^- configurations, which should be associated with 1_2^- and 2_1^- levels. The corresponding Nordheim number is 1, and therefrom it follows $E(1_2^-) > E(2_1^-)$. These two states might be associated with the experimental levels (1,2) at 718 keV and 2^- at 652 keV in ^{96}Y, respectively.

The lowest positive-parity state in ^{96}Y is 1^+ at 931 keV. If we associate $\tilde{\pi g}_{9/2}$ with the $(9/2^+)$ 1087 keV level in ^{95}Y and $\tilde{\nu g}_{7/2}$ with the $(7/2^+)$ 556 keV level in ^{95}Sr, we have in the zeroth-order the proton-neutron multiplet ($\tilde{\pi g}_{9/2}$, $\tilde{\nu g}_{7/2}$) 1^+, 2^+, ..., 8^+

in ^{96}Y at 1642 keV. In this case the occupation number [32] is $\bar{v}$≈1 (both quasiparticles obviously are particle-like) and the Nordheim number is zero. Thus, the parabolic rule [32] predicts that the levels 1^+, 2^+, ..., 8^+ lie on a parabola open down, with 1^+ and 8^+ members being shifted down, with $E(1^+) < E(8^+)$. Thus, the 1^+ member of the $\pi\tilde{g}_{9/2}$ $\nu\tilde{g}_{7/2}$ multiplet is predicted to be the lowest positive-parity level, in accordance with experiment. Adopting such an identification, the 1_2^+ level is predicted to be associated with the one d-boson state based on $[\pi\tilde{g}_{9/2} \nu\tilde{g}_{7/2}]$ 1. Its zeroth-order energy is $E(1_1^+:^{96}Y) + E(2_1^+:^{94}Sr) = 1.77$ MeV. This state might correspond to the observed 1^+ level in ^{96}Y at 1984 keV. Employing the information on the quasiparticle positions from the previous discussion, we have performed the IBFFM calculation for ^{96}Y [31].

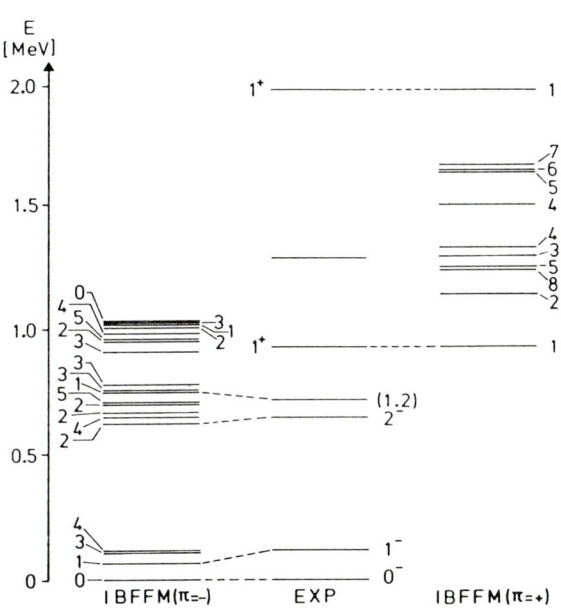

Since we consider the low-lying levels, with rather small boson admixtures, and the core in that region is close to SU(5) boson limit, we have employed the harmonic IBM core with d-boson energy given by the position of 2_1^+ level in ^{94}Sr. In the calculation three bosons have been included; in IBM this corresponds to protons and neutrons outside of Z=38, N=50 core. The quasiparticle parameters employed in the calculation are:

Fig.3: IBFFM positive- and negative-parity energy spectrum of ^{96}Y in comparison to the available experimental data.

$\varepsilon(\pi\tilde{p}_{3/2}) - \varepsilon(\pi\tilde{p}_{1/2}) = 0.686$ MeV, $\varepsilon(\pi\tilde{f}_{5/2}) - \varepsilon(\pi\tilde{p}_{1/2}) = 0.827$ MeV, $\varepsilon(\pi\tilde{g}_{9/2}) - \varepsilon(\pi\tilde{p}_{1/2}) = 1.087$ MeV, $v^2(\pi\tilde{p}_{1/2}) = 0.617$, $v^2(\pi\tilde{p}_{3/2}) = 0.924$, $v^2(\pi\tilde{f}_{5/2}) = 0.944$, $v^2(\pi\tilde{g}_{9/2}) = 0.044$,
$\varepsilon(\nu\tilde{g}_{7/2}) - \varepsilon(\nu\tilde{s}_{1/2}) = 0.556$ MeV, $\varepsilon(\nu\tilde{d}_{5/2}) - \varepsilon(\nu\tilde{s}_{1/2}) = 1.15$ MeV, $v^2(\nu\tilde{s}_{1/2}) = 0.1$, $v^2(\nu\tilde{g}_{7/2}) = 0.05$, $v^2(\nu\tilde{d}_{5/2}) = 0.9$. These values, adjusted to the experimental properties of ^{96}Y, are rather close to the standard BCS parametrization [33]. The ex-

ception is the $\nu \tilde{d}_{3/2}$ quasiparticle; on the basis of previous discussion it is assumed to be higher-lying and, accordingly, is omitted from the calculation. The interaction strengths employed in the present calculation, determined in order to get an overall agreement with the experimental properties of ^{96}Y are $\Gamma_0^{\pi}= 0.3$, $\Gamma_0^{\nu}= 0.62$ and the strengths of the residual force $V_\delta= -0.16$ (surface delta interaction strength), $V_{\sigma\sigma}= 0.1$ (spin-spin interaction strength). The calculated IBFFM levels are presented in Fig. 3 in comparison with the available experimental data. In Fig. 4 we present the calculated levels in the E/I plot, together with the classification of some states into multiplets on the basis of largest components in the wave functions.

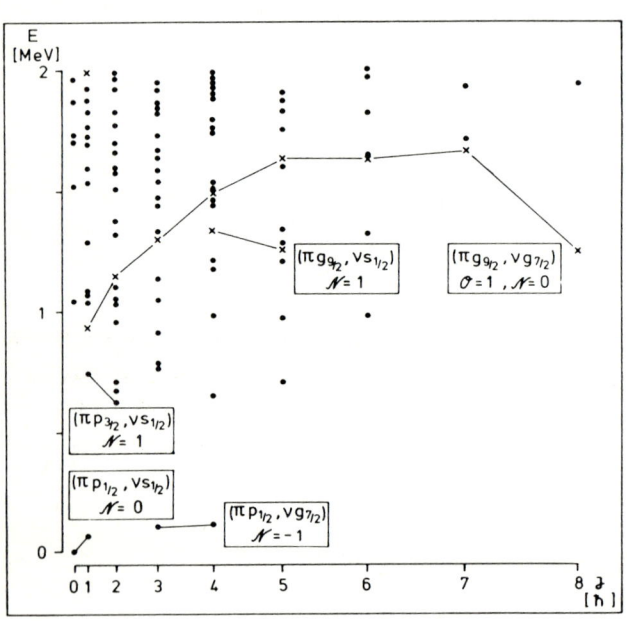

Fig. 4: Calculated energy spectrum of ^{96}Y in the E/I plot. Negative-parity levels are denoted by full circles and positive-parity by crosses.

In Ref. [31] the electromagnetic properties of ^{96}Y have been also calculated in IBFFM and compared to scarce experimental data. In particular, the calculated $2\bar{1} \to 1\bar{1}$ transition is M1+2% E2, in good agreement with experimental result M1+1% E2.

2.2.2. ^{98}Y

A particularly interesting nucleus is ^{98}Y. Recent data [34] on ^{98}Y indicate the coexistence of spherical and deformed shapes, in accordance with the expectation from its position between the spherical ^{97}Y and the good rotor ^{99}Y. The available experimental levels of positive and negative parity which are suggested to belong to the spherical shape are presented in Fig. 5 (spectra labelled EXPS). However, the spin parity assignment has been given only to the ground-state ($I^{\pi}= 0^-$). On the other hand, the energy sequence of the levels at 496, 596, 726, 884 and 1070 keV and the existence of a complete set of the transitions among them fit to a rotational band; therefore these states have been associated with a deformed shape. In Fig. 5 these states are presented in the spectrum labelled EXPD.

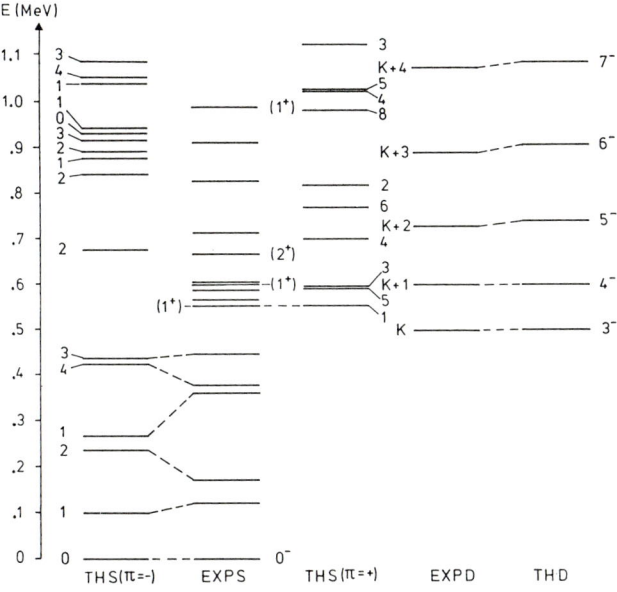

Fig. 5: IBFFM claculation for the energy spectrum of ^{98}Y in comparison to the experimental data. The lowest positive-parity theoretical state 1^+ is only tentatively assigned to the (1^+) 548 keV level.

In such a challenging situation, with scarce spin/parity assignment, we have attempted to describe this nucleus in IBFFM [35] using the experience with the abovementioned studies and, on this basis, to predict possible spin/parity assignments.

We proceed as follows.
In the first step we have performed the IBFM calculation for the negative-parity states in the odd-even neighbour ^{97}Y. The boson core has been taken the same as in the IBFM calculation for the positive-parity states in ^{97}Y (Fig. 1). The occupation probabilities for proton quasiparticles ($\tilde{p}_{1/2}$, $\tilde{p}_{3/2}$, $\tilde{f}_{5/2}$) are taken the same as in the IBFFM calculations for ^{96}Y (Fig. 3). The quasiparticle positions $\varepsilon(\tilde{p}_{3/2}) - \varepsilon(\tilde{p}_{1/2}) = 0.78$, $\varepsilon(\tilde{f}_{5/2}) - \varepsilon(\tilde{p}_{1/2}) = 1.15$ MeV are rather similar to the values used for ^{96}Y; some stretching of the quasiparticle spectrum improves the agreement with the energy spectrum. The boson-fermion interaction strengths Γ_0^π and χ are taken the same as in the IBFM calculation for positive-parity states in ^{97}Y (Fig. 1). The Λ_0^π value from calculation of positive parity states in ^{97}Y has been renormalized with reduction factor 10, in accordance with the IBFM calculation for

^{91,93}Rb [22]. It was found there that Λ_0 for normal-parity states is by a factor 10 smaller than for the unique parity.

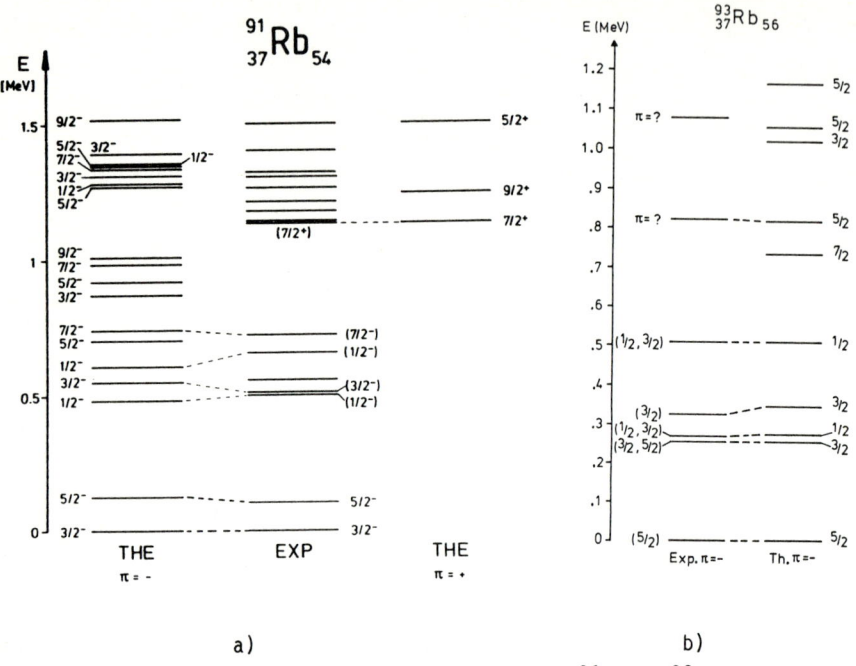

Fig. 6: IBFM and experimental energy spectrum of: a)^{91}Rb, b)^{93}Rb

The monopole interaction strength, employed to stretch somewhat the negative-parity spectrum in ^{97}Y, is $A_0^\pi = -0.25$. The resulting IBFM spectrum for negative-parity states in ^{97}Y is presented in Fig. 7a.

In the second step we have performed the IBFM calculation for the normal (positive) parity states in the even-odd neighbour ^{97}Sr. With 9 neutrons in the valence shell, $\tilde{d}_{5/2}$ quasiparticle is lying above $\tilde{g}_{7/2}$, $\tilde{d}_{3/2}$, $\tilde{s}_{1/2}$ and therefore is omitted from the calculation. The proton quasiparticle parameters are adjusted to the levels in ^{97}Sr: $\varepsilon(\tilde{g}_{7/2}) - \varepsilon(\tilde{s}_{1/2}) = 0.9$ MeV, $\varepsilon(\tilde{d}_{3/2}) - \varepsilon(\tilde{s}_{1/2}) = 1.2$ MeV, $v^2(\tilde{s}_{1/2}) = 0.5$, $v^2(\tilde{g}_{7/2}) = 0.2$, $v^2(\tilde{d}_{3/2}) = 0.15$. We note that $\tilde{d}_{3/2}$ is included above $\tilde{g}_{7/2}$, in accordance with the calculation for ^{96}Y (Fig. 3). The boson-fermion interaction strengths are $A_0^\nu = 0.18$, $\Gamma_0^\nu = 0.5$, $\Lambda_0^\nu = 0.4$. As a core we have taken ^{96}Sr, that we have already discussed, see section 2.1.1. The corresponding IBFM spectrum is presented in Fig. 7b.

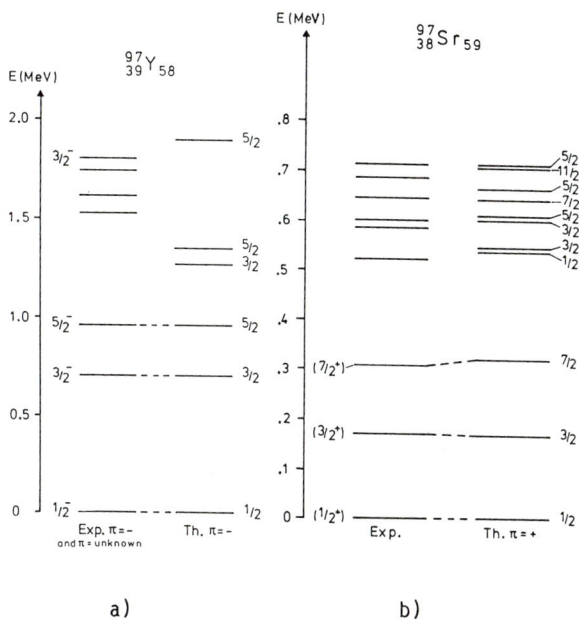

Fig. 7: IBFM calculations in comparison to experiment for: a) negative-parity energy spectrum of ^{97}Y b) positive-parity energy spectrum of ^{97}Sr.

In the third step, we have performed the IBFFM calculation for the states in ^{98}Y which are associated with the spherical shape. For the boson core we use the same U(5) parameters as in the IBFM calculations for ^{97}Y and ^{97}Sr. For quasiparticle parameters and boson-fermion interaction strengths we use the values from the IBFM calculations for the normal-parity states in ^{97}Y (Fig. 7a) and ^{97}Sr (Fig. 7b); only for Γ_0^ν we use the value from the IBFFM calculation for ^{96}Y, which is somewhat higher. For the spin-spin residual interaction strength we employ the standard estimate $V_{\sigma\sigma} = \frac{20}{A} = 0.2$. For the surface-delta interaction strength we start with the value used in the calculation for ^{96}Y (Fig. 3). Requiring that the diagonal matrix element for the $\lfloor \pi \tilde{g}_{9/2}, \nu \tilde{g}_{7/2} \rfloor$ configuration is the same, this gives the estimate $V_\delta = -0.24$ for ^{98}Y. In our calculation we use a slightly smaller value $V_\delta = -0.22$.

In this way, we have determined the parameters for ^{98}Y using the parameters corresponding to odd-even neighbours ^{97}Y, ^{97}Sr and odd-odd neighbour ^{96}Y. Thus, we perform the IBFFM calculation for ^{98}Y without fitting parameters to this nucleus. The resulting energy spectrum is presented on the l.h.s. in Fig. 5. Since the experimental spin/parity assignments, except for the ground-state, are not known, the attribution of the calculated to the experimental levels is only tentative. It should be noted that conversion coefficient data ⌊34⌋ suggest spin/parity 1^- and 2^- for the 119 and 170 keV levels, respectively.

In order to test these assignments of levels, we have performed the IBFFM calculation of the branching ratios and lifetimes for the low-lying levels of ^{98}Y. In this calculation we have employed the same values of effective charges and gyromagnetic ratios as used in the calculation for ^{96}Y, except $e^{vib} = 0.25$ and $g_1^\nu = \frac{1}{90} g_s^{free}$ $\langle r^2 \rangle = -0.78$, that were fited to the properties of ^{97}Sr. The calculated results, presented in table 1, are in very good agreement with the experiment.

Table 1

Electromagnetic properties of the low-lying levels of ^{98}Y calculated in IBFFM. Comparison to experiment is given under assumption that the spin/parity assignment of experimental levels corresponds to the IBFFM predictions from Fig. 5. We note that no parameter was fitted to ^{98}Y.

State (keV)	Transition	I_γ EXP	I_γ THE	$T_{1/2}$ EXP	$T_{1/2}$ THE
119.2	$1_1^- \to 0_1^-$	1	1	0.41ns	0.12ns
170.5	$2_1^- \to 0_1^-$	1	1	0.62μs	0.4 μs
	$\to 1_1^-$	0.19	0.22		
358.2	$1_2^- \to 0_1^-$	0.11	3.9		
	$\to 1_1^-$	1	1		0.03ns
	$\to 2_1^-$	0.32	0.96		
374.3	$4_1^- \to 2_1^-$	1	1	35.8 ns	34.6 ns
445.6	$3_1^- \to 1_1^-$	-	0.000003		
	$\to 2_1^-$	1	1	< 0.7 ns	0.05ns
	$\to 1_2^-$	-	0.00001		
	$\to 4_1^-$	0.07	0.02		

In the fourth step we perform the IBFFM calculation for the states in ^{98}Y which are associated with deformed shape. To this end we use the SU(3) boson core instead of SU(5) core. The spin/parity of the band-head is not known, but the probable multipolarity of the 121 keV transition, leading to the theoretical 4_1^- state, is E1(M2) or M1 [34], that gives $K^\pi = 3^\pm$. In order to obtain a completely regular rotational band, given by $E(I) = E_0 + \delta I(I+1)$, in the presence of a prolate core, both fermions (i.e. the proton and the neutron) have to be of hole-type [36,37]. The only Nilsson configurations that derive parentage from hole states and that lie close to the Fermi surface in this region are $\pi\ 3/2^-\ [301]$, $\pi\ 5/2^-\ [303]$ and $\nu\ 9/2^+\ [404]$. The coupling of $\pi\ 5/2^-\ [303]$ to $\nu\ 9/2^+\ [404]$ would give $K^\pi = 2^-$, with $\delta = 16.8$ keV, in good agreement with the condition that the moment of inertia is smaller than the rigid-rotor value (in this region the rigid-rotor value is $\delta \sim 15$ keV [38]). This K^π assignment, however, violates γ-transition data. The $\pi\ 3/2^-[301]\ \nu\ 9/2^+[404]$ configuration for the band head 3^- gives $\delta = 12.6$, i.e. the value for the moment of inertia that is bigger than the rigid-rotor value. In order to test this effect in

IBFFM we proceed as follows. The core nucleus ^{98}Sr we describe in IBM as a rotor with N=7, and $\delta = \delta^{EXP}$ = 20.65 keV. The boson-fermion interaction strenghts we take according to the SUSY limit [39], that for $v^2(\tilde{\pi p}_{3/2})$ = 0.924 and $v^2(\tilde{\nu g}_{9/2})$ = 0.99 gives Γ_0^{π} = 0.243 MeV, Γ_0^{ν} = 0.350 MeV. Adding $A_0^{\pi} = A_0^{\nu}$ = 0.42 MeV the IBFM calculation for $\tilde{\pi p}_{3/2}$ and $\tilde{\nu g}_{9/2}$ fermions coupled to ^{98}Sr give 3/2$^-$ and 9/2$^+$ rotational bands, respectively, with $\delta = \delta$ (RIGID ROTOR) = 15 keV. The same parametrization in IBFFM gives K^{π} = 3$^-$ with δ = 12.6 keV. From these calculations it follows that the rotational band head in ^{98}Y is probably 3$^-$ with the configuration π 3/2$^-$ [301] ν 9/2$^+$ [404], but the increase of the moment of inertia requires further investigation.

3. The N=59 Isotones

Another interesting application of the boson-fermion approach to the nuclei in A $\simeq$ 100 region is the IBFM calculation for a sequence of N=59 nuclei. In Figs. 8a-c we present the calculated energy spectra for ^{105}Pd [40], ^{103}Ru [41] and ^{101}Mo [42], in comparison to experiment.

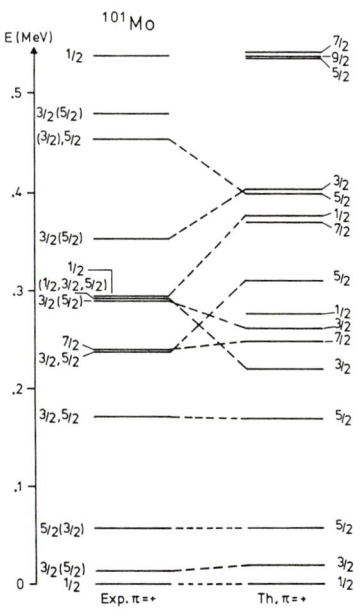

a)

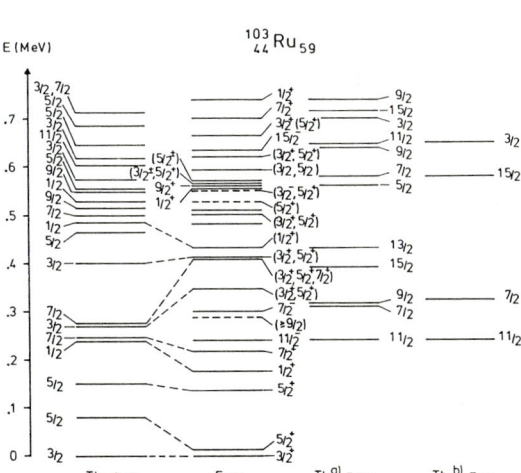

b)

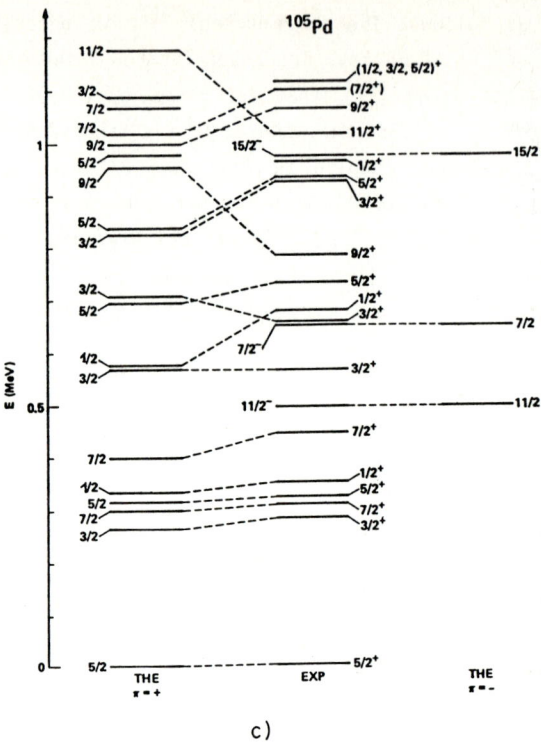

Fig. 8: IBFM calculations for the sequence of N=59 isotones in comparison to experimental data a) ^{101}Mo, b) ^{103}Ru, c) ^{105}Pd

In Refs. [40-42] we have also calculated the electromagnetic and one-particle transfer properties, and rather good agreement with experiment was obtained. In IBFM this sequence is associated with a rapid change of the quasiparticle pattern (in particular, the lowering of $\tilde{s}_{1/2}$ quasiparticle) and with a boson core of transitional character.

Conclusions

Concluding, our investigations indicate that the boson-fermion approach seems to be appropriate to account for many features of the complex nuclei in A ≃ 100 region. An interesting aspect is the fact that the properties of the Y isotopes with N<59 are well reproduced assuming individual particle character and using a core which shows no special softness. In ture a good rotor can be used as a core for the decription of ^{99}Y. Thus it appears that both spherical and deformed nuclear configurations are rigid and that the transition (in the Y chain) does not proceed via soft intermediate nuclei. Going away from Zr isotopes (in Mo, Ru, Pd) the softness start to play an important role. However, some of the features indicate that, in order to obtain better quantitative agreement with experiment, more complex correlations should be included.

References

[1] A. Arima and F. Iachello, Phys. Rev. Lett. 35 (1975) 1069
[2] D. Janssen, R.V. Jolos and F. Dönau, Nucl. Phys. A224 (1974) 93
[3] A. Arima and F. Iachello, Adv. Nucl. Phys. 13 (1984) 139
[4] F. Iachello and O. Scholten, Phys. Rev. Lett. 43 (1979) 679
[5] V. Paar, Verhl. Dtsch. Phys. Ges. 3 (1979) 683
    V. Paar, Inst. Phys. Conf. Ser. 49 (1980) 53
[6] O. Scholten, Thesis, University of Groningen (1980)
[7] S. Brant, V. Paar and D. Vretenar, Z. Phys. A319 (1984) 355
[8] T. Hübsch, V. Paar and D. Vretenar, Phys. Lett. 151B (1985) 320
[9] V. Paar, in "Proceedings of the International Symposium on Capture Gamma-Ray Spectroscopy and Related Topics", ed. S. Raman (American Institute of Physics, New York, 1985)
[10] P. Van Isacker, J. Jolie, K. Heyde and A. Frank, Phys. Rev. Lett. 54 (1985) 653
[11] V. Lopac, S. Brant, V. Paar, O.W.B. Schult, H. Seyfarth and A.B. Balantekin, Z. Phys. A323 (1986) 491
[12] A.B. Balantekin and V. Paar, Phys. Lett. 169B (1986) 9
[13] S. Brant, V. Paar, D. Vretenar, G. Alaga, H. Seyfarth, O. Schult and M. Bogdanović, Phys. Lett. 195B (1987) 111
[14] J. Blomquist, P. Kleinheinz and P.J. Daly, Z. Phys. A312 (1983) 27
[15] R.A. Meyer, E. Monnand, J.A. Pinston, F. Schussler, I. Ragnarsson, B. Pfeiffer, H. Lawin, G. Lhersonneau, T. Seo and K. Sistemich, Nucl. Phys. A439 (1985) 510
[16] K. Sistemich, G. Sadler, T.A. Khan, J.W. Grüter, W.D. Lauppe, H. Lawin, H.A. Selič, F. Schussler, J. Blachot, J.P. Bocquet, E. Monnand and B. Pfeiffer, Proc. 3rt Int. Conf. on nuclei far from stability, Cargese 1976, Report CERN 76-13 (1977) 495
[17] G. Lhersonneau, D. Weiler, P. Kohl, H. Ohm and K. Sistemich, Z. Phys. A323 (1986) 59
[18] G. Alaga and V. Paar, Phys. Lett. 61B (1976) 129
[19] S. Brant and V. Paar, Z. Phys. A329 (1988) 151
[20] S. Brant, K. Sistemich, G. Lhersonneau and V. Paar, Z. Phys. A, in print
[21] Y. Tokunaga, H. Seyfarth, O.W.B. Schult, S. Brant, V. Paar, D. Vretenar, H.G. Börner, G. Barreau, H. Faust, Ch. Hofmeyr, Ch. Schreckenbach and R.A. Meyer, Nucl. Phys. A430 (1984) 269
[22] K. Sistemich, K. Kawade, H. Lawin, G. Lhersonneau, H. Ohm, U. Paffrath, V. Lopac, S. Brant and V. Paar, Z. Phys. A325 (1986) 139
    S. Brant et al., to be published
[23] E. Monnand, J.A. Pinston, F. Schussler, B. Pfeiffer, H. Lawin, G. Battistuzzi, K. Schiruma and K. Sistemich, Z. Phys. A306 (1982) 183
[24] R.E. Petry, H. Dejbakhsh, J.C. Hill, F.K. Wohn, M. Shmid and R.L Gill, Phys.

Rev. C31 (1985) 634
[25] F.K. Wohn, J.C. Hill and R.F. Petry, Phys. Rev. C31 (1985) 634
[26] B. Pfeiffer, S. Brant, K.-L. Kratz, R.A. Meyer and V. Paar, Z. Phys. A325 (1986) 487
[27] S. Brant, Thesis, University of Zagreb (1984)
[28] D.K. Sunko and V. Paar, Phys. Lett. 146B (1984) 279
[29] F.K. Wohn, Phys. Rev. C36 (1987) 1204
[30] D. Vretenar, K.-L. Kratz and V. Paar, submitted for publication
[31] S. Brant, G. Lhersonneau, M.L. Stolzenwald, K. Sistemich and V. Paar, Z. Phys. A329 (1988) 301
[32] V. Paar, Nucl. Phys. A331 (1979) 16
[33] B.S. Reehal and R.A. Sorensen, Phys. Rev. C2 (1970) 819
[34] K. Sistemich et al., to be published
[35] S. Brant et al., to be published
[36] S. Brant, V. Paar, D.K. Sunko and D. Vretenar, Phys. Rev. C37 (1988) 830
[37] D. Vretenar, S. Brant and V. Paar, to be published
[38] K. Sistemich, G. Lhersonneau, R.A. Meyer and T. Seo, Proc. Int. Symp. on In-Beam Nuclear Spectroscopy, Debrecen, Hungary, Zs. Dombradi and T. Fenyes eds. (1984) 51
[39] S. Brant and V. Paar, Fizika 18 (1986) 279
[40] R.A. Meyer, S.V. Jackson, S. Brant and V. Paar, submitted to Phys. Rev. C
[41] H. Seyfarth, A.M. Hassan, W. Delang, P. Göttel, B. Kardon, K. Schreckenbach, R.A. Meyer, S. Brant and V. Paar, to be published
[42] K. Sistemich, W.D. Lauppe, N. Kaffrell, R.A. Meyer, S. Brant, V. Paar and D. Vorkapić, to be published

Single-Particle Excitations and Collective Vibrational Modes in ^{96}Zr

G. Molnár[1,2], T. Belgya[1], B. Fazekas[1], A. Veres[1], S.W. Yates[3],
E.W. Kleppinger[3], R.A. Gatenby[3], H. Mach[4], R. Julin[5], J. Kumpulainen[5],
A. Passoja[5], and E. Verho[5]

[1] Institute of Isotopes of the Hungarian Academy of Sciences,
 H-1525 Budapest, Hungary
[2] Institut für Kernphysik, Kernforschungsanlage Jülich,
 Postfach 1913, D-5170 Jülich, Fed. Rep. of Germany
[3] University of Kentucky, Lexington, KY 40506, USA
[4] Brookhaven National Laboratory, Upton, NY 11973, USA
[5] University of Jyväskylä, SF-40100 Jyväskylä, Finland

New results from (n,n'γ) and (p,p'γ) reactions, as well as from β^- decay confirm the double subshell closure in ^{96}Zr and provide a detailed level scheme for this nucleus. The 1750 keV first 2^+ state is described as a neutron particle-hole excitation, while the positive-parity band built on the 1582 keV excited 0^+ state is characterised by quadrupole vibrational collectivity at low spins. Candidates for two-phonon octupole, quadrupole-octupole and higher multipole states are suggested on the basis of fast E1 and E2 decays to the 3^- octupole state. The recently observed large octupole strength, its inverse relationship with the filling of the $\nu 1h_{11/2}$ orbital, and the onset of quadrupole deformation are also discussed.

1. Introduction

The doubly closed subshell structure of ^{96}Zr, while of great interest in itself, becomes especially important in view of the hindrance to the onset of deformation in the A$\sim$100 mass region. Despite the early recognition of its nearly magic character, which is ascribed to the simultaneous closure of the Z=40 and N=56 subshells, this nucleus has not been studied in detail.

The nucleus ^{96}Zr and its immediate neighbours form an island of spherical stability in a region where the nuclear shape changes abruptly from spherical near the N=50 neutron shell closure to axially deformed towards the middle of the 50-82 neutron shell. The reinforcement of stable spherical shape of the ground state is accompanied by the occurrence of 0^+ first excited states in ^{96}Zr and ^{98}Zr, as well as in ^{98}Mo. Shape coexistence has long been proposed [1] in these nuclei, although experimental attempts [2-5] to locate coexisting deformed band structures in these nuclei have not been successful.

Only recent spectroscopic studies utilizing the (n,n'γ) reaction [6,7], the (t,pγ) reaction [8,9] and β decay of low-spin [10,11] and high-spin [12,13] ^{96}Y have revealed the existence of a band built on the 0_2^+ state in ^{96}Zr. Comparisons with similar band structures in doubly closed shell ^{16}O and ^{40}Ca have led to the assumption [6] of its four-particle, four-hole nature. Assessing the degree of quadrupole collectivity of this intruder band has been one of the major objectives of the present study.

On the other hand, octupole softness, evinced by low-lying 3^- octupole states in the region around ^{96}Zr [14-16], suggests similarity with another class of doubly closed shell nuclei, such as ^{146}Gd and ^{208}Pb. Recent observation [8,9] of a number of E1 transitions to the 3^- octupole state in ^{96}Zr has motivated our search for two-phonon octupole multiplets of states.

In the present review, we first discuss the single-particle structure underlying the double subshell closure at ^{96}Zr on the basis of existing particle transfer data, complemented by our β decay results [10,11]. Then the systematics of octupole excitations are examined in the context of the recently observed [17] increase of octupole excitation strength in ^{96}Zr. In section 4, results of an extended study of ^{96}Zr levels using (n,n'γ) and (p,p'γ) reactions are presented. The next section deals with the classification of states according to particle-hole excitations, collective quadrupole, octupole and possible higher multipole excitations. Finally, conclusions are drawn concerning the underlying microscopic structure and its relationship with the earlier proposed mechanism [18] of the onset of quadrupole deformation in the A$\sim$100 mass region.

2. Double Subshell Closure

The double subshell closure at Z=40 and N=56 has emerged as one of the most fascinating features of the A$\sim$100 mass region. This double closure is evinced by the sudden increase of proton, neutron [19-21] and α-particle [21] binding energies, as well as by the high energy and low collectivity of the first 2^+ state of ^{96}Zr [22,23].

Single-proton transfer studies on ^{96}Zr have shown [24-26] that the $\pi 2p_{1/2}$ subshell is nearly filled and that the $\pi 1g_{9/2}$ single-particle state lies higher by at least 0.75 MeV, thus creating a proton subshell gap at Z=40, as shown in Fig. 1. The purity of the ground-state proton configuration has been confirmed [10,11] by the recent observation of an extremely fast first-forbidden β transition from the 0^- ground state of ^{96}Y to the ^{96}Zr ground state.

On the other hand, single-neutron stripping reactions indicate [27-29] a filled $\nu 2d_{5/2}$ neutron orbital at N=56 and an energy separation of at least 1 MeV, with respect to the next single particle orbital, the $\nu 3s_{1/2}$ orbital (Fig. 1). Moreover, an additional gap of similar size exists at N=58, making the neutron subshell closure at ^{96}Zr even stronger.

Thus, mutual reinforcement of the proton and neutron subshell closures leads to the highly stable spherical ground state structure of the ^{96}Zr nucleus and, to a lesser extent, of ^{98}Zr. The basic idea of de-Shalit and Goldhaber [30], that increased stabilisation results from the strong

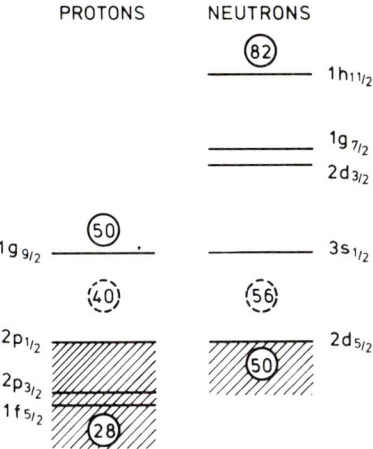

*Fig.*1 : Single-particle level structure at ^{96}Zr.

monopole interaction between protons and neutrons occupying highly overlapping shell-model orbitals, provides an explanation. Indeed, the radial overlap integral value, calculated by Zeldes [31] using harmonic oscillator wave functions and the delta interaction, amounts to about 430 keV for the $\pi 2p_{1/2}$ - $\nu 2d_{5/2}$ orbital pair while the similar quantity for the $\pi 1g_{9/2}$ - $\nu 2d_{5/2}$ combination is only about 200 keV. The situation is similar for the $\nu 3s_{1/2}$ neutron orbital which also has a good overlap with the $\pi 2p_{1/2}$ proton orbital. Since the radial overlap does not depend on the total angular momentum of the single-particle orbitals, similar conclusions can be drawn for the corresponding Sr isotopes where the $\pi 2p_{3/2}$ orbital is nearly filled.

3. Enhanced Octupole Collectivity

In closed shell nuclei, the octupole vibrations have relatively low excitation energies and compete successfully with the quadrupole mode. Microscopically, they are coherent superpositions of $1\hbar\varpi$ particle-hole excitations involving orbitals of opposite parity. It is obvious from Fig. 1 that the $2p_{3/2} \to 1g_{9/2}$ proton and the $2d_{5/2} \to 1h_{11/2}$ neutron excitations across the corresponding subshell gaps into the essentially empty high-j orbitals are of stretched E3 type. Hence, strong octupole collectivity is expected in ^{96}Zr in analogy with doubly closed shell nuclei such as ^{40}Ca and ^{208}Pb, and the closed proton subshell nucleus ^{146}Gd [32].

Indeed, the recently measured [33] lifetime of the 1897 keV 3^- octupole state in ^{96}Zr yields 39^{+49}_{-15} single particle (Weisskopf) units for the $B(E3; 3^- \to 0^+))$ transition probability. Even though the uncertainties are large, new preliminary lifetime data obtained at the TRISTAN isotope separator and comparisons with known $B(E3)$ rates in the region confirm the established inverse relationship between the octupole energies and transition rates for Z=40-42 and N=56-58, as shown in Fig. 2. It is interesting that the isoscalar transition rates obtained from (α,α') scattering [38, 39] also exhibit a rapid increase from ^{90}Zr to ^{96}Zr, although the absolute values depend on the way those rates have been extracted from the measured scattering cross sections.

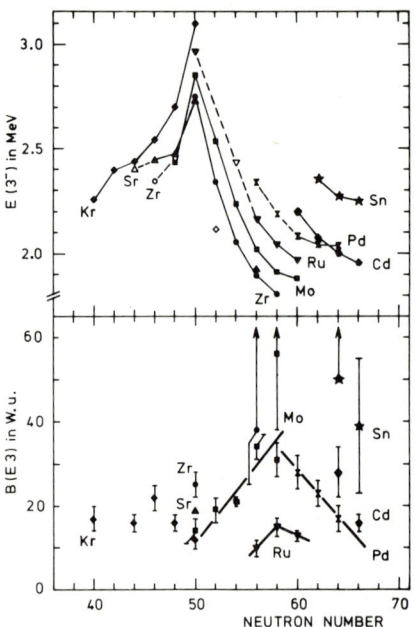

$\underline{Fig}.2$: Energies [34] and $B(E3)$ ground state transition probabilities [35,36(^{100}Mo),37(^{106}Pd)] for 3^- octupole states in the A$\sim$100 region. Unfilled symbols represent tentative assignments.

It is clear that filling of the $\nu 2d_{5/2}$ neutron orbital brings the energies of the 3^- states down rapidly due to the increasing contribution of the $2d_{5/2} \to 1h_{11/2}$ neutron excitation. Since neutrons do not have electric charge, the maximum of the E3 transition strength at N=56-58 should be associated with the enhancement of the proton octupole transition rate due to polarisation of the proton core. High spatial overlap of the $\pi 1g_{9/2}$ proton and $\nu 1h_{11/2}$ neutron orbitals makes their simultaneous population energetically favorable due to the strong isoscalar interaction, as long as saturation is not reached. Indeed, the octupole strength does not show this increase in the $_{44}$Ru nuclei where the $\pi 1g_{9/2}$ orbital is already half-filled. By the same token, decrease of the octupole transition strength beyond N=60, represented by the $_{46}$Pd data, should be ascribed to the rapid filling of the $\nu 1h_{11/2}$ orbital.

4. Experimental Studies of ^{96}Zr Levels

Until recently, the level structure of ^{96}Zr was poorly known. Most of the data included in recent compilations [40,41] came from early charged-particle studies. These data are, however, often in conflict with more recent high resolution (p,p') scattering [42] and (α,α') scattering [43] results. Moreover, when our work was initiated, the only source of γ-spectroscopic information was from the β decay of the 10 s ^{96}Y high-spin isomer [3]. Nuclear level lifetimes had been determined only for the lowest two states, i. e. for the 1582 keV 0_2^+ state [44] and the 1750 keV 2_1^+ state [22].

Application of the nonselective inelastic neutron scattering (INS) or (n,n'γ) reaction method [45,46] and reinvestigation of the β decay of 0^- $^{96}Y^g$ at the TRISTAN on-line fission product mass separator at BNL led to the discovery [6,7,10,11] of a coexisting band, associated with the 0_2^+ state of ^{96}Zr. This result has been confirmed by subsequent in-beam (t,pγe^-) reaction studies [8,9] and by reinvestigation of the β decay of high-spin $^{96}Y^m$ [12,13] where additional band members, up to spin 8^+, have been found.

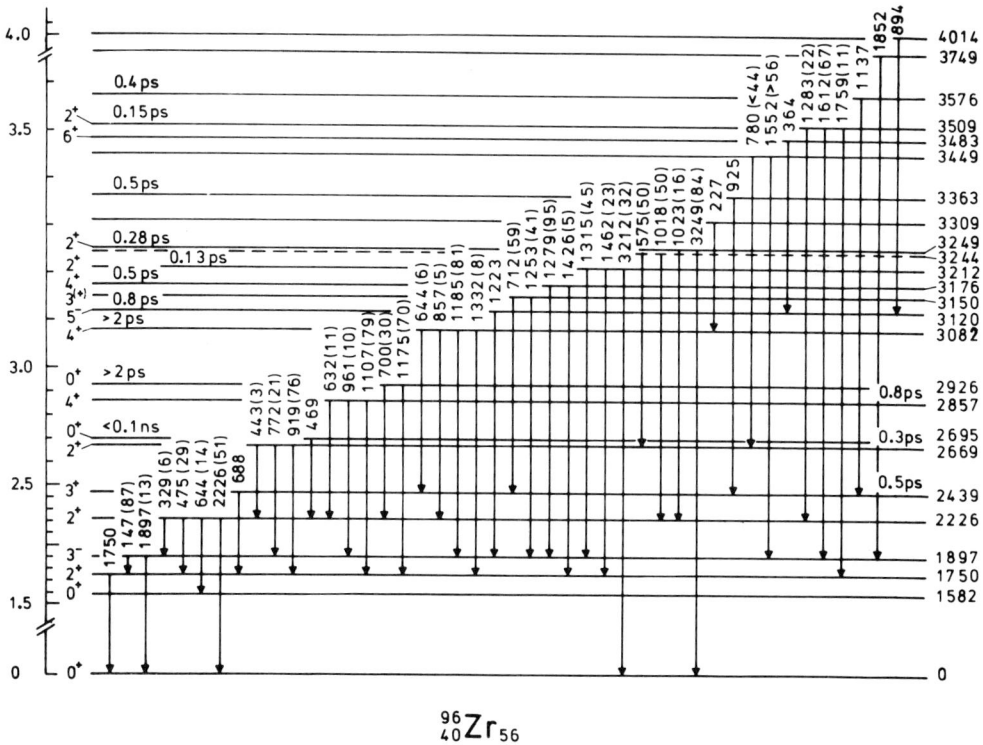

Fig.3: Partial level scheme of ^{96}Zr obtained from (n,n'γ) and (p,p'γ) reaction studies. More precise level lifetime and γ-ray energy values are given in Ref. [48].

Completion of our (n,n'γ) reaction studies at the University of Kentucky 7 MV Van de Graaff accelerator and of new (p,p'γ) coincidence experiments at the University of Jyväskylä cyclotron have resulted [47] in a fairly detailed scheme of low and intermediate spin states. This is presented in Fig. 3 where eight additional levels, decaying only to the ground state, have been omitted from the scheme. Spin assignments were made on the basis of (n,n'γ) reaction cross sections and γ-ray angular distributions, which also yielded multipole mixing ratio values. For the spin-parity assignments, results from β decay studies [10-13] and from conversion electron measurements [8] had to be used occasionally. Level lifetimes were determined from the Doppler shifts of γ rays following the (n,n'γ) reaction [48] and from electronic timing measurements with the (p,p'γ) reaction.

Detailed analysis of ^{96}Zr excitations on this basis will be given in the next section. Here we only mention that two additional 0^+ states were found at 2695 keV and 2926 keV energy, respectively. Both assignments have been confirmed by observing the corresponding E0 decays [8,9].

5. Classification of ^{96}Zr Excitations

The new spectroscopic results have enabled us to recognize the unique pattern followed by the low-energy excitations of the doubly closed subshell ^{96}Zr nucleus. This is shown in Fig. 4 where only the most important decay paths are indicated for each level chosen. In addition to the (n,n'γ) and (p,p'γ) reaction data [47], as well as information from low-spin decay [10,11], high spin states observed in the β decay of 8^+ $^{96}Y^m$ [12,13] and (t,pγ) in-beam studies [8,9] are also included. Stronger (filled symbols) or weaker (open symbols) population of the states in (p,p') [42], (α,α') [43], (t,p) [2] and (d,^{6}Li) [49] reactions is also characteristic of their origin, therefore this information is also added when available.

Quadrupole Intruder Band

The already mentioned intruder band is shown on the left side of Fig. 4. The 1582 keV 0^+ state has recently been interpreted [10,11] as a shape isomer with a quadrupole deformation parameter of 0.2 and a 96 percent $\pi 1g_{9/2}^2$ shell model configuration for protons. The sequence of levels decaying to this shape isomeric state by stretched E2 transitions has been identified up to spin 4^+ in Ref. [6] where four-particle, four-hole character has been suggested on the basis of strong excitation of the 0^+ and 2^+ members in the α-pickup reaction [49]. It is worth mentioning that in the latter study the 2226 keV 2^+ state could not be recognised as such because, as it became clear afterwards, the particle angular distribution was distorted by the collective coupling to the 0^+ bandhead [50].

The spin 6^+ and 8^+ members of the band were subsequently identified in the β decay of the $(\pi 1g_{9/2}, \nu 1g_{7/2})8^+$ isomer of ^{96}Y [12,13]. In the latter study, a pure $\pi 1g_{9/2}^2$ configuration was

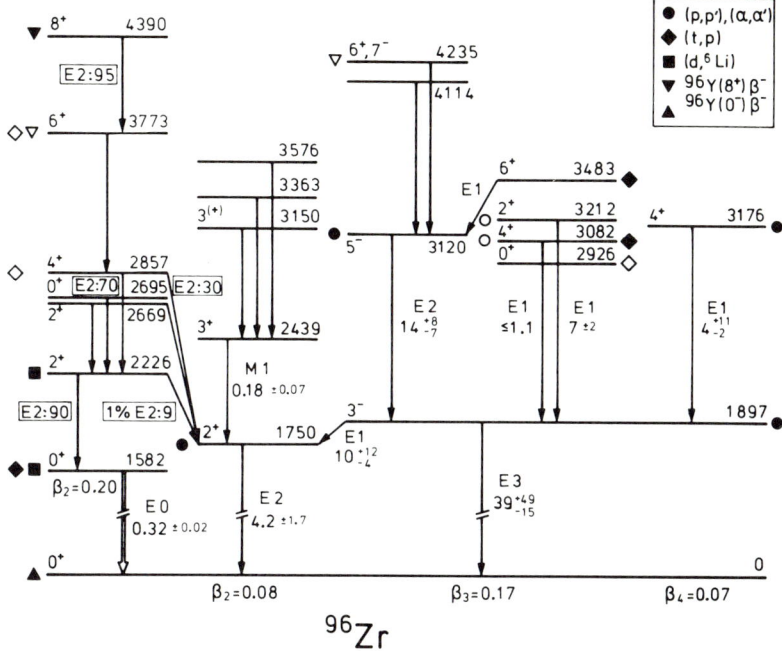

Fig.4 : Classification of ^{96}Zr levels according to their γ decay properties and populations in different reactions. Relative $B(E2)$ percentages (in boxes) or reduced transition probabilities are given beside the arrows in Weisskopf units, $B(E1)$ in 10^{-4} W.u., $\varrho(E0)$ in Bohr-Mottelson units.

suggested for the entire band on the basis of the observed Gamow-Teller decay to the 8^+ member. Comparison with the presumably pure $\pi 1g_{9/2}^2$ ground state band of ^{92}Mo certainly shows excellent agreement as far as the $0^+ - 8^+$ energy spacing is concerned, as indicated in Fig. 5. Strong disagreement is observed, however, for lower spins, since already the 6^+ state does not show the expected compression of level spacings [13].

On the other hand, good agreement is obtained up to spin 6^+ (see Fig. 5) with the ground state vibrational band of ^{98}Ru having four valence neutrons and four valence protons. This correspondence, based on the near equality of the deformation parameters and the level spacings has led to the suggestion [10] that the intruder band has collective, vibrational character at low spins. Indeed, the new $B(E2; 4^+ \to 2^+)$ value supports this view. Moreover, candidates for two-phonon 0^+ and 2^+ states could also be found, although the latter case is less clear. Hence, in addition to the proposed two-proton, two-hole configuration two-neutron, two-hole (or even more complex) excitations also have to be considered for the low spin states, as suggested in Ref. [6] where similar conclusions were drawn on the basis of multiparticle transfer data.

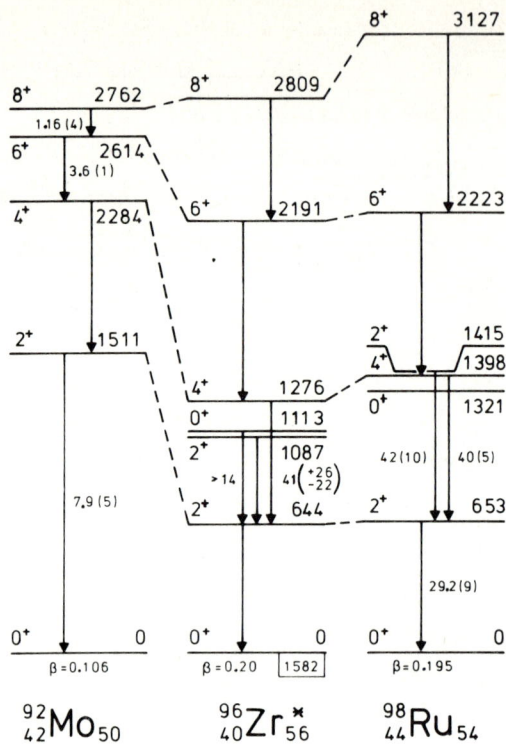

Fig.5 : Comparison of the band built on the 1582 keV 0_2^+ state of ^{96}Zr with the $\pi 1g_{9/2}^2$ terminating band of ^{92}Mo (left) and the ground state vibrational band of the ^{98}Ru intruder analog (right). Data are from Refs. [34,35].

Neutron Particle-Hole States

The first 2^+ state at 1750 keV is believed to be the lowest particle-hole state in this nucleus. The only state which decays exclusively into it is the new 2438 keV 3^+ state. The connecting pure M1 transition is one of the fastest in the region [35]. In analogy with proton excitations [51] in doubly closed shell ^{146}Gd, it is straightforward to interpret these two states as members of the $\nu(3s_{1/2}, 2d_{5/2}^{-1})$ neutron particle-hole doublet which is expected to be the lowest energy excitation of this kind. This is in accordance with the low collectivity of the 1750 keV first 2^+ state and, consequently, with the nearly magic character of the ^{96}Zr nucleus. Other states, decaying to the first 3^+ level are probably also neutron particle-hole states.

Other Collective Modes

The third class of levels decays to the strongly collective 1897 keV 3^- octupole state by fast E1 or E2 transitions. The 3120 keV 5^- level, strongly populated in (p,p') and (α,α') scattering [42,43], decays to this one-phonon octupole state by a collective E2 transition. It could be a quadrupole-octupole coupled state, similar to that observed [52] in the nearby vibrational nucleus ^{112}Cd. On the other hand, it is probably a superposition of $\pi(1g_{9/2}, 2p_{1/2}^{-1})$, $\pi(1g_{9/2}, 2p_{3/2}^{-1})$ proton, and $\nu(1h_{11/2}, 2d_{5/2}^{-1})$ neutron particle-hole excitations and, as such, can be a dyotriacontapole excitation. More sophisticated inelastic scattering experiments are needed to clarify this interesting issue. If the 4235 keV state decaying to the 5$^-$ level has spin 7$^-$, it probably also has similar neutron particle-hole structure.

One of the states decaying to the 3$^-$ octupole level by a fast E1 transition is the 3176 keV level. It is populated with large cross section in the (p,p') and (α,α') reactions, yielding a deformation length comparable to the first 2$^+$ state value. In analogy with a similar case in ^{90}Zr [53] it is suggested as a hexadecapole excitation. This conjecture also has to be checked, however, by microscopic analysis of the inelastic scattering data.

Finally, the 0$^+$, 2$^+$, 4$^+$, 6$^+$ quartet of levels is distinguished either by E1 decay or by population in the (t,p) two-neutron transfer reaction or by both features. The 3082 keV 4$^+$ and 3483 keV 6$^+$ levels have recently been suggested [54,9] to be two-phonon octupole states on the basis of predominant E1 decay to the 3$^-$ octupole state and to the 5$^-$ possible quadrupole-octupole state, respectively. In case of the former our lifetime limit [48] does not exclude this possibility, although population of such a state in the (t,p) reaction is hard to explain unless strong mixing with the 2857 keV intruder band member is assumed. The 2926 keV 0$^+$ state, on the other hand, could be excited in this reaction via coupling to the neutron pairing vibration [55]. The 3212 keV 2$^+$ level is probably the best candidate for a two-phonon octupole state since its E1 decay to the 3$^-$ state is very fast.

6. Conclusions

Reinvestigation of doubly closed subshell ^{96}Zr with complementary spectroscopic techniques has brought a much better understanding of the structure of this important nucleus. Simultaneous application of the (n,n'γ) and (p,p'γ) reactions, including Doppler shift [48] and direct timing measurements, has been particularly successful, and detailed information on the level scheme, including γ-ray transition rates, has been obtained.

The closure of the $\pi 2p_{1/2}$ proton subshell at N=40 is confirmed by the observation [10,11] of one of the fastest known first-forbidden β transitions, from 0$^-$ ^{96}Y^g to the 0$^+$ ground state of ^{96}Zr. On the other hand, identification of the first 2$^+$ level as a $\nu(3s_{1/2}, 2d_{5/2}^{-1})$ neutron particle-hole state supports the idea of a neutron subshell closure at N=56. This double subshell closure can be understood as a result of the strong monopole interaction between protons and neutrons occupying the highly overlapping $\pi 2p_{1/2}$ and $\nu 2d_{5/2}$ spherical shell model orbitals.

Reinforcement of the spherical subshell gaps blocks the high-j proton and neutron orbitals, hence quadrupole deformation can occur only as a result of cross-subshell excitations of proton and neutron pairs. This has been demonstrated here by observing a weakly collective quadrupole vibrational-like band built on the 1582 keV 0_2^+ state. A quadrupole deformation parameter of 0.2 has been deduced [10,11] for the shape isomer. On the other hand, fast E1 or E2 transitions to the strongly collective 3^- octupole state distinguish candidates for two-phonon octupole, quadrupole-octupole and higher multipole excitations.

The observed increase of octupole excitation strength at the N=56 neutron subshell closure is attributed to the increasing contribution of the $2d_{5/2} \rightarrow 1h_{11/2}$ neutron excitation. By the same token, a decrease of octupole collectivity beyond N=60 signals rapid filling of the $\nu 1h_{11/2}$ orbital. This decrease coincides with the onset of quadrupole deformation of the ground state, hence illustrating the importance [56-58] of the $\nu 1h_{11/2}$ orbital. Since the $\nu 1h_{11/2}$ orbital has nearly as good spatial overlap with the $\pi 1g_{9/2}$ proton orbital as the $\nu 1g_{7/2}$ spin-orbit partner, the arguments put forward [18] for the decisive role of the $\pi g_{9/2} - \nu g_{7/2}$ interaction as a deformation-driving force remain applicable.

We wish to thank J. Blomqvist, S. Brant, R. F. Casten, E. A. Henry, K. Heyde, P. Kleinheinz, R. A. Meyer, K. Sistemich and M. L. Stolzenwald for most fruitful discussions. This work was supported by the Hungarian Academy of Sciences and the US National Science Foundation under Grants No. INT-8512479 and No. PHY-8403278 as well as by the US Department of Energy under Contract No. DE-AC02-76CH00016.

References

1 R. K. Sheline, I. Ragnarsson and S. G. Nilsson, Phys. Lett. 41 (1972) 115

2 E. R. Flynn, J. G. Beery and A. G. Blair, Nucl. Phys. A218 (1974) 285

3 G. Sadler, T. A. Khan, K. Sistemich, J. W. Gueter, H. Lawin, W.-D. Lauppe, H. A. Selic, M. Shaanan, F. Schussler, J. Blachot, E. Monnand, G. Bailleul, J. P. Bocquet, B. Pfeiffer, H. Schrader and B. Fogelberg, Nucl. Phys. A252 (1975) 365

4 T. A. Khan, W.-D. Lauppe, K. Sistemich, H. Lawin, G. Sadler and H.A. Selic, Z. Phys. A283 (1977) 105

5 R. A. Meyer, J. Lin, G. Molnár, B. Fazekas, A. Veres and M. Sambataro, Phys. Rev. C29 (1984) 1839

6 G. Molnár, S. W. Yates and R. A. Meyer, Phys. Rev. C33 (1986) 1843

7 G. Molnár, S. W. Yates, E. W. Kleppinger, T. Belgya, B. Fazekas, A. Veres and R. A. Meyer, in: *Symmetries and Nuclear Structure*, R. A. Meyer and V. Paar, eds. (Harwood Pub. Co., New York, 1987) p. 237

8 E. A. Henry, in: *Nuclear Structure, Reactions and Symmetries*, A. Meyer and V. Paar, eds. (Word Scientific, Singapore, 1986) p. 616

9 E. A. Henry et al., contribution to this Workshop

10 H. Mach, G. Molnár, S. W. Yates, R. L. Gill, A. Aprahamian and R. A. Meyer, Phys. Rev. C37 (1988) 254

11 H. Mach et al., contribution to this Workshop

12 M. L. Stolzenwald, G. Lhersonneau, S. Brant, G. Menzen and K. Sistemich, Z. Phys. A327 (1987) 359

13 M. L. Stolzenwald et al., contribution to this Workshop

14 O. Nathan and S. G. Nilsson, in: *Alpha, Beta and Gamma Ray Spectroscopy*, K. Siegbahn, ed. (North-Holland, Amsterdam, 1965)

15 A. Bohr and B. Mottelson, *Nuclear Structure*, vol. 2 (W. A. Benjamin, New York, 1975)

16 W. Nazarewicz, P. Olanders, I. Ragnarsson, J. Dudek, G. A. Leander, P. Möller and E. E. Ruchowska, Nucl. Phys. A429 (1984) 269

17 G. Molnár, H. Ohm, G. Lhersonneau and K. Sistemich, submitted to Z. Phys, A

18 P. Federman and S. Pittel, Phys. Rev. C20 (1979) 820

19 M. Graefenstedt, U. Keyser, F. Munnich, F. Schreiber, H. R. Faust and H. Weikard, Z. Phys. A327 (1987) 383

20 U. Keyser, contribution to this Workshop

21 K. Bos, G. Audi and A. H. Wapstra, Nucl. Phys. A432 (1985) 140

22 Yu. P. Gangrskii and I. Kh. Lemberg, Yad. Fiz. 1 (1965) 1025; Soviet J. Nucl. Phys. 1 (1965)731

23 S. Raman, C. H. Malarkey, W. T. Milner, C.W. Nestor, Jr. and P.H. Stelson, Atomic Data Nucl. Data Tables 36 (1987) 1

24 M. R. Cates, J. B. Ball and E. Newman, Phys. Rev. 187 (1969) 1682

25 B. M. Preedom, E. Newman and J. C. Hiebert, Phys. Rev. C166 (1968) 1156

26 E. R. Flynn, R. E. Brown, F. Ajzenberg-Selove and J. A. Cizewski, Phys. Rev. C28 (1983) 575

27 B. L. Cohen, Phys. Rev. 125 (1962) 1358

28 B. L. Cohen and O. V. Chubinsky, Phys. Rev. 131 (1963) 2184

29 C. R. Bingham and G. T. Fabian, Phys. Rev. C7 (1973) 1509

30 A. de-Shalit and M. Goldhaber, Phys. Rev. 92 (1953) 1211

31 N. Zeldes, Nucl. Phys. 2 (1956/57) 1

32 P. Kleinheinz, Phys. Scripta 24 (1981) 236

33 H. Ohm, G. Lhersonneau and K. Sistemich, Ann. Rep. 1987, IKP, KFA Jülich (1988)

34 M. Sakai, Atomic Data Nucl. Data Tables 31 (1984) 399

35 P. M. Endt, Atomic Data Nucl. Data Tables 23 (1979) 547; 26 (1981) 47

36 S. J. Mundy, W. Gelletly, J. Lukasiak, W. R. Phillips and B. J. Varley, Nucl. Phys. A441 (1985) 534

37 D. De Frenne, E. Jacobs, M. Verboven and G. De Smet, Nucl. Data Sheets 53 (1988) 73

38 A. M. Bernstein, in: *Advances in Nuclear Physics*, M. Baranger and E. Vogt, eds., 3 (Plenum Press, New York, 1969) p. 325

39 D. Rychel, R. Gyufko, B. van Kruchten, M. Lahanas, P. Singh and C. A. Wiedner, Z. Phys. A326 (1987) 455

40 *Table of Isotopes*, 7th ed., C. M. Lederer and V. S. Shirley, eds. (Wiley, New York, 1978)

41 H.-W. Moeller, Nucl. Data Sheets 35 (1982) 281

42 M. Fujiwara, Y. Fujita, S. Y. Hayakawa, H. Ikegami, I. Katayama, K. Katori, S. Morinobu, Y. Tokunaga and T. Yamazaki, Ann. Rep. 1983, RCNP Osaka University (1983) p. 59

43 M. Lahanas, D. Rychel, P. Singh, R. Gyufko, D. Kolbert, B. van Krüchten, E. Madadakis and C. A. Wiedner, Nucl. Phys. A455 (1986) 399

44 D. Burch, P. Russo, H. Swanson and E.G. Adelberger, Phys. Lett. 40B (1972) 357

45 S. W. Yates and G. Molnár, in: *Nuclear Structure, Reactions and Symmetries*, R. A. Meyer and V. Paar, eds. (World Scientific, Singapore, (1986) p. 624

46 G. Molnár, B. Fazekas, L. Dabolczi, T. Belgya and A. Veres, Sci. Instrum. (Poland) 1 (1986) 63

47 G. Molnár, T. Belgya, B. Fazekas, A. Veres, S. W. Yates, E. W. Kleppinger, R. A. Gatenby, R. Julin, J. Kumpulainen, A. Passoja and E. Verho, submitted to Nucl. Phys. A

48 T. Belgya et al., contribution to this Workshop

49 A. M. van den Berg, A. Saha, G. D. Jones, L. W. Put and R. H. Siemssen, Nucl. Phys. A429 (1984) 1

50 R. H. Siemssen, private communication

51 S. W. Yates, R. Julin, P. Kleinheinz, B. Rubio, L. G. Mann, E. A. Henry, W. Stöffl, D. J. Decman and J. Blomqvist, Z. Phys. A324 (1986) 417

52 R. De Leo, M. Pignanelli, W. T. A. Borghols, S. Brandenburg, M. N. Harakeh, H. J. Lu and S. Y. van der Werf, Phys. Lett. 165B (1985) 30

53 Y. Fujita, M. Fujiwara, S. Morinobu, I. Katayama, T. Yamazaki, T. Itahashi and H. Ikegami, Phys. Lett. 98B (1981) 175

54 R. A. Meyer et al., Int. Conf. on Nuclear Structure through Static and Dynamic Moments, Melbourne, Australia, Aug. 25-28, 1987, H. H. Bolotin, ed., to be published

55 R. Julin, J. Kantele, J. Kumpulainen, M. Luontama, A. Passoja, W. Trzaska, E. Verho and J. Blomqvist, Phys. Rev. C36 (1987) 1129

56 B. Pfeiffer and K.-L. Kratz, contribution to this Workshop

57 K. Heyde, contribution to this Workshop

58 W. Nazarewicz and T. Werner, contribution to this Workshop

Doppler-Shift Lifetime Measurements in ^{96}Zr

T. Belgya[1], G. Molnár[1], B. Fazekas[1], A. Veres[1], S.W. Yates[2], and R.A. Gatenby[2]

[1] Institute of Isotopes of the Hungarian Academy of Sciences, H-1525 Budapest, Hungary
[2] University of Kentucky, Lexington, KY 40506, USA

Introduction

While differing types of collective structures have been proposed [1-3] in ^{96}Zr, γ-ray transition rates in this nucleus were not known. To assess the degree of collectivity of excited states of ^{96}Zr, we have measured the Doppler shifts of γ rays emitted following inelastic neutron scattering (INS) of monoenergetic accelerator-produced neutrons and determined lifetimes of many levels of this doubly closed subshell nucleus.

Elenkov and coworkers [4] have demonstrated that the DSAM following INS of reactor fast neutrons can provide lifetimes for levels in light nuclei despite the low recoil velocities involved. In medium-mass nuclei, the recoil velocities are even smaller; nonetheless, it is still possible to observe Doppler shifts of γ rays, and comparisons with known lifetimes indicate that the methods of evaluation are reliable.

Experimental procedures and results

The experiments were performed with the 7 MV Van de Graaff accelerator at the University of Kentucky where the ^{3}H(p,n)^{3}He reaction was used to produce fast neutrons. A ZrO$_2$ powder target, enriched to 60% in ^{96}Zr, was placed in the pulsed neutron flux, and a coaxial HPGe detector of 20% efficiency and 1.8 keV energy resolution was employed to observe γ rays following INS. Excitation function, angular distribution and Doppler-shift measurements were performed. Details of the experimental arrangements and the data reduction procedures can be found elsewhere [5], while the Doppler-shift measurements are described here.

The Doppler-shift lifetime measurements were carried out at neutron energies of 3.8 MeV and 4.3 MeV. At the lower energy, γ-ray spectra at detector angles of 90° and 145° with respect to the incident neutrons were acquired, while at the higher energy, spectra at 35°, 90° and 145° were recorded in 8k channels of analyzer memory. A ^{56}Co source was always present to obtain an internal energy calibration. The peak positions were corrected for the nonlinearity of the spectrometer system which was measured with a ^{226}Ra source in a separate experiment. The energies of the peaks were then calculated using a linear fit to the ^{56}Co peaks.

The Doppler shifts were determined using the approximation described as method I in Ref. [6]. The procedure consisted of a maximum likelihood fit of the γ-ray energies to the formula

$$E_\gamma(\Theta) = E_\gamma (1 + F_{exp}(\tau) v_{cm} \cos(\Theta)/c) \tag{1}$$

where $E_\gamma(\Theta)$ is the Doppler-shifted γ-ray energy measured at a detector angle of Θ with respect to the incident neutron. E_γ is the energy of the γ ray emitted by a nucleus at rest, v_{cm} is the velocity of the center of mass in the inelastic neutron scattering collision with the ^{96}Zr atom, and c is the

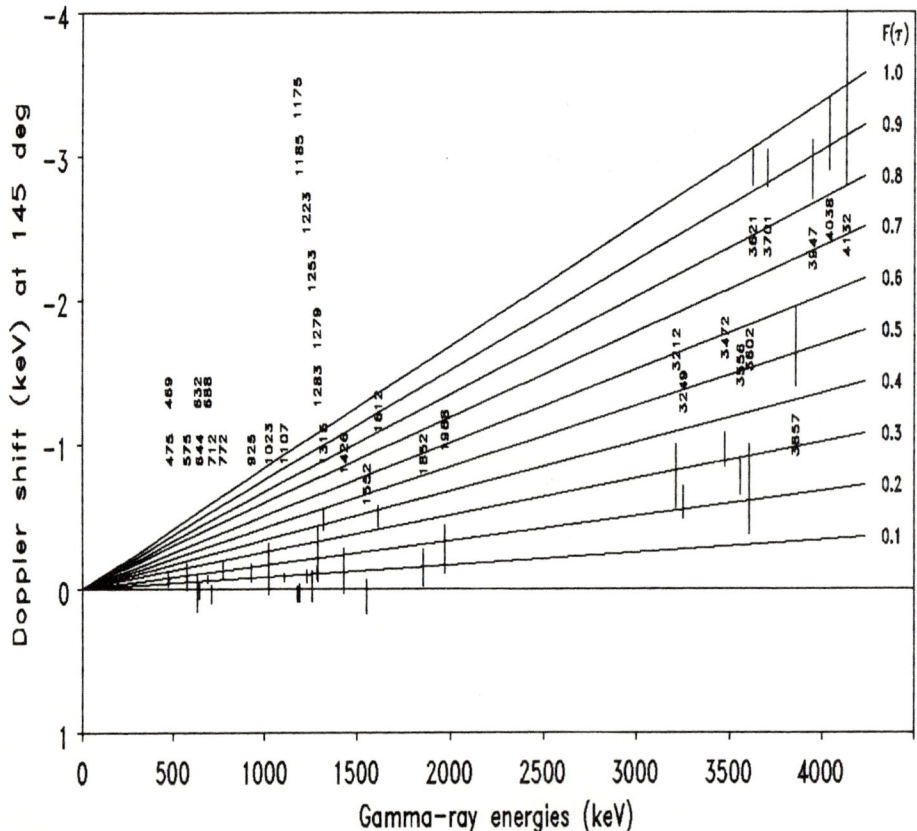

Figure 1: The Doppler shifts of ^{96}Zr γ rays at 145° detector angle and 4.3 MeV neutron energy. Vertical bars indicate the measured shifts (with ±1σ uncertainty) for γ rays with energies indicated in keV. Straight lines represent Doppler shifts expected for the attenuation factors F(τ) shown on the right.

velocity of light. In the present case v_{cm}/c is about 0.1 %. Finally, $F_{exp}(\tau)$ is the experimental attenuation factor, to be determined from the measured Doppler shift. The Doppler shifts observed at an angle of 145° are shown in Figure 1 where the shifts for constant $F_{exp}(\tau)$ values are represented as a function of γ-ray energy, by straight lines. The lifetime for each level was determined with the following averaging procedure. First, the F_{exp} values of the γ rays deexciting each individual level were averaged at a given incident neutron energy. Then the F_{exp} values obtained from the evaluation of 3.8 MeV and 4.3 MeV measurements were averaged for the same level.

To determine the lifetimes of levels from F_{exp} values, the theoretical attenuation factors were calculated using the method of Blaugrund [7] with standard LSS stopping powers and the geometrical density of 1.026×10^{22} ^{96}Zr atoms/cm^3, instead of the crystalline density, of the enriched ZrO_2 powder target. The validity of the theoretical procedure could be checked by determining two known lifetimes of levels of ^{90}Zr, also present in the target. Comparison of our results with previously measured lifetimes, determined by the resonance fluorescence method and not affected by stopping power uncertainties, shows good agreement (Table 1). The use of geometrical density instead of the crystalline density has been justified by examination of the fine structure of grains in the powder sample with an electron microscope.

Table 1: Comparison of known ^{90}Zr level lifetimes with the present results.

$E_{lev.}$ (keV)	τ (fs)	
	Present work	Ref. [8]
3308.3	140±20	107±20
3842.4	15±5	17.2±0.6
		20±3

The $F(\tau)$ curve calculated for ^{96}Zr at 4.3 MeV incident neutron energy is shown in Figure 2. The 3.8 MeV $F(\tau)$ curve is only slightly different from the 4.3 MeV one and, in this way, the average of the two curves could be used to calculate level lifetimes from the averaged F_{exp} values. This procedure was used because the F_{exp} values have symmetric errors, while the lifetimes have nonsymmetric errors and a more complicated averaging procedure would have been required.

The most important mean lifetime values obtained for ^{96}Zr levels [9] are collected in Table 2 where the reduced electromagnetic transition

octupole coupled state may be evinced by the enhanced 1223 keV E2 transition to the same 3⁻ level. More detailed discussion of these conjectures based on the present lifetime results is given in a separate contribution [9].

References

[1] G. Molnár, S. W. Yates, and R. A. Meyer, Phys. Rev. **C33**, 1843 (1986)
[2] H. Mach et al., Phys. Rev. **C37**, 254 (1988)
[3] R. A. Meyer et al., Int. Conf. on Nuclear Structure through Static and Dynamic Moments, Melbourne, Australia, Aug. 25 Sep. 1 1987
[4] D. Elenkov, D. Lefterov and G. Toumbev, Nucl. Instr. Methods **228**, 62 (1984)
[5] S. W. Yates and G. Molnár, Nuclear Structure, Reactions and Symmetries, Eds. R. A. Meyer and V. Paar (World Scientific, Singapore, 1986) p. 632
[6] C. Moazed, T. Becker, P. A. Assimakopoulos and D. M. Van Patter, Nucl. Phys. **A169**, 651 (1971)
[7] A. E. Blaugrund, Nucl. Phys. **88**, 501 (1966)
[8] C. M. Lederer and . S. Shirley, Table of Isotopes, 7th ed. (Wiley, New York, 1978)
[9] G. Molnár et. al., Contribution to this Workshop.

In-Beam Studies of ^{96}Zr and ^{98}Zr: Collective Excitations

E.A. Henry[1], R.A. Meyer[1], A. Aprahamian[1], K.H. Maier[2], L.G. Mann[1], and N. Roy[1]

[1]Lawrence Livermore National Laboratory, Livermore, CA 94550, USA
[2]Hahn-Meitner-Institut für Kernforschung,
  Glienicker Str. 100, D-1000 Berlin, Germany

Introduction

Nearly two decades ago signatures of deformation in the ground state bands of ^{100}Zr and ^{102}Zr were identified, and the rapid change in the deformation of heavy zirconium nuclei noted.[1] It is now well accepted that the short-range proton-neutron interaction between the $1g_{9/2}$ and $1g_{7/2}$ spin-orbit partners plays an important role in producing ground state deformation in this region.[2] Nevertheless, recent studies of zirconium nuclei, including those in the transition region,[3-10] continue to refine our understanding of the interplay between single-particle and collective degrees of freedom. In this report we discuss some aspects of the level structures of ^{96}Zr and ^{98}Zr with emphasis on collective excitations.

Cheifitz et al.[1] first observed the rapid decrease in the energy of the 2_1^+ level as neutron pairs were added to ^{98}Zr, and measured large collectivities for the $2_1^+ \rightarrow 0_1^+$ transitions in ^{100,102}Zr. They concluded that there was a strong indication of deformation in these nuclei. From recent inelastic scattering experiments, Rychel et al.[6] obtained $0_1^+ \rightarrow 2_1^+$ transition rates for $^{90-96}$Zr which indicate that the collectivity of these transitions decreases by a factor of up to 50 compared to the same transitions in ^{100,102}Zr. They deduced further that the $2_1^+ \rightarrow 0_1^+$ transition strength in $^{90-96}$Zr is dominated by neutron matrix elements. For the $0_1^+ \rightarrow 3_1^-$ transitions, Rychel et al. obtained B(E3) enhancements of approximately 20 compared to single particle estimates for all $^{90-96}$Zr nuclei.

The monopole matrix elements provide additional information on deformation and mixing of 0^+ states. The ρ_{21}^2 value is about 0.01 for all the $^{92-98}$Zr nuclei, but increases to 0.15 for ^{100}Zr reflecting the coexistence of spherical and deformed (β = 0.35) shapes and the mixing of these states in that nucleus.[3] Recently a modest deformation of β = 0.20 has been deduced by Mach et al.[10] for the 0_2^+ level of ^{96}Zr by assuming that the first two 0^+ levels can be described as mixtures of $(p_{1/2})^2$ and $(g_{9/2})^2$ proton configurations, and that these configurations represent spherical and deformed structures, respectively. Using the same model and new data from Mach and Gill[9], a value of β = 0.14 can be calculated for the 0_2^+ level of ^{98}Zr. This counterintuitive decrease in the model dependent deformation for ^{98}Zr undoubtedly results from the simplifying assumption that the 0_1^+ and 0_2^+ levels are mixtures of the $p_{1/2}$ and $g_{9/2}$ proton states only. Wave functions calculated by Federman and Pittel[2] indicate that the 0_2^+ level in ^{98}Zr is highly admixed and, hence, deformed. Examination of those wave functions suggests that the admixture of deformed and spherical components is approximately the same for the two lowest 0^+ levels of ^{96}Zr and ^{98}Zr; thus, the β-values for the 0_2^+ levels of ^{96}Zr and ^{98}Zr are probably similar.

In ^{96}Zr, Molnar et al.[4] have identified 2^+ and 4^+ levels that decay most strongly to the 0_2^+ level. They proposed that these levels constitute a four-particle, four-hole band, coexisting with the spherical shell model states in ^{96}Zr. Stolzenwald et al.[7] have identified the 6^+ and 8^+ levels that belong to this band structure. They indicated that though the band might be considered the complete set of states from the $(g_{9/2})^2$ proton configuration, it is not compressed as is the analogous band in ^{90}Zr. They suggested that the level energy sequence of this band in ^{96}Zr is due to a mixture of vibrational and shell-model properties. Similarly, we have found a set of nine levels in ^{98}Zr that decay most strongly to the 0_2^+ level and appear to form a fully developed O(6) excitation structure.[5] Levels were identified that are reasonable candidates for 0^+ and 2^+ "two-phonon" states in ^{98}Zr. Juxtaposition of the excited band structures built on the 0_2^+ levels in $^{90-98}$Zr with the ground state band structures of $^{100-102}$Zr suggests a smooth change in collective character from vibrational through gamma-soft to symmetric-rotor structure for these nuclei.

In order to provide experimental information on the nature of the transitional nuclei, we have employed in-beam spectroscopy using the (t,p) reaction to study ^{96}Zr (gamma rays and conversion electrons) and ^{98}Zr (gamma rays). Photopeaks and conversion-electron multiplets belonging to these nuclei are identified by requiring coincidences with outgoing protons. The usual techniques of in-beam spectroscopy, including gamma-gamma coincidences, were used to establish level schemes for these two nuclei.

Quadrupole Excitations

We have tried to identify candidates for the two-phonon 0^+ and 2^+ levels in order to characterize the nature of the suggested vibrational band in ^{96}Zr. Such states will be found near the 4^+ level at 2857 keV, among and probably mixed with shell model states. Consider first possible 2^+ levels. The 2_1^+ level decays with approximately single particle strength. The dominance of neutron matrix elements[6] indicates that a major component of its wavefunction is the $d_{5/2}s_{1/2}$ neutron configuration. We can expect to find the 3^+ level from this configuration at a low excitation energy. We identify two new levels in this work at 2439 and 2669 keV. The most intense transition from both levels is to the 2_1^+ level. Both transitions can be characterized as M1, E2, or M1/E2 multipolarity from our conversion electron data, requiring a positive parity for the two levels. However, the 2439-keV level is not populated in the beta decay of 0^- ^{96}Y (refs. 4, 12), suggesting a spin of 3 or 4 for this level. This level is a reasonable

Table 1. A summary of ρ^2- and X-values for 0^+ levels in ^{96}Zr.

0^+ level energy (keV)	Measured quantity	Value (Δ%)
1582	ρ^2_{21}	0.0107
2695	X_{312}	0.0039(22)[a]
	X_{322}	0.037-0.043(16)[b]
	ρ^2_{32}/ρ^2_{31}	9.4(28)
2925	X_{412}	0.067(40)[a]
	X_{422}	<0.119 (2 sigma)
	X_{432}	<2.8 (2 sigma)
	ρ^2_{42}/ρ^2_{41}	<3.0

[a]Statistical uncertainty only. A calibration uncertainty of 50% for E_{e^-} > 1600 keV is not included.

[b]The lower (higher) value applies if the 1114-keV gamma ray is M1 or E2 (E1).

----

candidate for the 3^+ member of the $d_{5/2}s_{1/2}$ doublet. The level at 2669 keV is populated by first forbidden unique beta decay with a log $f_1 t$ of 9.6 (ref. 10) and is close in energy to the two-phonon 4^+ level, making it a reasonable candidate for the two-phonon 2^+ level. A transition to the one-phonon 2^+ level is not observed. The enhancement of such an E2 transition would have to be greater than 1000 to compete with a single-particle M1 transition to the 2^+_1 level.

In addition to the two-phonon 0^+ state, a 0^+ state arising from the two-neutron configuration $(s_{1/2})^2$ undoubtedly exists at about 2.68 MeV also.[13] Two 0^+ levels are identified at 2695 keV (refs. 13, 14) and 2925 keV (this work and ref. 11), near the 4^+ level at 2857 keV. We have observed E0 transitions from the 2695-keV level to the 1582-keV level and the ground state, and from the 2925-keV level to the ground state. The intensity of the 1175-keV gamma-ray that depopulates the 2925-keV level is enhanced by nearly a factor of two in the 1750-keV coincidence gate compared to its singles intensity. This is consistent with an assignment of 0^+ for the 2925-keV level and a $0^+ \to 2^+ \to 0^+$ level sequence. Only upper limits are obtained for E0 transitions from the 2925-keV level to the 1582- and 2695-keV levels. X-values and ratios of ρ^2-values deduced from these data are summarized in Table 1.

Based on the evidence of a strong E2 transition to the 2225-keV 2^+ level, the level at 2695-keV is a good candidate for the two-phonon 0^+ state. There is some mixing between the levels at 2695 and 2925 keV since the strongest E2 transition from the higher level is also to the 2225-keV level. However, the E0 data do not lend themselves to a clear interpretation of either level as the two-phonon 0^+ level when discussed in terms of simple models. Since the "ground state" of the collective band has a deformation of β = 0.20, the model for spheroidal nuclei might apply: $X_{\beta 11} = 4\beta_0^2$. From this model β_0 is 0.1 for the 2695-keV level, in significant

This work was performed under the auspices of the US Department of Energy by the Lawrence Livermore National Laboratory under contract number W-7405-ENG-48.

1. E. Cheifetz, R. C. Jared, S. G. Thompson, and J. B. Wilhelmy, Phys. Rev. Lett. 25, 38 (1970).
2. P. Federman and S. Pittel, Phys. Lett. 69B, 385 (1977). Phys. Rev. C20, 820 (1979).
3. F. K. Wohn, J. C. Hill, C. B. Howard, K. Sistemich, R. F. Petry, R. L. Gill, H. Mach, and A. Piotrowski, Phys. Rev. C33, 677 (1986).
4. G. Molnar, S. W. Yates, and R. A. Meyer, Phys. Rev. C33, 1843 (1986).
5. R. A. Meyer, E. A. Henry, L. G. Mann, and K. Heyde, Phys. Lett. 177B, 271 (1986).
6. D. Rychel, R. Gyufko, B. van Kruchten, M. Lahanas, P. Singh, and C. A. Wiedner, Z. Phys. A - Atoms and Nuclei 326, 455 (1987).
7. M. L. Stolzenwald, G. Lhersonneau, S. Brant, G. Menzen, and K. Sistemich, Z. Phys. A - Atoms and Nuclei 327, 359 (1987).
8. M. Graefenstedt, U. Keyser, F. Munnich, F. Schreiber, H. R. Faust, and H. Weikard, Z. Phys. A - Atoms and Nuclei 327, 383 (1987).
9. H. Mach and R. L. Gill, Phys. Rev. C36, 2721 (1987).
10. H. Mach, G. Molnar, S. W. Yates, R. L. Gill, A. Aprahamian, and R. A. Meyer, Phys. Rev. C37, 254 (1988).
11. G. Molnar (private communication, 1987).
12. H. Mach, R. L. Gill, J. A. Winger, and A. Wolf, Bull. Am. Phys. Soc. 32, No. 8, 1570 (1987); and H. Mach (private communication 1987).
13. J. B. Ball and K. H. Bhatt, private communication cited in E. R. Flynn, J. G. Beery, and A. G. Blair, Nucl. Phys. A218, 285 (1974).
14. S. W Yates, L. G. Mann, E. A. Henry, D. J. Decman, R. A. Meyer, R. J. Estep, R. Julin, A. Passoja, J. Kantele, and W. Trzacka, Phys. Rev. C36, 2143 (1987).
15. R. Julin, J. Kantele, J. Kumpulainen, A. Passoja, W. Trzaska, E. Verho, and J. Blomquist, JYFL Annual Report 1987, University of Jyvaskyla, Jyvaskyla, Finland, p. 81.
16. S. Lunardi, P. Kleinheinz, M. Piiparinen, M. Ogawa, M. Lach, and J. Blomquist, Phys. Rev. Lett. 53, 1531 (1984).
17. W. Nazarewicz, P. Olanders, I. Ragnarsson, J. Dudek, G. A. Leander, P. Moller, and E. Ruchowska, Nucl. Phys. A429, 269 (1984).
18. W. Nazarewicz, Proceedings of the International Conference of Nuclear Structure through Static and Dynamic Moments, Melbourne, August 25-28, 1987, H. H. Bolotin, ed. (Conference Proceedings Press, Melbourne, 1987) p. 180.

Band Structures in ^{96}Zr and ^{98}Zr Studied Through the High-Spin Beta Decays of the Y Parents

M.L. Stolzenwald[1], S. Brant[1,*], H. Ohm[1], K. Sistemich[1], and G. Lhersonneau[2]

[1]Institut für Kernphysik, Kernforschungsanlage Jülich,
  Postfach 1913, D-5170 Jülich, Fed. Rep. of Germany
[2]Institut für Kernchemie, Universität Mainz,
  D-6500 Mainz, Fed. Rep. of Germany

The isotopes ^{96}Zr and ^{98}Zr are presently subject to intensive studies since their properties play a key role in the understanding of the rapid transition from spherical to deformed nuclear shapes at N $\sim$ 60. An interesting question is the strength of the subshell closures at $Z = 38, 40$ and $N = 56$ and to which extent the structure of ^{96}Zr, ^{98}Zr and of their immediate neighbours is influenced by these shell effects. There is evidence, although limited, for closed-shell-or individual-particle character of these nuclei, but more information on the nature of the excited states, especially those with sizeable spins is desired. This is expected to help clarifying the situation with respect to the possible coexistence of spherical shapes with collective structures including deformed configurations.

In order to improve the knowledge of ^{96}Zr and ^{98}Zr and to search for band structures, angular correlation measurements have been performed. A way to reach high-spin states in ^{96}Zr and ^{98}Zr, is the study of the high-spin β decaying isomers of the ^{96}Y and ^{98}Y parents [1,2] which are populated in fission. These isomers are accessible with the recoil separator JOSEF of the Kernforschungsanlage Jülich [3] which has no chemical selectivity. At fission-product separators with ion sources one can usually study only the low-spin decays of the Y isotopes. This is so since Y is not separated directly but is only produced in the decay of the Rb and Sr precursors with low spin.

Through the analysis of $\gamma\gamma$-coincidences and $\gamma\gamma$-angular correlations the decay schemes of ^{96}Zr (Fig.1) and ^{98}Zr (Fig.3) have been considerably extended compared to early publications[1,2] and in particular information on the spins and parities of levels in ^{96}Zr and ^{98}Zr has been obtained.

Band Structures in ^{96}Zr

The nucleus ^{96}Zr with closed proton and neutron subshells $Z = 40$ and $N = 56$ has magic properties. This is evident from its 2_1^+ energy of 1751 keV which is about 1 MeV larger than the typical values in this mass region, and from the observation of single-particle configurations in neighbouring nuclei. The results of the angular correlation measurement [4] reveal the spins of 8 levels and limit the spin assignment to a few alternatives in 5 cases. In particular, the identification of the most intense γ-ray cascade of 617, 915, 1107 and 1751 keV as strechted E2 transitions characterizes the 4390 keV level as an 8^+ state which is fed strongly by β^--decay. Some results are compiled in Fig.2. The transitions with the largest relative B(E2) values are those of 617, 915, 631 and 644 keV which lead into the first excited 0^+ state at 1583 keV. The 631 keV ($4^+ \to 2_2^+$) and 644 keV ($2_2^+ \to 0_2^+$) transitions have been previously proposed [5] to form members of a band based on the 0_2^+ level. It should be pointed out that the fast β-decay of the high-spin isomer of ^{96}Y to the 8$^+$ level in ^{96}Zr can best be understood in terms of a $[\pi g_{9/2} \cdot \nu g_{7/2}]8^+ \xrightarrow{GT} [\pi g_{9/2}^2]8^+$ transition [4]. Since it has been shown that the 0_2^+ level of ^{96}Zr also contains a strong fraction of $[\pi g_{9/2}^2]$ [6] it is reasonable to assume basically $[\pi g_{9/2}^2]$ structure for this band.

Then the most remarkable feature of the band is the lack of compression of the energy levels with increasing spin since the $2^+, 4^+, 6^+$ and 8^+ levels are almost equidistant. This seems to establish a clear difference to other nuclei with two-nucleon configurations, e.g. the isotope ^{90}Zr, where the energies of the corresponding levels are at 0, 2186, 3077, 3448 and 3589 keV.

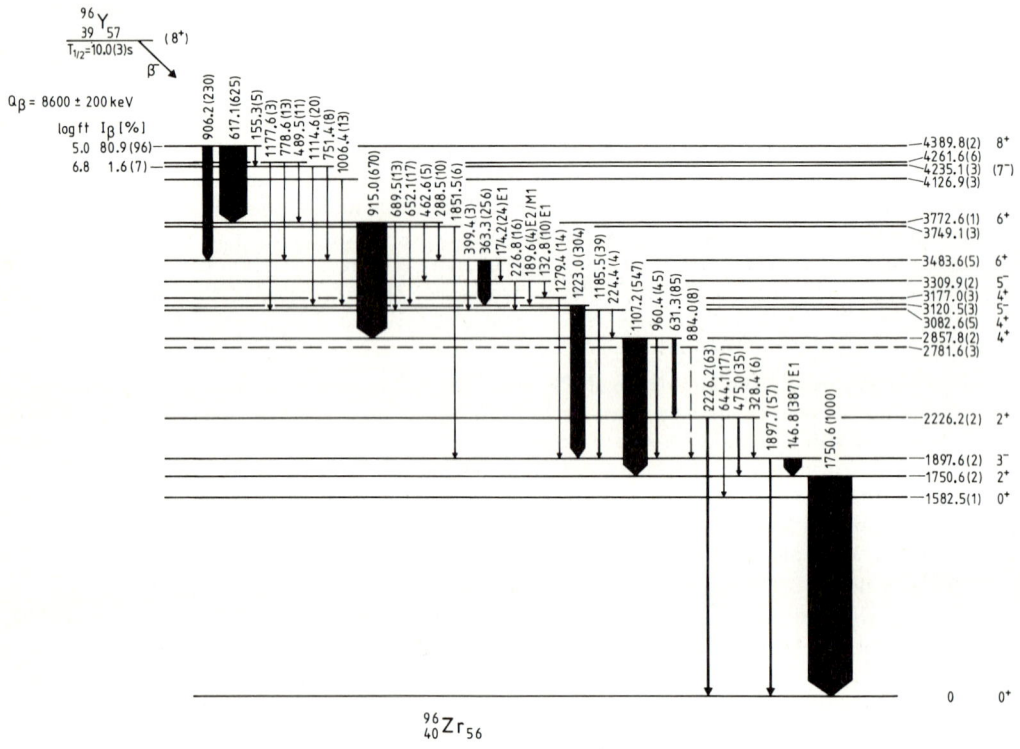

Fig.1 : Partial Level Scheme of ^{96}Zr. Selected logft values are given. They have been determined with I_{rel}^{exp}, $T_{1/2}(^{98}Y^m) = 10s$, $Q_\beta = 8.6(2) MeV$.

This raises the question, whether the identified 6^+ and 4^+ states are, indeed, the members of the corresponding band or whether there are other levels with this nature. There is another candidate for the 6^+ level : the 4235 kev state for which the angular correlation studies allow $I = 6, 7$. The 155 keV transition from the 8^+ level into this state is comparable to the $8^+ \rightarrow 6^+$ energy of 141 keV in ^{90}Zr. Moreover, if this transition has multipolarity E2 then its strenght is larger than that of the 617 keV transition. If the 4235 keV level is the 6^+ member of the $[\pi g_{9/2}^2]$ band then one would expect its decay to be similar to the one of the corresponding 6^+ states in other 2-particle or 2-hole nuclei. Hence, in order to prove this alternative, the hindrance factors of the $(6^+ \to 4^+)$ E2 transitions and of competing $(6^+ \to 5^-)$ E1 transitions have been deduced for known two-particle or two-hole nuclei in order to determine the expected relative intensity of the ("6^+" $\to 4^+$) E2 transition from the 4235 keV to the 2858 keV level in ^{96}Zr, which has not been observed. The result is that the ("6^+" $\to 4^+$) transition should have $I_r = 505$ relative units which is half of the intensity of the 1751 keV transition $(2_1^+ \to 0_1^+)$. The transitions from the 4235 keV level to the 4^+ states at 3177 and 3083 keV which might be other candidates for beeing band members would yield $I_r = 135$ and $I_r = 211$ relative units. Such intense transitions would have been observed in the coincidences. Hence it is concluded that the 4235 keV level is not the 6^+ state of the $[\pi g_{9/2}^2]$ configuration and that the lack of compression in ^{96}Zr is an experimental fact. Anyhow, the results of the angular correlation and the logft- value slightly favour spin 7^- for the 4235 keV level .

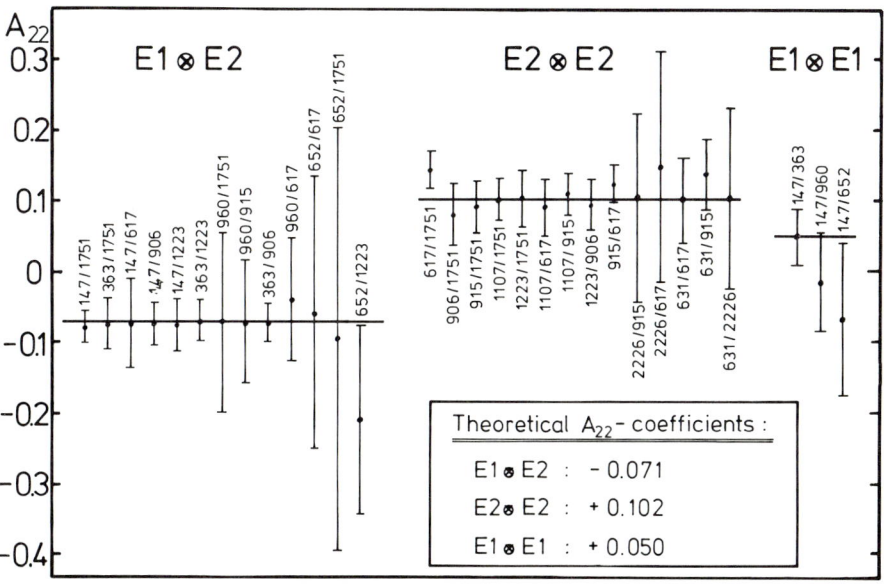

Fig.2 : Experimental A_{22} coefficients in ^{96}Zr compared to the expected values for stretched $E1 \otimes E2, E2 \otimes E2, E1 \otimes E1$ transitions. The actual cascades are labeled through their initial and final transition energy. The inset shows the theoretical A_{22} coefficients.

The Nucleus ^{98}Zr

A similar band as in ^{96}Zr has been proposed to exist in ^{98}Zr [7]. Since part of the members of this band have been observed [2] in the β^- decay of the high-spin isomer of ^{98}Y, here again angular correlation studies at JOSEF can be used to check this suggestion.

The levels at 2491 and 1844 keV have been assumed to be the 6^+ and 4^+ members of this band [7]. Their spins can be deduced from the analysis of the angular correlation of the cascade 1801, 647, 621 and 1223 keV. The six possible correlations favour strongly a spin sequence of $(6 \to 4 \to 3 \to 2 \to 0)$ for the levels at 4292, 2491, 1844, 1223 and 0 keV. This is in contradiction to the proposal of the intruder band (Fig.4), since the levels at 1844 and 2491 keV have spin 3 and 4, respectively, and not 4^+ and 6^+. Thus the proposed band is not reproduced in ^{98}Zr by the present results, which do, however, not exclude the existence of such a band. The 4^+ and 6^+ members are then still unknown. It should be mentioned that there is evidence for another 4^+ state at 2277 keV, which could be a candidate for the missing band member. The analysis of the coincidence data reveals a β feeding of the 6^+ and 4^+ states at 4292 and 2491 keV, respectively.

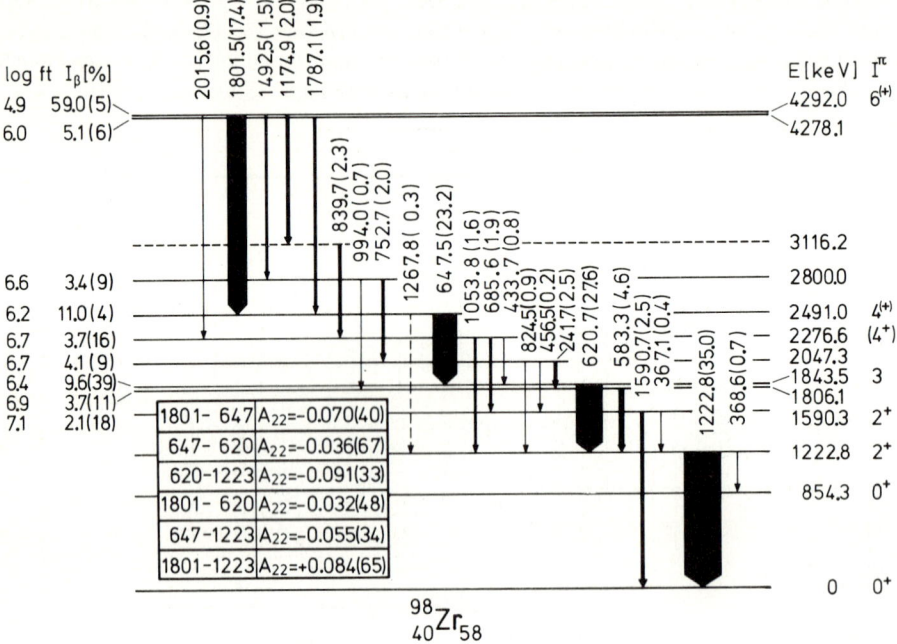

Fig.3 : Level Scheme of ^{98}Zr. Selected logft values are given. They have been determined with I_{rel}^{exp}, $T_{1/2}(^{98}Y^m) = 2s$, $Q_\beta = 9.8(2) MeV$. The inset shows the experimental A_{22} coefficients.

Nature of the High-Spin Isomers in ^{96}Y and ^{98}Y

The new data on the spins of levels in ^{96}Zr and ^{98}Zr lead to the assumption that the high-spin isomers in ^{96}Y and ^{98}Y have spin and parity 8^+ and 5^+, respectively. These isomers can be understood in terms of two-quasiparticle configurations which demonstrates the efficient subshell closures near ^{96}Zr. In fact, the nature of the β decaying isomeric states in the Y parents can well be revealed from the decay properties with the aid of IBFFM calculations.

The contributing nuclei for the IBFFM calculations for ^{96}Y are ^{95}Y and ^{95}Sr, in order to determine the quasiparticle states and energies, and the core nucleus ^{94}Sr, which is a harmonic vibrator with 3 bosons, since Z = 38 and N = 50 shell closures are taken into consideration. For ^{98}Y the quasiparticle states and energies of ^{97}Y and ^{97}Sr are used as well as the core nucleus ^{96}Sr, which is a vibrator, too, with 4 bosons.

The details are discussed in the contribution of S.Brant et al. to these proceedings. Fig. 5 shows the comparison of calculated and experimental levels in ^{96}Y and ^{98}Y. In ^{96}Y an isomerism of an 8^+ state based on the $[\pi g_{9/2} \cdot \nu g_{7/2}]_{8^+}$ configuration is predicted in accordance with the data

[4]. There is no chance for the 8+ state to decay through γ-emission into a lower lying level. In
98Y the isomeric state should be 5+ based on the $[\pi g_{9/2} \cdot \nu s_{1/2}]_{5+}$ configuration. The energies
of the calculated levels in 98Y have been normalized by fitting the one of the $[\pi g_{9/2} \cdot \nu g_{7/2}]_{1+}$
band member to the 548 keV level. This interpretation is not unambiguous since the 548 keV
level may be a deformed state [8]. But, although the absolute energies of the calculated levels
are subject to uncertainties, the relative positions, in particular the one of the 5+ state below
the 8+ member of the $[\pi g_{9/2} \cdot \nu g_{7/2}]_{1+}$ band, are firm.

The basic reason, why in one case the 8+ and in the other case the 5+ state is the isomeric state,
is that in 98Y the two additional neutrons will influence the occupation probabilities, considerably
lowering the energy of the 5+ level in 98Y respective to the one of the 8+ [9].

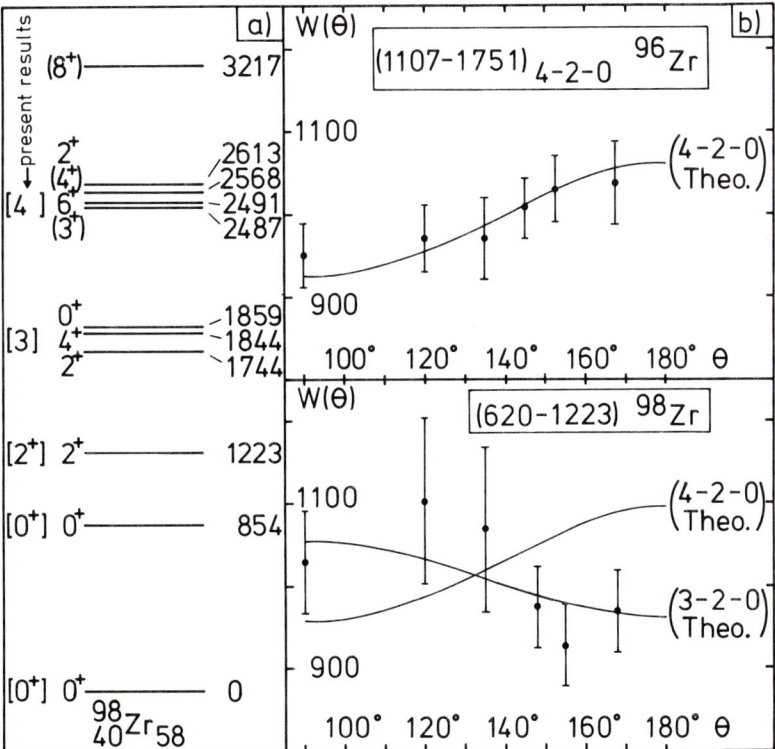

Fig.4a: Proposed Intruder Band in 98Zr in comparison with the present results.

Fig.4b: Angular Correlation for the (620/1223 keV) cascade in 98Zr in comparison with a (4-2-0)
sequence in 96Zr. The data show that the proposed interpretation [7] of the 1844 keV level as a
4+ state is not supported by the present results.

High Collectivity of the 3_1^- Level

Another interesting result in context with the study of the shell effects at A ∼ 100 is the half life
of the 3− level at 1898 keV in 96Zr recently obtained at JOSEF [10] from γγ delayed coincidence

measurements. This value of $T_{1/2} = 84(44)$ ps indicates a large octupole strength which is expected [11] because of the subshell closures at ^{96}Zr and the particular single-particle orbitals near the Z= 40 and N=56 gaps.

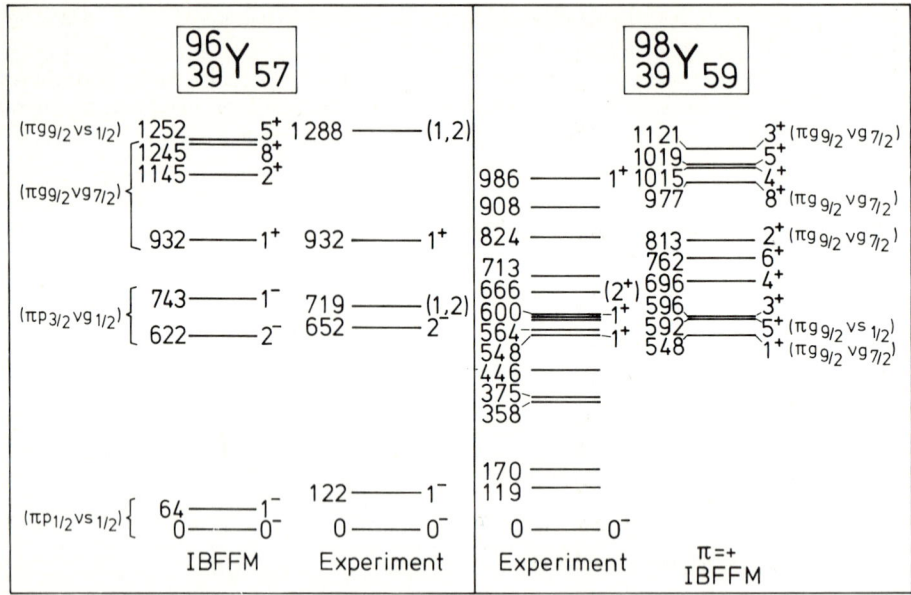

Fig.5 : Comparison of IBFFM and Experimental Levels in ^{96}Y and ^{98}Y [9,12]

*) Alexander von Humboldt fellow, on leave of absence from the University of Zagreb, Yugoslavia

References :

[1] G.Sadler et al., Nucl.Phys. A252 (1975) 365
[2] K.Sistemich et al., Z.Physik A281 (1977) 169
[3] H.Lawin et al., Nucl.Instr.Meth. 137 (1976) 103
[4] M.L.Stolzenwald et al., Z.Physik A327 (1987) 359
[5] G.Molnàr et al., Phys.Rev. C33 (1986) 1843
[6] H.Mach et al., Phys.Rev. C37 (1988) 254
[7] R.A.Meyer et al., Phys.Lett. B177 (1986) 271
[8] G.Lhersonneau et al., Contribution to this conference
[9] S.Brant et al., Contribution to this conference
[10] H.Ohm et al., Annual Report 1987, IKP, KFA Jülich
[11] G.Molnàr et al., Contribution to this conference
[12] S.Brant et al., Z.Physik A329 (1988) 301

Possible Evidence for Subshell Closures at N = 38, 40 and 56

N. Severijns, E. van Walle, D. Vandeplassche, J. Wouters, J. Vanhaverbeke, W. Vanderpoorten, and L. Vanneste

Instituut voor Kern- en Stralingsfysika, Leuven University, Celestijnenlaan 200 D, B-3030 Leuven, Belgium

1. Introduction.

The nuclei ^{69,70}As and ^{102,104}Ag were cryogenically oriented in an iron host. From the β-anisotropy of ^{69}As the magnetic moment of the $5/2^-$ ground state was deduced, while the γ-anisotropies in the decay of ^{70}As and ^{102,104}Ag yielded δ(E2/M1) mixing ratios for transitions in the respective daughter isotopes. All these results are discussed here in terms of subshell effects in the As, Ge and Pd isotopes.

2. Experimental procedure.

The experiments were done with the KOOL low temperature nuclear orientation facility which is coupled directly to the LISOL mass separator on-line with the cyclotron CYCLONE at Louvain-la-Neuve. Mass separated activities are implanted into a cold (mK region) and magnetized iron foil (soldered onto the cold finger of a ^{3}He-^{4}He dilution refrigerator) and subsequently oriented. The anisotropies of the γ- and β-rays emitted are measured by taking data points as a function of temperature at both 0° (or 180°) and 90°. The β-rays are detected with high purity Ge particle detectors working at 4K and with no absorbing material between them and the source. To avoid corrections for short lifetimes and production instabilities, our analyses are based on the quantity

$$\frac{W(0)}{W(90)} = \frac{[N(0)/N(90)]_{cold}}{[N(0)/N(90)]_{warm}},$$

where $W(\theta) = 1 + \alpha \sum_k B_k A_k U_k Q_k P_k(\cos\theta)$ is the normalized angular distribution [1]. In this expression, the parameters B_k describe the orientation (depending on μH/IT), A_k are the angular distribution coefficients which depend on initial spin, final spin and multipolarity of the observed radiation, while the deorienting effects of unobserved preceding transitions are entered into the U_k parameters. The $P_k(\cos\theta)$ are the Legendre polynomials and the Q_k coefficients correct for finite solid angle detection. The correction factor α, finally, is introduced to allow for a fraction of the nuclear ensemble not being oriented.

3. The nuclear magnetic moment of ^{69}As.

^{69}As was produced by bombarding a ^{54}Fe-target with a 70 MeV Ne^{6+} beam. From the β-anisotropy data (fig. 1) the magnetic moment of the 5/2$^-$ ground state was fitted as μ = +1.58(16)μ$_N$ for pure Gamow-Teller, and covering practical limits of Fermi admixture. Our result is significantly larger than the previously reported one (see table 1) and removes the need to postulate a proton configuration that is different from ^{71,73}As. As can be seen from table 1, the only drastic change in the moment for this 5/2$^-$ state occurs between ^{73}As and ^{75}As. This could be well explained by a rearrangement of proton pairs while passing the magic neutron number 40 [2]).

Table 1: Theoretical and experimental magnetic moments of the lowest 5/2$^-$ state in the odd As isotopes. Theoretical values are from refs. 2 and 3 and were calculated in the core polarization model of Noya, Arima and Horie [4]), except for the value +1.75 for ^{69}As, which is the single particle moment calculated with the effective gyromagnetic ratios of ref. 5. The experimental values are from ref. 2 ($^{71-77}$As) and from ref. 3 and this work (^{69}As).

Isotope	E(5/2$^-$) (keV)	π-configuration	ν-configuration	μ$_{cal}$ (n.m.)	μ$_{exp}$ (n.m.)
^{69}As	0.	$(1f_{5/2})^5$		+0.96	1.2(2)
		$(2p_{3/2})^2(1f_{5/2})^3$		+1.34	
		$(1f_{5/2})^1$		+1.75	+1.58(16)
^{71}As	0.	$(2p_{3/2})^2(1g_{9/2})^2(1f_{5/2})^1$	$(1f_{5/2})^6(2p_{3/2})^4$	+1.81	(+)1.6735(18)
^{73}As	67.	$(2p_{3/2})^2(1g_{9/2})^2(1f_{5/2})^1$	$(1f_{5/2})^6(2p_{1/2})^2$	+1.72	+1.63(10)
^{75}As	280.	$(1f_{5/2})^5$	$(2p_{3/2})^4(1g_{9/2})^4$	+0.96	+0.88(10)
^{77}As	264.	$(1f_{5/2})^5$	$(2p_{3/2})^4(2p_{1/2})^2 (1g_{9/2})^4$	+0.91	+0.83(8)

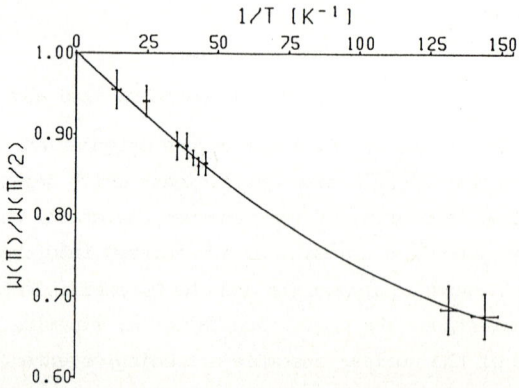

Fig. 1. Anisotropy of the β-spectrum of ^{69}As between 1.40 and 2.95 MeV.

4. E2/M1 mixing ratios in ^{70}Ge.

Using the same reaction as was used for ^{69}As, and during the same run, nuclear orientation was performed also on ^{70}As. From the anisotropies of the γ-rays (fig. 2), δ(E2/M1) mixing ratios were deduced for ten transitions in ^{70}Ge (table 2). The sign convention of Krane and Steffen [6] is used for these and all other δ-values quoted here. In table 3 the δ-values for the $2^+_2 \to 2^+_1$, $3^+_1 \to 2^+_1$ and $3^+_1 \to 2^+_2$ transitions in the even Ge isotopes are listed. A remarkable change of sign shows up between ^{70}Ge and ^{72}Ge for all three δ-values. It should be noted that for the $2^+_2 \to 2^+_1$ transition in ^{72}Ge we preferred the result of Chung et al. [7] instead of the value -10.3(13) quoted by Chen et al. [8]. Indeed, not only is the value of ref. 7 consistent with two previous results [9,10], but also was the change of sign for this δ-value between ^{70}Ge and ^{72}Ge observed in the experiments of both Chung et al. [7] and Mohindra and Van Patter [10]. Our result for $\delta(2^+_2 \to 2^+_1)$ in ^{70}Ge is moreover consistent with the values quoted in these two references. This sudden sign inversion could be related to a subshell closure at N = 38 in the Ge isotopes. The energies of the 2^+_1 state indeed indicate a subshell closure at N = 38 for Ge, Zn and Ni nuclei, i.e. for nuclei with Z near 28. Also, in the Nilsson scheme an important gap is present at N = 38 for small deformations. The even-A Ge nuclei are indeed known to be only weakly deformed [11].

Table 2: δ(E2/M1)-mixing ratios for transitions in ^{70}Ge.
The underscored values are preferred on the basis of literature values or arguments inherent to the analysis.

γ-transition			δ(E2/M1)-values	literature	
668.4keV	2^+_2 →	2^+_1	-0.97(45) or <u>-2.8(8)</u>	$-(5.0 ^{+4.0}_{-1.5})$	10)
				$-(6^{+10}_{-3})$	7)
1412.5	3^+_1 →	2^+_1	0.50(8) or 4.5(4)		
744.8	3^+_1 →	2^+_2	<u>-0.24(7)</u> or -1.93(14)	-0.14(20)	16)
				-0.05(8)	17)
				0.04(8)	18)
1339.4	3^+_2 →	2^+_1	<u>-25.(6)</u> or 0.21(6)		
889.2	3^+_2 →	2^+_2	-1.2(3) or -0.50(24)		
595.2	3^+_2 →	3^+_1	<u>-0.036(40)</u> or 1.42(7)		
1781.3	3^+ →	2^+_2	<u>-21.(8)</u> or 0.20(9)		
607.6	4^+_3 →	3^+_1	5.2(6) or 0.40(7)		
905.7	4^+_3 →	4^+_1	1.23(6) or -0.13(3)		
252.3	4^+_3 →	4^+_2	-0.10(11) or 1.2(2)		

Table 3: δ(E2/M1) mixing ratios in the even Ge isotopes.
Numbers for ^{70}Ge are from this work; for the other isotopes from ref. 14 and several volumes of the Nuclear Data Sheets.

Isotope	$\delta(2_2^+ \to 2_1^+)$	$\delta(3_1^+ \to 2_1^+)$	$\delta(3_1^+ \to 2_2^+)$
^{66}Ge	$-(3.3^{+2.6}_{-1.8})$		$-2.2(2)$
^{68}Ge	$-0.09(2)$	$0.16(8)$	$-0.02(2)$
^{70}Ge	$-2.8(8)$	$0.50(8)$ or $4.5(4)$	$-0.24(7)$
^{72}Ge	5^{+3}_{-1}	$-(2.0^{+1.5}_{-3.5})$	$0.25(10)$
^{74}Ge	$3.3(3)$	$-0.34(5)$	$1.3(4)$
^{76}Ge	$3.5(15)$		

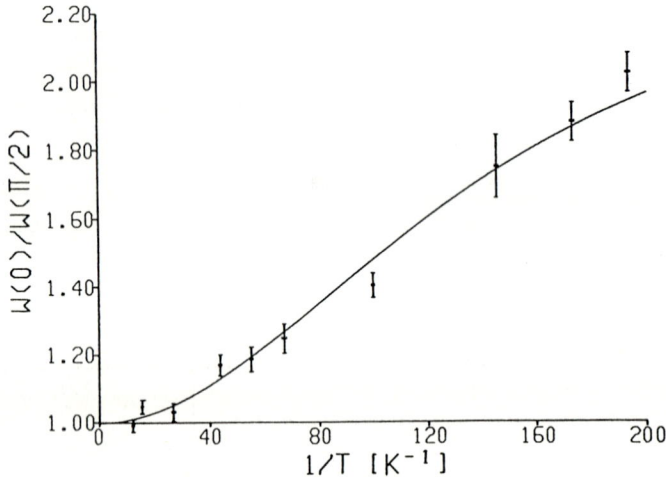

Fig. 2. Anisotropy of the mixed $(3_1^+ \to 2_2^+)$ 744.8 keV γ-transition in ^{70}Ge.

5. E2/M1 mixing ratios in ^{102,104}Pd.

Nuclear orientation measurements on ^{102,104}Ag, produced on a Mo-target with a 92 MeV N^{4+} beam, yielded δ-values for several transitions in ^{102,104}Pd [12]. In table 4 the $\delta(2_2^+ \to 2_1^+)$, $\delta(3_1^+ \to 2_1^+)$ and $\delta(3_1^+ \to 2_2^+)$-values in $^{102-106}$Pd are listed. Data for ^{102,104}Pd are from ref. 12, for ^{106}Pd from ref. 13. A change of sign occurs for $\delta(2_2^+ \to 2_1^+)$ and $\delta(3_1^+ \to 2_1^+)$ between ^{102}Pd and ^{104}Pd. Comparison with the Ru isotopes [14] shows that the same phenomenon occurs between ^{100}Ru and ^{102}Ru, thus also between N = 56 and 58. Although no subshelleffect can be inferred from the 2_1^+-energies, the Nilsson scheme for neutrons nevertheless shows an important gap at N=56 for the deformation values around 0.15 which are assumed for these nuclei [15].

Table 4: $\delta(E2/M1)$ mixing ratios in the even Pd isotopes.
The values between brackets are limits for extreme U_k coefficients.

Isotope	$\delta(2^+_2 \to 2^+_1)$	$\delta(3^+_1 \to 2^+_1)$	$\delta(3^+_1 \to 2^+_2)$
^{102}Pd	2.8(2)	[4.4, 9.]	-
^{104}Pd	-25.(5)	[-14.,-11.]	[-29.,-22.]
^{106}Pd	-14.1(15)	-3.8(3)	-9.(1)

6. Conclusion.

Inspecting the tabulation of Lange et al. [14], 16 cases are found where the $\delta(2^+_2 \to 2^+_1)$ value is known for both an isotope with a closed neutron-and/or proton-shell and its (A+2)-isotope (table 5 and 6). For eleven of these, the δ-values differ in sign, while in three cases the sign remains the same and for two the accuracies are not sufficient to make any definite conclusion. Taking into account the difficulty for several techniques to determine precise δ-values we tentatively suggest that a (sub)shell-closure causes an inversion of sign for at least the $\delta(2^+_2 \to 2^+_1)$ value, but in view of our results for Ge and Pd nuclei perhaps also for others. There is at present no theoretical explanation.

Table 5: δ-values for nuclei with a closed neutronshell and their respective (A+2)-isotope.

Isotopes	δ-values	
$^{18,20}_{10}$Ne$_{8,10}$	-0.6(6)	$8.4^{+1.5}_{-1.0}$
$^{40,42}_{20}$Ca$_{20,22}$	13^{+6}_{-3}	-0.18(2)
$^{50,52}_{22}$Ti$_{28,30}$	0.26(17)	0.03(10)
$^{52,54}_{24}$Cr$_{28,30}$	-6.25(150)	$-(0.22^{+0.15}_{-0.10})$
$^{54,56}_{26}$Fe$_{28,30}$	0.11(4)	-0.179(11)
$^{92,94}_{42}$Mo$_{50,52}$	$0.69^{+0.54}_{-0.38}$	-2.0(3)
$^{140,142}_{58}$Ce$_{82,84}$	0.34(2)	-0.10(5)

Table 6: δ-values for nuclei with a closed protonshell and their respective (A+2)-isotope.

Isotopes	δ-values	
$^{18}_{8}$O$_{10}$ $^{20}_{10}$Ne$_{10}$	-0.12(4)	$8.4^{+1.5}_{-1.0}$
$^{42}_{20}$Ca$_{22}$ $^{44}_{22}$Ti$_{22}$	-0.18(2)	$-(7.5^{+8.0}_{-2.5})$
$^{44}_{20}$Ca$_{24}$ $^{46}_{22}$Ti$_{24}$	-0.123(16)	1.28(18)
$^{48}_{20}$Ca$_{28}$ $^{50}_{22}$Ti$_{28}$	0.04(3)	0.26(17)
$^{60}_{28}$Ni$_{32}$ $^{62}_{30}$Zn$_{32}$	0.82(15)	$-(1.2^{+0.5}_{-0.4})$
$^{62}_{28}$Ni$_{34}$ $^{64}_{30}$Zn$_{34}$	3.2(1)	-1.6(4)
$^{120}_{50}$Sn$_{70}$ $^{122}_{52}$Te$_{70}$	-1.43(25)	-3.48(7)
$^{122}_{50}$Sn$_{72}$ $^{124}_{52}$Te$_{72}$	4^{+2}_{-1}	-3.55(7)
$^{124}_{50}$Sn$_{74}$ $^{126}_{52}$Te$_{74}$	$3.0^{+1.2}_{-0.6}$	$-(5.6^{+0.5}_{-0.4})$

References.

1) K.S. Krane in 'Low Temperature Nuclear Orientation', eds. N.J. Stone and H. Postma, North-Holland, Amsterdam (1986), ch. 2
2) P. Herzog, N.J. Stone and P.D. Johnston, Nucl. Phys. A259 (1976) 378
3) W. Hogervorst, H.A. Helms, G. Zaal, J. Bouma and J. Blok, Z. Phys. A294 (1980) 1
4) H. Noya, A. Arima and H. Horie, Prog. Theor. Phys. Suppl. 8 (1958) 33
5) K. Kumar, Phys. Scr. 11 (1975) 179
6) K.S. Krane and R.M. Steffen, Phys. Rev. C2 (1970) 724
7) K.C. Chung, A. Mittler, J.D. Brandenberger and M.T. McEllistrem, Phys. Rev. C2 (1970) 139
8) H. Chen, P.L. Gardulski and M.L. Wiedenbeck, Nucl. Phys. A219 (1974) 365
9) R.G. Arns and M.L. Wiedenbeck, Phys. Rev. 112 (1958) 229
10) R.K. Mohindra and D.M. Van Patter, Bull. Am. Phys. Soc. 10 (1965) N°1, 38, CC13
11) R. Lecomte, M. Irshad, S. Landsberger, P. Paradis and S. Monaro, Phys. Rev. C22 (1980) 1530
12) E. van Walle, Ph. D. thesis, Leuven (1985)
13) R. Eder, E. Hagn and E. Zech, Phys. Rev. C31 (1985) 190
14) J. Lange, K. Kumar and J.H. Hamilton, Rev. Mod. Phys. 54 (1982) 119
15) H.A. Smith and F.A. Rickey, Phys. Rev. C14 (1976) 1946
16) C. Morand, M. Agard, J.F. Bruandet, A. Giorni, J.P. Longequeue and Tsan Ung Chan, Pys. Rev. C13 (1976) 2182
17) L. Cleemann, U. Eberth, J. Eberth, W Neumann and V. Zobel, Phys. Rev. C18 (1978) 1049
18) R.L. Robinson, H.J. Kim, R.O. Sayer, J.C. Wells, R.M. Ronningen and J.H. Hamilton, Phys. Rev. C16 (1977) 2268

Subshell Closure at N = 56 in Very Neutron-Rich Bromine Isotopes[*]

M. Graefenstedt[1], U. Keyser[1], F. Münnich[1], F. Schreiber[1],
The LOHENGRIN and OSTIS Collaboration[2],
and The ISOLDE Collaboration[3]

[1]Institut für Metallphysik und Nukleare Festkörperphysik,
Technische Universität, Mendelssohnstr. 3,
D-3300 Braunschweig, Fed. Rep. of Germany, and
[2]Institut Laue-Langevin, F-38046 Grenoble, France
[3]CERN, CH-1211 Geneva 23, Switzerland

1. Introduction

The existence of a nuclear subshell closure at N = 56 has been derived by us several years ago from the systematics of nuclear binding energies, which were calculated from experimental Q_β-values |1|. This effect was shown to exist for nuclei with $37 \leq Z \leq 42$. For nuclei with proton number $Z \geq 43$, no such subshell closure is observed. In a later measurement of the Q_β-value of $^{93}_{36}Kr_{57}$ it was found that for Z = 36, the submagic behaviour at N = 56 is still present |2|. But according to a recent theoretical calculation, this effect should vanish for the bromine isotopes (Z = 35) |3|.

In order to check this prediction, Q_β-measurements of $^{87-90}$Br isotopes have been performed at the mass separators LOHENGRIN and OSTIS of the ILL in Grenoble |4|. But there, the production rate of the heavier bromine isotopes with mass number A > 90 is too low. At CERN-ISOLDE II, however, it has been demonstrated, that the high electron affinities of the halogens allow the use of negative surface ionization for the production of intense beams of heavy bromine isotopes |5, 6|. In fact, we were able to measure for the first time the total beta-decay energies of ^{91,92}Br from βγ-coincidences at this separator.

2. Experimental Techniques

The beta-spectra of the samples were recorded by a 720 cm³ plastic scintillator telescope in coincidence with a 265 cm³ (HP)Ge-detector, which had an energy resolution of 2.26 keV for 1,33 MeV γ-ray energy, a relative efficiency of about 64% and a peak-to-Compton ratio of 61.5. A 1 mm thick NE 104-scintillator in front of the large E_β-scintillator served as a ΔE-detector with very low γ-ray sensitivity ($\bar{Z} \approx 3$). It was applied for further reduction of the γ-sensitivity of the E_β-scintillator and for the backscattering effect of the electrons. The linearity of the telescope was tested with the conversion electron spectrometer BILL at the Institut Laue-Langevin in Grenoble.

---

[*] This work has been funded by the German Federal Minister of Research and Technology (BMFT) under the contract number 06 BS 452 I.

Before and during the decay energy measurements, the beta-emitters shown in Table 1 were used as reference sources for the energy calibration.

Table 1: Beta-endpoint energies used for energy calibration of the plastic-scintillator telescope.

Nuclide	^{38}Cl	^{89}Br	^{90}Sr	^{91}Kr	^{92}Kr	^{106}Ru
E_β/keV	4917±1	8125±50	2282±3	6460±70	4695±50	1978±9; 2406±9; 3028±9; 3540±9

The value for ^{91}Kr is the weighted mean of the one given in |7| and the value from own measurements. The uncertainty in the endpoint energy, resulting from this calibration so far, is of the order of 110 keV and is included in the total error of the measured decay energies given later in Table 2.

The βγ-coincidence events were stored on-line in a matrix of 256 β-channels and 2048 γ-channels of the memory of a Hewlett Packard HP 1000/A 600 computer. In addition, β- and γ-singles spectra have been recorded in two multichannel analyzers with 2048 and 8192 channels respectively.

3. Experimental Results

In this section, we will mainly report the results obtained for the isotopes ^{91}Br and ^{92}Br. More details concerning the isotopes $^{87-90}$Br will be given elsewhere |4|.

For $^{91}_{35}$Br ($T_{1/2}$ = 0.53±0.03 s |6|), no decay scheme has been published so far, but preliminary information was obtained by B.Pfeiffer |10|. In the decay of this nuclide, 10 γ-gates with energies from 1122 keV to 3970 keV have been evaluated, depopulating energy levels up to 4452 keV, and all of them directly fed by β-decay. The correct placement of these levels was checked in our measurement by the comparison of the intensities obtained from the γ-singles spectrum and the β-gated γ-spectrum.

For the decay of $^{92}_{35}$Br ($T_{1/2}$ = 0.31±0.02 s |5|), again only preliminary information is available |10|.

In Figure 1, the γ-singles spectrum for energies up to 6 MeV is given. From this picture it is clear, that a determination of the total β-decay energy from a β-singles spectrum is practically impossible. In addition, many of the γ-lines belong to the decay of daughter nuclei in the decay chain of ^{92}Br and to contaminations undergoing β-decay, cf. Figure 2a. In Figure 2b, an isometric plot of the first part of

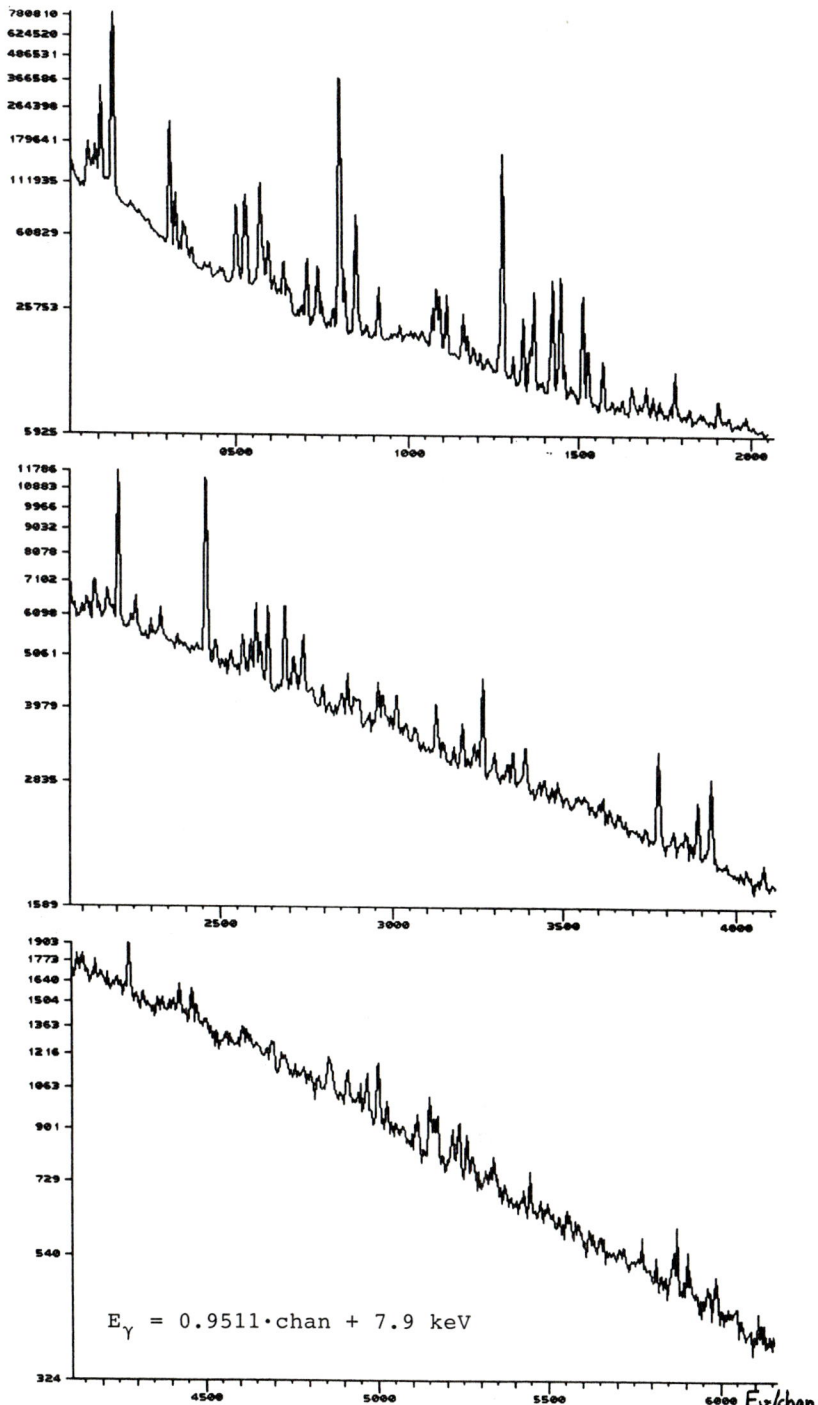

$E_\gamma = 0.9511 \cdot \text{chan} + 7.9 \text{ keV}$

Figure 1: γ-singles spectrum of ^{92}Br-decay.

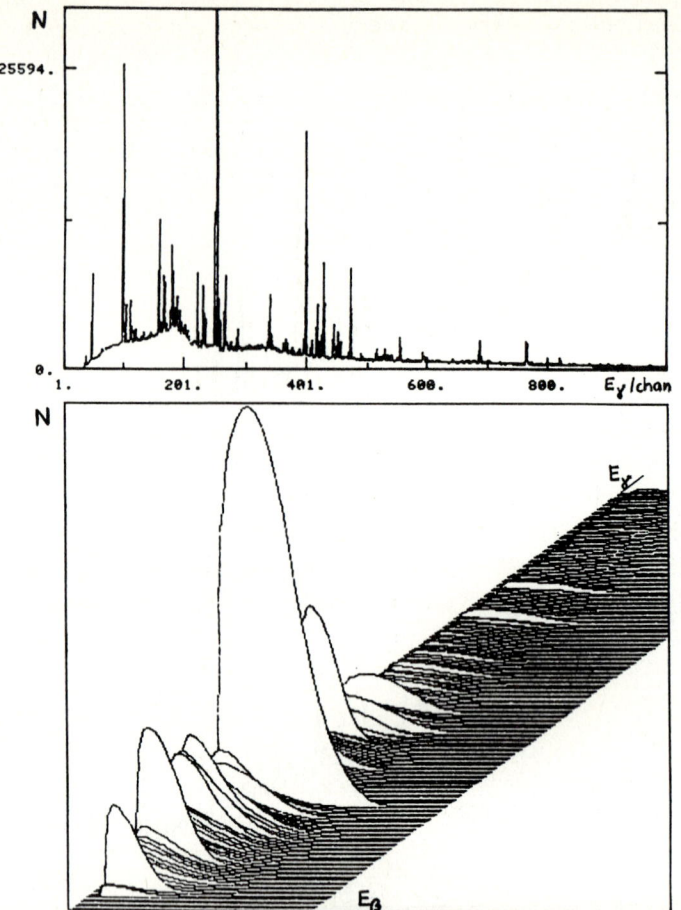

Figure 2: a) β-gated γ-spectrum of ^{92}Br-decay;
b) isometric plot of the βγ-coincidence matrix of ^{92}Br-decay.

the βγ-coincidence matrix is shown. The β-spectra are well separated and differ in height, length (e.g. intensity and endpoint energy) and shape. From this matrix, β-spectra coincident with different γ-lines are obtained by setting digital windows on the γ-lines of interest. Due to the good time resolution of the coincidence electronics, the contribution of chance coincidences to these spectra are very low in general. But for the determination of the endpoint energies from a Fermi-Kurie analysis, the Compton background from γ-lines of higher energy has to be subtracted to increase the length of the fit interval. The number of gates, depopulating all known strongly β-fed energy levels above 2 MeV, is given in Table 2 together with the Q_β-values of the lighter isotopes.

Table 2: Q_β-values of heavy Br isotopes; preliminary results.

Isotope	$T_{1/2}$	γ-gates evaluated	Present value	Other authors	Wapstra |7|	Adopted value
^{87}Br	55.7s	15	6760±120	6830±120|8|	6830±120	6795± 85
^{88}Br	16.3s	18	8950±100	8970±120|9|	8970±130	8960± 80
^{89}Br	4.5s	25	8125±100	8140±140|9|	8300±400*	8130± 80
^{90}Br	1.9s	7	10280±120	9800±400|9|	10700±400*	10280±120
^{91}Br	0.53s	10	9710±120	-	-	9710±120
^{92}Br	0.31s	11	12155±130	-	-	12155±130

A detailed discussion of the subshell closure effect at N = 56 will be given in another contribution to this workshop |11|. The existence of this effect also for bromine with proton number Z = 35 can be derived from a plot of the Q_β-values as a function of mass number, which is shown for odd-odd nuclei in Figure 3. The subshell closure manifests itself in the increase of the Q_β-value for the isotope $^{92}_{35}$Br$_{57}$, similar to the increase observed for $^{94}_{37}$Rb$_{57}$, $^{96}_{39}$Y$_{57}$ and $^{98}_{41}$Nb$_{57}$.

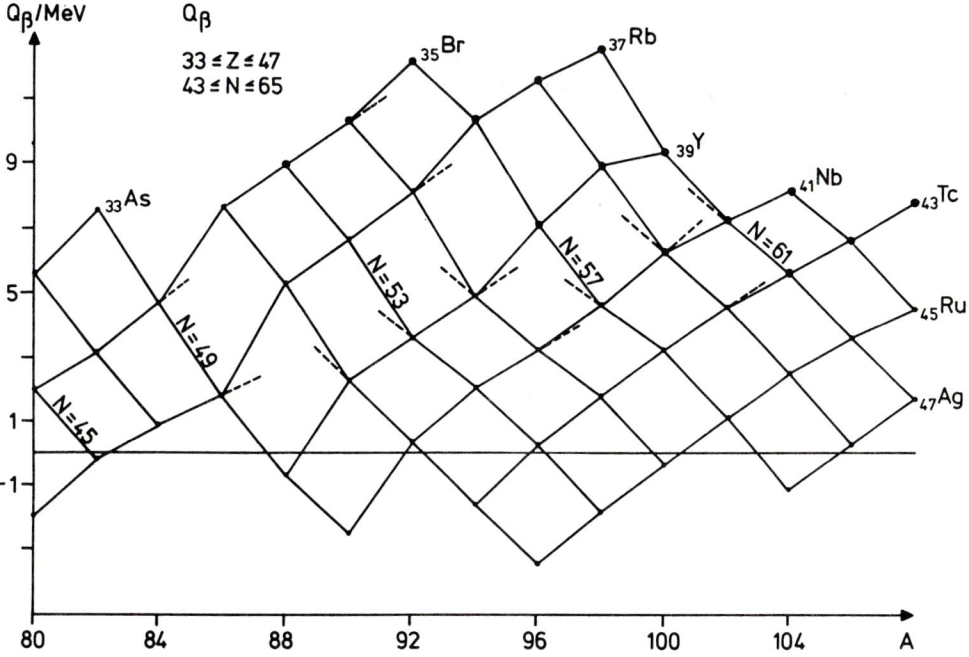

Figure 3: Way-Wood diagram for odd-odd nuclei; heavy dots: Q_β-values from own measurement.

Acknowledgement

We are indebted to the staff members of the CERN-ISOLDE, especially to the Drs.: H. Haas, O. Jonsson, E. Kugler, G. Olesen, H. L. Ravn, B. Vosicki and H. Gabelmann, and to K. Balog, P. Jürgens and T. Winkelmann from our group as well, for their valuable assistance before and during the experiment. Thanks are due to Mrs. Laupheimer for their help in the preparation of the manuscript.

References

|1| U. Keyser, F. Münnich and L. J. Weigert: J.Phys.G4,L119(1978)
|2| H. Berg, U. Keyser, F. Münnich, K. Hawerkamp, H. Schrader and B. Pfeiffer: Z.Phys.A288,59(1978)
|3| R. Bengtsson, P. Møller, J. R. Nix and Jing-Ye Zhang: Phys.Scripta 29,402(1984)
|4| F. Schreiber: Thesis TU Braunschweig 1988
|5| G. T. Ewan, P. Hoff, B. Jonson, K.-L. Kratz, P. O. Larsson, G. Nyman, H. L. Ravn and W. Ziegert: Z.Phys.A318,309(1984)
|6| CERN-ISOLDE USERS' GUIDE: CERN Yellow Report 86-05
|7| A. H. Wapstra and G. Audi: Nucl.Phys.A432,1(1985)
|8| K. Aleklett, E. Lund and G. Rudstam: Z.Phys.A290,173(1979)
|9| P. Hoff, K. Aleklett, E. Lund and G. Rudstam: Z.Phys.A300,289(1981)
|10| B. Pfeiffer: priv. comm.
|11| M. Graefenstedt, U. Keyser, F. Münnich and F. Schreiber: this workshop

Recent Results of Beta-Decay-Energies of Very Neutron-Rich Nuclei Around Mass Number A = 100*

M. Graefenstedt[1], U. Keyser[1], F. Münnich[1], F. Schreiber[1],
The LOHENGRIN and OSTIS Collaboration[2],
and The ISOLDE Collaboration[3]

[1]Institut für Metallphysik und Nukleare Festkörperphysik,
 Technische Universität, Mendelssohnstr. 3,
 D-3300 Braunschweig, Fed. Rep. of Germany, and
[2]Institut Laue-Langevin, F-38046 Grenoble, France
[3]CERN, CH-1211 Geneva 23, Switzerland

1. Introduction

The determination of beta-decay energies is a very useful method to explore the nuclear mass surface outside the region of stability. Such experiments are simple in principle, but they are difficult to evaluate due to the high decay energies, the large number of beta-transitions and the great complexity of the decay schemes of nuclei far away from the valley of β-stability |1|. These decay schemes must be known in their main features in order to derive the correct Q_β-value from the measured β-endpoint energies. A reliable Q_β-value can only be deduced in such experiments, if as many β-spectra as possible are measured in coincidence with γ-transitions, which depopulate different known levels in the daughter nucleus. In this way, also the reliability of the decay scheme of the nucleus studied can be checked. In addition, the final experimental error of the Q_β-value is reduced to about 100 keV for $Q_\beta \geq 10$ MeV; this gives a relative error of about 10^{-6} for the nuclear masses (for A ≅ 100).

From the study of the decay energies - or of nuclear binding energies derived from them - it is possible to investigate the systematics of nuclear structure effects as a function of mass number A, proton number Z or neutron number N. This will be done in a later section of this paper.

2. Experimental Results

Measurements of beta-decay energies of very neutron-rich nuclei have been performed by us since several years at the mass separator LOHENGRIN and OSTIS of the ILL (Grenoble) and recently also at ISOLDE II CERN (Geneva). A short discussion of the experimental details of these experiments is found in another contribution of this workshop |2|.

---

* This work has been funded by the German Federal Minister of Research and Technology (BMFT) under the contract number o6 BS 452 I.

In our latest measurements at the LOHENGRIN separator, neutron-rich isotopes of the elements $_{42}$Mo, $_{43}$Tc, $_{44}$Ru and $_{45}$Rh with mass numbers $107 \leq A \leq 109$ have been studied, using targets of ^{239}Pu with high independent yields. Preliminary results of these investigations have been reported in a recent publication |3|, together with a summary of all our earlier results of Q_β-measurements.

The beta-decay energies of the heavy bromine isotopes with A = 87, 88 and 89 have been measured at the OSTIS separator and those with A = 91 and 92 at ISOLDE II; the Q_β-values obtained are given in |2|.

These experimental results will be discussed in the following section in connection with the nuclear structure effects which have been found in the mass region around A = 100.

3. Nuclear Structure Effects

The measurement of nuclear binding energies or masses is a very suitable method to investigate nuclear structure effects such as main shell or subshell closures or regions of deformations, because the slope of the mass surface depends very sensitively upon these effects. In a plot of the nuclear binding or decay energies as a function of Z, N or A, large changes of the slope of these curves are observed at main shell closures for all known nuclei with a main magic proton or neutron number. At the semi-magic numbers or at the onset of a deformation region, the change in slope of the curves is normally much smaller and it is a local effect on the mass surface, i.e. it occurs only for nuclei within a small part of the nuclear chart. The knowledge of the extension of these regions of subshell closures and of deformations is of great importance for the theoretical understanding and interpretation of nuclear properties in the framework of the IBM-model |4, 5|.

The binding and decay energies of a single particle, however, are strongly influenced by the proton and neutron pairing energies. Therefore, to display the systematics of these energies it is necessary to plot four diagrams for each energy (for even-even, even-odd, odd-even and odd-odd nuclei). A single diagram can be drawn for two-particle decay and binding energies, because these properties are only little influenced by the even-odd straggling due to the pairing energies.

In Figure 1, the change in slope of the $Q(2\beta)$-values for all known nuclei with $35 \leq Z \leq 43$ and $48 \leq N \leq 64$ is shown as a function of the neutron number in order to demonstrate the shell closures as N = 50 and N = 56. The subshell closure at N = 56 is only observed for nuclei with $Z \leq 42$; a fact which can be explained by shell model arguments |6|.

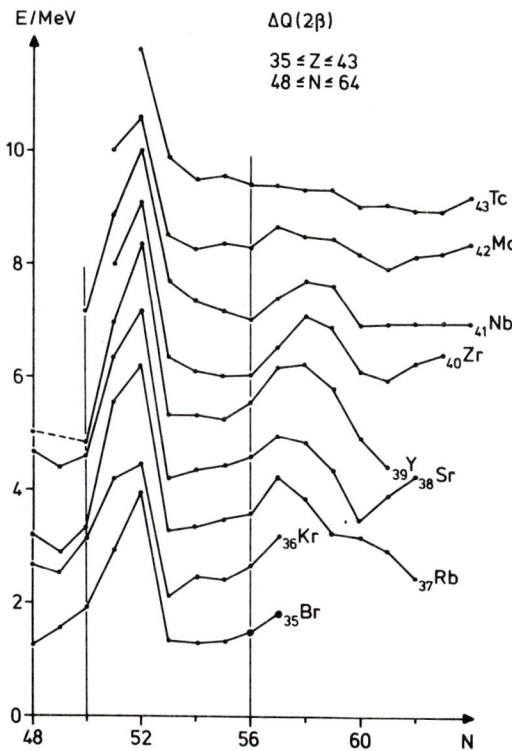

Figure 1: Difference of Q(2β)-values,
$\Delta Q(2\beta) = Q_\beta(A,Z) - Q_\beta(A-2,Z)$,
as a function of neutron number N.

The effect is still present for bromine with Z = 35, cf. also |2|, contrary to theoretical predictions |7|.

The behaviour of the $\Delta Q(2\beta)$-values as a function of N is quite different for the main shell closure at N = 50 and the subshell closure at N = 56. At the main shell closure, the $\Delta Q(2\beta)$-values are about the same at N = 50 and N = 53 for each nucleus, i.e. the main shell closure is "sharp" and quite similar for all nuclei. The subshell closure, however, is broader and much more irregular; it depends sensitively upon the proton number Z.

These shell closures are easily observed also in the more familiar plot of the two-neutron separation energies S(2n) shown in Figure 2. In addition, the onset of nuclear deformation at N = 60 can easily be derived from this figure; it manifests itself in the rise of the S(2n)-curves at this neutron number for the nuclei with Z ≤ 45. The deformation is seen to increase with decreasing proton number and becomes quite strong for $_{40}$Zr. For lower proton numbers, the behaviour of the S(2n)-values as a function of N is seen to be quite irregular. The strong rise of this value for $^{100}_{39}$Y and $^{100}_{38}$Sr indicates that also these nuclei are strongly deformed.

The extension of the deformation region cannot be derived from Figure 2, because the S(2n)-values of nuclei with Z ≤ 45 and N > 66 are not known. The S(2n)-value of $^{109}_{43}$Tc, deduced from our experimental Q_β-value of this nucleus |3| indicates that also the isotones with N = 66 are still deformed.

In order to investigate the proton subshell closures in this mass region, the two-proton separation energies S(2p) are plotted as a func-

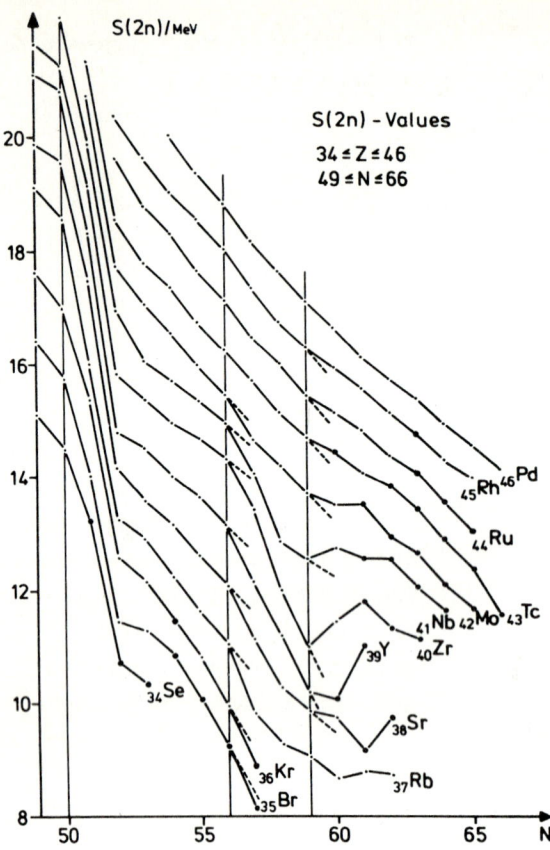

Figure 2: Two-neutron separation energies S(2n) for nuclei with $82 \leq A \leq 112$. Heavy dots: values calculated from our own Q_β-values only.

tion of mass number A in Figure 3. One observes a discontinuity in the slopes of these curves around the proton number Z = 40, which is limited to nuclei with neutron numbers $50 \leq N \leq 60$ and depends sensitively upon N. Contrary to Figure 2, where the strong change in the slope of the S(2n)-curves associated with the semi-magic number N = 56 begins at this neutron number, the discontinuity in Figure 3 is observed in the range of the proton numbers $39 \leq Z \leq 41$ for nuclei with neutron numbers $50 \leq N \leq 54$. The range of Z then increases and the slope of the S(2p)-curves changes between Z = 38 and Z = 42 for nuclei with $N \geq 56$. This fact is due to a considerably smaller pairing energy of the last $2p_{1/2}$ proton pair in Zr |8|. So it seems that for neutron numbers $N \geq 56$, the proton number Z = 38 as well as Z = 40 is semi-magic.

4. Outlook

In our future exeriments, we will concentrate our efforts upon two areas of investigations. One is the existence of isomerism for quite a large number of nuclei in the mass region around A = 100. We have already studied the isomers of several of these nuclei |1|, but most of these measurements have been performed several years ago. They should be repeated again under improved experimental conditions, because no other measurements have been reported in the meantime. Most of these experiments can be performed at the LOHENGRIN separator with the sources available there.

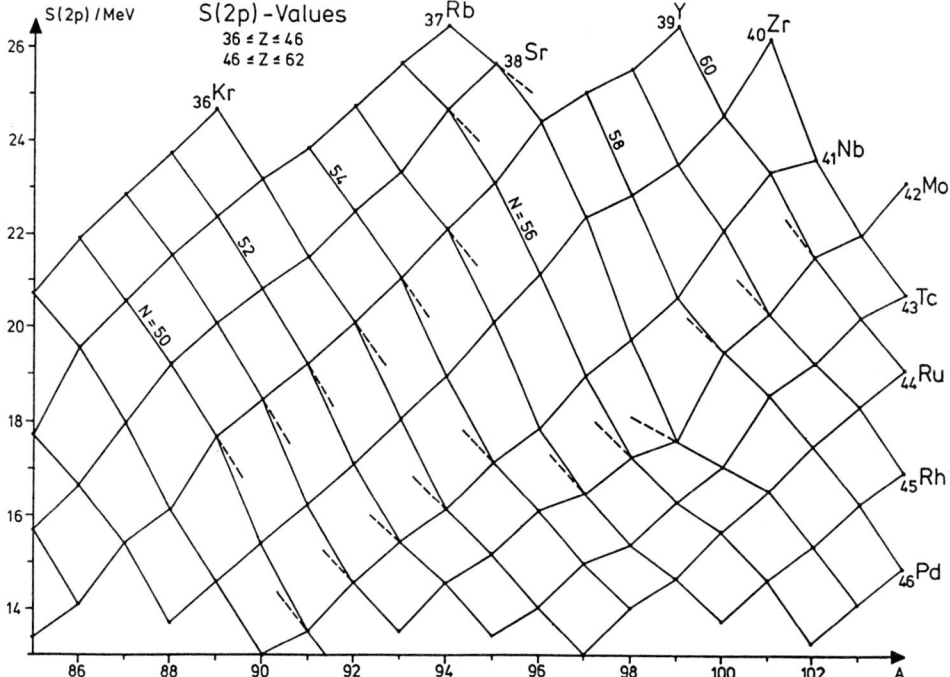

Figure 3: Two-proton separation energies S(2p) as a function of mass number A for nuclei with $82 \leq A \leq 108$.

And then, of course, we want to perform Q_β-measurements of nuclei still farther away from stability. Looking at Figure 2 it is evident that no simple extrapolation of the nuclear masses is possible for the very neutron-rich isotopes of Y, Sr, Rb, Kr and Br. At least several of these nuclei can be produced at ISOLDE II and III with a source strength sufficiently strong for a reliable determination of their Q_β-value with our equipment. For some of these nuclides, the decay schemes are already known well enough to allow the derivation of their Q_β-value from the βγ-coincidences measured.

In order to determine also the beta-decay energies of still more neutron-rich nuclei in this mass region, however, it will be necessary to investigate also the main features of unknown decay schemes. Therefore, we plan to extend our experimental equipment in order to measure also γγ-coincidences, in addition to the βγ-coincidences recorded already now. A sketch of the βγγ-coincidence system we are going to built is shown in Figure 4. By adding another Ge-detector of medium size to our existing system it will be possible to obtain two different βγ-matrices simultaneously, together with a γγ-matrix registred in list-mode and the β- and γ-singles spectra. This system will be quite compact and self-contained and can be installed at every mass separator, which can deliver the sources we are interested in.

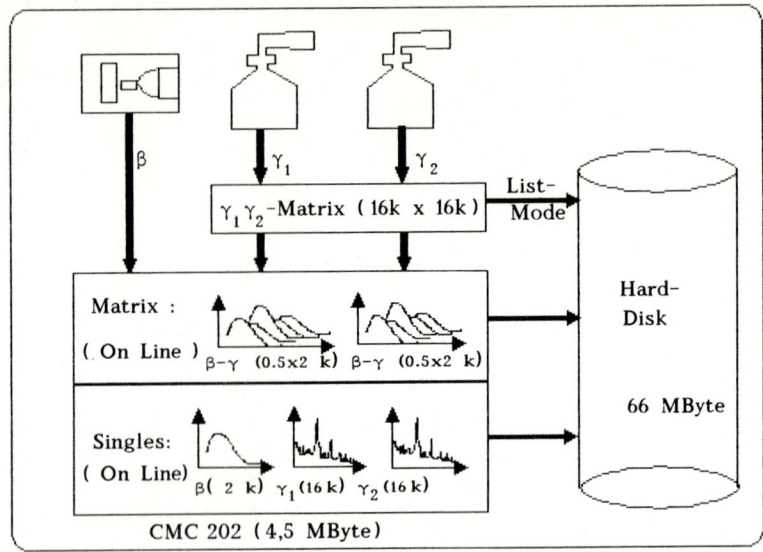

Figure 4: Sketch of a planned βγγ-coincidence system for the measurement of Q_β-values of nuclei very far from stability.

Acknowledgement

We want to thank the staff members of LOHENGRIN/ILL and ISOLDE/CERN for their valuable assistance. Thanks are due to Mrs. Ch. Laupheimer for her help in the preparation of the manuscript.

References

|1| F. Münnich, U. Keyser and B. Pahlmann: Proc.AMCO-7; Schriftenreihe Wiss.u.Technik, Vol.26,pp.155, Darmstadt: THD 1984
|2| M. Graefenstedt, U. Keyser, F. Münnich and F. Schreiber: this workshop
|3| M. Graefenstedt, U. Keyser, F. Münnich and F. Schreiber: Nuclei far from Stability; AIP Conference Proceedings 164,30(1988)
|4| R. F. Casten, W. Frank and P. von Brentano: Nucl.Phys.A444,133(1985)
|5| M. Tönhardt: Dissertation TU Braunschweig 1987
|6| N. Zeldes: Proc.Int.Workshop on "Gross Properties of Nuclei and Nuclear Excitations VII", Hirschegg 1979; INKA-Conf.79-001-046
|7| R. Bengtsson, P. Møller, I. R. Nix and Jing-ye Zhang: Phys.Scripta 29,402(1984)

Level Structure of $^{89}_{39}Y$, N = 50 Isotone

T. Batsch[1], J. Kownacki[1], Z. Zelazny[1], M. Guttormsen[2], T. Ramsøy[2], and J. Rekstad[2]

[1]Institute of Experimental Physics, University of Warsaw,
ul. Hoza 69, 00681 Warsaw, Poland
[2]Institute of Physics, University of Oslo, N-0315 Oslo, Norway

The $^{87}Rb(\alpha,2n)$ reaction has been used to populate states in ^{89}Y with spin values up to around 31/2.

The observed energy states have been arranged into two sequences with opposite parities. Up to about 7 MeV of excitation energy they show a level systematics typical for nuclei lying in the vicinity of closed shells. The levels lying higher in spin and energy may be arranged into a rotational-like sequence suggesting a change of the nuclear shape.

Using our and previously known data [1,2,3] we have tried to locate the members of the $1g_{9/2}$ core multiplets. The present results are compared with the semi-empirical shell model and the particle-core weak coupling description. The E = f[I(I+1)] dependence in N = 50, odd-A isotopes is also displayed.

Our starting point to understand the structure of the states in ^{89}Y is based on the former descriptions [4-8] of the low-spin excitations. Thus, the four lowest-lying states seen in the reaction [1,2,3] and radioactive decay [3] works are supposed to be mainly the single-particle shell-model states lying between the Z = 38 and Z = 40 gaps ($p_{3/2}$, $f_{5/2}$, $p_{1/2}$ and $g_{9/2}$). In ref. [7] the core-coupling model [6] has been used to calculate the level energies in ^{89}Y below 3.2 MeV (up to $13/2^+$ state). In this calculation the basis states are given by the coupling of the first 2^+ and 3^- states of the ^{88}Sr core to the shell-model orbits $p_{1/2}$ and $g_{9/2}$. Then the states with $17/2 \leq I \leq 23/2$ can arise as suggested by Arnell et al [8] from coupling of a $g_{9/2}$ proton to higher negative parity excitations ($5^-,6^-,7^-$) in the ^{88}Sr core.

It is known that the semi-magic nucleus $_{38}Sr_{50}$ has the $f_{5/2}$ and $p_{3/2}$ proton subshells filled up, while in $_{40}Zr_{50}$ also the $p_{1/2}$ orbital is filled. It turns out [9], however, that the best fit to pertinent experimental data is obtained by assuming that about 40% only of the ground state wave function of ^{90}Zr represents the closed-shell configuration (i.e. $\pi(p_{1/2})^2$).

The shell closure in ^{88}Sr is more pronounced [10]: the probability of finding the outermost proton pair outside the $p_{3/2}$ and $f_{5/2}$ subshells being only 20-30% about evenly distributed over the $p_{1/2}$ and $g_{9/2}$ orbitals.

In view of more data now available we would like to find the members of possible multiplets formed due to the ^{88}Sr core-particle couplings, using the weak-coupling basis [11] as a first order approximation.

Inspecting those results one can conclude that the energy levels of ^{89}Y excited in the present work combined with those seen in refs. [1,2,3] and [9] can be divided into a few groups: low-lying levels containing a large part of the single particle (single hole) strength [7], levels produced mainly by the coupling of a $g_{9/2}$, $p_{1/2}$ proton (or proton-hole) to the core excited states (up to 8^+) of ^{88}Sr and high energy levels (above 6 MeV) which are beyond the scope of the particle-vibration coupling and may represent entirely different configurations. The three uppermost transitions observed (402.8, 429.5 and 456.3 keV) are found in a kind of rotational-like cascade suggesting possible shape changes at spin 31/2 and at approximately 8 MeV excitation energy.

In a further attempt to describe the higher-spin states in ^{89}Y a semi-empirical shell-model calculation along the line developed by deShalit and Talmi [12] and Blomqvist [13] has been performed. It would be rather complicated to calculate the energies of the 2p-1h and 3p-2h states by a matrix diagonalization in the full valence particle space. One can try then an extreme simplification assuming for the yrast states only one configuration with the lowest unperturbed single-particle energy. We use empirical (if available, otherwise calculated with a δ-force) single-particle and 2-body matrix elements derived from level energies in the neighbouring nuclei assuming pure shell-model configurations. The matrix elements we have used in our calculations are listed in Table 1. Some of the effective 2-body interactions have been taken from the compilation of Bałanda [14] and of Blomqvist and Rydström [15].

Table 1

Two-particle matrix elements Δ_J

Nucleus					
^{90}Zr	^{88}Sr	^{88}Sr	^{88}Sr	^{88}Y	^{88}Y
Configuration					
E_0 (keV) - zero order energy					
$\pi(1g_{9/2})^2 J$	$\pi(g_{9/2}p_{3/2}^{-1})J$	$\pi(g_{9/2}p_{1/2}^{-1})J$	$\pi(g_{9/2}f_{5/2}^{-1})J$	$\pi p_{3/2}^{-1}\nu g_{9/2}^{-1}$	$\pi p_{1/2}^{-1}\nu g_{9/2}^{-1}$
-12332	4447	5292	4849	5544	4037
J^π Δ_J^a(keV)	J^π Δ_J^b(keV)	J^π Δ_J^b(keV)	J^π Δ_J^b(keV)	J^π Δ_J^b(keV)	J^π Δ_J^b(keV)
0^+ 1643.2	3^- -1713	4^- (-2292)$^{c)}$	3^- -2115	3^- -1500	4^- 367
2^+ 561.8	4^- (-1447)$^{c)}$	5^- -1707	4^- (-1849)$^{c)}$	4^- -1140	5^- 599
4^+ 170.0	5^- -866		5^- -1264	5^- -908	
6^+ 431.6	6^- -427		6^- -829	6^- -300	
8^+ 528.2			7^- -421		

a) Ref. [15]   b) Ref. [16]   c) δ-force estimation

The excitation energies of states observed in the $(\alpha,2n)^{89}$Y reaction have been calculated relative to the ^{88}Sr core with the following configurations:

$\pi(g_{9/2}^2 p_{3/2}^{-1})$ $13/2^- \leq I^\pi \leq 19/2^-$

$\pi(g_{9/2}p_{1/2}p_{3/2}^{-1})$ $5/2^+ \leq I^\pi \leq 13/2^+$

$\pi(g_{9/2}p_{3/2}f_{5/2}^{-1})$ $9/2^+ \leq I^\pi \leq 17/2^+$

Basing on the empirical effective two-body interactions a qualitative correspondence of the calculated and the measured excitation energies is obtained. The negative parity states, $13/2^- \leq I \leq 19/2^-$ are thought to be the $\pi(g_{9/2}^2 p_{3/2}^{-1})$ multiplet. The positive parity states $5/2^+ \leq I \leq 17/2^+$ are most probably due to the $\pi(g_{9/2} p_{1/2} p_{3/2}^{-1})$ and $\pi(g_{9/2} p_{3/2} f_{5/2}^{-1})$ configurations. The $17/2^+$ state is reproduced by the latter one in its maximal alignment, the $15/2^+$ state is not observed in the experiment, while both the $13/2^+$ and $11/2^+$ states of the $\pi(g_{9/2}\ p_{3/2}\ f_{5/2}^{-1})$ configuration come too high compared to experiment, suggesting that they are mainly due to the $\pi(g_{9/2}\ p_{1/2}\ f_{5/2}^{-1})$ configuration (fig.1).

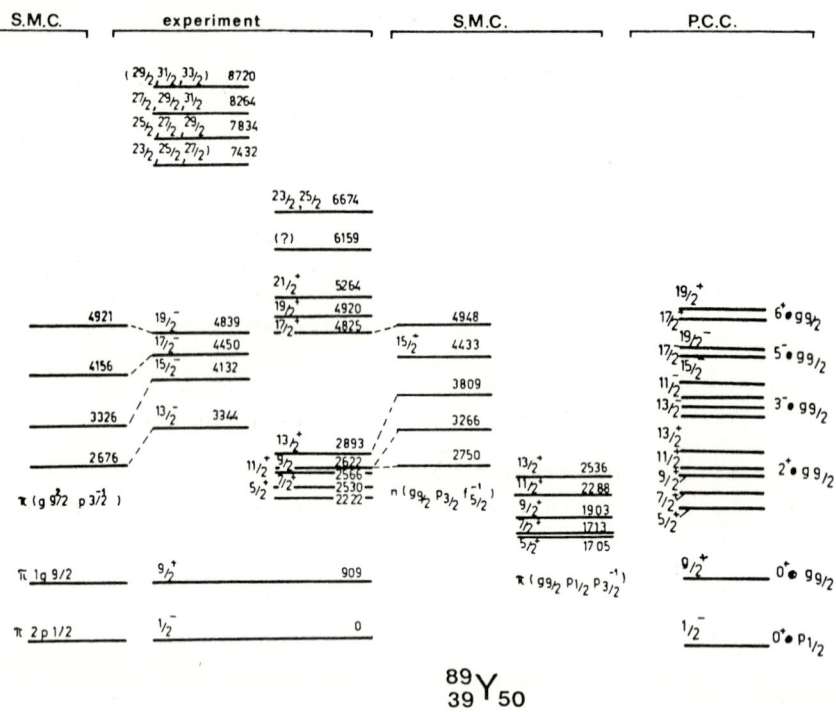

Fig.1. Comparison between the ^{83}Y experimental levels and states calculated within the shell-model (S.M.C.) and particle-core weak coupling (PVC).

Finally, looking for systematical features, the yrast states for the $N = 50$, odd-A nuclei $^{89}_{39}$Y, $^{91}_{41}$Nb, $^{93}_{43}$Tc and $^{95}_{45}$Rh are drawn as a function of $I(I+1)$ (Fig.2). Whereas, the three latter nuclei display signature dependent values, the ^{89}Y case gives quite smooth curves. Also, the values for ^{89}Y are larger, indicating a smaller effective moment of inertia.

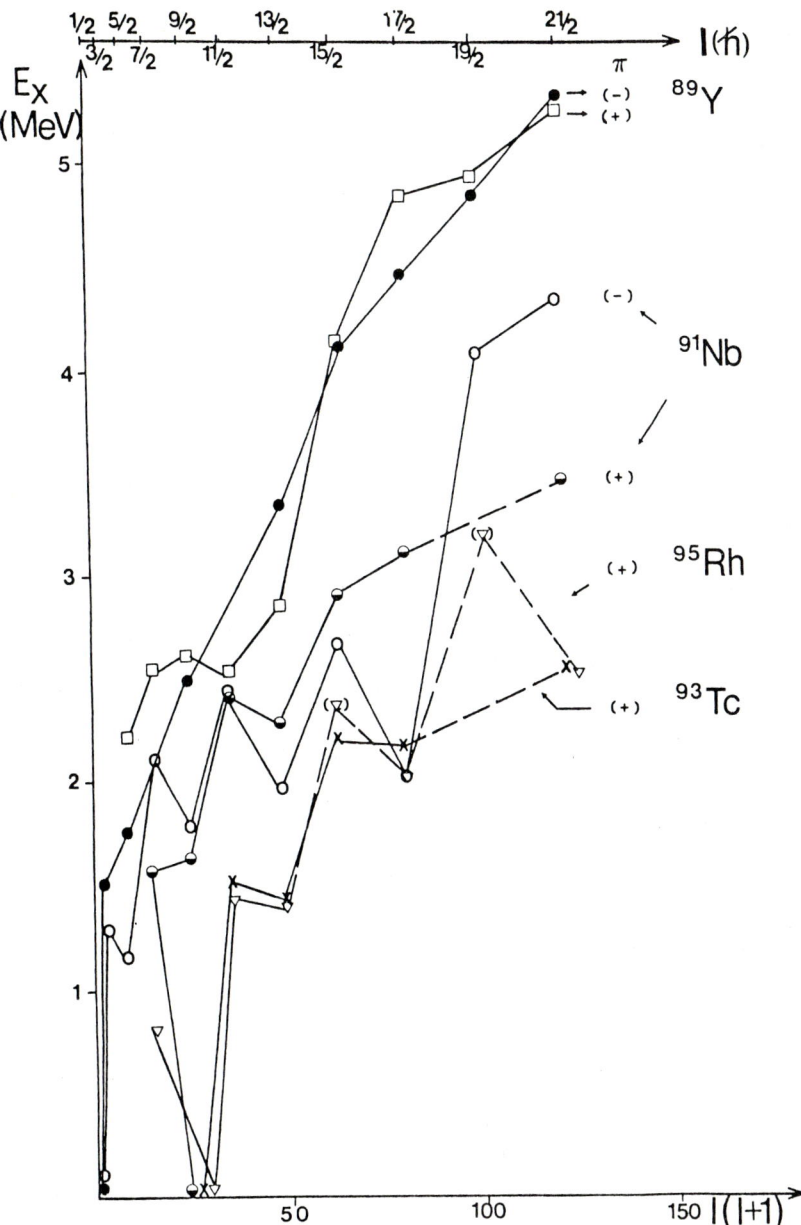

Fig.2 The excitation energies of the $^{95}_{45}$Rh, [17], $^{93}_{43}$Tc [19], $^{91}_{41}$Nb [18] and $^{89}_{39}$Y yrast states versus I(I+1). The two point marked (△) show estimated positions of $15/2^+$ and $19/2^+$ states in ^{95}Rh nucleus.

In conclusion, the states with spin lower than 21/2 are rather satisfactorily described within the particle-core coupling model. The semi-empirical shell-model approach, however needs more data on 2-body matrix elements around the ^{88}Sr core nucleus. The high-spin levels above 7 MeV excitation energy have been tentatively interpreted as aligned many-quasiparticle states with possible rotational-like bands built upon them. The more extensive description of the present results is prepared [20], however a further experimental studies could clarify if a transition into a deformed nuclear shape takes place at this spin and excitation energy.

References

1. M. Davidson, J. Davidson, M. Behar, G. Garcia Bermudez and M.A.J. Mariscotti, Nulc.Phys. A306 (1978) 113.
2. C.A. Fields and L.E. Samuelson, Phys.Rev. C20 (1979) 2442
3. D.C. Kocher, Nulc.Data Sheets 16 (1975) 445
4. S.M. Shafroth, N.P. Trehan and D.M. Van Patter, Phys.Rev. 129 (1963) 704
5. B.J. Dalton and D. Robson, Nucl.Phys. A210 (1973) 1
6. V.K. Thankappan and W. True, Phys.Rev. 137 (1965) 8793
7. P. Hoffmann-Pinther and J.L. Adams, Nucl.Phys. A229 (1974) 365
8. S.E. Arnell, A. Nilsson and O. Stankiewicz, Nucl.Phys. A241 (1975) 109
9. W.J. Courtney and H.T. Fortune, Phys.Rev.Lett. 41B (1972) 4
10. J. Picard and G. Bassani, Nucl.Phys. A131 (1969) 636
11. B.R. Mottelson, Nordita Publications No.288, Nikko Summer School 1967
12. A.de-Shalit and I. Talmi, Nuclear Shell Theory, Academic Press, N.Y. 1963
13. J. Blomqvist, I. Bergstrøm, C.J. Herrlander, C.G. Liden and K. Wickstrøm, Phys.Rev.Lett 38 (1977) 534
14. A. Batanda, Acta Phys.Pol. 88 (1977) 501
15. J. Blomqvist and L. Rydstrøm, Phys.Scripta 31 (1985) 31
16. E.K. Warburton, J.W. Olness, C.J. Lister, R.W. Zurmuhl and J.A. Becker, Phys.Rev. C31 (1985) 1184
17. P. Komnios, E. Nelte, P. Blasi, Z.Phys.A. - Atoms and Nuclei 314 (1983) 135
18. A. Balanda, R. Kulessa, W. Waluś and J. Sieniawski, Acta Phys.Pol. B7 (1976) 355
19. B.A. Brown, O. Hausser, T. Faestermann, D. Ward, H.R. Andrews and D. Horn, Nulc.Phys. A306 (1978) 242
20. T. Batsch, J. Kownacki, Z. Zelazny, M. Guttormsen, T. Ramsøy and J. Rekstad, to be published in Physica Scripta.

Nuclear Structure Close to ^{100}Sn: Excitations of the Shell Model Core

H. Grawe[1], D. Alber[1], H. Haas[1], H. Kluge[1], K.H. Maier[1], B. Spellmeyer[1], and X. Sun[2]

[1] Hahn-Meitner-Institut für Kernforschung,
   Glienicker Str. 100, D-1000 Berlin, Germany, and
   Freie Universität Berlin, D-1000 Berlin, Germany
[2] Institute of Modern Physics, Acádemia Sinica, Lanzhou, P.R. China

Introduction

The main topics of experimental β-decay and in-beam studies in the ^{100}Sn region are the shell model structure [1-5], the residual interaction [6-9] and spin response phenomena as the quenching of M1 [3,10] and Gamow-Teller strengths [11]. Shell model calculations [3,6-9,12] in a small configuration space are generally in good qualitative agreement with the experiment. There are, however, a few unsolved problems, like e.g. the failure to reproduce the ^{94}Ru and ^{95}Rh B(E2)-values in the middle of the $\pi g_{9/2}$ shell [1,6] or the spin gap isomerism in ^{95}Pd [5]. Whether this is due to an inadequate residual interaction and/or configuration space can be studied best as close as possible to ^{100}Sn.

Experimental Set-up and Results

In a series of experiments we have studied the nuclei ^{96}Pd, ^{97}Ag and ^{100}Cd, using the nuclear reactions

$$^{36}\text{Ar} + {}^{64}\text{Zn} \rightarrow {}^{100}\text{Cd}^* \text{ at 130 MeV}$$
$$^{58}\text{Ni} + {}^{46}\text{Ti} \rightarrow {}^{104}\text{Sn}^* \text{ at 230 MeV}$$
$$^{36}\text{Ar} + {}^{70}\text{Ge} \rightarrow {}^{106}\text{Sn}^* \text{ at 135 MeV}$$

Multiple coincidences between γ-rays and evaporation neutrons, protons and α-particles were measured to clean up γ-ray spectra and to identify weakly populated neutron deficient exit channels. The γ-rays were detected in up to 10 Compton suppressed Ge-detectors from the OSIRIS spectrometer [13], a 16 segment $\Omega = 2\pi$ NE 213 scintillator was used as a multiplicity filter for neutron detection [14] and a four-segment large area (4 x 300 mm^2, $\Omega_{\text{eff}} = 2\pi$) surface barrier ΔE-detector-lantern [15] served as a charged particle identification and multiplicity device. Magnetic moments were measured for several isomers in a standard PAD set-up. Spins and parities were deduced from systematics of neighbouring nuclei and anisotropies measured in the PAD experiment. To demonstrate the power of the filter detectors, the identification of ^{97}Ag, populated with only 0.9 mb or 0.2 % of the total residue cross section, is shown in Fig. 1.

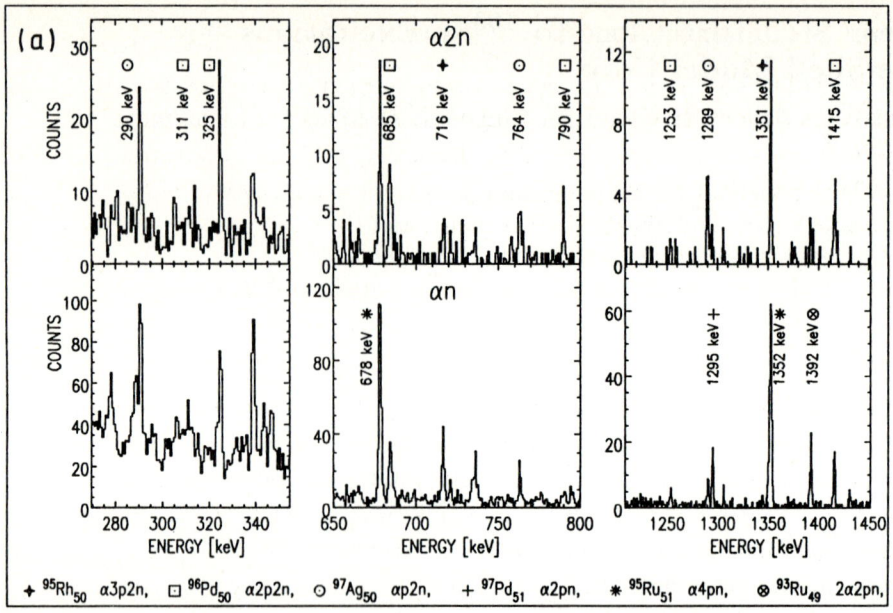

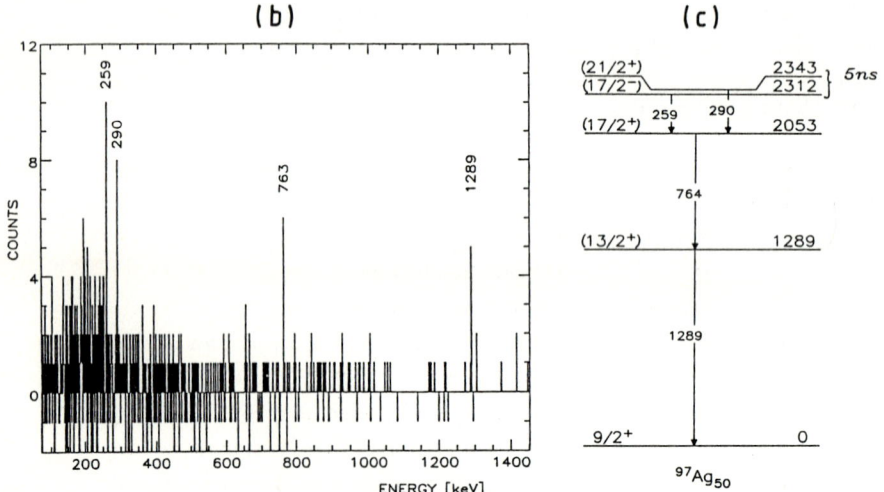

Fig. 1a. Delayed (10-60ns) αnγ and α2nγ coincidence spectra from the reaction ^{58}Ni + ^{46}Ti. Note the enhancement of the 2n exit channels (N=50 nuclei) ^{95}Rh, ^{96}Pd, ^{97}Ag.

b. Delayed (7-26ns) nγγ sum spectrum for ^{97}Ag. Gates on 1289, 764 and 290 keV γ-rays.

c. Preliminary level scheme of ^{97}Ag.

Discussion

We have performed shell model calculations in the spirit of the approaches in Refs. [6,9,12] using a ^{100}Sn core with single particle energies calculated from the values for ^{88}Sr and ^{90}Zr cores by means of the chosen empirical two-body matrix elements. Particles (holes) were allowed in the $2p_{1/2}$, $1g_{9/2}$ orbitals for protons (π) and neutrons (ν) and in the $2d_{5/2}$ shell for neutrons. For N = 50 nuclei a seniority

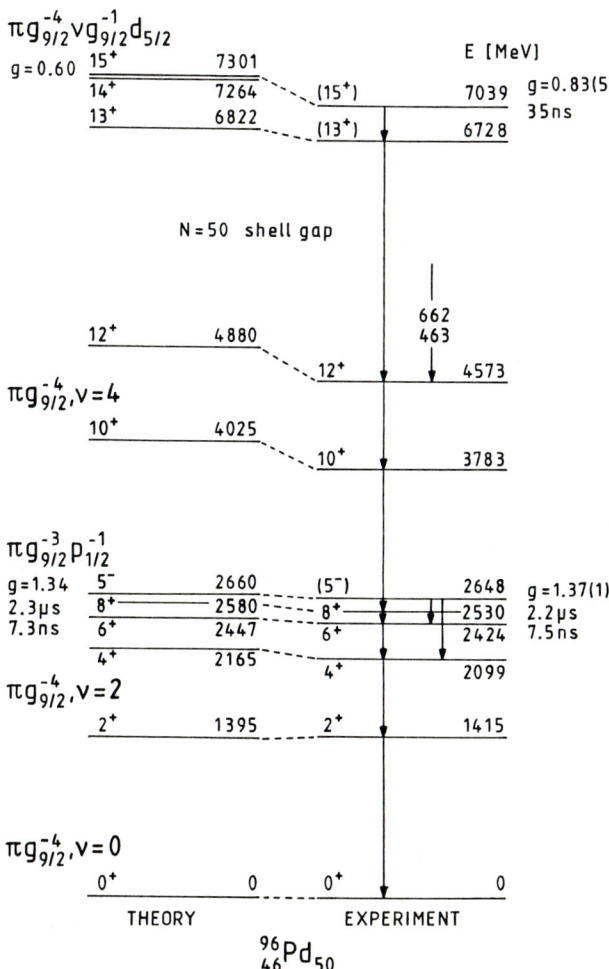

Fig. 2 Experimental and theoretical level scheme and electromagnetic properties for ^{96}Pd.

conserving interaction [6] was used for protons, and the πν interaction was taken from Ref. [12]. For N=52 nuclei only the π1g$_{9/2}$ and ν2d$_{5/2}$, 1g$_{7/2}$ orbitals were used. For the unknown interaction involving the ν1g$_{7/2}$ orbital realistic two-body matrix elements derived with the Sussex and Yale codes [16] were used.

N=50 nuclei and the ππ interaction

The merits and drawbacks of the shell model approach can be demonstrated in the well studied [3,4] N=50 nucleus ^{96}Pd (see Fig. 2). Excellent agreement is found for the v=2 states. Note, that the lifetimes of the $I^\pi = 8^+$ and 6^+ states cannot be reproduced with a seniority mixing interaction [3] that has been invoked to explain the E2 discrepancy in ^{94}Ru [6]. In Fig. 3 the B(E2) values for the seniority v=2, v=3, I=I$_{max}$ states are compared to the expectations for pure seniority B(E2, j^n,v,I → I-2) = [(2j+1-2n) / (2j+1-2v)]^{2}B(E2, j^v,I → I-2). Obviously the g$_{9/2}$ shell is half filled already for ^{94}Ru, which is not reproduced in the present shell model calculation. Even more pair scattering from the lower (p,f) shell than obtained in a recent calculation [17] is needed to improve the discrepancies for ^{94}Ru and ^{95}Rh. This is supported by the observed discrepancy in excitation energy for the v=4 states (Fig. 2) and the v=3 states in ^{97}Ag. Their energies do not contain the (g$_{9/2}^2$,I=0) matrix element, which empirically is found to be bound too strongly, as it has to simulate the inadequacy of the 2p$_{1/2}$, 1g$_{9/2}$ model space.

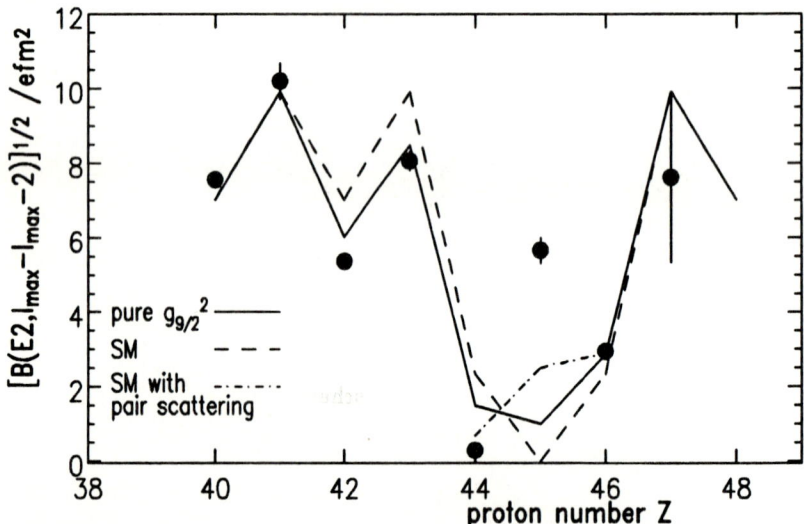

Fig. 3 Experimental E2 strength of the lowest seniority excited states with maximum spin (8$^+$, v=2 and 21/2$^+$, v=3, respectively) in N=50 nuclei in comparison to theoretical predictions.

πν interaction in hole-hole and particle-hole configurations

With the $I^\pi = 12^+$ state in ^{96}Pd the proton model space $\pi(p_{1/2}, g_{9/2})$ is exhausted and in the N=50 neighbour ^{94}Ru the $I^\pi \geq 13^+$ states have been interpreted as $\nu g_{9/2}^{-1} d_{5/2}$ core excitations [12]. Experimentally, this is proven by our g-factor measurement for the $I^\pi = (15^+)$ isomer $g = 0.83(5)$ showing a large reduction as compared to the pure proton state $g(8^+) = 1.37(1)$. The shell model theory, though in contrast to ^{94}Ru indicating a trend towards isomerism, cannot reproduce the level order needed for an E2 isomer. As found previously for the $I^\pi = 21/2^+$ isomer in ^{95}Pd a more strongly bound $(\pi g_{9/2}^{-1} \nu g_{9/2}^{-1})_{I=9}$ matrix element would reproduce the level order [5,12]. The common reason for the isomerism is the πν interaction, which is strong in hh and pp configurations, but weak in ph configurations as in the lower $\pi g_{9/2}$ shell in ^{94}Ru. However, a more serious discrepancy is observed for the g-factor indicating that the πν interaction needs further improvement, i.e. more admixture from the v=4 proton states $I^\pi = 10^+, 12^+$ in the $I^\pi = 15^+$ wave function.

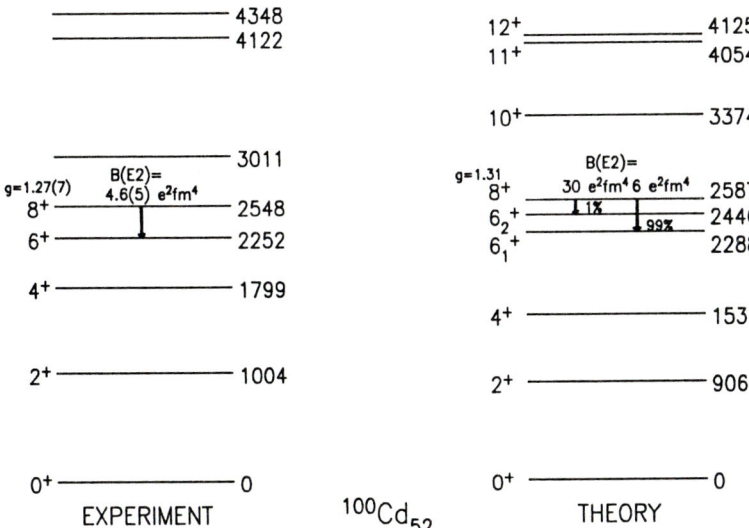

Fig. 4   Experimental and theoretical level scheme and electromagnetic properties for ^{100}Cd.

For the N=52 nucleus ^{100}Cd little mixing of proton and neutron configurations is expected in high spin states due to the ph-character of the interaction. This is corroborated by the almost pure proton g-factor measured for the $I^\pi = 8^+$ isomer in ^{100}Cd $g = 1.27(7)$. In Fig. 4 the preliminary results of a shell model calculation are compared to the experimental data [18]. The level scheme and the lifetime of the

$(\pi g_{9/2}^{-2})_{8^+} \nu^2 0^+$ isomer is well reproduced. The position of the two 6^+ states with predominant configurations $(\pi g_{9/2}^{-2})_{6^+} \nu^2 0^+$ and $\pi^2 0^+ (\nu d_{5/2} g_{7/2})_{6^+}$ depends sensitively on the $\pi g_{9/2}^{-2} \nu g_{7/2}$ interaction, which empirically had to be reduced by 25 % from the strength deduced from the Sussex/Yale interaction, to reproduce the lifetime of the $I^\pi = 8^+$ state.

Conclusion

In nuclei close to ^{100}Sn shell model calculations in a limited configuration space using empirical residual interactions are found to be in qualitative good agreement with the now available experimental information. Discrepancies arise from two effects:

(i) excitation and/or pair scattering from below the ^{88}Sr proton core is responsible for the problems of B(E2) values in the middle of the $\pi g_{9/2}$ shell (^{94}Ru, ^{95}Rh) and the deviations of the seniority $v = 4$ and $v = 3$ excitation energies;

(ii) the empirical $\pi\nu$ interaction being strong in pp and weak in hh configurations needs further refinement to yield quantitative agreement for neutron core excited states.

[1] E. Nolte, G. Korschinek, U. Heim, Z. Physik A298 (1980) 191
[2] W. Kurcewicz, E.F. Zganjar, R. Kirchner, O. Klepper, E. Roeckl, P. Komninos, E. Nolte, D. Schardt, D. Tidemand-Petersson, Z. Physik A308 (1982) 21
[3] H. Grawe, H. Haas, Phys. Lett. 120B (1983) 63
[4] W.F. Piel Jr, G. Scharff-Goldhaber, C.J. Lister, B.J. Varley, Phys. Rev. C28 (1983) 209
[5] E. Nolte, H. Hick, Z.Physik A305 (1982) 289
[6] D.H. Gloeckner, F.J.D. Serduke, Nucl. Phys. A220 (1974) 477
[7] F.J.D. Serduke, R.D. Lawson, D.H. Gloeckner, Nucl.Phys. A256 (1976) 45
[8] R. Gross, A. Frenkel, Nucl. Phys. A267 (1976) 85
[9] J. Blomqvist, L. Rydström, Phys. Scripta 31 (1985) 31
[10] O. Häusser, I.S. Towner, T. Faestermann, H.R. Andrews, J.R. Beene, D. Horn, D. Ward, Nucl. Phys. A293 (1977) 248
[11] K. Rykaczewski, I.S. Grant, R. Kirchner, O. Klepper, V.T. Koslowski, P.O. Larsson, E. Nolte, G. Nyman, E. Roeckl, D. Schardt, L. Spanier, T. Tidemand-Pertersson, E.F. Zganjar, J. Zylicz, Z. Physik A322 (1985) 263
[12] K. Muto, T. Shimano, H. Horie, Phys. Lett. 135B (1984) 349
[13] R.M. Lieder, A. Jäger, A. Neskakis, T. Venkova, C. Michel, Nucl. Instr. and Meth. 220 (1984) 363
[14] D. Alber, H. Grawe, H. Haas, B. Spellmeyer, Nucl. Instr and Methods in Physics Research A263 (1988) 401
[15] D. Alber, H. Grawe, H. Haas, B. Spellmeyer, Annual Report 1986, HMI-B 441, p. 127
[16] L.D. Skouras, C. Dedes, Phys. Rev. C15 (1977) 1873
[17] J. Dobaczewski, W. Nazarewicz, A. Plochocki, K. Rykaczewski, J. Zylicz, Z. Physik A329 (1988) 267
[18] D. Alber, H. Grawe, H. Haas, B. Spellmeyer, X. Sun, Z. Physik A327 (1987) 127

Part VI

Deformations

A ≈ 80 and A ≈ 100 Nuclei Studied Within the Mean Field Approach

W. Nazarewicz[1] and T. Werner[2]

[1]Institute of Physics, Warsaw Institute of Technology,
ul. Koszykowa 75, 00662 Warsaw, Poland

[2]Institute of Theoretical Physics, University of Warsaw,
ul. Hoza 69, 00681 Warsaw, Poland

> *The mean field theory is applied to nuclei from the zirconium region. Special attention is given to the shape variations with particle number and angular momentum, shape coexistence effects, changes in the pairing field, and the octupole correlations.*

1. Introduction

It has been shown experimentally ([1-4] and earlier papers quoted therein) that shape coexistence, unusually large deformations, the presence of well-deformed intruder orbitals, quenching of pairing correlations, and dramatic shape changes induced by rotation are quite common phenomena in the zirconium region.

The richness of different structural effects and the degree of their dependence on particle number and/or angular momentum make this region an ideal area for testing various theoretical approaches. The microscopic reason for such a strong variation of collective properies is the relatively low single-particle (s.p.) level density in these medium-mass nuclei. While in the heavier systems equilibrium deformations vary smoothly from nucleus to nucleus, in the zirconium region shape properties essentially depend on the proton and neutron numbers: the onset and termination of deformation are much more abrupt here.

In the mean field approach (or its simple version, the deformed shell model) the shape effects are automatically accounted for through the deformation of the mean field. The nuclear deformation depends essentially on the individual motion of the valence nucleons which polarize the nuclear core. The underlying shell structure of the A≈80 and A≈100 nuclei is analyzed in sect. 2. The different mean field calculations in the zirconium region are briefly reviewed in sect. 3. In sect. 4 a comment is given on the apparent reduction of pairing correlation in a number of A≈80 and A≈100 isotopes. It is expected that due to the γ-softness, the aligned $g_{9/2}$ orbitals strongly polarize the core, triggering the shape change to different areas in the deformation plane, depending on the actual position of the Fermi level. Also, the well-deformed (superdeformed) structures predicted for N=43 and 44 can be explicitly related to the large shell gaps and the alignment properties of high-j intruder orbitals, like $g_{7/2}$ or $h_{11/2}$. Selected aspects of the interplay between deformation, rotation, and pairing in the A≈80 and A≈100 nuclei are discussed in sect. 5 and 6. Medium-mass nuclei from the zirconium region are not expected to be octupole unstable. However, as one approaches the "optimal" particle numbers 32 and 56 the octupole coupling becomes enhanced (sect. 7). Finally, the role played by different terms in the nuclear Hamiltonian in creating the nuclear deformation, especially the effect of neutron-proton interaction, is commented in sect. 8.

2. Shell effects in the zirconium region

Calculations employing the shell correction method applied to the A=80 [5-12] and A=100 [5-9,13-17] nuclei suggest an interpretation of the experimental data in terms of the well-deformed, elongated (prolate) shapes, the structures with flattened (oblate) shapes, and the shell-model (spherical) configurations. The microscopic origin of strong shape variations in these nuclei may be attributed to the sizeable gaps in the single-particle spectrum.

The basic assumption of the shell correction approach is that the total energy of a nucleus, E_{tot}, can be composed of two parts [18,19]:

$$E_{tot} = E_{macr} + \delta E_{shell} \tag{1}$$

where E_{macr} is the macroscopic energy (smoothly depending on the number of nucleons) and δE_{shell} is the shell correction term. In the framework of the deformed shell model, the shell correction term can be calculated from the single-particle spectra of the deformed mean field, while the macroscopic energy is replaced by the phenomenological energy of the liquid drop or droplet model. The macroscopic energy always increases with the deformation of nuclear shape. Therefore, the spontaneous breaking of the spherical symmetry can be directly associated with the shell correction term. More details about the shell correction method and the related configuration concept can be found in [10] and references quoted therein.

To gain some understanding of the dominant shape-polarisation forces it is very instructive to plot δE_{shell} versus the number of particles and selected deformation parameters. The total shell energy is a sum of corresponding proton and neutron contributions, therefore the latter ones can be treated separately. The shell energy landscapes representative for the zirconium region can be found e.g. in [7,16] (Nilsson model) and [6] (folded-Yukawa model). Fig. 1 presents the neutron Woods-Saxon shell energy landscape (without pairing) as a function of neutron number and quadrupole deformation β_2.

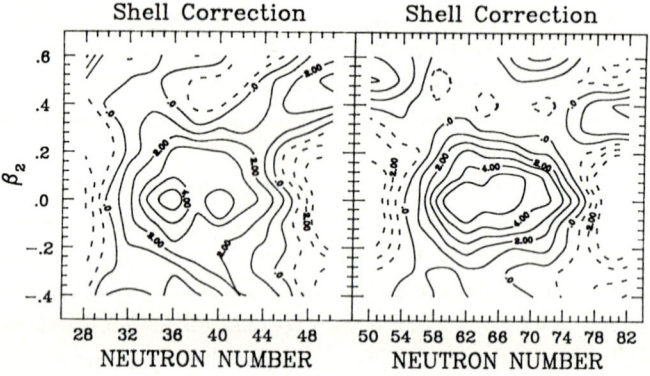

Fig. 1 Woods-Saxon neutron shell correction landscapes for the $A \approx 80$ (left) and $A \approx 100$ (right) regions plotted versus neutron number and quadrupole deformation β_2. The hexadecapole deformation, β_4, minimizes the liquid-drop-model energy for each value of β_2. The diagrams of the proton shell energy are very similar.

The topology of δE_{shell} is fairly similar in both regions of zirconium nuclei. The lowest shell energy is expected at spherical shapes for the magic numbers, N=28, 50 and 82. The transition to a deformed region occurs around N=32, 44, 58 and 76. The strongest shell-polarisation towards deformed shapes are predicted for N=36-42 and N=60-70. For these particle numbers the shell energy slopes down towards both positive and negative values of β_2. This is a microscopic origin of the observed prolate-oblate isomerism or even the γ-instability. The best prospects for well deformed shapes are found for the particle numbers 38-44 and 60-62. In fact, the nuclei around ^{76}Sr and ^{100}Sr are among the best deformed systems known. The best candidates for superdeformed shapes ($\beta_2 \approx 0.6$) are the nuclei with particle numbers 44 and 60-64, like ^{83}Zr, ^{84}Zr and $^{104-108}$Ru (see sect. 6 and [14]).

3. Mean field calculations in the A≈80 and A≈100 regions

There already exist a number of calculations in the zirconium region which employ the mean field approach. The differences between various models are clearly reflected by their predictions concerning (i) prolate-oblate energy differences and deformation energies; (ii) particle numbers at which the shape-transition from spherical to deformed systems takes place; (iii) the nature of shape-coexisting states.

For the A≈80 nuclei the probably most systematic calculations are those of [6,8-10] (see also refs. quoted therein). An astonishingly close agreement between predicted prolate-oblate energy differences, deformation energies, and equilibrium deformations has been found for the folded-Yukawa model of [6] and the Woods-Saxon model of [10] (see fig. 2 of [10]). The sefconsistent Hartree-Fock calculations with the Skyrme III interaction [8] yield similar results. Fig. 2 presents a summary of the calculated equilibrium deformations for Sr and Zr isotopes with $40 \leq N \leq 48$. The N=40 isotopes are predicted to be prolate-deformed with a large quadrupole deformation. For N=42 the Potential Energy Surfaces (PES) have a more complex structure with coexisting prolate and oblate minima. The transition from deformed to spherical shapes is predicted in all models to occur at N=44, although triaxial or prolate local minima have also been found. Finally, the $N \geq 46$ isotopes are expected in all models to be spherical in their ground states. A more detailed discussion of the low-spin shape effects in the A≈80 nuclei is given in [11,12].

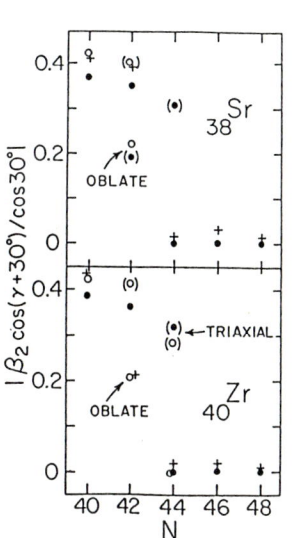

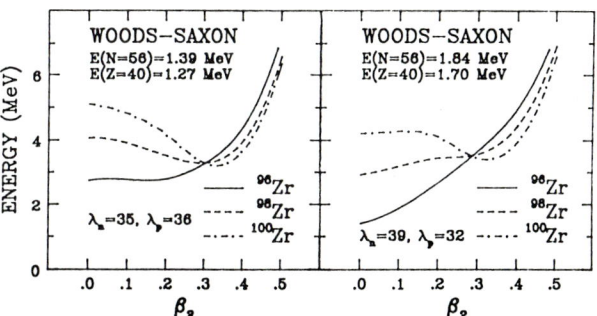

Fig. 3 The total energy calculated with the Woods-Saxon potential for $^{96,98,100}Zr$. The results shown in the left diagram were calculated using the standard values of the spin-orbit strength (λ_n, λ_p). In the right diagram, λ_n and λ_p were slightly modified. The values of the N=56 and Z=40 shell gaps in ^{96}Zr are given in both cases.

Fig. 2 Theoretical quadrupole equilibrium deformations: $\beta_2(eff)=\beta_2 \cos(\gamma+30°)/\cos(30°)$ for the even-even Sr and Zr with $40 \leq N \leq 44$ [34]. Results of three theoretical models are presented: the Woods-Saxon model [10] (•), the Skyrme III model [8] (o), and the folded-Yukawa model [5] (+). Encircled symbols represent local (excited) energy minima.

For the A≈100 nuclei the Skyrme-HF model [8] turned out to be quite successful in reproducing large deformations around ^{102}Zr. The heavy isotopes of Mo were predicted by this model to be triaxial, as in [15]. Let us also mention the Nilsson-Strutinsky calculations [9] where, after careful adjustment of the Nilsson parameter, the authors were able to reproduce the main experimental trends in both the A≈80 and A≈100 mass regions.

The main problem which one encounters in mean field calculations in the $A\approx 100$ region is the correct description of the shape transition around N=58. Here, a key nucleus is ^{96}Zr which is a semi-magic spherical system (Z=40 and N=56 are semi-closures at spherical shape). A transition from sphericity to the deformed regime takes place at N=58. Experimental data for ^{98}Zr indicate a spherical ground state and a deformed intruder state with an excitation energy of 853 keV (however, see [21]). This configuration becomes a ground state in ^{100}Zr which has been found to be fairly well deformed [22], with the spherical excited state at 331 keV. Both in ^{98}Zr and ^{100}Zr the spherical and the deformed configurations are strongly mixed [22].

In the language of the geometrical model the onset of quadrupole deformation at N=60 can be understood in the following way. At large deformations ($\beta_2 \approx 0.4$) two large shell closures at Z=40 and N=60 open up. Compared to the spherical configuration there are three $g_{9/2}$ protons, and two $h_{11/2}$ neutrons occupied at the deformed minimum. The intruder configuration in ^{98}Zr can thus be viewed as a two-quasiparticle excitation to the $h_{11/2}$ subshell. As soon as $h_{11/2}$ levels are filled the nuclear deformation starts to develop.

Any model which aims to reproduce the experimetal data in the $A\approx 100$ region should be able to correctly predict the balance between spherical and deformed subshell closures discussed above. In [8] the authors have noticed the great sensitivity of the model predictions for ^{96,98}Zr to the choice of Skyrme parameters used. In [5,6] the deformed shell model calculations based on the folded-Yukawa average field were quite successful in reproducing experimental masses and large quadrupole deformations in the whole Zr-region. However, the transition from a spherical to a deformed shape was predicted too early, around N=55. The authors correlated this discrepancy with too small theoretical gaps at Z=40 and N=56. The Nilsson model calculations with standard parameters have also had difficulties reproducing the suddenness of the shape transition, see [17] and refs. quoted therein. The agreement with experiment was eventually achieved, but only after some modifications of the Nilsson model paremeters [9,17].

In order to illustrate the great sensitivity of calculated PES around ^{96}Zr to the choice of model parameters, we performed the calculations with the deformed Woods-Saxon potential. Fig. 3 shows the total energy versus quadrupole deformation for ^{96,98,100}Zr. In the left portion the standard set of parameters was used, the same which was employed for nuclei in the $A\approx 80$ region [10]. In this case PES for ^{96}Zr was calculated to be extremely shallow around the spherical shape while both ^{98}Zr and ^{100}Zr were predicted to be well deformed with $\beta_2 \approx 0.3$. The experimental trend was restored after small modification in the spin-orbit strength for protons and neutrons, right portion of fig. 3. The new parameters increase the spherical Z=40 and N=56 gaps and therefore stabilize the spherical configuration. It can be seen that ^{96}Zr is now well spherical, there are two minima in the PES for ^{98}Zr, one spherical and one deformed, and there is only one deformed minimum in the PES for ^{100}Zr. Fig. 4 shows corresponding PES in the (β_2,β_4)-plane.

4. Pairing correlations

In the deformed shell model theory the second building block, after the single-particle mean field, is the short range residual interaction. This force is often approximated by means of the state-independent monopole pairing interaction. The main effect of pairing is the smearing out the occupation probabilities around the Fermi level. A direct measure of the smearing width is the gap parameter, Δ. The macroscopic part of the total energy, E_{macr}, already contains the average pairing energy which accounts for the main part of the even-odd mass difference. Therefore it is the fluctuating part of the pairing energy, δE_{pair}, that gives an additional contribution to the total shell energy. This so-called pairing correction is shown in fig. 5 as a function of neutron number and quadrupole deformation, β_2. It is worth emphasizing that the pairing correction is anticorrelated with the shell correction shown in fig. 1. This is

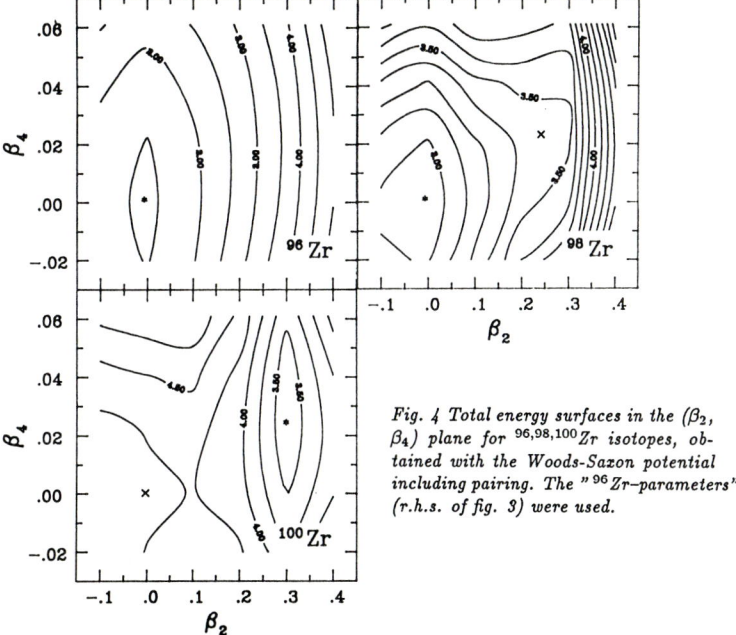

Fig. 4 Total energy surfaces in the (β_2, β_4) plane for $^{96,98,100}Zr$ isotopes, obtained with the Woods-Saxon potential including pairing. The "^{96}Zr-parameters" (r.h.s. of fig. 3) were used.

because the pairing gap depends on the level density around the Fermi level, $g(\lambda_F)$, as

$$\Delta \propto exp\left[-\frac{1}{Gg(\lambda_F)}\right] \tag{2}$$

where G is the monopole pairing strength. In the case of a large single-particle level density δE_{shell} is positive but δE_{pair} is large and negative. The opposite is true for systems with small $g(\lambda_F)$, i.e. around the large shell gaps. This tendency is clearly seen in fig. 5. The regions with positive δE_{pair} can easily be associated with the large spherical openings at N=28, 50 and 82 and deformed gaps at N=38-42, 64-68. The strongest pairing is predicted at spherical shape (due to large degeneracy between single-particle orbitals) for N=36 (the middle of the $p_{3/2}$, $f_{5/2}$, $p_{1/2}$ area), N=44 (the middle of the $g_{9/2}$ subshell), N=62 (the middle of the $d_{5/2}$, $g_{7/2}$ and $d_{3/2}$ area) and N=74 (the middle of the $h_{11/2}$ subshell).

In a number of odd-A and odd-odd A≈80 and A≈100 nuclei, rotational bands with unusually large moments of inertia have been found. The best examples are: the [312]3/2 and $g_{9/2}$ band in ^{77}Rb [23], the $g_{9/2}$ bands in ^{81}Y [24], ^{83}Y [25] and ^{99}Y [26], the three-quasiparticle band in ^{99}Y [26] and two-quasiparticle bands in ^{100,102}Y and ^{102,104}Nb [27,28]. These large moments of inertia have usually been taken as evidence of reduced pairing correlations [29].

Fig. 6 shows calculated values of the pairing gap versus β_2 in selected one-quasiparticle configurations in ^{77}Rb, ^{79}Y, ^{99}Y, ^{101}Nb. A quite dramatic reduction of proton pairing is seen in the [422]5/2 and [301]3/2 configurations in Y as well as in the [431]3/2, [312]3/2 and [310]1/2 states in Rb. The former states lie just above the well-deformed Z=38 gap while the latter ones are just below this gap. Therefore, by blocking some of those orbitals one effectively increases the Z=38 gap, which leads to a reduction of the pairing gap, or even, as in the case of [422]5/2 level in Y, to a collapse of the static pairing. In the N=61 and N=63 systems, fig. 7, the neutron pairing is reduced most strongly in the [411]3/2 and [532]5/2 configurations. However, this effect is not as strong as in odd-Z systems.

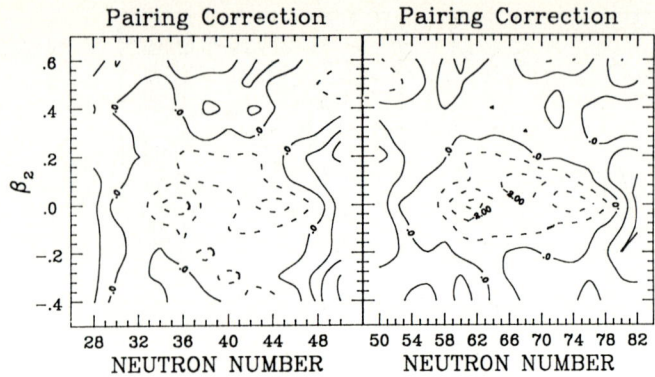

Fig. 5 Similar to fig. 1, but for the pairing correction.

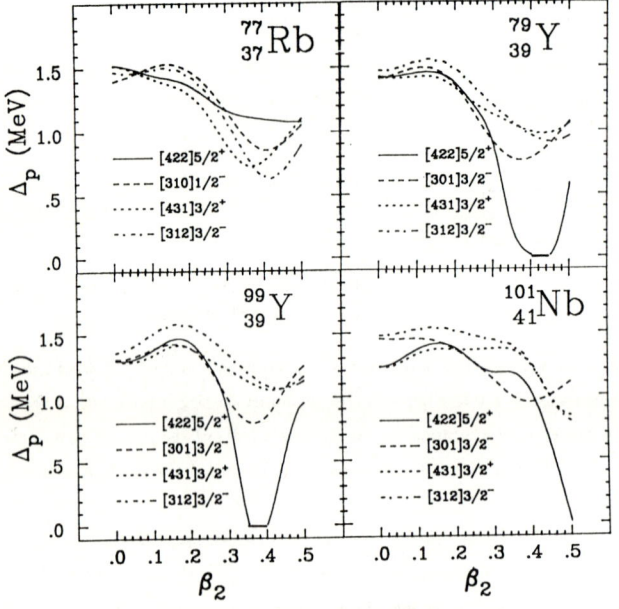

Fig. 6 Proton pairing gap versus β_2 for selected one-quasiproton configurations in ^{77}Rb, ^{79}Y, ^{99}Y, and ^{101}Nb calculated with the blocked BCS approach.

A particularly interesting example is the high spin spectrum of ^{77}Rb. In this nucleus both [312]3/2 and $g_{9/2}$ bands are known up to relatively high rotational frequencies [23]. The calculated equilibrium deformations for [312]3/2 and [431]3/2 band-heads are $\beta_2 \approx 0.36$ and $\beta_2 \approx 0.42$, respectively. At these deformations the calculations predict strong reduction in the proton pairing, see fig. 6. Experimentally, the [312]3/2 band behaves very regularly. A very weak bump in the dynamical moment of inertia, $J^{(2)}$, can be seen around $\hbar\omega=0.55$ MeV (cf. fig. 15 below). This weak irregularity can immediately be associated with the $g_{9/2}$ proton crossing, which is also expected to take place, at similar frequencies in neighbouring even-even nuclei. However, in the even-even neighbours the $g_{9/2}$ proton crossing

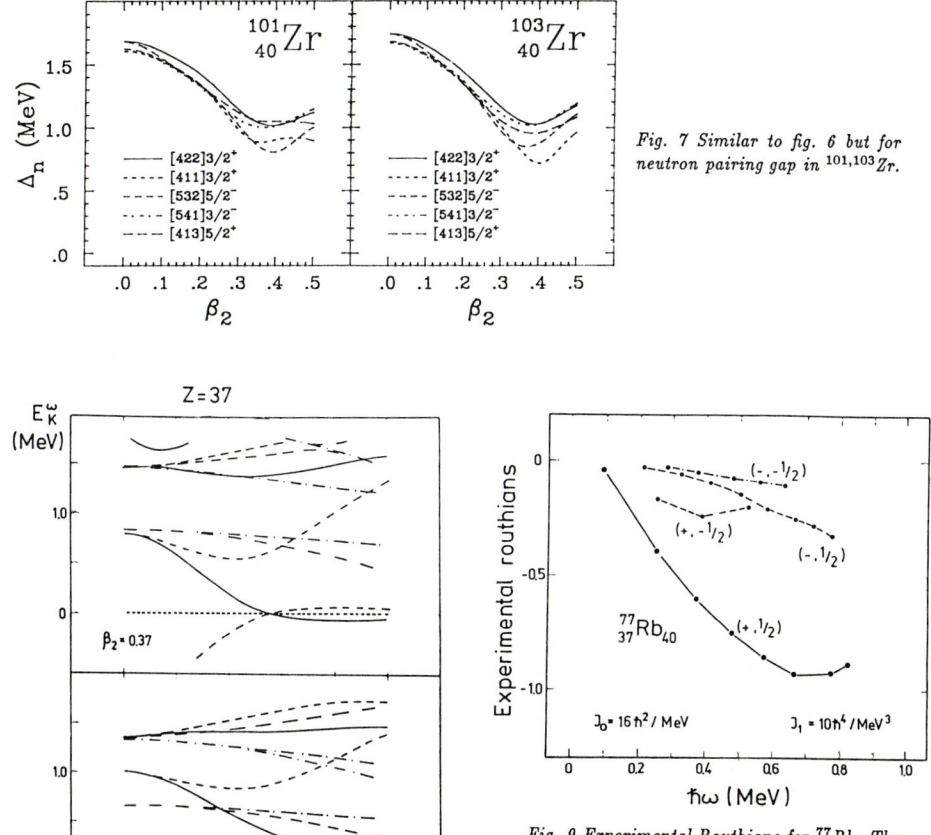

Fig. 7 Similar to fig. 6 but for neutron pairing gap in ^{101,103}Zr.

Fig. 9 Experimental Routhians for ^{77}Rb. The experimental data were taken from [23].

Fig. 8 Proton quasiparticle Routhians in ^{77}Rb calculated at $\beta_2=0.42$ (representative of the [301]3/2 configuration) and at $\beta_2=0.38$ (representative of the $g_{9/2}$ configuration).

is much more pronounced: it influences to a much greater extent the angular momentum alignment and moments of inertia. The predicted reduction in the proton pairing offers some explanation of this puzzle. Fig. 8 shows the calculated quasiparticle routhian diagrams for Z=37. It is seen that, with the proton pairing reduced, the first crossing of $g_{9/2}$ protons is almost washed out, i.e. only very gradual proton alignment with a very large interaction strength is expected. The experimental data for ^{77}Rb are presented in fig. 9 in the form of the experimental routhian diagram. The agreement between figs. 8 and 9 is worth noting.

Concluding this section: the blocked BCS calculations confirm the previous speculations about the reduction of pairing at certain quasiparticle configurations. This is consistent with the analysis of beta-decay properties of selected nuclei from the A≈100 region [30]. The results of RPA calculations for the Gamow-Teller decay suggest a reduction of pairing by about 50 % in ^{100,102}Y and ^{100,102}Nb.

5. Shape transitions induced by rotation in the A≈80 nuclei

Probably the most interesting feature of Ge/Zr nuclei is the richness of various structural effects which occur at high angular momenta. Many of these effects have a straighforward interpretation in terms of the interplay between the long-range quadrupole ineraction responsible for the nuclear deformation (cf. sect. 8), the short-range pairing interaction, and the Coriolis force.

The deformed mean field theory contains, by definition, the two first building blocks, i.e. deformation and pairing. The third element, rotation, can easily be accounted for by means of the cranking approximation. Within this approach a transformation is made to the intrinsic system of coordinates which rotates with a fixed angular velocity, ω. Such an extension of the mean field theory, the Hartree-Fock-Bogolyubov-Cranking (HFBC) method, is widely employed in the description of high spin states. The simplest approximation of the HFBC approach is the Cranking Shell Model (CSM) [31] which was mentioned in sect. 4 when discussing the high spin properties of ^{77}Rb (figs. 8,9). However, in the case of strong shape variations with angular momentum the CSM analysis cannot be applied directly because the band crossing frequencies and the angular momentum alignment can be influenced by changes in deformation and the pairing field.

Unlikely in rare earth nuclei where the first band crossing can be associated with $i_{13/2}$ neutrons and the second band crossing is caused by the alignment of $h_{11/2}$ protons, in the A≈80 region both protons and neutrons lie in the same $g_{9/2}$ subshell. A similar situation is kown in the region around ^{128}Ba, where proton and neutron Fermi surfaces lie within the $h_{11/2}$ intruder subshell. Consequently, proton and neutron band crossings are expected at similar rotational frequencies. For γ-soft systems such configuration changes induce strong shape variations caused by aligned quasiparticles which polarize the core.

Results of cranking calculations can be conveniently presented in a form of Total Routhian Surfaces (TRS) at a fixed rotational frequency, for selected many-quasiparticle configurations labeled by means of two quantum numbers, parity and signature. For more details of the Woods-Saxon cranking (WSC) model used we refer reader to [10,11,32-35]. Let us briefly discuss some results of the WSC model.

(i) Interplay between shape transition and quasiparticle alignment in Se-isotopes.

The recent experimental data for ^{70,72}Se [36] and ^{74}Se [37] reveals a very regular yrast band above spin I=10. Previously, the irregularity observed at low spins was attributed to the crossing between the weakly deformed oblate band and the excited prolate band with large moment of inertia (see e.g. [1]). The calculations [11,36] reproduce this situation: a near oblate ground state with $\beta_2 \approx 0.2$ and an excited 0_2^+ near prolate state with $\beta_2 \approx 0.32$. At higher frequencies the prolate minimum becomes lowest in energy up to $\hbar\omega$=1.2 MeV. It is worth emphasizing that the alignment of the first pair of $g_{9/2}$ protons (neutrons) is predicted to occur already at $\hbar\omega$=0.45 MeV (0.50 MeV) in the near-prolate configuration. Such a low value of band crossing frequency is, in fact, nothing unusual in the region around Z=34, N=40. For example, the first proton crossing in the neighbouring ^{75}Br has been observed [38] at $\hbar\omega$= 0.35 MeV. It can thus be concluded that the yrast configuration in discussed Se-isotopes has a four-quasiparticle nature above I=10.

(ii) Competition between prolate and oblate shapes in odd-A nuclei.

The best examples of the ground-state shape isomerism in odd-A nuclei from the light zirconium region are ^{69,71}Se [39], and ^{81}Sr [40,41]. While in the former case the Fermi surface penetrates the bottom of the $g_{9/2}$ subshell, ^{81}Sr is

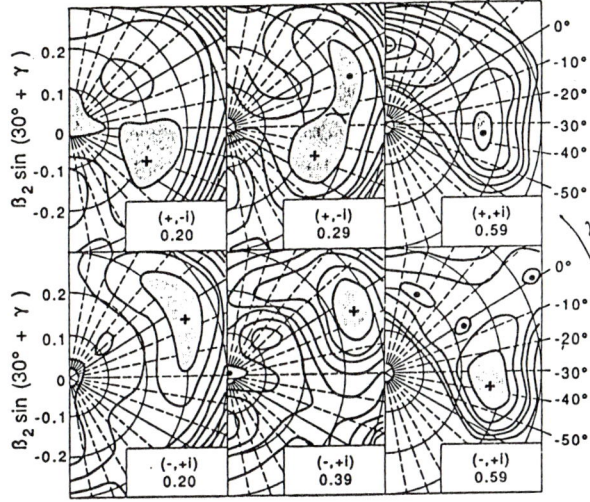

Fig. 10 Total Routhian surfaces (with pairing) in the (β_2, γ) plane for various quasiparticle configurations in ^{81}Sr [41]. The numbers give values of rotational frequency (in MeV). The distance between contour lines is 0.25 MeV. Quasiparticle configurations are labeled using the parity and signature quantum numbers (π, r).

an example of a heavier, γ-soft nucleus which lies in the upper border of the deformed region. Fig. 10 shows selected TRS for ^{81}Sr. At low spins the $g_{9/2}$ configuration corresponds to an oblate shape with $\beta_2 \approx 0.2$ while the negative parity bands can be associated with nearly-prolate shapes. A well deformed prolate configuration seen in the (+,-i) diagram for $\hbar\omega$=0.29 MeV has a strong component of the $\nu[431]1/2$ intruder orbital. At higher frequencies the $g_{9/2}$ proton and neutron crossings are expected. They trigger a shape change to a triaxial region.

(iii) Well-deformed intruder bands.

At sufficiently large deformations (or high spins), the states originating from the region above the Z(N)=50 gap may become occupied. One example of such an intruder state is the [431]1/2 orbital in ^{81}Sr. Even more interesting are the high-j low-K members of the $h_{11/2}$ subshell. These are expected to approach the Fermi surface at large deformations, $\beta_2 \approx 0.42$, for N$\geq$44 [10]. The best candidates to find the $h_{11/2}$ states are ^{81}Sr, ^{83}Zr and ^{84}Zr. The high-spin spectrum of ^{83}Zr has recently been measured [34] and the positive and negative parity bands in this nucleus are predicted to be triaxial, even at high rotational frequencies, see fig. 11.

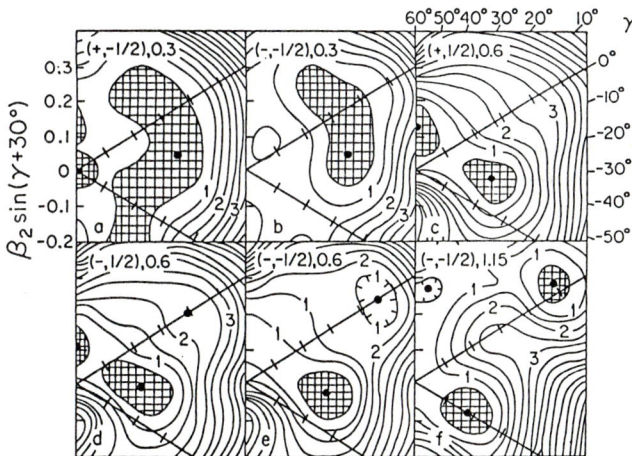

Fig. 11 Similar to fig. 10, but for ^{83}Zr [34]. The distance between contour lines is 0.5 MeV. Quasiparticle configurations are labeled using the parity and signature exponent quantum numbers (π, α).

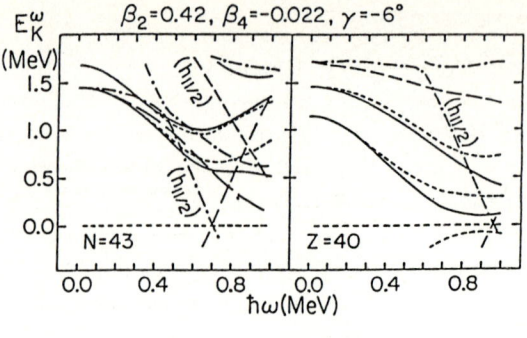

Fig. 12 Quasiparticle Routhians for ^{83}Zr at a deformation characteristic of the $\nu h_{11/2}$ intruder band [34].

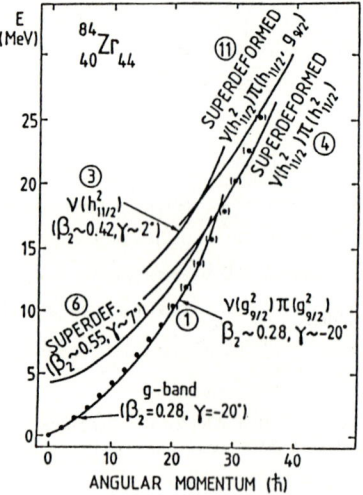

Fig. 13 Total energy calculated using the particle number projection method versus angular momentum for near-yrast rotational bands in ^{84}Zr [33]. Calculated bands are labeled by means of aligned quasiparticles and by deformation parameters.

Note that already at $\hbar\omega=0.6$ MeV a secondary minimum at TRS $(-,-1/2)$ ($= (-,i)$) can be seen. This well-deformed minimum becomes yrast at $\hbar\omega \geq 1$ MeV. A close inspection of the quasiparticle diagram shown in fig. 12 indicates that this configuration contains one $h_{11/2}$ neutron. The well deformed $h_{11/2}$ band in ^{83}Zr is analogous to the observed $i_{13/2}$ band in ^{131}Ce [42] (see also [43]). The analogon of the well deformed intruder band in ^{132}Ce [44] would be the $\nu(h_{11/2})^2$ configuration in ^{84}Zr. A detailed theoretical analysis of the high spin behaviour of ^{84}Zr is given in [33]. As in ^{83}Zr, the near yrast bands in ^{84}Zr correspond to the triaxial shape. Calculations explain two irregularities seen in the ground band of ^{84}Zr [45] as being caused by the alignment of $g_{9/2}$ protons and neutrons at similar frequencies. The moment of inertia of the highest (tentative) part of the yrast band is in very good agreement with the calculated well deformed $\nu(h_{11/2})^2$ structure shown in fig. 13. In the calculations, however, this band lies too high in energy. It is even higher than the superdeformed band 4 involving four $h_{11/2}$ particles. Moreover, if the highest part of the yrast band really contains $h_{11/2}$ particles one would expect a very sharp band crossing with the four-quasiparticle $g_{9/2}$ band, i.e. exactly the same situation as in ^{132}Ce (cf. also [46]). Does this mean that there are some low energy transitions missing experimentally?

6. Rigid rotors in the zirconium region

At low angular momenta one can expect many differences between rotational patterns of A=70-110 nuclei. First, they have very distinct shapes in their ground states. Secondly, there is a lot of configuration mixing at low spins, mainly due to the interaction between structures with different shapes. At very high rotational frequencies these effects are expected to play a much less important role. Rotational bands built on the less deformed configurations coexisting at low spins are very far from the yrast line as they have much smaller moments of inertia. Due to the large Coriolis force the pairing field is seriously diminished after the alignment of the first or second pair of quasiparticles [10,33]. Of course, the weak static pairing or dynamical pair vibrations will slightly modify (reduce) the total moment of inertia. However, it will be mainly the geometry of the nuclear shape (deformation) and the underlying single-particle structure (single-particle alignment) that will determine the high-spin behaviour of a nucleus.

In the best investigated high-spin area, i.e. in the rare earths, many nuclei exhibit very "rigid" rotational pattern at high rotational frequencies (see e.g. [47]). Such a limit is usually defined by a condition:

$$J^{(2)} = J^{(1)} \quad (3)$$

where $J^{(1)}$ and $J^{(2)}$ are the kinematical ($=I_x/\omega$) and dynamical ($=dI_x/d\omega$) moments of inertia, respectively. This condition (3) means that a nucleus rotates like a perfect rotor with a constant moment of inertia. It is instructive to relate eq. (3) to the identity:

$$J^{(2)} = J^{(1)} + \omega \frac{dJ^{(1)}}{d\omega} \quad (4)$$

In the region of structural changes (band crossings, deformation changes, variations in the pairing field) the second term in eq. (4) exhibits rather violent fluctuations and is generally positive and large in magnitude, thus $J^{(2)} > J^{(1)}$. Consequently, changes in the intrinsic configuration will lead to deviations from the "rigid" rotational behaviour. On the other hand, for a fixed configuration which contains a number of high-j strongly aligned quasiparticles, eq. (4) can be rewritten as:

$$J^{(2)} = J^{(1)} - \frac{i}{\omega} \quad (5)$$

where i is the quasiparticle alignment. In this limit $J^{(2)} < J^{(1)}$.

How rigid is the rotation in the zirconium region? In order to illustrate this point the correlation diagrams between $J^{(2)}$ and $J^{(1)}$ are presented in figs. 14,15. Moments of inertia are given in units of a rigid body moment of inertia at spherical shape, $J_{rig} = A^{5/3}/72 \; \hbar^2/MeV$, to make it possible to compare different nuclei on the same plot. The line $J^{(2)}=J^{(1)}$ indicates the "rigid" rotation limit, see above.

Fig. 14 shows the experimental data for the best A≈80 even-even rotors known today, i.e. ^{72}Se [36], ^{74}Se [37], ^{80}Sr [48] and ^{84}Zr [38]. At low spins $J^{(2)}$ is larger than $J^{(1)}$ which reflects changes in deformation (or interaction between structures with different shapes) and /or pairing field. Each $J^{(2)}$-curve exhibiths several characteristic bumps. The first two irregularities, which are most pronounced in ^{84}Zr, can be attributed to the $g_{9/2}$ proton and neutron crossings while the nature of the remaining bumps is not fully clear. They can, for example, be caused by the collapse of the static pairing correlations [33] or by the weak alignment of negative-parity nucleons. It is interesting to observe that after the second $g_{9/2}$ crossing $J^{(2)}$ is reduced due to the large gain in alignment (eq. (5)) and sometimes it may even become locally equal to $J^{(1)}$. Of course, this effect cannnot be related to the rotation of a rigid system. However, at highest spins, the kinematical moment of inertia doesn't change too much. In ^{74}Se, ^{80}Sr and ^{84}Zr the dynamical moment of inertia oscillates in the vicinity of the "rigid" rotation limit. In ^{80}Sr and ^{84}Zr this limit ($J^{(1)} = 1.14 \; J_{rig}$)

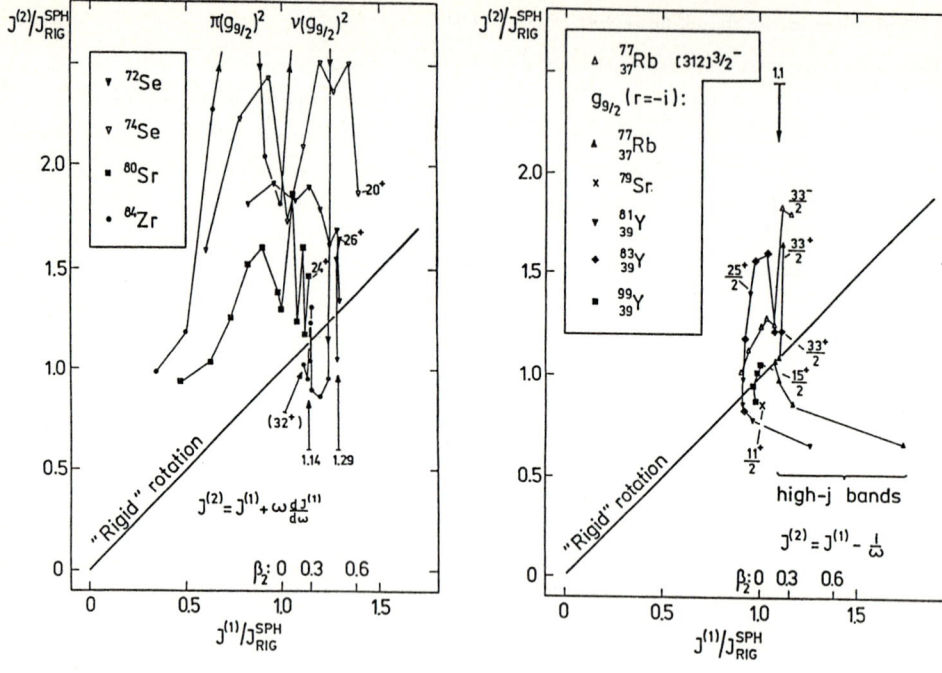

Fig. 14 $J^{(2)}$ vs. $J^{(1)}$ correlation diagram for $^{72,74}Se$, ^{80}Sr and ^{84}Zr.

Fig. 15 Similar to fig. 14 but for ^{77}Rb, ^{79}Sr, $^{81,83,99}Y$.

corresponds to the rigid-body moment of inertia with $\beta_2 \approx 0.3$ - the value which is expected from calculations [10,33]. On the other hand, in ^{72}Se the quadrupole deformation deduced from the ratio $J^{(1)}/J_{rig}$=1.29 is $\beta_2 \approx 0.5$, i.e. much larger than the predicted value ($\beta_2 \approx 0.33$). The situation is even more striking in the case of ^{74}Se, although the plateau in $J^{(1)}$ has not yet been reached there.

The experimental data for selected odd-A nuclei are presented in fig. 15. In this case we have chosen the [422]5/2 bands and $K^\pi = 3/2^-$ bands in ^{77}Rb [23], ^{79}Sr [49], ^{81}Y [24], ^{83}Y [25] and ^{99}Y [26], i.e. the very configurations in the Z (N) = 37 and 39 systems which have been expected to have seriously quenched pairing correlations, see sect. 4. Two clear bumps can be seen in the $J^{(2)}$-curve for the [312]3/2 band in ^{77}Rb. They can be given the same ieterpretation as in the previous case of even-even nuclei. In the $g_{9/2}$ bands of Rb and Y isotopes the first (proton) crossing is blocked and the first irregularity in $J^{(2)}$ has to be associated with the alignment of $g_{9/2}$ neutrons. In all the cases displayed, apart from the region of band crossing, $J^{(1)}$ and $J^{(2)}$ are very close to each other. In ^{77}Rb and ^{83}Y $J^{(1)}$ reaches the limit of $J^{(1)} = 1.1 - 1.15 J_{rig}$, which corresponds to $\beta_2=0.2-0.3$ for the rigid rotor. However, in ^{79}Sr, ^{81}Y and ^{99}Y the experimental data are known at much lower spins where the neutron pairing still plays a very improtant role. Indeed, in these nuclei $J^{(1)}/J_rig = 1$. A very special case is ^{99}Y. It is close to the rigid rotation limit already at lowest spins, which indicates a very low spin alignment in the $g_{9/2}$ band. Such a decrease at low spins is typical of systems with weak pairing.

7. Octupole correlations

Medium-mass transitional nuclei from the zirconium region are not expected to be octupole unstable, in contrast to the nuclei from light actinides and the heavy barium region which exhibit various fingerprints of static octupole deformation. The reason for this is the large energy separation between the spherical subshells with $\Delta l = \Delta j = 3$, i.e. the ones which interact strongly through the octupole field. These are the $g_{9/2}$ and $p_{3/2}$ proton (neutron) subshells and the $h_{11/2}$ and $d_{5/2}$ neutron subshells. At non zero quadrupole deformation, orbitals belonging to the unique parity subshells approach the strongly quadrupole-mixed normal parity states which have the same K-quantum number. The small number of active subshells in the zirconium region makes the octupole effect more sensitive to quadrupole distortion than in heavier nuclei.

The deformed shell model indicates that Z(N)=34 as well as N=56 are the optimal particle numbers for enhanced octupole correlations [50,51]. Fig. 16 displays the octupole stiffness, C_3, for nuclei from the zirconium region. The stiffness C_3 is defined from the equation:

$$E = E_o + \frac{1}{2} C_3 \beta_3^2. \tag{6}$$

The C_3-values of the macroscopic energy, C_3^{macr}, are indicated in fig. 16 by dashed lines. If the C_3 value is smaller than C_3^{macr} the nucleus is said to be soft towards the octupole deformation.

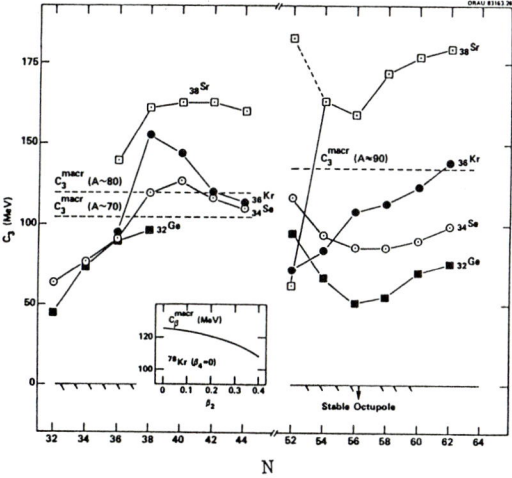

Fig. 16 Octupole stiffness for nuclei from the zirconium region [51].

According to the results presented in fig. 16 in the A≈80 region only the Ge and Se isotopes with N≤36 are octupole-soft. Recent experimental study of the very neutron deficient isotopes ^{64}Ge and ^{65}Ge [52] indeed indicate the very low - 2119 keV - spacing between $15/2^-$ ($\nu g_{9/2} \otimes 3^-$) and $9/2^+$ ($\nu g_{9/2}$) states. In the heavier Ge-isotopes this spacing closely follows the 3^- octupole excitation in the even-even cores. The predicted low energy of 3^- state in ^{64}Ge is consistent with the simple estimate based on the results of fig. 16 [52].

In the A≈100 region there are more nuclei with octupole softness. One of the best candidates for strong octupole correlations in the ground state is ^{88}Ge. The PES for this nucleus is shown in fig. 17 as a function of β_2 and β_3. The analogous diagram for ^{94}Zr is also presented in fig. 17. In this nucleus the ground state corresponds to the spherical shape, but the excited prolate minimum associated with the secondary shape-isomeric 0_2^+ state can be seen. The results presented in figs. 16, 17 lead to the following general observation [51]: the octupole softness should be

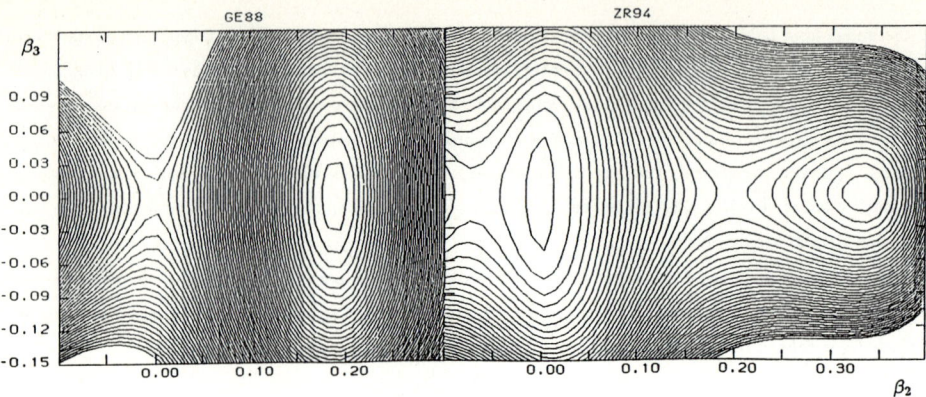

Fig. 17 Total energy for ^{88}Ge and ^{94}Zr as a function of β_2 and β_3. The energy is optimized with respect to β_4 at each point. The contour lines are 0.1 MeV apart.

enhanced in spherical and transitional nuclei, i.e. there is an anticorrelation between quadrupole and octupole effects. As a consequence, in the well deformed Kr, Sr and Zr isotopes, C_3 is much larger than the corresponding C_3^{macr} value.

8. Role of neutron-proton interaction

A different explanation of the onset of stable deformations in atomic nuclei (and, particularly, in the A≈100 region) has been suggested by Federman and Pittel [53]. According to them, it is the T=0 neutron-proton (n-p) interaction between the spin-orbit partners, i.e. between the $g_{9/2}$ protons and the $g_{7/2}$ neutrons, that determines the nuclear deformation.

Recently, this attractive suggestion was analyzed within the Hartree-Fock method and the harmonic oscillator model [54]. The total nuclear energy was decomposed with respect to isospin and angular momentum. An example of such a procedure is illustrated in fig. 18, which shows different contributions to the total energy of ^{72}Ge as a function of total quadrupole moment Q_o. The total energy can be viewed as a subtle balance between two terms: the monopole term, δE_o which inreases with Q_o and the quadrupole term, E_2, which decreases with Q_o. The latter can be closely related to the quadrupole-quadrupole interaction responsible for the nuclear deformation. It can be seen in fig. 18 that the n-p quadrupole interaction is about five times stronger than the sum of the unlike-particle components. From this point of view one can indeed conclude that the quadrupole deformation is caused by the n-p quadrupole interaction. The quadrupole-quadrupole isoscalar and isovector coupling constants are presented in fig. 19 as functions of mass number for molybdenum isotopes. They are compared to the harmonic oscillator estimates of [55]. The fair agreement between the two models strengthens the role played by the n-p interaction as the main mechanism for nuclear collectivity.

Another interesting question is the relation between the shell structure and the n-p interaction. According to [54] δE_o changes much faster with the number of particles than E_2^{np}, therefore the actual deformation of a nucleus mostly depends on the monopole part of the energy.

Although in [53] the authors emphasized the importance of the $\pi g_{9/2}$-$\nu g_{7/2}$ coupling, they also pointed out the significance of the $\nu h_{11/2}$ subshell in stabilizing nuclear deformation in Mo-isotopes. In our opinion, the increased occupancy of the $h_{11/2}$ intruder orbitals *is the main mechanism* for the onset of nuclear deformation in Zr and Mo

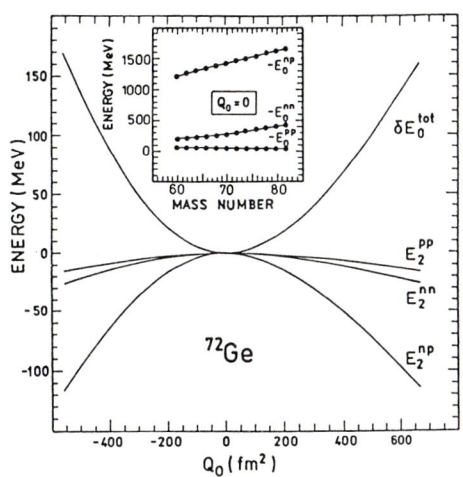

Fig. 18 Different contributions to the total energy of ^{72}Ge as functions of total quadrupole moment [54].

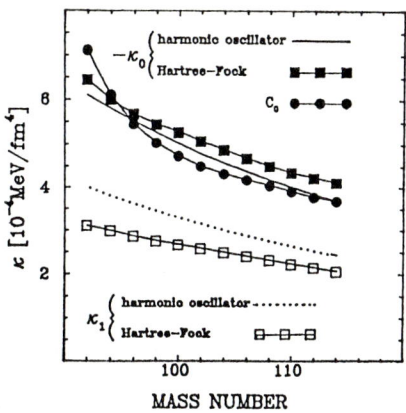

Fig. 19 Quadrupole T=0 and T=1 coupling constants, κ_0 and κ_1, calculated using the Skyrme III model as a function of mass number for Mo-isotopes [54]. They are compared to the harmonic oscillator values of ref. [55]. The stiffness of the monopole part of the energy, C_o, is also shown. The variation in C_o around A=96 reflects the shape transition to the deformed region.

isotopes when going from N=96 to N=100. The [550]1/2 and [541]3/2 Nilsson states play exactly the same role here as the [431]1/2 orbital which is responsible for the shape-isomeric states in Sn and In isotopes, the [514]9/2 orbital which is responsible for the shape isomers in neutron-deficient Pb isotopes, or the [660]1/2 state which becomes occupied in the highly deformed intruder bands in nuclei around ^{132}Ce. All those states belong to the unnatural parity high-j intruders. They are particle-like at spherical shape but due to their large quadrupole moment they become occupied at large deformations.

This work was supported in part by the Polish Ministry of National Education under Contract CPBP 01.09

[1] Hamilton J.H., Lecture Notes in Physics **168**, Heavy Ion Collisions (Springer-Verlag, 1982), p. 287 (1982) and Refs. therein
[2] Lieb K.P., in Proc. Int. Symposium on Weak and Electromagnetic Interactions in Nuclei, Heidelberg ed. by H.V. Klapdor (Springer-Verlag, Berlin, 1986), p. 106 and Refs. therein
[3] Sistemich K. et al., Nuclear Structure, Reactions and Symmetries, ed. by R.A. Meyer and V. Paar, (World Scientific, Singapore, 1986), vol. 2, p.706 and Refs. therein
[4] Meyer R.A., Hyperfine Interactions **22**, 385 (1985) and Refs. therein
[5] Möller P. and Nix J.R., At. Data Nucl. Data Tables **26**, 165 (1981); Nucl. Phys. **A361**, 117 (1981)
[6] Bengtsson R., et al., Phys. Scr. **29**, 402 (1984)
[7] Ragnarsson I. and Sheline R.K., Phys. Scr. **29**, 385 (1984)
[8] Bonche P. et al., Nucl. Phys. **A443**, 39 (1985)
[9] Galeriu D., Bucurescu D. and Ivascu M., J. Phys. **G12**, 329 (1986)
[10] Nazarewicz W. et al., Nucl. Phys. **A435**, 397 (1985)
[11] Bengtsson R. and Nazarewicz W., Proc. XIX Winter School on Physics, ed. Z. Stachura, Report IFJ No. 1268 (1984), p. 171; to be published.
[12] Bengtsson R., these proceedings.
[13] Arseniev D.A., Sobiczewski A. and Soloviev V.V., Nucl. Phys. **A139**, 269 (1969)
[14] Sheline R.K., Ragnarsson I. and Nilsson S.G., Phys. Lett. **41B**, 115 (1972)

[15] Faessler A. et al., Nucl. Phys. **A230**, 302 (1974)
[16] Ragnarsson I., Proc. Int. Conf. on Atomic Masses, East Lansing, 1979; CERN preprint TH.2766-CERN.
[17] Azuma R.E. et al., Phys. Lett. **86B**, 5 (1979)
[18] Myers W.D. and Swiatecki W.J., Nucl. Phys. **81**, 1 (1966)
[19] Strutinsky V.M., Yad. Fiz. 3, **614** (1966); Nucl. Phys. **A122**, 1 (1968)
[20] Nazarewicz W., Nuclear Structure, Reactions and Symmetries, ed. by R.A. Meyer and V. Paar, (World Scientific, Singapore, 1986), vol. 2, p. 1079.
[21] Kawade K. et al,. Z. Phys. **A304**, 293 (1982)
[22] Wohn F.K. et al., Phys. Rev. **C33**, 677 (1986)
[23] Lühmann L. et al., Europhys. Lett. **1**, 623 (1986)
[24] Lister C.J. et al., J. Phys. **G11**, 969 (1985)
[25] Rapaport M.S., Liang C.F. and Paris P., Phys. Rev. **C36**, 303 (1987); Lister C.J., private communication 1985
[26] Meyer R.A. et al., Nucl. Phys. **A439**, 510 (1985); Wohn F. K., Phys. Rev. **C36**, 1204 (1987)
[27] Wohn F.K. et al., Phys. Rev. Lett. **51**, 873 (1983)
[28] Wohn F.K. et al., Phys. Rev. **C36**, 1118 (1987)
[29] Peker L.K. et al. Phys. Lett. **169B**, 323 (1986)
[30] Kratz K.L. et al., preprint 1986.
[31] Bengtsson R. and Frauendorf S., Nucl. Phys. **A314**, 27 (1979); **A327**, 139 (1979)
[32] Gross C.J. et al, Phys. Rev. **C36**, 2601 (1987)
[33] Dudek J., Nazarewicz W. and Rowley N., Phys. Rev. **C35**, 1489 (1987)
[34] Hüttmeier U.J. et al., Phys. Rev. **C37**, 118 (1988)
[35] Nazarewicz W. et al., to be published
[36] Mylaeus T. et al., submitted to Phys. Lett. B; these proceedings
[37] Gross C.J., Phys. Rev. **C36**, 2127 (1987)
[38] Lühmann L. et al., Phys. Rev. **C31**, 828 (1985)
[39] Wiosna M. et al., Phys. Lett. **B200**, 255 (1988)
[40] Arnell S.E. et al., J. Phys. **G9**, 1217 (1983)
[41] Moore E.F. et al., submitted to Phys. Rev. C (1988); Tabor S.L., these proceedings
[42] Love D.J.G. et al., to be published
[43] Wyss R. et al., Lund preprint Lund-MPh-87/16
[44] Kirwan A.J. et al., Phys. Rev. Lett. **58**, 467 (1987)
[45] Price H.G. et al., Phys. Rev. Lett. **51**, 1842 (1983)
[46] Ragnarsson I. and Bengtsson T. , contribution to this conference
[47] Garrett J.D. in Nuclear Structure 1985, ed. by R. Broglia, G. Hagemann and B. Herskind, (North-Holland, 1985), p. 111
[48] Davie R.F. et al., Nucl. Phys. **A463**, 683 (1987)
[49] Lister C.J. et al., Phys. Rev. Lett. **49**, 308 (1982)
[50] Leander G.A. et al., Nucl. Phys. **A388**, 425 (1982).
[51] Nazarewicz W. et al., Nucl. Phys. **A429**, 269 (1984)
[52] Görres J. et al., Phys. Rev. Lett. **58**, 662 (1987)
[53] Federman P. and Pittel S., Phys. Rev. **C20**, 820 (1979)
[54] Dobaczewski J. et al., submitted to Phys. Rev. Lett., 1988; to be published
[55] A. Bohr and B. Mottelson, Nuclear Structure, vol. 2 (Benjamin, New York, 1975)

On the Solution to the Large Amplitude Collective Motion of Finite Interacting Fermi Systems[*]

F. Dönau[1], Jing-ye Zhang[2], and L.L. Riedinger[2]

[1]Zentralinstitut für Kernforschung Rossendorf,
DDR-8051 Dresden, Postfach 19, G.D.R.
[2]Joint Institut of Heavy Ion Research, Oak Ridge, TN 37831, USA, and
Dep. of Physics, University of Tennessee, Knoxville, TN 37996, USA

[*] Preprint UT Knoxville 87

1 Introduction

Large amplitude collective motion in nuclei is a very fascinating subject for studying the properties of finite quantum systems. Examples include oscillations between coexisting deformed shapes or the interplay of the rotating deformed field and the alignment of individual particles. In the experimental data the related phenomena manifest themselves by the characteristic pattern of level schemes and transition rates as found, for example, in the research on properties of high-spin states. Concerning the theoretical description of those features, the construction of Potential Energy Surfaces (PES) by the Strutinsky shell-correction method or by constrained Hartree-Fock calculations provides a profound background for understanding and interpreting the observed features in a microscopic way. The knowledge of such PES maps gives us detailed information on the statics of the system under study, because they predict the stability of selected quasiparticle configurations which correspond to the minimal points in the potential-energy landscape. The so-called Total Routhian Surfaces [1] describe the spatial deformation of the nuclear density distribution as function of rotational frequency. These are examples of the succesful application of theoretical PES maps and are important for the interpretation of experimental data. However, to describe, in addition, possible transitions between coexisting quasiparticle configurations, one finds it necessary to account for the full dynamics of the collective motion treated for the case of large amplitudes.

Our conception for describing the collective dynamics is based on the Generator Coordinate Method (GCM) which is well-known for more than thirty years [2-4]. However, in spite of numerous successful applications some important features have not been treated with such attention to develop all the power of this formalism. Our preference for the GCM results from the fact that this method is based on collective variables that have in most cases a classical counterpart and hence enables one a transparent physical understanding.

2 The "horizontal" expansion of the Hamilton kernel in GCM

Here is presented a particular expansion for the Hamilton kernel of the Hill-Wheeler equation defining the quantized collective states within GCM. We start the considerations assuming that the many-body Hamiltonian H involving effective two-body interactions is given by

$$H = \Sigma t_{ij} c_i^\dagger c_j^\dagger + \Sigma V_{ijkl} c_i^\dagger c_j^\dagger c_l c_k \qquad (1)$$

To this Hamiltonian external auxiliary fields are added and written in short-hand notation as

$$- q \cdot Q = -(q_1 Q_1 + q_2 Q_2 + \cdots). \qquad (2)$$

The quantities $q = (q_1, q_2, ...)$ entering in eq.(2) as external strength parameters specify what we call below collective variables. The operators $Q_1, Q_2, \ldots$ are appropriate single particle fields serving as auxiliary potentials to shape the determinantal solutions of the "Routhian"

$$R(q) = H - q.Q \qquad (3)$$

to the collective properties one wants to study. We recall the deformed potential or the cranking term as examples of well-known external fields.

The Routhian (3) is utilized for executing an HFB procedure. Assuming that self-consistency can be achieved, one is able to form at any point (q) a determinantal state later on called the local vacuum $|q\rangle$. If one is studying slow collective motion, the state $|q\rangle$ is taken to be the energetically lowest HFB determinant satisfying

$$a|q\rangle = 0 \qquad (4)$$

where both the destruction operator $a = a(q)$ and its vacuum are defined locally for each point (q). The above procedure means an adiabatic preparation of a sequence of states labeled by $q, q', q'', \ldots$. In the "adiabatic" regime (which can be altered for describing other processes) the local vacuum $|q\rangle$ follows in a case of level crossing merely the envelope of the energy of pure configurations instead of continuing in the previous configuration which increases in energy.

According to the GCM, the traveling through the sequence $q, q', q'', \ldots$ is to be considered as a fermion realization of the collective motion. This means that a controllable manifold of states $|q\rangle$ is used as a basis for the diagonalization of the Hamiltonian H (eq.(1)). Thus, formally one arrives at the well-known Hill-Wheeler integral equation

$$\int dq' (\langle q|H|q'\rangle - E\langle q|q'\rangle) C(q') = 0 \qquad (5)$$

for the collective eigenstate

$$\Psi = \int dq\, C(q)|q\rangle. \tag{6}$$

Here the eigenvalue E is the collective energy and $C(q)$ is the unknown distribution function.

Now we benefit from the fact that $|q\rangle$ is a "local" self-consistent HFB solution. The unity operator is then written in terms of the quasiparticle excitations with respect to the vacuum $|q\rangle$ (note that only 2 qp, 4 qp, ... are possible)

$$1 = |q\rangle\langle q| + \sum (a_i^\dagger a_j^\dagger |q\rangle\langle q|a_j a_i) + \sum (a_i^\dagger a_j^\dagger a_k^\dagger a_l^\dagger |q\rangle\langle q|a_l a_k a_j a_i) + \cdots \tag{7}$$

and likewise with respect to the local vacuum $|q'\rangle$. In this manner one obtains the expansion of the Hamilton kernel as

$$\langle q|H|q'\rangle = 1/2(\langle q|H|q\rangle + \langle q|H|q'\rangle)\langle q|q'\rangle$$
$$+ 1/2\left[(q+q')\langle q|Q|q'\rangle - (\langle q|q.Q|q\rangle + \langle q'|q'.Q|q'\rangle)\langle q|q'\rangle\right]$$
$$+ 1/2 \sum \left[\langle q|a_i^{\dagger\prime} a_j^{\dagger\prime} a_k^{\dagger\prime} a_l^{\dagger\prime}|q'\rangle\langle q'|a_i' a_j' a_k' a_l' H|q'\rangle + \langle q|H a_i^\dagger a_j^\dagger a_k^\dagger a_l^\dagger|q\rangle\langle q|a_i a_j a_k a_l|q'\rangle\right]. \tag{8}$$

Due to the self-consistency of the HFB determinants, no terms of 2qp structure apppear in this exact expansion. The remaining 4qp contribution is expected to have a convenient convergency property, since it is combined with an overlap $\langle q|aaaa|q'\rangle$ which obviously measures the probability for the local vacuum q' to be a pure 4qp excitation on top of the vacuum q. However, it should be natural that all the members $q, q', q'', \ldots$ of the collective family correspond to a coherent rearrangement of all particles which does usually not favor a particular 4 qp state. As a consequence we have a good argument to obtain a reasonable approximation of the Hamilton kernel by neglecting 4qp contributions.

The remaining expression has a quite interesting structure. The first line stems from the zero-quasiparticle contribution and is the direct coupling between the various points (q). Hence, in the lowest order approximation the Hamilton kernel for the quantized collective motion is obtained by combining the static potential $\langle q|H|q\rangle$ with the overlap matrix $\langle q|q'\rangle$ between all the couples of states $|q\rangle$. It is interesting to mention that any finite overlap in the generator basis leads to a spread of the final collective wave function (6). Furthermore, because we are left only with a diagonal term originating from the Hamiltonian, we can approximate it by a PES constructed in terms of the Strutinsky shell-correction method or even by an appropriate phenomenological potential, like a Nilsson plus pair field potential in case of quadrupole deformation. This option would maximally employ the information present in the average fields and its dependence on collective parameters like shapes, pair field, cranking frequencies, etc.

The 2 qp term in the kernel is due to the auxiliary fields $-q.Q$ that shifts the system away from the minimal points in the PES. These contributions resemble formally the cranking terms which would remain also if the generator basis $|q\rangle$ would be orthogonal. Therefore, these terms give rise to additional mixings of the generator states. Their presense is, however, not genuinely affected from the fermion interaction but introduced externally with the generator coordinate under study.

Why is the expansion (8) called "horizontal"? The family (q) obtained by an adiabatic preparation of HFB determinants is quite different compared to the family of multi-quasiparticle excitations with respect to a fixed vacuum, which is usually taken at the minimum of the PES. Such quasiparticle excitations are the basis of phonon states obtained in RPA. They obviously form a "vertical" basis with respect to the energy. We believe that the horizontal expansion is the natural basis of the large amplitude collective motion, since each member $|q\rangle$ of the generator basis is tailored to follow, like instantaneous snap shots, the desired type of collective motion.

3 On the role of the overlap matrix

The "horizontal" expansion (8) shows explicitly the important role of the overlap matrix for the coupling of the generator states. In most calculations using the GCM, this overlap or norm matrix has been treated in a Gaussian approximation. The application of this approximation has been a quite fruitful step towards a wide exploration of the GCM, including the ambitious development of their specific techniques ([4] and the references therein). Other extensive studies performed in the seventies [5] applied the GCM without the Gaussian overlap approximation, and they have lead to encouraging results.

The presently available computational techniques enable one to calculate practically the overlap matrices in an acceptable precision even for heavy-mass systems, and thus there is no reason to keep the Gaussian approximation. In contrast, based on the expansion (8), the detailed study of the overlap matrix provides us as a new tool, i.e. a qualitative survey of the collective motion. This is because a sizeable overlap between two points q and q' tells us immediatly that a direct transition $q- > q'$ is favored, whereas a small overlap forbids such a transition. Thus, in rigorous calculations of overlap "trees", structural details which would be hidden in a Gaussian shape are revealed. This holds in particular for simultaneous considerations of several degrees of freedom as, for instance, when the pairing gap, triaxial shapes, and rotational frequency are of importance.

There is another important observation in order. The "natural" orthogonal basis [4] of the collective subspace is obtained by diagonalizing the overlap matrix and subsequently removing those eigenstates, for which the eigenvalue is lower than a given limit (these small-norm states are identified as redundant for defining the collective subspace).

Our experience in this direction demonstrates that the dimension of the collective subspace spanned by collective generator states is indeed surprisingly small compared with the number of available multi-quasiparticle excitations which form the above mentioned "vertical" basis. This is not only a great practical advantage. The use of such a discretized basis needs moreover no re-quantization like the continuous generator functions treated with the Gaussian overlap approximation.

The application of the "vertical" expansion opens a new field for the exploration of the dynamics of shape transitions, pair-field fluctuations, quantal rotational motion, and tunneling phenomena. The proposed method is considered as the logical extension of the potential energy calculations to obtain a quantized collective motion.

References:

1. R.Bengtsson, T.Bengtsson, J.Dudek, G.Leander, W.Nazarewicz and Jing-ye Zhang,Phys.Lett. 183B (1987) 1
2. D.L.Hill and J.A.Wheeler, Phys. Rev. 89(1953)1102
3. J.J.Griffin and J.A.Wheeler, Phys. Rev. 108(1957)311 J.J.Griffin, Phys. Rev. 108(1957)329
4. P.Ring and P.Schuck, "The Nuclear Many-Body Problem",Springer,1980 and references therein.
5. A.Faessler,F.Gruemmer,A.Plastino and F.Krmpotic,
Nucl. Phys. A217 (1973) 420
K.Goeke, K.Allaart,H.Muether and A.Faessler, Z.f. Physik 271 (1974) 377
H.Muether,K.Goeke,k.Allaart and A.Faessler, Phys. Rev. C15 (1977) 1467 and references therein

Nuclear Structure Effects along the N = Z Line

C.J. Lister[1], A.A. Chishti[2], B.J. Varley[2], W. Gelletly[3], and A.N. James[4]

[1] Wright Nuclear Structure Laboratory, Yale University,
  P.O. Box 6666, 272 Whitney Avenue, New Haven, CT 06511, USA
[2] Department of Physics, Schuster Laboratory,
  University of Manchester, M13 9PL, UK
[3] SERC Lodge Laboratory, Warrington, WA4 4AD, UK
[4] Oliver Lodge Laboratory, University of Liverpool,
  Liverpool L69 3BX, UK

INTRODUCTION

The nuclear states discussed in this paper bear considerable resemblance to the high spin superdeformed states which have aroused recent interest in γ-spectroscopy. They have a highly collective nature, shapes are established by shell effects, and they have high moments of inertia (close to their rigid body values) which reflects a considerable reduction in pairing effects compared to neighboring nuclei. However, there are significant differences. In actinine and rare earth nuclei the superdeformed 'second minimum' states can be seen to reflect the extra binding achieved by certain shell structures over classical Coulomb and centrifugal barriers. In lighter nuclei this distinction is not so easily made as quantum effects begin to dominate all modes of excitation, though the large collectivity can perhaps be seen to be shell structures overcoming the pairing field which stabilizes spherical shapes.

In N = Z nuclei the quantum effects may be anticipated to be especially distinct as extra binding and shape polarization influences occur simultaneously for protons and neutrons so changes in the Fermi surface may cause rapid variation of the shapes predicted to be most bound. Thus an experimental determination of nuclear shapes can be directly related to the underlying shell effects which stabilize them.

Experimental access to the N = Z nuclei is difficult. The fusion of stable nuclei near the Coulomb barrier to form the lightest possible compound nuclei requires further neutron evaporation to reach the N = Z line. This process is extremely unlikely as proton separation energies are small (< 5 MeV) compared to that of neutrons (> 15 MeV) so charged particle evaporation dominates. However, the evaporation process is statistical so a small fraction of exotic nuclei are produced (< 1 in 10^4). To study these nuclei, in this case by γ-ray spectroscopy, requires a reliable method of channel selection which was achieved by use of the Daresbury Recoil Separator [1].

THE EXPERIMENTS

A series of studies were performed using the reactions listed in table 1. The reactions were inverse (using a heavy beam) to enhance the focusing of the reaction products into the electromagnetic separator. Beam energies were chosen to maximize two nucleon evaporation. The recoil products were separated from the beam using a double Wien Filter and dispersed in A/q with a 60° dipole magnet. The ions were stopped in a split anode ionization chamber which provided Z information through their relative dE/dx energy loss. The lighter N = Z nuclei recoiled at V/c $\gtrsim$ 6.0 and exhibited nearly complete Z-separation so were relatively easily identified. However, the heavier systems were produced with a lower recoil velocity and smaller cross-section so identification was extremely difficulty requiring experiments of up to four days with 20 pnA of beam. Gamma rays were detected in an array of up to 15 BGO suppressed Ge detectors surrounding the target (typically 2 - 500 $\mu g/cm^2$). Data were collected for all identified recoil-residue coincidences.

RESULTS

The predictions of several theoretical groups [2-5] indicate that the ^{64}Ge - ^{68}Se - ^{72}Kr nuclei should exhibit competition between oblate and prolate shapes. Indeed, evidence for such behavior has been seen in previous studies of more neutron rich isotopes. However, the N = Z nuclei are predicted to hve stiffer potential energy surfaced with less triaxiallity. These predictions are consistent with our new data on these nuclei. Fig. 1 shows our results for ^{68}Se; the high 2^+ energy and indications of states above it suggest a decay pattern distinctly different from ^{70,72}Se where an oblate-prolate shape change at low spin [6,7] results in a vibration-like low spin sequence. ^{72}Kr has a γ-ray pattern [8] similar to that of its isobar ^{72}Se and two level mixing calculations indicate ^{72}Kr may be one of the few nuclei which have an oblate ground state.

The nuclei ^{76}Sr and ^{80}Zr show an abrupt change from their lighter couterparts. The first excited state energies drop drastically and a clear rotational-like γ-ray sequence is observed as for ^{76}Sr in fig. 2. The theoretical predictions from Hartree-Fock [9] and Strutinsky calculations [2-5] for these nuclei are that the potential energy surfaces should have a deep minima at $\varepsilon_2 \sim 0.4$. Most nuclei with N Z 40 are predicted to be γ-soft with the exception of ^{76}Sr which is predicted to be a stiffer, more axial rotor. The experiments indeed appear to verify the prediction of large deformation, although the spectrum of ^{76}Sr appears almost identical to its neighbors ^{80}Zr (ref.10) and ^{78}Sr with an $E(4^+)/E(2^+)$ ratio of 2.83, compared to 2.86, and 2.81 respectively, a feature which does not seem to support the suggestion that ^{76}Sr is a better rotor than its neighbors. The E(2) energy and E(4)/E(2) ratio do appear to be a remarkable verification of the P-scheme of Casten **et al.** [11] in two respects. Firstly, ^{76,78}Sr and ^{80}Zr lie exactly in mid-shell so have the same valve of P and would be expected to have the same spectra. Secondly, because the nucleon number 28-50 shell is small, P cannot exceed 5 for even-even nuclei, a value which is insufficient for the residual quadrupole interactions to polarize the nuclei into an axial shape as is the case in the middle of the larger shells of the Rare Earth and Actinide regions.

PROSPECTS

Beyond midshell at ^{80}Zr, the size of the groundstate quadrupole deformation and the dominance of prolate shapes are expected to be reduced. Recent calculations [2,12] have indicated that nucleon number 44 presents an ideal situation for nuclear hyperdeformation $\left(\varepsilon_2 \sim 1.0\right)$ to occur at high spin. Finally the spherical gap at N = Z = 50 is predicted to suppress collective effects. We have studied the ^{50}Cr(^{56}Fe, xn) and ^{50}Cr(^{54}Fe, xn) reactions in order to reach light Tin isotopes. Preliminary analyses indicate that ^{103,104}Sn can be studied but the lightest and most interesting isotopes present a tremendous experimental challenge.

CONCLUSIONS

New spectroscopic data on N = Z nuclei from Germanium to Zirconium have been collected using the Daresbury Recoil Separator. Evidence has been found for strong shell effects stabilizing oblate deformation in ^{68}Se and ^{72}Kr and strong prolate deformation ($\varepsilon_2 \sim 0.4$) ^{76}Sr and ^{80}Zr. Access to the spherical doubly magic ^{100}Sn appears extremely difficult.

REFERENCES

[1] A.N. James **et al.** Daresbury study weekend, SERC (UK) report #DL/NUC/R20 1979 p. 84 (unpublished); A.N. James et al., Nucl. Inst. Meth. **212** (1983) 545.
[2] W. Nazarewicz et al., Nucl. Phys. **A435** (1985) 397.
[3] I. Ragnarsson and R.K. Sheline, Phys. Scr. **29** (1984) 385.
[4] P. Moller, J.R. Nix et al., Nucl. Phys. **A361** (1981) 117; Phys. Scr. **29** (1984) 402.
[5] S. Aberg, Phys. Scr. **23** (1982) 23.
[6] J.H. Hamilton et al., Phys. Rev. Letts. **32** (1974) 239.
[7] J. Heese **et al.**, Z. Phys. **A325** (1986) 45.
[8] B.J. Varley et al., Phys. Lett. **B194** (1987) 463.
[9] P. Bonche et al., Nucl. Phys. **A443** (1985) 39
[10] C.J. Lister et al., Phys. Rev. Letts. **59** (1987) 1270.
[11] R. Casten, D.S. Brenner and P. Haustein, Phys. Rev. Letts. **58** (1987) 658 and references therein.
[12] I. Ragnarsson and T. Bengtsson: Contributors to this Conference.

Table 1

A list of reactions used to produce N = Z nuclei. The inverse two neutron evaporation reaction is ideally suited to the Daresbury Recoil Separator.

Isotope	Reaction	E_{beam} MeV	$E(2^+)$ (keV)	σ_{prod} (μb)	
$^{64}Ge_{32}$	$^{12}C(^{54}Fe,2n)^{64}Ge$	155	902	500	300
$^{68}Se_{34}$	$^{12}C(^{58}Ni,2n)^{68}Se$	175	854	22	5
$^{72}Kr_{36}$	$^{16}O(^{58}Ni,2n)^{72}Kr$	170	709	60	25
$^{76}Sr_{38}$	$^{24}Mg(^{54}Fe,2n)^{76}Sr$	175	261	10	5
$^{80}Zr_{40}$	$^{24}Mg(^{58}Ni,2n)^{80}Zr$	190	290	10	5

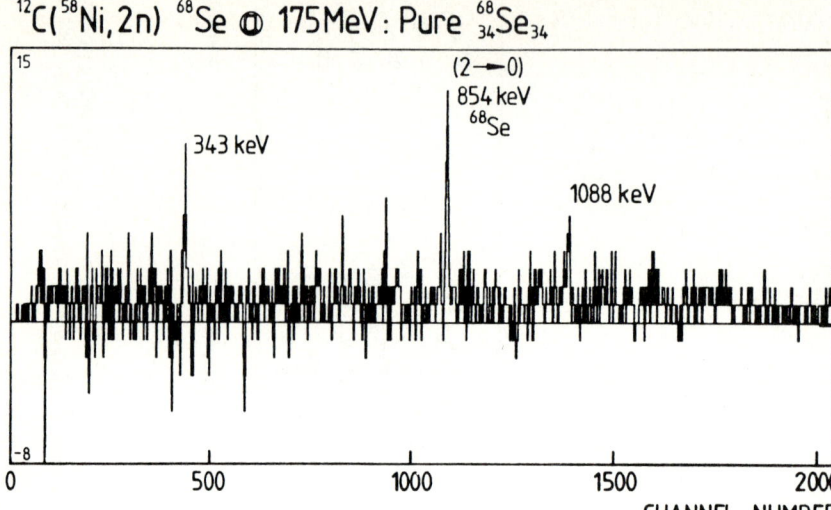

Fig. 1 The gamma ray spectrum associated with ^{68}Se.

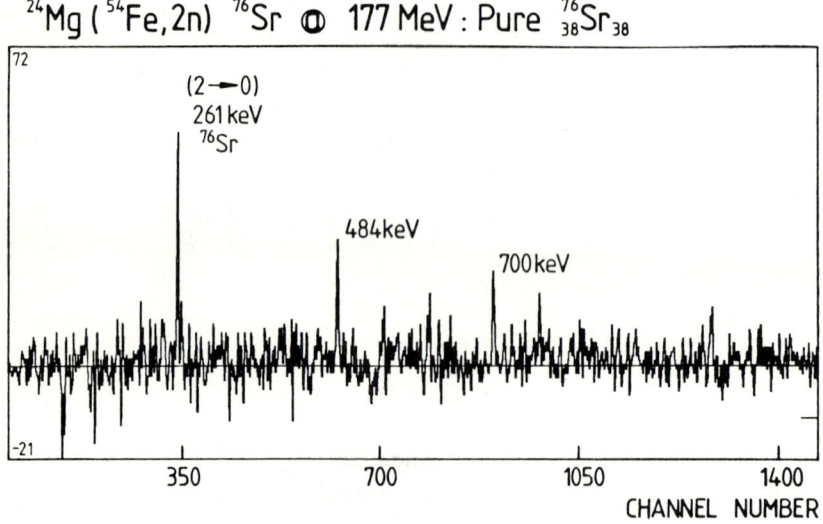

Fig. 2 The gamma ray spectrum associated with ^{76}Sr.

Core Polarization Effects in Odd A ≈ 80 Nuclei

S.L. Tabor[1], P.D. Cottle[1], C.J. Gross[1], D.M. Headley[1], U.J. Hüttmeier[1], E.F. Moore[1], and W. Nazarewicz[2]

[1] Physics Department, Florida State University,
Tallahassee, FL 32306, USA

[2] Institute of Physics, Warsaw Institute of Technology,
ul. Koszykowa 75, 00662 Warsaw, Poland, and
Supercomputer Computations Research Institute,
Florida State University, Tallahassee, FL 32306, USA

The nucleus ^{81}Sr provides one of the best examples of the effects of an unpaired nucleon in polarizing the nuclear core. Four different rotational bands can be seen in the level scheme [1-3] of ^{81}Sr, shown in Fig. 1. Each is built upon a different single-particle configuration and each exhibits distinctive properties. The large signature splitting of the strongly decoupled $g_{9/2}$ Yrast rotational band is consistent with the nearly oblate shape (β_2=0.23, γ=-50°) predicted by Woods-Saxon cranking model calculations [1,4]. The measured B(E2) strengths (See Fig. 2) also imply $|\beta_2| \approx 0.22$ in this band. The nuclear shape changes substantially after the proton quasiparticle alignment at $\hbar\omega \approx 0.5$ MeV. The signature splitting nearly vanishes and ΔJ=1 transitions become strong. A similar band crossing has been seen [5] in ^{81}Kr, and a nearly identical one has recently been reported [6] in ^{79}Kr. Theoretically, the alignment polarizes the core towards a triaxial or more nearly prolate shape with $\gamma \geq$ -30°.

In contrast, the negative parity bands associated with the $\nu[301]1/2$ and $\nu[303]5/2$ Nilsson orbitals show almost no signature splitting and no dramatic changes above the proton crossing at $\hbar\omega \approx$ 0.5 MeV. Theoretically these are expected to have near-prolate shapes with β_2 = 0.35 and -10° $\leq \gamma \leq$ -5°. The proton crossing drives the core triaxial toward $\gamma\approx$-30°. Experimentally, $|\beta_2|\approx$0.26 for the [303]5/2 band.

It is clear from the level spacings in Fig. 1 that the $K^\pi = 1/2^+$ band represents a considerably more deformed rotor. The lifetime measurements imply substantially higher transition quadrupole moments Q_t (See Fig. 2), averaging about 3.5 eb. These are as high as any that have been seen in the A ≈ 80 region and correspond to $\beta_2 \geq$ 0.4. Theoretically [2,4], the core is polarized to an extremely large prolate value of $\beta_2 >$ 0.37 by a $\nu[431]1/2$ particle. This intruder orbital originates from the higher $d_{5/2}$ subshell and plunges rapidly downward in energy with increasing ω, due to its large quadrupole moment. This $K^\pi = 1/2^+$ band provides perhaps the most dramatic evidence of the polarizing power of a single particle.

The example of ^{81}Sr shows that a single odd particle in different

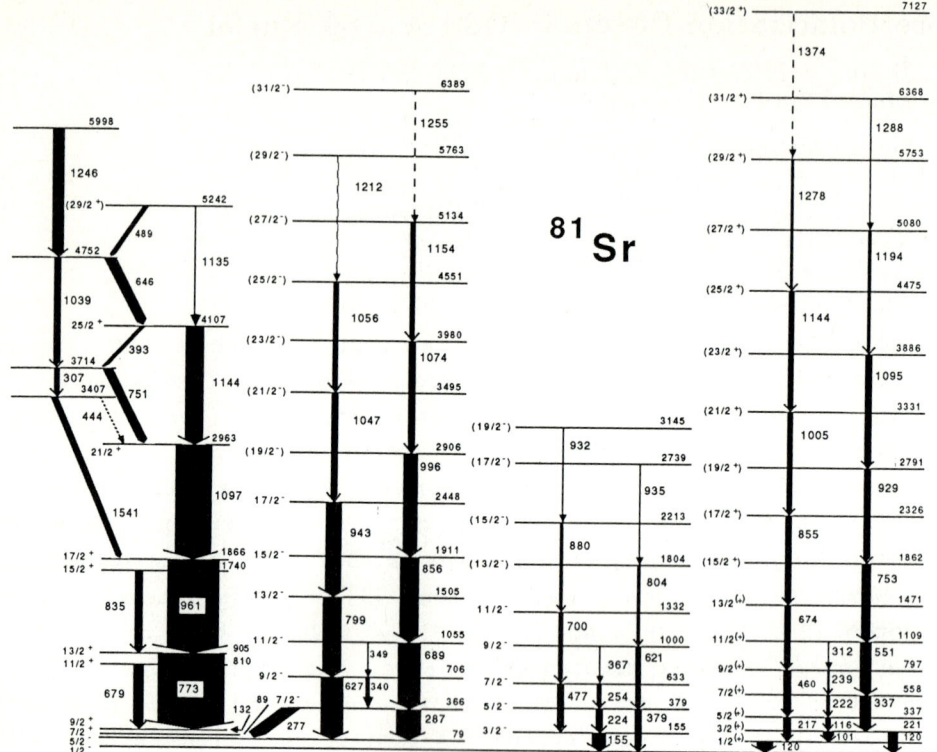

Figure 1. The high-spin level scheme of ⁸¹Sr.

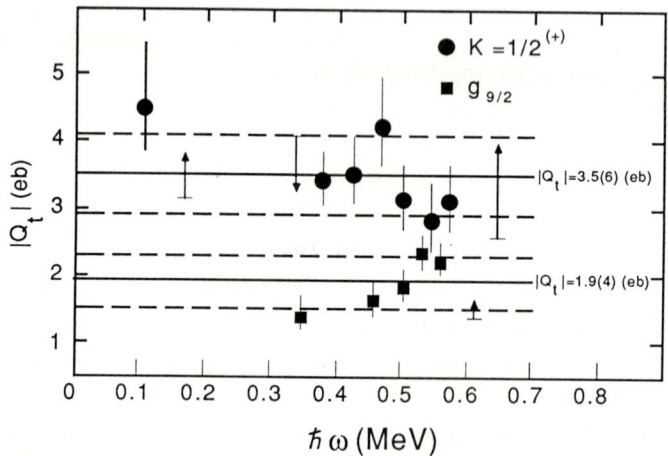

Figure 2. Transition quadrupole moments in the positive parity rotational bands of ⁸¹Sr.

orbitals can polarize the core very differently. The corollary to this statement is that the shape may remain rather constant if the odd particle stays in the same orbital while some other property, such as the number of even particles, changes. An example is the $g_{9/2}$ bands in the N=43 isotones [1,6-8] shown in Fig. 3. Below the proton alignment at J ≥ 21/2, the level schemes are quite similar. The only difference is a slow, smooth increase in deformation [9] as Z approaches the middle of the shell at Z=40. There is even a close similarity between ^{79}Kr and ^{81}Sr after the band crossing, as discussed earlier.

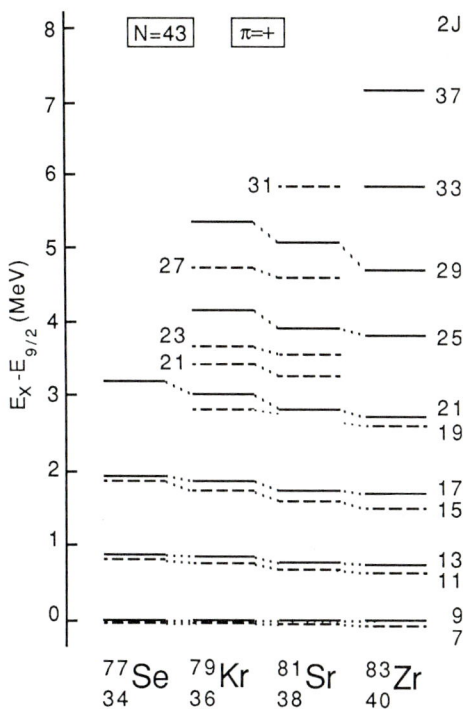

Figure 3. A comparison of the $g_{9/2}$ bands in the N=43 isotones. Levels of signature α=+1/2 are drawn as solid lines (except for the second 21/2 state) and those with α=-1/2 as dashed lines.

No higher states of signature α = -1/2 have been seen [7] in ^{83}Zr. This suggests that there is probably no reduction in signature splitting, which would favor decays to α = -1/2 states. Different shapes after the proton alignments do not violate the corollary because the active protons may be in different orbitals. A decay sequence has been found in ^{83}Zr which may arise from the analog of the $K^{\pi}=1/2^+$ band in ^{81}Sr, but spin assignments and lifetime measurements have not yet been made. The Woods-Saxon cranking calculations [7] for ^{83}Zr suggest that the lowest $h_{11/2}$ orbital will plunge to the Fermi surface at a rotational frequency ℏω somewhat above 0.6 MeV. This polarization mechanism would be similar to that observed for the $d_{5/2}$ intruder orbital, but would lead to a larger β_2 deformation of about 0.5.

^{77}Kr [10,11] provides another example of the shape polarization caused by rotationally aligned quasiparticles. As can be seen in Fig. 4, the alignment in the unfavored α=-1/2 band at ℏω = 0.51MeV is rather sharp, indicating a small interaction matrix element. On the other hand, the favored band (α=+1/2) undergoes a gradual alignment in the frequency interval 0.53 < ℏω < 0.64 MeV which is associated with a large band interaction. Since both bands are built on the $\nu g_{9/2}$ orbital, the first crossing must be due to the alignment of a $g_{9/2}$ proton pair because of blocking. The lifetime measurements imply some difference in the shapes of the two bands, with an axial deformation of $|\beta_2|\approx 0.22$ in the $\alpha=-1/2$ band and $|\beta_2|\approx 0.28$ in the $\alpha=+1/2$ band.

One indication of the change in shape in ^{77}Kr after the proton alignment is the decrease in signature splitting. This is apparent from the level scheme and can be seen from the Routhians in Fig. 4, whose separation decreases around $\hbar\omega \approx 0.5$ MeV. There is also a substantial increase in the M1 strengths after the crossing. Theoretically the $g_{9/2}$ bands are expected to correspond to a nearly prolate shape with $-8° \leq \gamma \leq -2°$. The two aligned quasiparticles drive the deformation towards positive γ values ($\gamma\approx 10°$), which reduces the signature splitting.

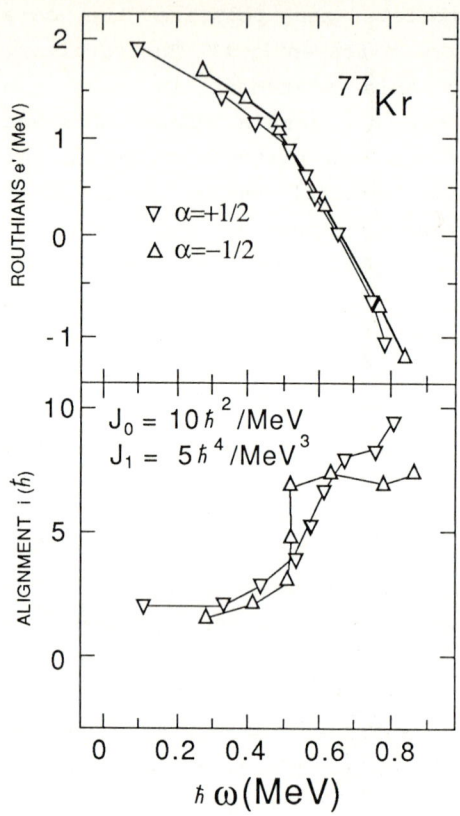

Figure 4. Experimental intrinsic frame excitation energies (Routhians) and alignments for the $g_{9/2}$ bands in ^{77}Kr.

Signature splitting is common among the $g_{9/2}$ bands in this mass region. In the Rb isotopes (Z=37) it is so large that the level ordering is often inverted. This is especially true of ^{81}Rb [12,13] whose level scheme is shown in Fig. 5. The negative parity bands exhibit much less signature splitting, as is also common. However, there is a rather sharp backbend around the $17/2^-$ state, a rarer phenomenon among these nuclei.

The backbend in ^{81}Rb shows up as a large excursion at $\hbar\omega = 0.4$ MeV in the dynamical moment of inertia $J^{(2)}$, plotted in Fig. 6b. A considerably smaller bump is seen in ^{77}Rb. Among the positive parity bands, the largest peak in $J^{(2)}$ also occurs for ^{81}Rb, at a frequency $\hbar\omega \approx 0.54$ MeV. It is, however, smaller and broader than that in the negative parity band (Note the expanded ordinate scale), indicating a much larger band crossing interaction.

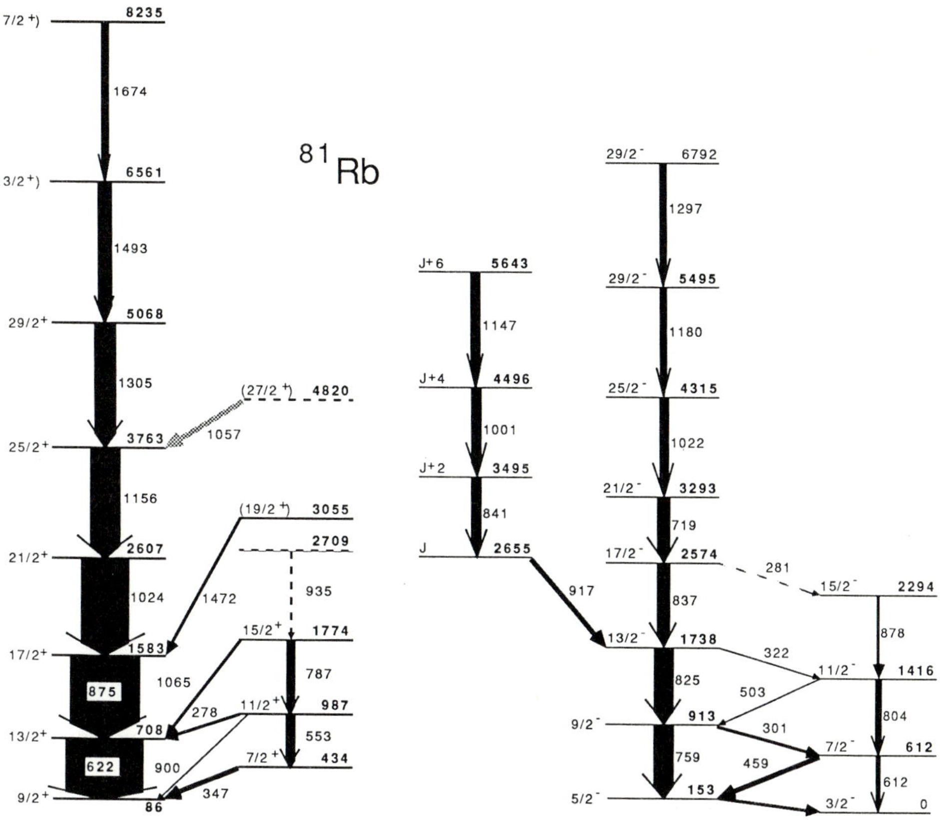

Figure 5. The high-spin level scheme of ^{81}Rb

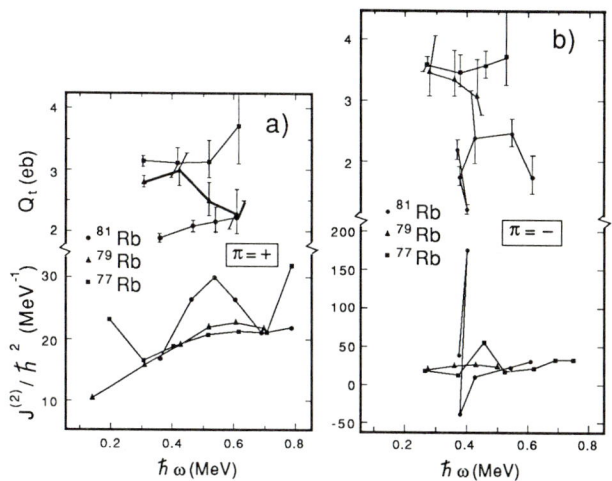

Figure 6. A comparison of the dynamical moment of inertia $J^{(2)}$ and the transition quadrupole moments Q_t as a function of rotational frequency among the well-deformed Rb isotopes.

The odd Rb isotopes constitute another series of nuclei with a constant number of odd particles and rather similar level structures. Deformation and collectivity increase smoothly as N decreases toward mid-shell near N=40. This trend can be seen both in the decreasing level spacings and in the increasing Q_t values [12,14-16] shown in Fig. 6. Although there is some variation with rotational frequency, the quadrupole transition strengths Q_t generally increase from ^{81}Rb through ^{79}Rb to ^{77}Rb, and the $\pi=-$ bands tend to be a little more deformed. The corresponding β_2 deformation values in the $\pi=-$ bands increase from somewhat below 0.3 for ^{81}Rb to somewhat above 0.4 for ^{77}Rb. In the negative parity band of ^{81}Rb there is a clear drop in the Q_t values at the backbend.

In summary, the shapes of the generally soft, shape-coexistent odd A≈80 nuclei are particlularly sensitive to polarization by an unpaired nucleon or by an aligned pair of nucleons. A combination of measurements and calculations shows that in ^{81}Sr a single odd neutron in different orbitals can polarize the shape of the core of 80 nucleons from oblate through triaxial to prolate to an extremely deformed value of $\beta_2 \geq 0.4$. An $h_{11/2}$ particle is predicted to polarize the ^{83}Zr core even more. Conversely, the shapes of other nuclei often change little if the odd particle remains in the same orbital. The alignment of a quasiparticle pair can also alter the nuclear shape, leading to substantial changes in signature splitting and in the relative strengths of $\Delta J=1$ and $\Delta J=2$ transitions.

This work was supported in part by the U.S. National Science Foundation. One of us (W.N.) was supported in part by the Polish Ministry of Science and Higher Education through Contract CPBP 01.09 and the U.S. Department of Energy through Contract No. DE-FC05-85ER250000.

References

1. E. F. Moore, et al., to be published.
2. S. E. Arnell, et al., J. Phys. G **9**, 1217 (1983).
3. G. C. Hicks, et al., Phys. Rev. C **29**, 1345 (1984).
4. W. Nazarewicz, et al., Nucl. Phys. **A345**, 397 (1985).
5. L. Funke, et al., Phys. Lett. **120B**, 301 (1983).
6. G. Winter, et al., J. Phys. G **14**, L13 (1988).
7. U. J. Hüttmeier, et al., Phys. Rev. C **37**, 118 (1988).
8. K. O. Zell, et el., Z. Phys. A **276**, 371 (1976).
9. S. L. Tabor, Phys. Rev. C **34**, 311 (1986).
10. C. J. Gross, et al., Phys. Rev. C **36**, 2601 (1987).
11. B. Wörmann, et al., Nucl. Phys. **A431**, 170 (1984).
12. S. L. Tabor, et al., to be published.
13. H. -G. Friederichs, et al., Phys. Rev. C **13**, 2247 (1976).
14. J. Panqueva, et al., Nucl. Phys. **A376**, 367 (1982).
15. J. Panqueva, et al., Nucl. Phys. **A389**, 424 (1982).
16. L. Lühmann, et al., Europhys. Lett. **1**, 623 (1986).

Deformation of Light Br Isotopes: New Ground State Spin and Moment Measurements

N.J. Stone[1], C.J. Ashworth[1], I.S. Grant[2], A.G. Griffiths[1], S. Ohya[1,3], J. Rikovska[1], and P.M. Walker[4]

[1]Clarendon Laboratory, Parks Road, Oxford, OX1 3RH, UK
[2]Schuster Laboratory, University of Manchester, Manchester, M13 9PL, UK
[3]Faculty of Science, Niigata University, Niigata, Japan
[4]Dept. of Physics, University of Surrey, Guildford, UK

INTRODUCTION

In a previous paper[1]) we have described and discussed measurements of the nuclear magnetic dipole moments of light Br isotopes ^{74m,75,77g}Br using the technique of low temperature nuclear orientation in both on- and off-line experiments. We now report the extension of these measurements to ^{73}Br and ^{72}Br. Improved data on ^{74m}Br are also given. The new results allow deduction of the ground state spin of ^{73}Br and of the magnetic dipole moment of ^{72}Br. As the data are very recent, the numerical moment values quoted may be subject to minor change before full publication.

EXPERIMENTAL DETAILS

The isotopes ^{74m}Br(41.5m), ^{73}Br(3.3m) and ^{72}Br(1.3m) were produced at the NSF accelerator of the SERC Daresbury Laboratory using 150 MeV ^{28}Si ions on a ^{54}Fe target which formed part of the FEBIAD ion source of the isotope separator of the on-line nuclear orientation facility DOLIS-COLD[2]). After acceleration to 60 keV the selected Br ions were implanted into a pure polycrystalline iron foil soldered to the copper cold finger of the on-line dilution refrigerator. Production rates were low but well resolved gamma spectra were recorded in four large Ge detectors, two on the axis of nuclear polarisation and two perpendicular to this axis. The axis was defined by a 0.7T magnetic field which polarised the iron foil in a horizontal direction parallel to the surface of the foil and perpendicular to the implanted beam direction.

RESULTS

The implanted Br nuclei are oriented through the large magnetic hyperfine interaction they experience in the iron lattice. For substitutional Br nuclei this field has been measured to be 81.31(8)T[3]). The observed gamma radiation angular distribution normalised to that from as randomly oriented sample, is given by

$$W(\theta) = 1 + \Sigma_i f_i \left(\Sigma_k (B_k(B_{HF}^i) U_k A_k Q_k P_k(\cos\theta) \right)$$

where f_i denotes the fraction of nuclei experiencing a magnetic hyperfine field B_{HF}^i so that $\Sigma_i f_i = 1$. The other symbols have their usual meaning[4]). The measured anisotropy is defined as

$\varepsilon = ([W(0)/W(90)] - 1)\%$

and is evaluated for two independent pairs of detectors.

74mBr. For this isotope large anisotropies were observed, as shown in Figure 1c for the 728 keV $4^+_1 \rightarrow 2^+_1$ transition. Table 1 gives the anisotropies found for all stronger transitions at 8.7mK. This isotope has a complex decay scheme and although the A_k coefficients for the pure E2 728 keV transition are known, the U_k coefficients are difficult to estimate. Accordingly fits to the temperature dependence of the anisotropies of the strongest transitions have been made using a simplified two site model in which the fraction f_1 experiences the substitutional site field, and the remainder zero field. The parameters $f_1 U_k(k=2,4)$ and the product (moment x field), μB_{HF}^1, are fitted. with the result

$\mu B_{HF}^1 = (+/-)134(5) \mu_N T$, which gives $\mu(^{74m}Br) = (+/-)1.63(6)\mu_N$.

This value is compatible with, and more accurate than, our previous result of $1.45(13)\mu_N$, being derived from data over a wider temperature range. Were we to use the more sophisticated model with two non-zero field sites described by Herzog et al. for BrFe[3]), we estimate the resulting moment would be changed by less that $0.05\mu_N$. A more thorough fitting will be done before final publication.

Table 1. Anisotropies in decay of 74mBr at 8.7mK

From (keV)	Transition	To (keV)		Energy (keV)	Anisotropy (%)
854	0^+_2	635	2^+_1	218	0.8(3.6)
1884	(3^+_1)	1269	2^+_2	615	-39.2(2.1)
1269	2^+_2	635	2^+_1		
635	2^+_1	0	0^+_1	635	-25.9(0.6)
1363	4^+_1	635	2^+_1	728	-42.3(0.9)
2108	4^+_2	1363	4^+_1	745	+23 (11)
2108	4^+_2	1269	2^+_2	839	-48.1(3.0)
1839	(2^+)	854	0^+_2	985	-29.5(4.6)
2564	(2,3,4)	1363	4^+_1	1200	-43.8(3.6)
1884	(3^+_1)	635	2^+_1	1250	-58.9(1.9)
1269	2^+_2	0	0^+_1	1269	-32.8(2.7)
2564	(2,3,4)	1269	2^+_2	1295	-45 (8)
3253	(2-5)	1884	(3^+_1)	1367	-38 (6)

73Br. For this isotope the lowest temperature reached (fully on-line) was 12.0 mK at which the anisotropies measured were as given in Table 2. All are consistent with zero orientation, as was true at all other temperatures. Although the spins of the levels populated in the daughter isotope 73Se are not well established, the fact that the 931 keV transition derives from a level at 1021 keV, the 848 keV transition from a level at 939 keV, the 700 keV from a 790 keV level, and the 400, 336 and 275 keV transitions depopulate the 426 keV level, and that all these levels are independently fed in the decay make it inherently

Fig 1. Angular Distribution Of Gamma Radiation From Oriented Nuclei As A Function Of Temperature.

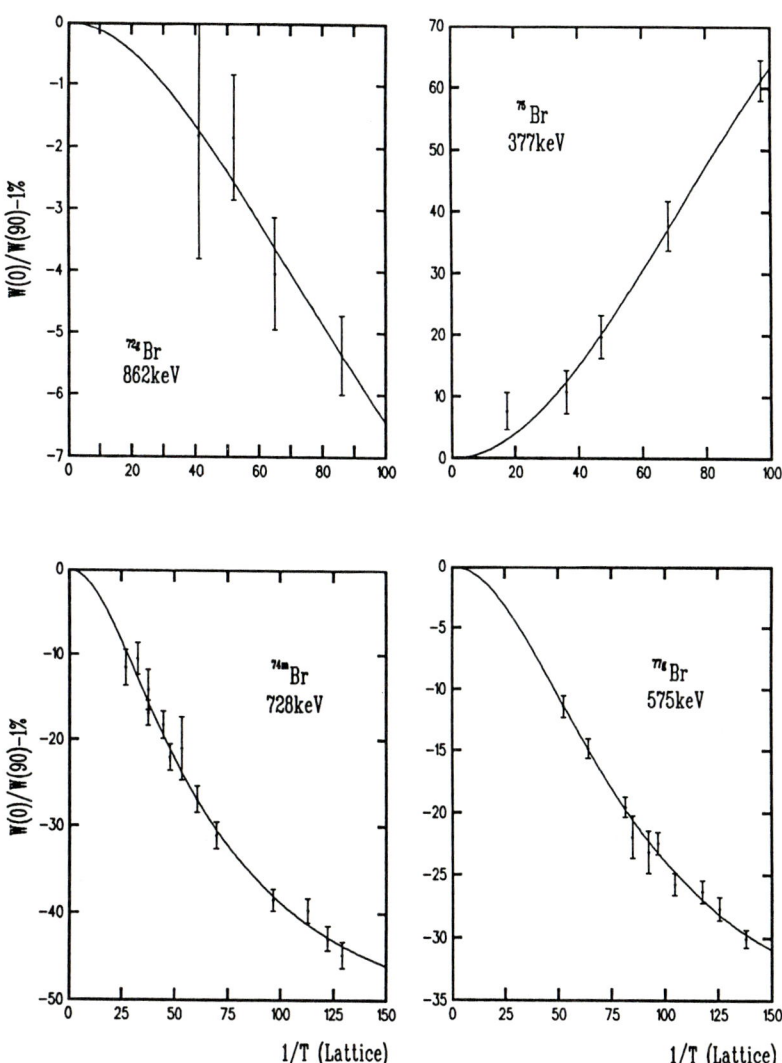

Table 2. Anisotropies in the decay of ^{73}Br at 12.0 mK.

Transition energy (keV)	Anisotropy (%)	Transition energy (keV)	Anisotropy (%)
275	-0.3(2.7)	336	-1.5(1.0)
401	-1.4(1.4)	700	-0.1(0.8)
848	-0.3(1.1)	931	0.0(1.2)

unlikely that they all have spin 1/2, or that in all cases the UA coefficients are close to zero, as required by the experimental observations. We therefore conclude that the parent state nuclear spin,

I (^{73}Br g.s.) = 1/2$^{(-)}$.

Although this value differs from the 3/2$^-$ spins found for heavier odd-A Br isotopes up to ^{85}Br it is fully consistent with all known features of the decay, and gives explanation for the absence of beta decay to the 5/2$^-$ level in ^{73}Se at 151 keV. Attenuation of anisotropy due to imperfect relaxation of the ^{73}Br prior to decay (as invoked below for ^{72}Br) is not taken to be the cause of the zero effects observed for ^{73}Br as the half life at 3.4m is 2.6 times longer than that of ^{72}Br and the moment of a 3/2$^-$ state would not be expected to differ strongly from the value of 0.8 μ_N found for ^{77,75}Br, which leads to relaxation in less than 30s.

^{72}Br. The anisotropies observed for the strongest transition in the decay of ^{72}Br, at 861 keV (pure E2, $2^+_1 \to 0^+_1$) are shown in Figure 1a. Their poor statistical quality reflects the very low production rate for this isotope. However the form of the temperature dependence is relatively well defined. In view of the limited precision of the data the same model as was used for ^{74m}Br is used here with the added complication of the shorter half life of ^{72}Br. The nuclear spin-lattice relaxation time and the thermal equilibrium nuclear orientation are both known functions of the nuclear magnetic dipole moment. Relaxation measurements on ^{82}BrFe have yielded the quantity g^2C_K = 0.017(1) for Br isotopes in iron lattice, where g is the isotope g-factor and C_K is the Korringa relaxation constant.[3] This quantity is isotope independent, which allows a self-consistent calculation to be made taking into account imperfect relaxation to thermal equilibrium with the Fe lattice of the ^{72}Br nuclei (implanted with random orientation) prior to their decay. The fitted line in Figure 1a is for a moment of 0.59μ_N. For this moment value the observed anisotropy at 12.5mK is attenuated by approximately 30%. The 861keV transition has a complex feeding, but if we take the 775 keV transition and accept the proposed spin of ^{72}Br (3) and decay ($3^{(+)} \to 4^+_1 \to 2^+_1$, pure E2), we have parameters U_2A_2 = -0.37 and a fit to the observed anisotropy of this transition (-6(2)% at 14 mK) yields

$\mu(^{72}$Br) = (+/-)0.6(2)μ_N.

This value contains a maximal allowance for uncertainty in the factor f_1. As the ^{72}Br and ^{74m}Br implantations were carried out under the same conditions, further analysis of the ^{74m}Br results may allow extraction of narrower limits on f_1 and hence a more precise moment for ^{72}Br.

Table 3. Spins and Magnetic Moments in the Light Br Isotopes

Nuclide	Spin I^π	Moment (expt.) μ_N	Config.	μ	Moment (calc.)[a] Config.	μ
^{79}Br	3/2$^-$	+2.106[6]	$\pi[301]3/2$	+2.0	$\pi[312]3/2$	+0.75
^{77}Br	3/2$^-$	+/- 0.845(16)[1]	$\pi[301]3/2$	+2.0	$\pi[312]3/2$	+0.75
^{76}Br	1$^-$	+/- 0.5482[6]	$\pi[301]3/2\nu[422]5/2$	-1.5	$\pi[301]3/2\nu[422]5/2$	-0.57
^{75}Br	3/2$^-$	+/- 0.73(9)[1]	$\pi[301]3/2$	+2.0	$\pi[312]3/2$	+0.75
^{74m}Br	4$^-$	+/- 1.63(6)	$\pi[301]3/2\nu[422]5/2$	+1.7	$\pi[301]3/2\nu[422]5/2$	-0.57
^{73}Br	1/2$^{(-)}$		$\pi[321]1/2$	+0.04		
^{72}Br	(3)	+/- 0.6(2)	$\pi[321]1/2\nu[422]5/2$	-0.81	$\pi[321]1/2\nu[303]5/2$	+0.87

[a] Calculated odd-A moments from ref 1) and present work; odd-odd moments adapted from Rb calculations[7]).

DISCUSSION

The magnetic moments measured in this and our previous paper are given in Table 3. We have argued [1]) that the reduction in moment of ^{77,75}Br in comparison with the moment of ^{79}Br (+2.1064μ_N) indicates increasing prolate deformation with the [301]3/2$^-$ low deformation orbital ($\varepsilon < 0.2$) being replaced as ground state in ^{77,75}Br by the [312]3/2$^-$ orbital, with deformation close to $\varepsilon = 0.35$. Calculations performed on the low lying excited state negative parity level structure of these odd-A Br isotopes, using a generalised particle asymmetric rigid rotor model[1]), gave support for these conclusions and good quantitative agreement with the measured magnetic moments (taking positive signs), see Table 3.

How then do we understand the spin I = 1/2 of ^{73}Br? As remarked previously[1]) the level structure calculations required no gamma deformation in ^{77,75}Br although the yrast negative parity prolate states were found to have wavefunctions which strongly overlap those of corresponding states on the oblate side allowing a great deal of oblate-prolate band mixing. Figure 2 shows how the single particle Nilsson orbitals vary with both beta and gamma deformation (gamma being varied for the case $\varepsilon = 0.3$). For $\varepsilon \approx 0.3$ and $\gamma > 25°$ an I = 1/2$^-$ level based on the [321]1/2$^-$ orbital becomes the ground state for Br (Z=35). It has been argued, based on transitional quadrupole moments[5]), that the predicted prolate to oblate transition in this region occurs via gamma deformation and triaxiality. The ground state of ^{73}Br is consistent with this prediction. The presence of a second low-lying negative parity state at 27keV suggests $\gamma \approx 30°$. At $\varepsilon \approx 0.3$ and $\gamma \approx 30°$, the calculated moment of this 1/2$^-$ state is 0.04μ_N.

For ^{72}Br the choice I = 3 is based on general decay scheme arguements, with no clear evidence on parity. Taking possible neutron g-factors from neighbouring Kr isotopes ([422]5/2$^+$) or ^{71}Se ([303]5/2$^-$), coupling of these neutron orbitals to the [321]1/2$^-$ proton orbital gives the moments in Table 3, both being compatible with the experimental value.

The authors acknowledge assistance from the crew of the NSF Daresbury and from J. Groves and J.R.H. Smith in the conduct of the experiments and from D. Denholm in data analysis.

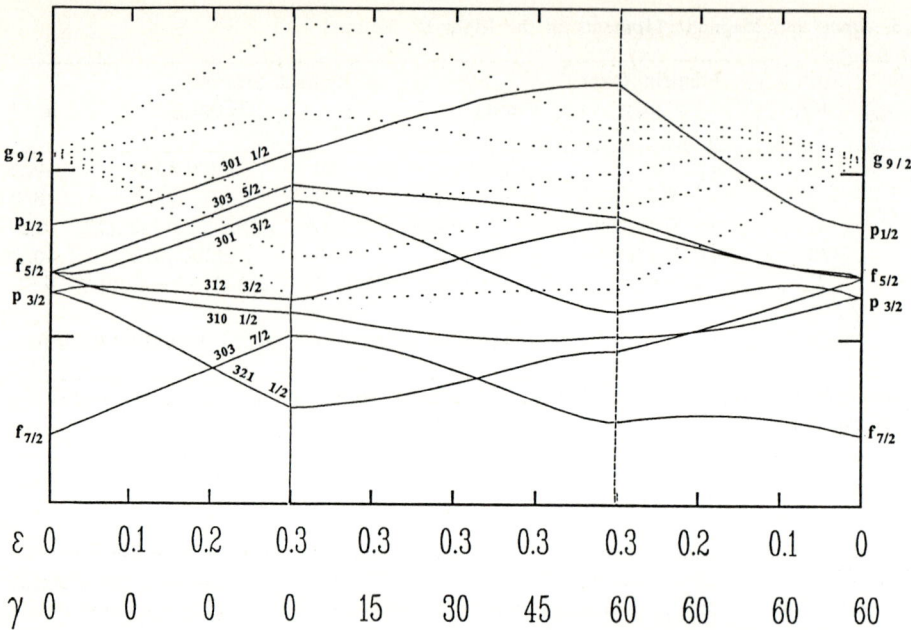

Figure 2. Single particle levels as a function of deformation. Note that the asymptotic quantum numbers have little meaning in the region of varying gamma in which the wavefunctions are strongly mixed. Thus, for instance, the state labelled [301]3/2 on the prolate side becomes, on the oblate side, the [321]1/2 state.

References.

1) A.G. Griffiths et al, Proc. 5th Int. Conf. on Nuclei Far From Stability, Rosseau Lake, (1987) p184.
2) I.S. Grant et al., Nucl. Inst. and Meth. B26 (1987) 95 and ibid p. 482.
3) P. Herzog et al., Z. Phys. B64 (1986) 853.
4) N.J. Stone and H. Postma, Low Temperature Nuclear Orientation, (North Holland) 1986.
5) K.R. Lieb et al., in "Recent Advances in Study of Nuclei Off the Line of Stability (eds. Meyer and Brenner) ACS Symposium Series 324, (1986) p. 233.
6) M. Lederer and V. Shirley, Table of Isotopes (1978) Wiley.
7) C. Ekstrom et al. Nucl. Phys. A311 (1978) 269, Phys. Scripta 22 (1980) 344.

Anomalous Behaviour of Transition Probabilities in ^{75}Kr *

S. Skoda, J. Busch, J. Eberth, M. Liebchen, T. Myläus, N. Schmal, R. Sefzig, W. Teichert, and M. Wiosna

Institut für Kernphysik der Universität zu Köln,
Zülpicher Str. 77, D-5000 Köln 41, Fed. Rep. of Germany

Abstract

The two collective bands of ^{75}Kr have been extended up to spin 21/2 using the compound reactions ^{64}Zn(^{14}N,p2n)^{75}Kr and ^{50}Cr(^{28}Si,2pn)^{75}Kr. Spins and parities were assigned from neutron-gated γ angular-distributions and excitation functions using the **OSIRIS** Anti Compton Spectrometer. The bands are interpreted to be built on well-deformed Nilsson states: [442]5/2 and [301]3/2. Energies and mixing ratios for both bands can be reproduced within the 'single particle and rotor' model, while the experimental $Q_0(I \to I-1)/Q_0(I \to I-2)$ ratios exhibit large deviations by a factor 4 to 6 from theoretical values.

*This work was supported by German Federal minister of Research and Technic grant contract No. 060K272

I. Introduction

In the mass region A≈70-80 and N≈Z we find strong shape variation as a function of particle number and spin and the shape-coexistence effect. Both features are a result of the shell structure in this mass-region. Due to the deformation-dependent splitting of the $1g_{9/2}$, $2p_{1/2}$, $1f_{5/2}$ and $2p_{3/2}$ orbitals, the nucleon-number N,Z=38 stabilizes highly-deformed prolate and N,Z=36 oblate configurations [1].

The first example of shape coexistence in the mass-region A≈70-80 was found in ^{72}Se [2], which has a less deformed ground state and a 0_2^+ state of large prolate deformation at 937 keV.

Theoretical calculations of Bengtsson et al. [3] delivered potential energy curves for isotopes of $_{36}$Kr and $_{38}$Sr with N≤40 displaying two competing minima at quite different deformations: at large prolate deformation ($\beta_2 \approx 0.35-0.40$) and at somewhat smaller oblate deformation ($\beta_2 \approx 0.30$).

In accordance with these calculations, we recently found in our investigation of ^{69,71}Se [4] clear experimental evidence for collective oblate deformation in the mass region A≈70. The results strongly support our interpretation of the level-structure of the neighboring even-even nuclei ^{70,72}Se and ^{74,76}Kr. Their low-spin (I≤8) level-structure is dominated by shape coexistence and mixing between prolate and oblate states.

In ^{74}Kr the shape coexistence between a strongly deformed prolate ground state and a less deformed oblate state causes, due to band mixing, large pertubations in the two bands built above these band heads. Since an unpaired nucleon is a sensitive probe to quadrupole field, the investigation of the even-odd nucleus ^{75}Kr should give more insight into the complex interplay between oblate and prolate structures.

II. γγ coincidences, γ excitation functions and angular distributions

17 new levels were established in ^{75}Kr from γγ coincidences via the reaction ^{14}N(54 MeV)-> ^{64}Zn with 4 Ge detectors.

The angular distributions of the ^{75}Kr transitions and the excitation functions of the ^{75}Kr levels were measured via the heavy ion reaction ^{50}Cr(^{28}Si(85-95 MeV),2pn)^{75}Kr with the 10 fold OSIRIS Anti Compton Spectrometer [5]. The Ge detectors were positioned at angles of 25°, 38°, 63°, 90°, 117° and 142°. We measured neutron-coincident γ-spectra to reduce the line intensities of the charged particle coincident reaction products, using two 4-fold neutron detectors. Fig. 1 shows the experimental setup.

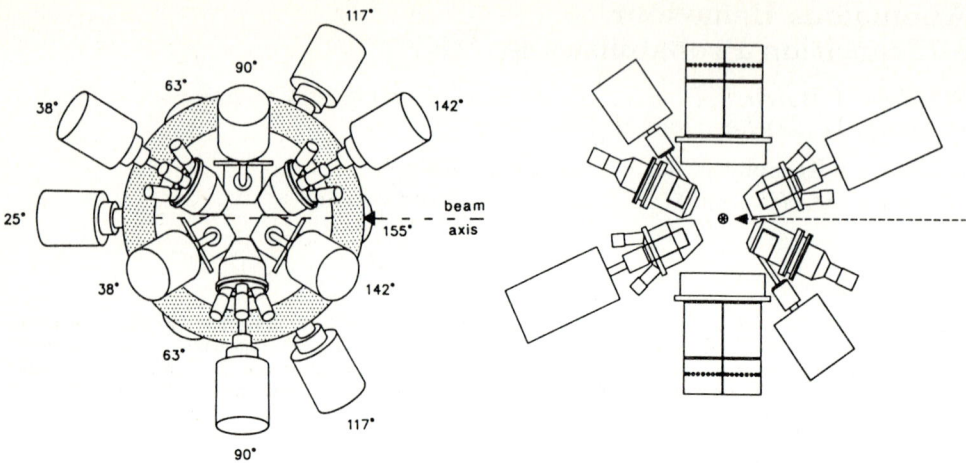

Fig.1   Experimental setup: left - from side (without neutron detectors), right - from above (schematic)

The angular distributions were measured at a beam energy of 95 MeV. In order to improve the reliability of the analysis, we fitted the angular distributions of transitions from the same level simultaneously, thus reproducing the same alignment for a given level. Beside the χ^2-test, the excitation functions together with the evaluated alignment coefficients α_2 were taken into consideration to choose the right spin hypothesis.

III. Proposed level scheme

Spin and parity of the ^{75}Kr ground state are assigned as $5/2^+$. The β^+ decay of ^{75}Kr populates the ^{75}Br levels at 132.4 keV (J =$5/2^+$, I(β^+)=40%) and at 154.5 keV (J =$3/2^+$, I(β^+)=37%) with log ft =5.4 [6,7], fixing a positive parity and allowing only the spins 5/2 and 3/2. The logft =5.9 value for the level 374.0 keV (J =$7/2^+$, I(β^+)=7.5%) fixes the spin 5/2. Yet, because the measured β^+-intensity may be attributed in part to γ-feeding, an assignment J^π =$3/2^+$ cannot be completly excluded.

Due to unresolved multiplets, the spin and parity of the bandhead of the second rotational band in ^{75}Kr at 179.1 keV could not be deduced. Nevertheless, various authors in accordance stated spin and parity of the first excited state in ^{75}Kr as J^π=$3/2^-$ using different models: Nazarewicz et al. calculating within the Cranked Shell model [1], Herath-Banda et al. using a triaxial Nilsson quasiparticle plus rotor model [8], and Ahalpara et al. working in the framework of the projected Hartree-Fock approximation [9]. Besides, the absence of interband transitions between the two rotational bands points to negative parity of the second band head.

The $\gamma\gamma$ coincidences, angular distributions, excitation functions and relative γ-ray yields lead to the level scheme displayed in Fig.2. Our results complement and confirm the work of Winter et al. [10], Garcia-Bermudez et al. [11] and Herath-Banda et al. [8]. Several transitions into the two rotational bands and many new yrast transitions have been identified. The connections of the weak transitions populating the yrast bands from the side are somewhat ambiguous. The highest yrast transitions seen in the cut of the considered line defined the connection level.

Fig. 2   Level scheme of ^{75}Kr

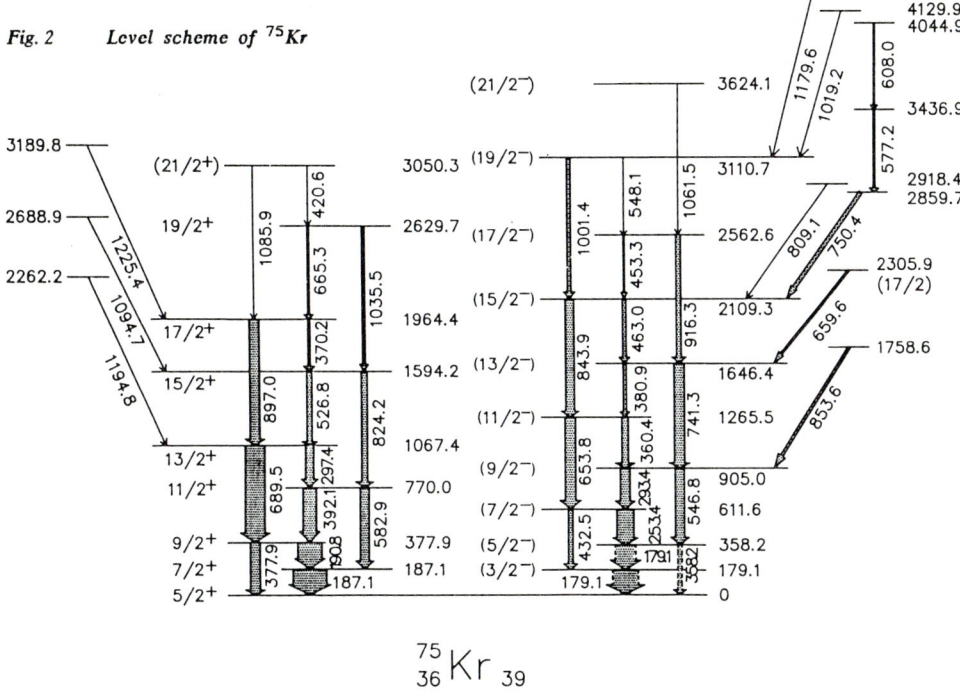

IV.1 Asymmetric Rotor and Single Particle

We interpreted ^{75}Kr within the model of a single particle in two mixing shells coupled to an asymmetric rotor with variable moment of inertia, using the program AROVM2 written by Toki and Faessler [12,13]. The ground state was described as $g_{9/2}$ with possible mixing of $d_{5/2}$, the first excited state as $f_{5/2}$ mixing with $p_{1/2}$. The core was chosen as ^{74}Kr. Whereas the assumed $f_{5/2}, \Omega=3/2^-$ band could be reproduced exceptionally well in both excitation energies and mixing ratios, there was a distinctive deviation of one order of magnitude in the mixing ratios of the $g_{9/2}, \Omega=5/2^+$ band with good agreement in excitation energies. This deviation in mixing ratios results from the Coriolis attenuation which was not included in AROVM2, as can be seen in [8], where a Coriolis attenuation of 0.5 had to be used. The model calculations in the asymmetric rotor model gave the best agreement at a triaxiality near zero. Fig. 3 shows the best fits of the two rotational bands of ^{75}Kr. Our calculations are in accordance with [8] reproducing ^{75}Kr

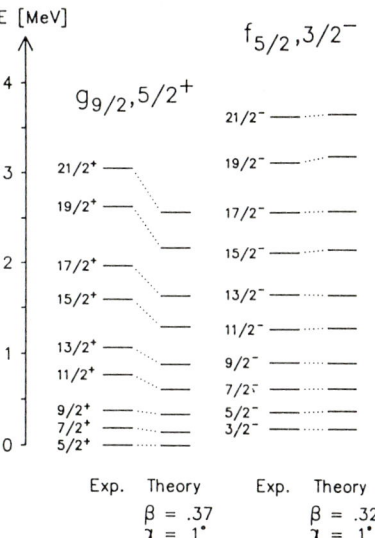

Fig. 3   Model calculations with AROVM2 compared to experimental data

with a triaxiality less then 10° and thus approaching axialsymmetric shape.

IV.2 Electromagnetic Properties

The electromagnetic properties of a nucleus can be described by the total transition probability $T(M\lambda, j_i \to j_f)$, with transition energy E_{if}, multipole order λ and reduced transition probability $B(M\lambda)$ [14]:

$$T(M\lambda, j_i \to j_f) = \frac{8\pi\,(\lambda+1)}{\lambda\,[(2\lambda+1)!!]^2} \cdot \frac{1}{h} \cdot \left[\frac{E_{if}}{hc}\right]^{2\lambda+1} \cdot B(M\lambda, j_i \to j_f).$$

The mixing ratio is defined by $\delta^2 = T(E2, I \to I-1)/T(M1, I \to I-1)$.

Since the AROVM2 model calculations give a strong indication of an axialsymmetric deformation of ^{75}Kr, we describe the reduced E2-transition probabilities as [15]:

$$B(E2) = 5/16\pi \cdot e^2 Q_0^2 \cdot \langle IK20|I'K\rangle.$$

Thus, we get the ratio of intrinsic quadrupole moments

$$\frac{Q_0(I \to I-1)}{Q_0(I \to I-2)} = \left[\frac{B(E2, I \to I-1)}{B(E2, I \to I-2)}\right]^{1/2} \cdot \frac{\langle IK20|I-2\,K\rangle}{\langle IK20|I-1\,K\rangle}$$

with E_{if} in MeV, Q_0 in barn. The relation between the B(E2) values can be deduced from the relation between the total transition probabilities $T(M\lambda, j_i \to j_f)$ [16]

$$\frac{B(E2, I \to I-1)}{B(E2, I \to I-2)} = \left[\frac{E_{if}(I \to I-2)}{E_{if}(I \to I-1)}\right]^5 \cdot \frac{\delta^2}{V_{int} \cdot (1+\delta)^2} \quad \text{with } V_{int} = \gamma\text{-Int.}(I \to I-2)/\gamma\text{-Int.}(I \to I-1).$$

Fig.4 displays the $Q_0(I \to I-1)/Q_0(I \to I-2)$ ratios extracted from experimental data compared with the theoretical ones. We find a rather disturbing discrepancy from the theory. The collective model of the asymmetric rotor as well as microscopic calculations by Ahalpara et al. [9] give both a ratio of nearly

Fig. 4
$Q_0(I \to I-1)/Q_0(I \to I-2)$ *ratios of the two rotational bands,* $g_{9/2}, \Omega=5/2^+$ *band (left) and* $f_{5/2}, \Omega=3/2^-$ *band (right)*

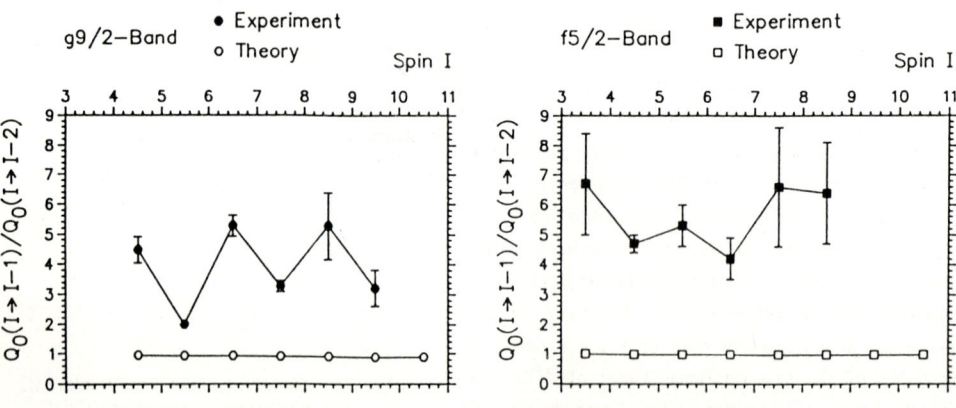

one with a small signature-dependent staggering of less than 5%. Therefore, with the models used so far, this surprising deviation could not be explained.

Since the Q_0-ratios depend strongly on the mixing ratios δ, the mixing ratios have to be evaluated unambiguously. We did not find obvious errors in our careful done data analysis, so possible explanations are undetected systematic errors, or unknown physical effects.

To get a comparable value for different nuclei of varying triaxiality, one defines the effective quadrupolemoment $Q_{\Delta I}(K)_{eff}$ as [17]

$$Q_{\Delta I}(K)_{eff} = [(16\pi/5)\, B(E2, I \rightarrow I-\Delta I)]^{1/2} / \langle IK20|I-\Delta I\,K\rangle .$$

The neighboring nuclei ^{77}Kr and ^{77}Rb show $(Q_1/Q_2)_{eff}$ values of around 2 [18,19] in the $g_{9/2}$ band; on the other hand for ^{77}Kr a new angular distribution measurement delivers systematically smaller mixing ratios [20], which lead to $(Q_1/Q_2)_{eff}$ values below 0.9. Both nuclei are supposed to have a triaxiality of about 20 to 30 degrees. To include the triaxiality-dependent mixing between different Ω-substates, we did AROVM2 model calculations covering all triaxialities between axialsymmetric prolate and oblate shapes. The results never reached deviations from the value one in the measured magnitude. If one looks into recent papers of I. Hamamoto [18,21], one sees that the calculations with triaxial deformation further reduce the $(Q_1/Q_2)_{eff}$ values relative to the axialsymmetric case. In the mass region considered by I. Hamamoto we find the nucleus ^{157}Ho investigated by G. Hagemann et al.. Here the $(Q_1/Q_2)_{eff}$ values reach 2, whereas the calculations with $\gamma = \pm 15$ both lie below 1, for positive triaxility nearer to 1 than for negative [16].

To summarize, up to now ^{75}Kr seems to be a unique nucleus with such extraordinary $(Q_1/Q_2)_{eff}$ values. Yet the crucial mixing ratios have to be reproduced by an independent measuerement, which is under investigation.

References:
[1] W. Nazarewicz, J. Dudek, R. Bengtsson, I. Ragnarsson; Nucl. Phys. A435, 397 (1985)
[2] J.H. Hamilton, A.V. Ramayya, W.T. Pinkston, R.M. Ronningen, G. Garcia-Bermudez, H.K. Carter, R.L. Robinson, H.J. Kim and R.O. Sayer; Phys. Rev. Lett. 32, 239 (1974)
[3] R. Bengtsson, P. Moeller, J.R. Nix and J. Zhang; Phys. Scri. 29, 402 (1984)
[4] M. Wiosna, J. Busch, J. Eberth, M. Liebchen, T. Myläus, N. Schmal, R. Sefzig, S. Skoda, W. Teichert; Phys. Lett. B200, 255 (1988)
[5] R.M. Lieder, H. Jäger, A. Neskakis, T. Venkova, C. Michel; Nucl. Instr. 220, 363 (1984)
[6] Nuclear Data Sheets 32 (1981)
[7] L. Lühmann, M. Debray, K.P. Lieb, W. Nazarewicz and B. Wörmann; Phys. Rev. C31, 828 (1985)
[8] M.A. Herath-Banda, A.V. Ramayya, L. Cleemann, J. Eberth, J. Roth, T. Heck, N. Schmal, T. Mylaeus, W. Koenig, B. Martin, K. Bethge and G.A. Leander; Jour. Phys. G, Nucl. Phys. 13, 43 (1987)
[9] D.P. Ahalpara, A. Abzouzi and K.H. Bhatt; Nucl. Phys. A445, 1 (1985)
[10] G. Winter, J. Döring, W.D. Fromm, L. Funke, P. Kemnitz and E. Will; Z. Phys. A309, 243
[11] G. Garcia Bermudez, C. Baktash and C.J. Lister; Phys. Rev. C30, 1208 (1984)
[12] H. Toki and A. Faessler; Nucl. Phys. A253, 231 (1975)
[13] H. Toki and A. Faessler; Phys. Lett. 63B, 121 (1976)
[14] H. Morinaga and T. Yamazaki: In-Beam Gamma-Ray Spectroscopy; North Holl.Publ.Comp. (1976)
[15] A. Bohr, B.R. Mottelson; Nuclear Structure Vol.II, W.A. Benjamin (1975)
[16] G.B. Hagemann, J.D. Garrett, B. Herskind, J. Kownacki, B.M. Nyako, P.L. Nolan, J. Sharply-Schafer and P.O. Tjom; Nucl. Phys. A424, 365 (1984)
[17] I. Hamamoto; Proc. Electromagn. Prop. of High Spin Levels, Annals of Israel Phys. Soc. 7, 209 (1984)
[18] B. Wörmann, K.P. Lieb, R. Diller, L. Lühmann, J. Keinonen, L. Cleemann and J. Eberth; Nucl. Phys. A431, 170 (1984)
[19] L. Lühmann, Doctor thesis, unpublished
[20] C.J. Gross, P.D. Cottle, D.M. Headly, U.J. Hüttmeier, E.F. Moore and S.L. Tabor; Phys. Rev. C36, 2601 (1987)
[21] I. Hamamoto; Phys. Lett. 194B, 399 (1987)

Backbending Behaviour in the Odd Neutron Nuclei ^{75}Kr and ^{79}Sr

A.A. Chishti[1], M. Campbell[1], W. Gelletly[2], L. Goettig[3], A.N. James[4], C.J. Lister[5], D.J.G. Love[2], J.H. McNeill[1], R. Moscrop[1], O. Skeppstedt[6], and B.J. Varley[1]

[1] Department of Physics, Schuster Laboratory,
University of Manchester, Manchester, M13 9PL, UK
[2] SERC Daresbury Laboratory, Daresbury, Warrington, WA4 4AD, UK
[3] Institute of Experimental Physics, University of Warsaw,
ul. Hoza 69, 00681 Warsaw, Poland
[4] Dept. of Physics, Oliver Lodge Laboratory,
The University of Liverpool, Liverpool, L69 3BX, UK
[5] Wright Nuclear Structure Laboratory, Yale University,
P.O. Box 6666, 272 Whitney Avenue, New Haven, CT 06511, USA
[6] Department of Physics, Chalmer University, Goteborg, Sweden

Abstract:

A study of high spins of odd neutron nuclei ^{75}Kr and ^{79}Sr, is presented in this report. The ^{75}Kr nucleus was populated through the inverse ^{24}Mg(^{54}Fe,2pn)^{75}Kr reaction at beam energies of 177 MeV and 190 MeV, and ^{79}Sr through the ^{24}Mg(^{58}Ni, 2pn)^{79}Sr reaction at 190 MeV beam energy. By using the data from these experiments we were able to extend the known bands to the point where alignment effects are observed. The levels in ^{75}Kr have been identified up to spin $37/2^+$ and $31/2^-$ in the positive and negative parity bands respectively. In ^{79}Sr, the level scheme has been extended to spins $37/2^+$ in the positive and $33/2^-$ in the negative parity bands.

Introduction:

The process of alignment has been found to be rather complicated in neutron deficient A $\simeq$80 rotors. This is for two main reasons: firstly, the neutron and proton Fermi surfaces lie close together, so the alignment effects occur at similar frequencies; secondly, the size of the deformation and the strength of the pairing interaction change rapidly with N and Z. The even-even and some odd proton Br, Rb and Y isotopes[1-5] have been studied in considerable detail but few data are available on the odd neutron nuclei with N~Z. The new measurements on ^{75}Kr and ^{79}Sr are important to redress this balance.

I. Experimental Techniques:

The results reported here are the outcome of three experiments, two for ^{75}Kr and one for ^{79}Sr. These experiments were performed at the NSF Daresbury using the Recoil Mass Separator operated in coincidence with an array of γ-ray detectors.

The Recoil Mass Separator[6] consists of a double Wien filter to reject the non-interacting beam particles and allow the selection of velocities of the reaction residues to be studied. Residues of suitable velocities are deflected by a 50° dipole magnet on to a focal plane which disperses them in mass but not in energy. The ions corresponding to the desired mass are then detected in the focal plane by a carbon foil and a position sensitive channel plate detector, thus providing a signal to detect γ-rays in coincidence with these ions. Gamma-rays from the reactions were detected in an array of 15 Compton suppressed germanium detectors located in rings at 143°, 117° and 101° with respect to the incident beam direction.

II. The ^{75}Kr Nucleus:

Two experiments were carried out to study high spin states in ^{75}Kr. In both cases the ^{24}Mg(^{54}Fe,2pn)^{75}Kr inverse nuclear reaction, using a 500 μg.cm^{-2} thick ^{24}Mg target was studied. In the reaction study at 190 MeV, only $\gamma-\gamma$ coincidence data were collected. From these coincidence events a two dimensional (2K×2K)

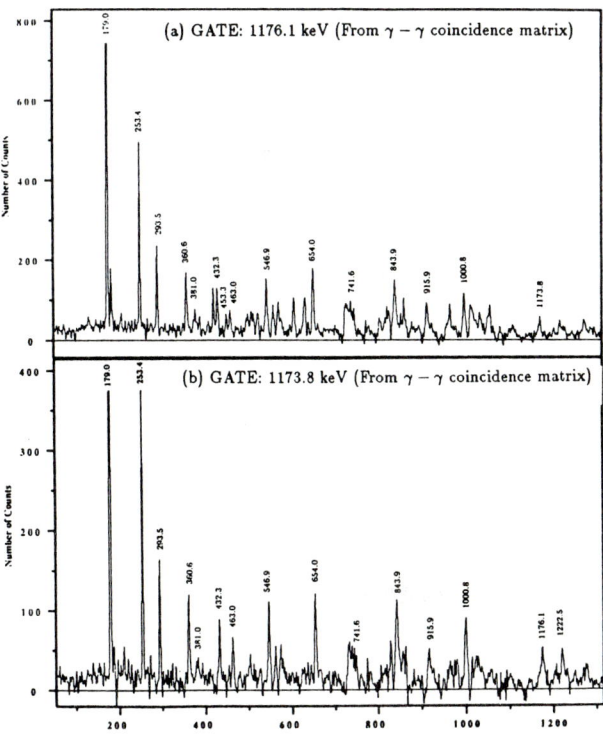

Fig.1 In the negative parity band of ^{75}Kr, the two transitions, 1173.8 keV and 1176.1 keV are in coincidence with each other, but it is difficult to assign spins in the right order as their energies are very close to each other and are seen as one broad peak when seen through a gate other than (a) or (b).

$\gamma - \gamma$ matrix was created with time random events subtracted. By gating on appropriate energy peaks and using careful background subtraction, a number of one dimensional energy spectra were extracted to develop the level scheme. Two examples of the one dimensional energy spectra extracted from the matrix are shown in Fig. 1.

In the reaction at 177 MeV, the Recoil Mass Separator was used to detect γ-rays in coincidence with mass 75 nuclei. By using the data from these two experiments considerable extensions were made in all the known bands[7] in the level scheme.

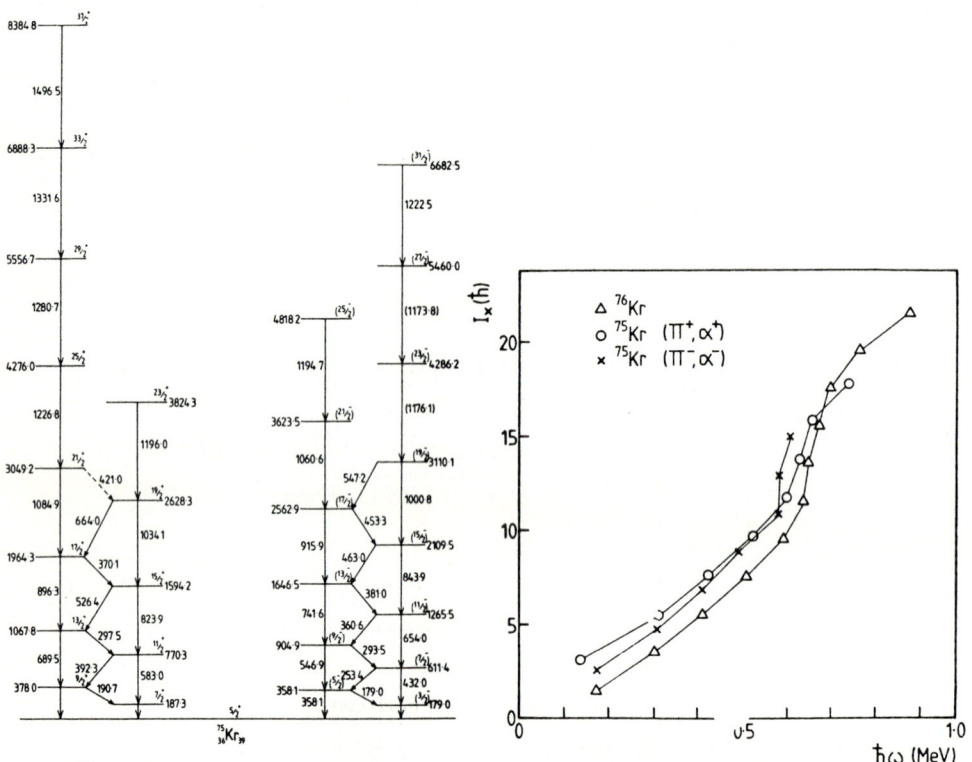

Fig. 2. Proposed level scheme for ^{75}Kr.

Fig. 3. Observed aligned angular momentum for ^{75}Kr and ^{76}Kr.

The decay scheme as shown in Fig.2 has been extended to spin and parity $I^\pi = 37/2^+$ and $31/2^-$ (the ground state spins and parities assigned in ref.7 have been used). The alignment effects are clear in both positive and negative parity bands. In Fig. 3 the total aligned angular momentum I_x is plotted as a function of rotational frequency. We have chosen to plot I_x because it is difficult to determine a set of Harris parameters to use as a reference. This is partly due to rapid changes of structure with proton and neutron number in this region. In ^{75}Kr alignments are observed at $\hbar\omega \simeq 0.56$ to 0.60 MeV in bands of both parities. On the other hand the alignment effects are not seen in the ^{75}Br positive parity bands at this frequency[3]. Blocking arguments suggest that the alignment we observe in the ^{75}Kr positive parity band is due to the protons. The amount of this alignment is $4.5\hbar$ in the positive parity band, which compares well with the expected values for the $g_{9/2}$ protons. The alignment in the negative parity band occurs at a lower frequency ($\hbar\omega \sim 0.57$ MeV) and may be due to protons or neutrons. The measured alignment i_x is more than $4\hbar$. Large alignment has also been observed in the ground state band of ^{76}Kr[1] where protons and neutrons appear to align at very similar frequencies.

Calculations using the model of Frauendorf et al[8] have been carried out to reproduce the observed signature splitting of ~ 100 keV in the positive parity band and ~ 60 keV in the negative parity band. The signature splitting in the positive parity band is reproduced by using the parameters $\epsilon_2=0.35$, $\Delta=1.5$ MeV, $V_{po}= -3$ MeV and Nilsson parameters of ref.9. This implies $\gamma \sim -5°$ to $-10°$, in line with triaxial rotor calculations of Herath-Banda et al[7]. However, it is not then possible to reproduce the signature splitting in the negative parity states ~ 20 keV at $\hbar\omega=0.4$ MeV, since these orbitals have a splitting of > 60 keV at all γ-values in a calculation with the above parameters. This indicates that the structure of ^{75}Kr nucleus is not same for the two configurations, [422]5/2 and [301]3/2.

III. The ^{79}Sr Nucleus:

The odd neutron ^{79}Sr nucleus was produced in the inverse ^{24}Mg(^{58}Ni, 2pn)^{79}Sr reaction at 190 MeV beam energy with a 500 μg.cm^{-2} thick ^{24}Mg target. During the experiment two types of coincidence data were recorded on the magnetic tape:(a) any two γ-rays detected in coincidence with each other and (b) any two γ-rays detected in coincidence with an event due to a mass 79 nucleus.

As discussed in section II, the level scheme was developed by making use of these coincidence measurements. The spectrum in Fig.4(a) is from the $Recoil - \gamma\gamma$ matrix and the one in Fig.4(b) from the $\gamma - \gamma$ matrix. As can be seen in these figures the number of events is greatly reduced when γ-rays are detected in coincidence with mass 79. In the $Recoil - \gamma\gamma$ matrix the number of events was less than 2% as compared to the $\gamma - \gamma$ matrix. However the $Recoil - \gamma\gamma$ data are much cleaner and are very useful in confirming or rejecting ambiguous transitions. The proposed level scheme is given in Fig.5.

The data collected from this experiment permitted the extension of all the known bands[10]. The two yrast bands have been extended to the point where it is possible to study the alignment effects. The observed backbending in ^{79}Sr is difficult to explain. We observe a clear upbending at a rotational frequency of $\hbar\omega \sim 0.6$ MeV in only the positive parity, positive signature band and there is no sign of alignment in the negative parity band. The alignments are plotted in Fig. 6.

If we examine the alignment in the neighbouring even-even ^{78}Sr and ^{80}Sr nuclei we observe that it is very slow[2], which indicates a very strong interaction, and it appears to be due to the $g_{9/2}$ protons. In ^{79}Sr the observed alignment apears to be anomalous in two respects. We would expect a gradual alignment due to the protons in both the positive and negative parity bands. In the case of the negative parity band we do not observe any alignment, but it should be noted that we do not have a good reference in this mass region and the effect may be masked. In the case of the positive parity band we see a very strong upbend, just as has been observed in ^{83}Zr[12], which must be due to a weak interaction. To date we do not understand this behaviour.

In conclusion, the study of high spin states of ^{75}Kr and ^{79}Sr is an important contribution toward our understanding of the alignment effects in the mass 80 region. The observed alignment in only one band and its absence in all the Sr isotopes, which have been studied in detail, apears anomalous.

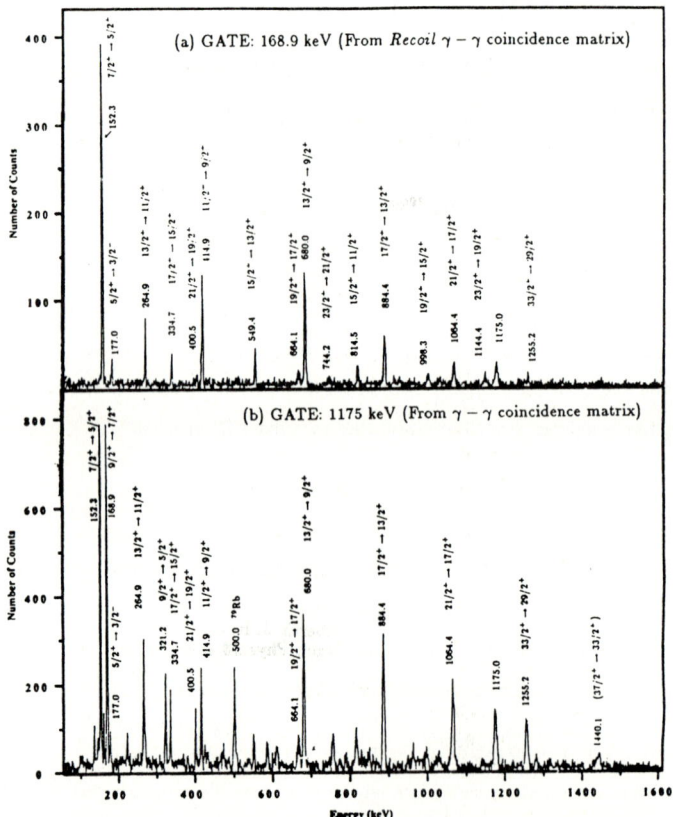

Fig. 4 .Energy spectra of ^{79}Sr positive parity band: (a) γ-rays in coincidence with 168.9 keV transition. The 1175 keV peak appears to be more intense when compared with 1064.4 keV transition which follows it in the decay pattern, (b) the 1175 keV is seen in coincidence with itself. The $T_{1/2}=20+/-1$ ns has been measured for 177 keV isomeric state by Lister et al[11] in a separate experiment.

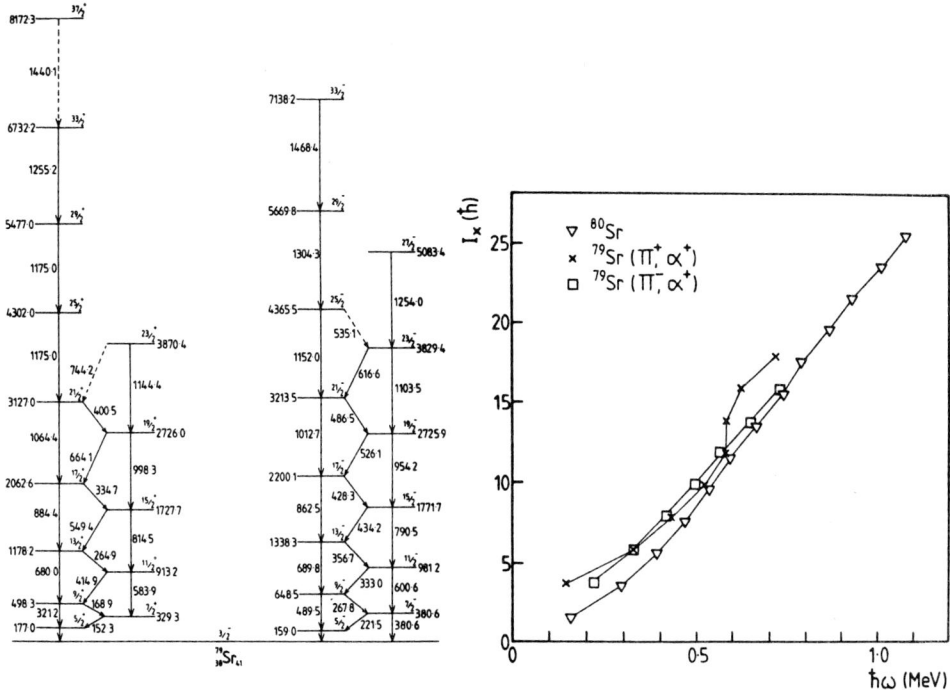

Fig. 5. Proposed level scheme for ^{79}Sr.　　Fig. 6. Observed aligned angular momentum for ^{79}Sr and ^{80}Sr

1. K. P. Lieb, Proc. of Conf. on Weak and Electromagnetic Interactions in Nuclei, Heidelberg, July 1986, edited by H. V. Klapdor (Springer-Verlag 1986).
2. C. J. Gross, J. Heese, K. P. Lieb, , C. J. Lister, B. J. Varley, A. A. Chishti, and W. Gelletly, Poster Session in the present meeting.
3. L. Luhmann, M. Debray, K. P. Lieb, W. Nazarewicz, B. Wormann, J. Eberth and T. Heck, Phys. Rev. **C31** (1985) 828.
4. L. Luhmann, K. P. Lieb, C. J. Lister, B. J. Varley, J. W. Olness and H. G. Price, Europhys. Lett. **1(12)** (1986) 623.
5. C. J. Lister, R. Moscrop, B. J. Varlely, H. G. Price, E. K. Warburton, J. W. Olness and J. A. Becker, J. Phys. G: Nucl. Phys. **11** (1985) 969.
6. A. N. James, T. P. Morrison, K. L. Ying, K. A. Connell, H. G. Price and J. Simpson, Nucl. Inst. Methods. **A267** (1988) 144.
7. M.A. Herath-Banda, A. V. Ramayya, L. Cleemann, J. Eberth, J. Roth, T. Heck, N. Schmal, T. Mylaeus, W. Koenig, B. Martin, K. Bethge and G. A. Leander, J. Phys. G: Nucl. Phys.**13** (1987) 43.
8. S. Frauendorf and F. R. May, Phys. Lett. **125B** (1983) 245.
9. D. Galeriu, D. Bucurescu and M. Ivascu, J. Phys. G: Nucl. Phys. **12** (1986) 329.
10. C. J. Lister, B. J. Varley, H. G. Price, and J. W. Olness, Phys. Rev. Lett. **49** (1982) 308.
11. C. J. Lister, Private communications.
12. U. J. Huttmeier, C. J. Gross, D. M. Headly, E. F. Moore, S. L. Tabor, T. M. Cormier, P. M. Stwetka and W. Nazarewicz, Phys. Rev. **C37** (1988) 118.

Proton-Neutron Interaction and the 1 g Orbital in the $33 \leq Z \leq 50$ Region

A.G. Griffiths and J. Rikovska

Clarendon Laboratory, Parks Road, Oxford, OX13RH, UK

Nuclei with Z<40 and 36<N<50 show large changes in deformation and coexistence both between prolate and oblate and spherical and deformed shapes. Theoretical interpretation of experimental data in this region is rather complicated by the presence of the N,Z ~ 38,40 subshell closures, which is deformation dependent and its manifestation is a function of the number of protons and neutrons involved. The region has been thoroughly studied in numerous works (see e.g. [1-17]) with special interest in features of rotational bands up to high-spins. Various nuclear models, including shell model, particle - rotor (axial, triaxial) with and without variable moment of inertia, cluster vibrational model and IBM have been used in efforts to understand experimental results in this area.

There is a striking similarity between A=80 and A=100 regions in occurence of considerably deformed nuclei. In the A=100 region, this phenomenon has been attributed either to the T=0 p-n $g_{9/2}$-$g_{7/2}$ interaction[18] or to increasing occupancy of the $h_{11/2}$ intruder neutron orbital[19]. In even-even Mo nuclei, intruder core excitations have been identified in [96-104]Mo which become ground states in [102-104]Mo[20]. In the A=80 region, there is an experimental evidence for increase in deformation of the $1g_{9/2+}$ proton state in As,Br,Rb and Y nuclei with growing occupancy of the neutron $1g_{9/2+}$ shell. In even-even nuclei with N close to the middle of the major $28 \leq N \leq 50$ shell, there is a coexistence of deformed and spherical ground states and first excited 0+ states. In this contribution we explore these phenomena using extended systematics of experimental data and results of the particle-triaxial rotor and IBM models calculation.

1. Odd- proton $1g_{9/2+}$ states in the A~80 region.

The proton $1g_{9/2+}$ states are strongly populated in heavy-ion reactions in most of odd-A isotopes in this region[1-17]. In most nuclei these states are band-heads of a rotationally-aligned band which resembles the ground state band in the adjacent even-even nucleus. The bands are considerably influenced by a strong Coriolis coupling, especially in cases of lower deformation ($\beta \leq 0.25$). A sensitive measure of the Coriolis coupling strength is the relative position of the $I^\pi = 5/2+$ and 9/2+ member of the bands; energy spacing and relative placement of favored (signature $\alpha=1/2+$) and unfavored ($\alpha=1/2-$) states has proved to yield information on deviation of shape of the core from axial symmetry. As a consequence of stronger deformation, bands built on $1g_{9/2+}$ states in [79,81]Y[11,12] and [77,79] Rb[7,8] exhibit level order characteristic of deformation alignment but with staggering still present, thus indicating the presence of Coriolis force even for deformation as high as $\beta \sim 0.4$. Some properties the 9/2+ states are summarized in Table 1, together with corresponding data for adjacent A-1 even-even core nuclei.

It may be first noted that in all cases except those with neutron number close to N=50, ΔE (13/2+ - 9/2+) is systematicaly lower then ΔE (2_1^+ - 0_1^+) for As and Ge isotopes, but almost equal for Rb and Y nuclei. B(E2, 13/2+ - 9/2+) value is almost always larger then B(E2, 2_1^+ - 0_1^+) . Available effective g-factors, deduced from B(M2) values for transitions from the $1g_{9/2+}$ state to a 5/2- ground state, indicate that these g-factors are much smaller than the single-particle estimate. The same information can be deduced from known magnetic dipole moments of the 9/2+ states. This 'excess of collectivity' in the $1g_{9/2+}$ proton states has been discussed in length in many papers (e.g.Ref.7,8) and attributed to the core polarization effect due to residual interaction of the rather soft core with a proton occupying the $1g_{9/2+}$ deformation driving orbital. However, most models underestimate this effect and do not reproduce fully all experimental data on the $1g_{9/2+}$ proton states and low members of the associated bands.

In Fig.1 we show systematics of energies of the $g_{9/2+}$ proton states as a function of neutron number. There is an apparent correlation between decrease of the 9/2+ state energy and filling of the $1g9/2+$ neutron orbital. For all Z the functions reach their minimum for $42 \leq N \leq 44$, i.e. higher then the expected middle of the full $28 \leq N \leq 50$ neutron shell. This effect cannot be explained by changes in deformation of the core and corresponding variation of the Fermi level.

Table 1. Electromagnetic properties of proton $g_{9/2+}$ states[a].

N	ΔE[MeV]			B(E2)[e^2fm^4]			R[b]	g_{eff}/g_{sp} [c]	μ[n.m.][d]
	($13/2^+$-$9/2$)	($9/2^+$-$5/2^+$)	(2_1^+-0_1^+)	($13/2^+$-$9/2^+$)	($17/2^+$-$13/2^+$)	(2_1^+-0_1^+)			
As(Ge)									
36	0.854		1.026	305(41)	280(42)		1.1(2)	0.197(5)	+4.73(5)
38	0.714	-0.129	1.039	775(80)	1074(535)	352(8)	2.2(2)	0.129(1)	+5.15(9)
40	0.609	-0.082	0.834	815(50)	445(13)		1.8(1)	0.097(3)	+5.234(14)
44		-0.156	0.563		536(16)				+5.525(9)
Br(Se)									
38	0.583	0.188	0.862	2505(346)	>1381	350(40)	7(1)		
40	0.563	0.086	0.635	2065(175)	1725(345)	774(16)	2.7(2)		
42	0.533	-0.024	0.559	1345(100)	3232(2000)	840(20)	1.6(1)		
46		-0.256	0.666		506(12)			0.073(1)	5.694(45)
Rb(Kr)									
40	0.502	0.185	0.424	2800(180)	3020(430)	1640(80)	1.7(1)		
42	0.501	0.097	0.455	2180(110)	2760(460)	1200(100)	1.8(1)		
44	0.623	-0.168	0.616	989(55)	1270(100)	740(40)	1.3(1)		+5.598(2)
48	0.779	-0.383	0.882	234(27)	45(10)	250(12)	0.9(1)	0.093(1)	+6.046(10)
Y(Sr)									
44	0.595		0.573	1420(260)	1790(790)	1026(40)	1.4(2)		
46	0.795	-0.416	0.793	757(44)	468(69)	560(80)	1.3(2)		6.2(5)
48		-0.772	1.077		212(32)				6.10(3)

a) Data on odd-A nuclei are taken from Refs.1-17 and on even-even from Ref.21,22.
b) R=B(E2,$13/2^+$-$9/2^+$)/B(E2,2^+-0^+)   c) Ref.1   d) Ref.23

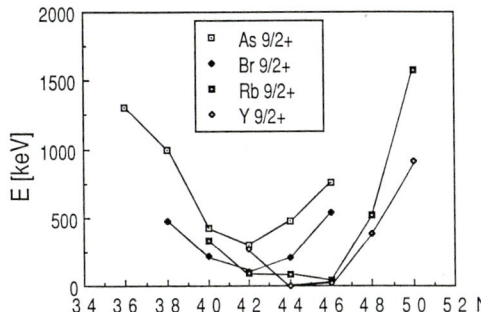

Fig.1 Enegy of $1g_{9/2+}$ states as a function of neutron number.

To establish characteristics of these bands, we have carried out a systematic calculation using the particle-triaxial rotor model[24] for all known proton bands in the A=80 region. Main emphasis was on calculation of electromagnetic properties and energies of states of $I^\pi \leq 17/2$, as our model did not include variable moment of inertia and was therfore not suitable for handling of the higher spin states. In contrast to the previous calculation of Toki and Faessler[25], we did not use effective charges different from 1e. For all isotopes κ=0.069, μ=0.420, g_s=0.7 have been used. A prolate triaxial core (γ=28°) has been chosen for all nuclides. The value of parameter γ is in a broad agreement with the experimental ratio of energies of the 2_2^+ and 2_1^+ states (assuming that the 2_2^+ level is indeed of a quasi-vibrational origin). Other parameters used in the calculation and results obtained are summarized in Table 2. The latter should be compared to experimental data in Tab.1 A very good overall agreement has been achieved to fit energy levels, B(E2 ,13/2+-9/2+) ,B(E2 ,17/2+-13/2+) and magnetic dipole moment. Both energy differences between 9/2+ and 5/2+ states and 7/2+ and 9/2+ states are very reasonably reproduced. These results give a supporting evidence that in all cases a triaxial shape with positive value of β_2 for the core state is a good approximation which is not in line with predictions of Moller and Nix[26] who take axially symmetrical ground states for nuclei in the A=80 region with some prolate and some oblate deformations.

Table 2. Parameters and results of a particle-triaxial rotor calculation

N	ϵ_2	ξ	E_{2+}	B(E2) [e^2fm^4]		μ [n.m.]	ΔE[MeV]	
				(13/2+-9/2+)	(17/2+-13/2+)		(13/2+-9/2+)	(9/2+-5/2+)
As(Z=33)								
36	0.17	0.93	0.58	318	358	5.64	0.861	-0.257
38	0.25	0.93	0.46	757	778	5.48	0.759	-0.132
40	0.26	0.93	0.35	857	857	5.04	0.602	-0.088
Br(Z=35)								
38	0.37	0.85	0.43	2652	3286	5.19	0.625	+0.188
40	0.34	0.88	0.37	1994	2673	5.37	0.571	+0.077
42	0.29	0.90	0.31	1332	1563	5.42	0.540	-0.014
44	0.25	0.98	0.35	964	979	5.48	0.511	-0.179
46	0.22	0.98	0.43	748	775	5.54	0.597	-0.250
48	0.20	0.98	0.44	626	654	5.57	0.597	-0.274
Rb(Z=37)								
40	0.34	0.80	0.36	2812	3276	5.06	0.500	+0.199
42	0.31	0.87	0.38	2172	2690	5.35	0.505	+0.091
44	0.23	0.95	0.45	956	1137	5.61	0.624	-0.180
46	0.18	0.96	0.58	571	672	5.88	0.753	-0.365
48	0.12	0.97	0.65	253	307	5.77	0.773	-0.456
Y(Z=39)								
40	0.44	0.78	0.38	4468	5838	4.59	0.659	+0.404
42	0.36	0.82	0.44	3444	4176	5.05	0.620	+0.221
44	0.25	0.93	0.50	1472	1864	5.59	0.600	-0.093
46	0.19	0.97	0.71	760	950	5.74	0.820	-0.398
48	0.12	1.00	0.95	288	355	5.84	0.981	-0.792
50	0.06	1.00	1.50	71		5.92	1.485	-1.328

Our calculation shows that to obtain a good fit , the parameter E_{2+} , which should be equal to the energy of the 2_1^+ state of the core must in fact be less than this for As and Br. As discussed later, this suggests that the $1g_{9/2+}$ particle is not coupled to the g.s. of Ge and Se, but to the 0_2^+ state. The same can be seen in Table 3, where a comparison of experimental deformations deduced from B(E2) values with calculated β_2 of the isomeric $1g_{9/2}$ states (particle-triaxial rotor model) and of corresponding ground states of both, odd-A and (A-1) even-even core nuclei , obtained from Q_2 values by Moller and Nix[26] is presented.

The rotation-aligned pattern of the positive parity bands in the $33 \leq Z \leq 39$ nuclei with N>42 suggests that $1g_{9/2+}$ states with lowest K^π (1/2+,3/2+) are the primary components of the $1g_{9/2+}$ wavefunctions. These states in As and Br can be excited by promotion of a 1p-1h proton pair from $2p_{3/2}$ or $1f_{5/2}$ orbitals across the Z=34 ($0.1 \leq \epsilon_2 \leq 0.3$) subshell gap. Thus $[1g_{9/2+}]^{(1)} + [2p_{3/2}, 1f_{5/2}]^{-k}$ structures are created, with k proton hole pairs below the gap. For Z=37 and 39 the situation is more complex. The sharp Fermi surface is very close to 1/2+[440] and 3/2+[431] orbitals for a wide range of deformations larger than 0.1. The $1g_{9/2+}$ states are much closer to the ground state or become ground states (^{83}Y). However, subshell gaps at Z=40 ($\epsilon_2 \sim -0.2$) and Z=38 ($0.2 \leq \epsilon_2 \leq 0.5$) exist and 1p-1h excitations cannot be excluded. These intruder states 1p-2h involve more active particles (holes) then the normal 1p ones and are therefore more deformed due to increased residual proton-neutron quadrupole interaction.

Table 3. Comparison of deformations of the core, the ground and the $g_{9/2+}$ states.

N	β_2 ($\gamma=28$)[a] core	β_2 ($\gamma=0$)[b] core	β_2 ($\gamma=0$)[c] core	I^π g.s	β_2 ($\gamma=0$)[c] g.s	β_2 ($\gamma=0$)[b] $g_{9/2+}$
As(Z=33)						
36	0.17	0.20	+0.22	5/2$^-$	-0.25	0.18
38	0.24	0.22	-0.20	5/2$^-$	-0.23	0.28
40	0.25	0.24	-0.18	3/2$^-$	-0.20	0.28
Br(Z=35)						
38	0.35	0.21	-0.23	(1/2$^-$)	-0.27	0.47
40	0.32	0.23	-0.21	3/2$^-$	-0.24	0.42
42	0.28	0.31	-0.21	3/2$^-$	-0.23	0.33
44	0.24	0.27	+0.16	3/2$^-$	+0.12	
46	0.22	0.23	+0.16	3/2$^-$	+0.14	
48	0.20	0.19	+0.14	3/2$^-$	+0.10	
Rb(Z=37)						
40	0.32	0.41	+0.43	3/2$^-$	+0.44	0.45
42	0.30	0.34	-0.19	5/2$^+$	+0.40	0.39
44	0.23	0.27	+0.10	3/2$^-$	-0.10	0.26
46	0.18	0.20	+0.10	5/2	+0.10	
48	0.12	0.15	+0.08	5/2$^-$	+0.07	0.12
Y(Z=39)						
40	0.44	0.43	+0.45			
42	0.34	0.39	+0.45	(3/2$^-$)	+0.46	
44	0.25	0.29	-0.03	9/2$^+$	-0.00	0.29
46	0.19	0.21	-0.03	(1/2$^-$)	-0.13	0.21
48	0.12	0.13	-0.03	1/2$^-$	-0.13	
50	0.06	0.12	-0.01	1/2$^-$	-0.13	

a) Present calculation (all β_2 are positive).
b) Absolute values of β_2 deduced from experimental B(E2) values
c) Deduced from calculation of Moller and Nix[26]. The sign of β_2 was taken equal to the calculated sign of Q_2.

The energy of these states is equal to the difference between the unperturbed single particle normal and intruder configurations, decreased due to pairing, monopole and quadrupole interactions[27]. In particular, the quadrupole interaction between active protons and neutrons is proportional to $2\kappa \Delta N_\pi N_v$, where κ is the proton-neutron quadrupole interaction strength, ΔN_π is the difference in number of proton bosons between the intruder and normal states and N_v is number of neutron bosons. This interaction is responsible for strong neutron number dependence of the intruder state energy. For a given number of protons, the quadrupole correction will reach its maximum in the middle of corresponding neutron shell where the number of active neutron bosons is highest. Thus one of the fingerprints of intruder states is their strong neutron number dependence as compared to normal states. This dependence of the $1g_{9/2+}$ proton state energy on the corresponding distribution of active neutron pairs over the $38,40 \leq N \leq 50$ region can be clearly seen in Fig.1.

The concept of of 2h+1p configurations in the 1g proton orbital would explain increased collectivity of the odd-A $g_{9/2+}$ states. A similar idea has been explored before in the shell-model calculation of ^{85}Rb by Luhmann et al.[10]. They assumed the lowest 1/2$^-$ and 9/2$^+$ states to be of (1p2h) character with two proton holes equally distributed over the $2p_{3/2}$ and $1f_{5/2}$ orbits. They found out that the occurence of the (1p2h) 9/2$^+$ state is the common character of all the wave functions involved.

2. Excited 0+ states in even-even Ge, Se, Kr and Sr nuclei

Heyde et al[27] showed recently that in the SU(3) limit there is a scaling between energies of proton (neutron) intruder states in odd-A and even-even A nuclei with the same number of neutrons (protons), namely the energy of a 2h-1p intruder state is about one half of the excitation energy of the corresponding 2h-2p 0+ intruder state in the neighbouring even-even nucleus (involving same orbitals). The validity of this result was succesfully demonstrated for nuclei around the Z=82 and Z=50 shell gaps. This scaling phenomenon can be invoked to assist identification of intruder states in either odd-A or even-even nuclei provided their position in the other is known.

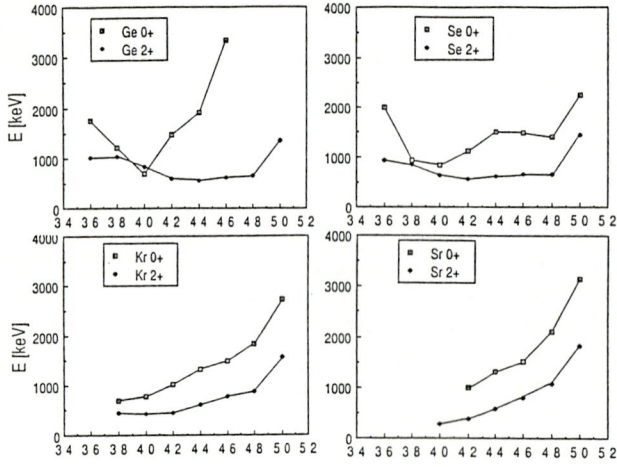

Fig.2 Neutron number dependence of the 2_1^+ and 0_2^+ states energy in Ge,Se,Kr and Sr nuclei

There is strong experimental evidence on different deformation of the ground state and the first excited 0+ states in Ge-Sr nuclei [28-30]. However, the nature and structure of the coexisting states is still unclear. In Fig.2 we display neutron number dependence of the 0_2^+ energy as compared to the energy of the 2_1^+ state. There is obviously a striking difference between Ge and Se isotopes. and Kr and Se nuclides. In the former, the 0_2^+ state energy changes rapidly with N and is minimal for N=40, very close to the middle of the 28-50 neutron shell, whilst the 2_1^+ state energy is more or less constant. In the latter, the 0_2^+ state energy together with that of 2_1^+ state monotonically increases as one approaches the N=50 shell.

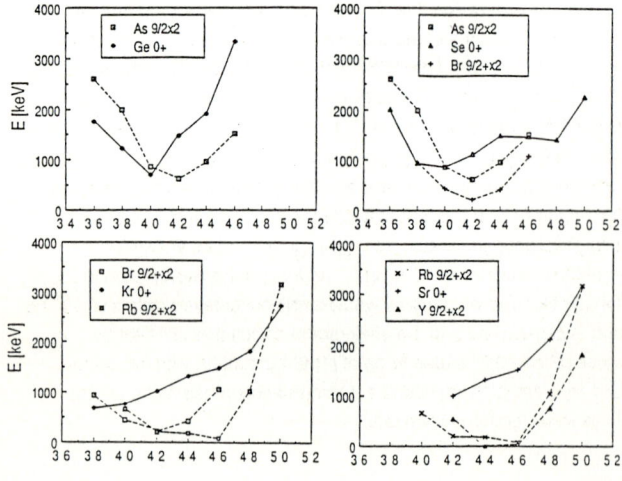

Fig.3 Scaling of intruder states energy in odd-A and even-even nuclei in the A=80 region (see text)

In Fig.3 we compare the 0_2^+ state energies to corresponding energies of the $g_{9/2+}$states in neighbouring nuclei scaled according to the scaling law discussed above. Again, there is a considerable difference between Ge and Se isotopes, and Kr and Sr isotopes. Because of complicated structure of the Z=40 region, we expect rather complex behaviour of the pairing and monopole corrections to the intruder state energy. It is not therefore surprising that the scaling is not perfect, as for example in the Z=82 region [27]. Nevertheless there is a definitive resemblance in trend of both the Ge and Se 0_2^+ state energy and the energy of adjacent odd-A $1g_{9/2+}$ proton states. This suggests that the 0_2^+ states contain an intruder component across the same gap as the $1g_{9/2+}$states. In contrast, the $1g_{9/2+}$ states energies in Rb and Y indicate that the 0_2^+ state in Kr and Sr belongs to the normal configuration and that the 2p-2h excitations contribute to the ground state. This conclusion is in line with experimental evidence[29,30] that in both Ge and Se a nearly spherical ground state coexists with a deformed 0_2^+, whilst in Kr and possibly Sr the situation is reversed. The $g_{9/2+}$ particle couples mostly to the core state involving the same orbital. According to our calculation these states are of triaxial shape with positive β_2. This is close to predictions of the deformed shell model[19] on the shape of 0_2^+ states in Se isotopes and experimental data on Kr and Sr ground states.

The present interpretation is in line with recent calculation of Duval et al. [31] who performed IBM2 plus configuration mixing calculations on $^{68-76}$Ge isotopes and showed, that the assumption that 0_2^+ states are 2p-2h excitations to $1f_{5/2}$ and $1g_{9/2}$ orbital mixed with states where the proton pairs are primarily distributed over the $2p_{3/2}$ orbital gives a very good agreement with experimental data, including two-neutron transfer reaction cross sections. Also, Wood [32] suggested similar nature of some excited 0+ states in the A=80 region.

Therefore we conclude that there is an experimental evidence supported by theoretical calculation that the deformation in A=80 region in both, odd-A and even-even nuclei, is mostly due to presence of the intruder 1g orbital and proton-neutron interaction betwen 1g proton and 1g neutron orbitals. Neverthelless, more experimental and theoretical effort is needed to give a conclusive answer to the interesting possibility of existence of intruder states in the Z=40 region, responsible for sudden increase of deformation.

References
1. H.P.Hellmeister,E.Schmidt,M.Uhrmacher,R.Rascher,K.P.Lieb and D.Pantelica, Phys.Rev.C17,2113,(1978)
2. B.Heits et al. Phys.Rev.C15,1742,(1975)
3. J.Heese et al.,Phys.Rev.C36,2409,(1987)
4. L.Luhmann,M.Debray,K.P.Lieb,W.Nazarewicz,B.Wormann,J.Eberth and T.Heck, Phys.Rev.C31,828,(1985)
5. H.Schafer,A.Devald,A.Gelberg,U.Kaup,K.O.Zell and P. von Brentano, Z.Phys. A293,85,(1979)
6. C.J.Lister et al. Phys.Rev.C28,2127,(1983)
7. J.Panqueva et al., Nucl.Phys.A389,424,(1982)
8. J.Panqueva et al., Nucl.Phys.A376,367,(1982)
9. W.Gast,K.Dey,A.Gelberg,U.Kaup,F.Paar,R.Richter,K.O.Zell and P. von Brentano, Phys.Rev.C22,469,(1980)
10. L.Luhmann,K.P.Lieb,A.Moussavi-Zarandim,J.Panqueva and J.Sau, Z.Phys.a313,297,(1983)
11. U.Lenz, private communication,1988
12. C.J.Lister et al., J.Phys.G11,969,(1985)
13. C.J.Lister et al.,submitted for publication in Z.Phys.A, 1988
14. D.Bucurescu,G,Constantinescu,M.Ivascu,N.V.Zamfir,A.Avrigeanu and D.Cutoiu, J.Phys.G7,667,(1981)
15. C.A.Fields et al., Z.Phys.A295,365,(1980)
16. N.J.Stone,C.J.Ashworth,I.S.Grant.A.G.Griffiths,S.Ohya,J.Rikovska and P.M.Walker, in this Proceedings
17. I.Berkes et al., in Nuclei far from stability, ed.I.S.Towner, Rosseau Lake, Ontario,Canada,1987
18. P.Federman and S.Pittel, Phys.Rev. C20,820,(1979)
19. W.Nazarewicz and T.Werner, in this Proceedings
20. M.Sambataro and G.Molnar, Nucl.Phys. A376,201,(1982)
21. M.Sakai, Atomic Data and Nuclear Data Tables, 31,399, (1984)
22. S.Raman et al. Atomic Data and Nuclear Data Tables, 36,1,(1987)
23. P.Raghavan, Table of Nuclear Moments, private communication and submitted to ADNDT, 1988
24. S.E.Larsson,G.Leander and I.Ragnarsson, Nucl.Phys.A261,77,(1976)
25. H.Toki and A.Faessler, Nucl.Phys.A253,231,(1975)
26. P.Moller and J.R.Nix, Atomic Data and Nuclear Data Tables, 26,165,(1981)
27. K.Heyde et al, Nucl.Phys.A466,189, (1987)
28. A.M.Van den Berg et al., Nucl.Phys. A379, 239, (1982)
29. S.Matsuki et al., Nucl.Phys. A370, 1, (1981)
30. R.B.Piercey et al., Phys.Rev.Lett. 47,1514,(1981)
31. P.D.Duval,D.Goute and M.Vergnes, Phys.Lett. 124B,297,1983
32. J.Wood in Proc.Int.Symp. on In-Beam Nuclear Spectroscopy, Debrecen,1984

Particle-Rotor Calculations in the A = 100 Region

P.B. Semmes

Dept. of Mathmatical Physics, Lund Inst. of Technology,
P.O. Box 118, S-221 00 Lund, Sweden

Abstract: An odd nucleon can be a sensitive probe of collective core structure, and thus studying the properties of odd-mass nuclei often give direct information on nuclear shapes. However, the available experimental information for the deformed odd-mass nuclei in the A=100 region is seriously limited in at least one respect: The lack of transition multipolarity measurements. As a result, spin and parity assignments have depended strongly on model considerations, and should be regarded with caution. Particle-rotor calculations (including E1, M1 and E2 transition rates) have been made for Y and Nb nuclei to study how much can be learned from the currently available data, and whether there is any evidence for triaxial shape.

Data for deformed odd-mass nuclei with Z~40, A~100 have become available in recent years through decay studies of separated fission products (reviewed in [Me85a]), and through such studies clear insights into the nuclear shape and single-particle structure are possible. The most direct measures of these quantities are provided by static and dynamic E2 and M1 moments, but such detailed information is not generally available. In many cases, only level energies and transition intensities are clearly established; spin and parity assignments are often uncertain and based on model dependent considerations.

For nearly pure rotational bands, the bandhead spin and the underlying Nilsson orbital can often be identified by a "first order" analysis. This consists of fitting the level energies with the expression [Bo75]

$$E(I,K) = E_0 + AI_x^2 + BI_x^4 + A_{2K}(-1)^{I+K}(I+K)!/(I-K)! \qquad (1)$$

and extracting the quantity $|(g_K-g_R)/Q_0|$ from the relative intensities of $\Delta I=2$ and $\Delta I=1$ transitions within the band. This quantity is related to the E2/M1 mixing ratio δ^2 by [Bo75]

$$\left|\frac{g_K-g_R}{Q_0}\right| = \frac{0.933 \, E_\gamma}{\sqrt{(I+1)(I-1)}} \frac{1}{|\delta|} \qquad (2)$$

which remains constant within a pure rotational band. The intrinsic g-factor g_K is characteristic of the particular Nilsson orbital and is given by

$$g_K = g_\ell + \frac{1}{K}(g_\ell - g_s) \langle s_z \rangle \tag{3}$$

A representative Nilsson diagram for protons in this mass region is shown in Fig. 1, and g_K parameters for a few orbitals expected near the Fermi level for $\varepsilon=0.30$ and $Z=39$ are given in Table 1. Such considerations have been extremely valuable, and a number of assignments have been made in this way (e.g., [Wo83], [Me85b], [Sh84]).

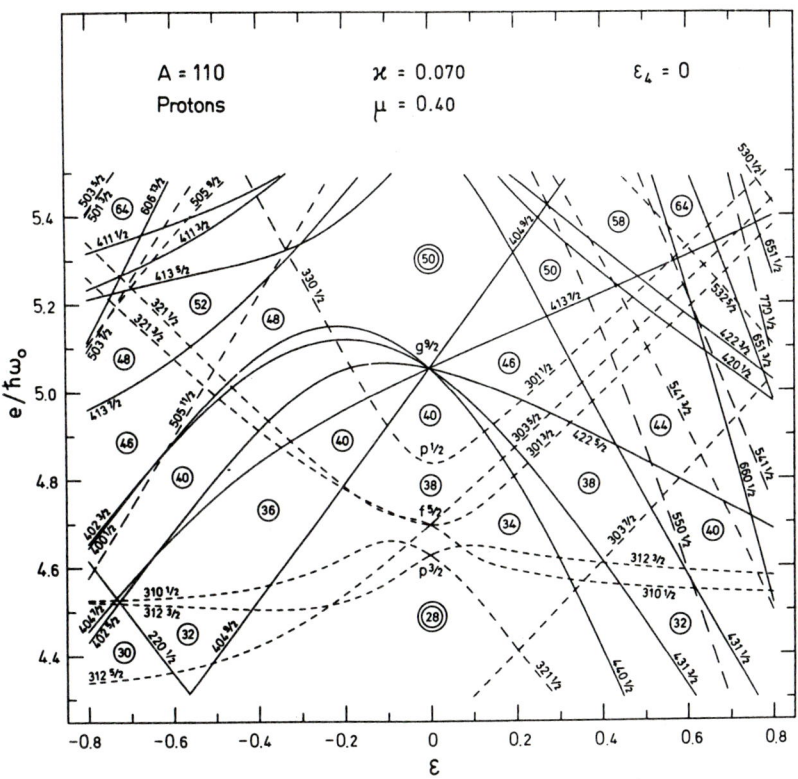

1. Representative Nilsson diagram for protons in this mass region.

Orbital	$\langle s_z \rangle$	g_K	$|(g_K - g_R)/Q_0|$ (b^{-1})
5/2+[422]	0.39	1.45	0.32
3/2+[431]	0.31	1.60	0.37
3/2-[301]	0.45	1.87	0.45
5/2-[303]	-0.44	0.48	0.03

Table 1. Intrinsic g_K factors calculated for $\varepsilon=0.30$ and $g_s=0.7 \times g_s$(free); Q_0 uses $Z=39$.

However, such first-order analysis is much less useful when strong interband transitions and strong perturbations of the rotational patterns are observed (e.g. through Coriolis mixing or axial asymmetry). In such cases, more detailed calculations based on the particle-rotor model are required. Furthermore, strong interband transitions have been observed in ^{99}Y [Pe85] (Fig. 2) and ^{101}Y [Pe87] (Fig. 3) that have been interpreted as E1, even though no multipolarity measurements were made. This contribution explores the reliability of E1 calculations in the particle-rotor model, and just how much can be reliably deduced from the currently existing data for deformed odd-Z nuclei in this region. The calculations described below use the triaxial particle-rotor model of Ref. [La78]. The details of the E1 calculations for this model are given in the Appendix.

Consider first the E1 rates, for which the lifetime measurements in ^{103}Nb [Se84] provide the most direct information. The data can be summarized as follows: three rotational bands were observed at low energy, and were assigned to the Nilsson orbitals 3/2[301], 5/2[303] and 5/2[422]. No interband transitions were reported. These assignments were based on first order considerations, i.e. by fitting the energies to the rotational formula and by comparing extracted and calculated values for $|(g_k-g_R)/Q_o|$ within each band. The sensitivity of calculated E1 rates on the details of the model is illustrated in Table 2, where the experimental halflives of the negative parity bandheads are compared with Nilsson + pairing estimates, i.e. no mixing between Nilsson orbitals is considered. The results marked "A" were obtained with $\kappa=0.070$, $\mu=0.40$ and $\varepsilon=0.30$ (as in Ref. [Se84]) while the "B" results used $\kappa=0.067$, $\mu=0.53$ and $\varepsilon=0.30$ as suggested for ^{99}Y [Wo85]. Both calculations correspond to a pairing strength $G \cdot A=19.2$ MeV (about 81% of the usual value) which gave a pairing gap of $\Delta=0.72$ MeV in each case. The intrinsic matrix elements are corrected for the stretching of the basis as described in the Appendix (the correction is ~ 10% here; for comparison, the matrix elements without the correction are 1.52×10^{-2} e·fm and 7.26×10^{-3} e·fm for set "A"). Comparison of the calculated results shows that the intrinsic matrix elements change by ~30% in the two cases, which disagrees with Ref. [Wo85] where the GE1 parameters were taken to be the same for ^{99}Y and ^{103}Nb. Furthermore, the pairing factors change tremendously and hence so do the calculated halflives. In set A, the two negative parity orbitals are on opposite sides of the Fermi level with the 5/2+ orbital slightly above λ, while for set B λ is nearly between the 5/2+ orbital (below λ) and the two negative parity orbitals (above λ). Thus for set B, a near cancellation occurs in the pairing factor, resulting in a tremendous sensitivity of the E1 rates on the details of the pairing calculation. This suggests that these calculated E1 transition rates can be quite unstable with respect to uncertainties in the single-particle and pairing parameters, and probably should not be trusted as much more than order of magnitude

335

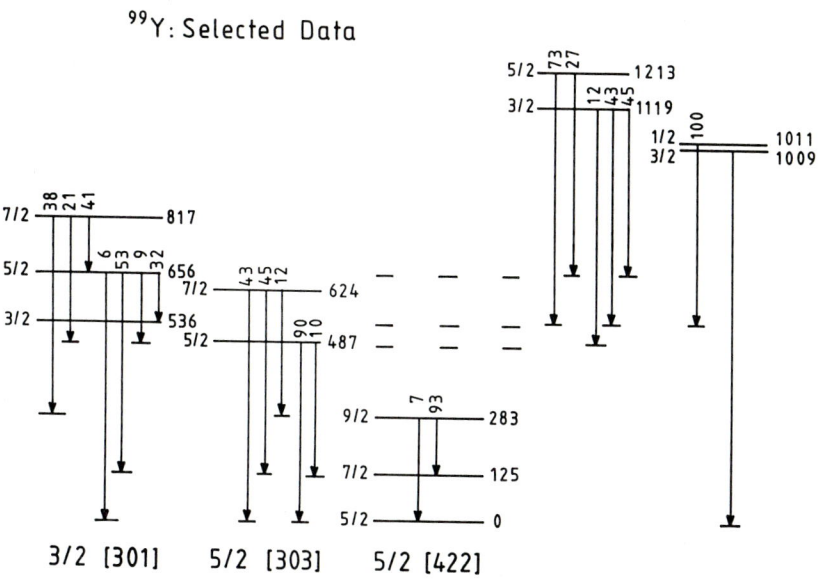

2. Main features of the experimental level energies and transition intensities for ^{99}Y [Me85] [Pe85].

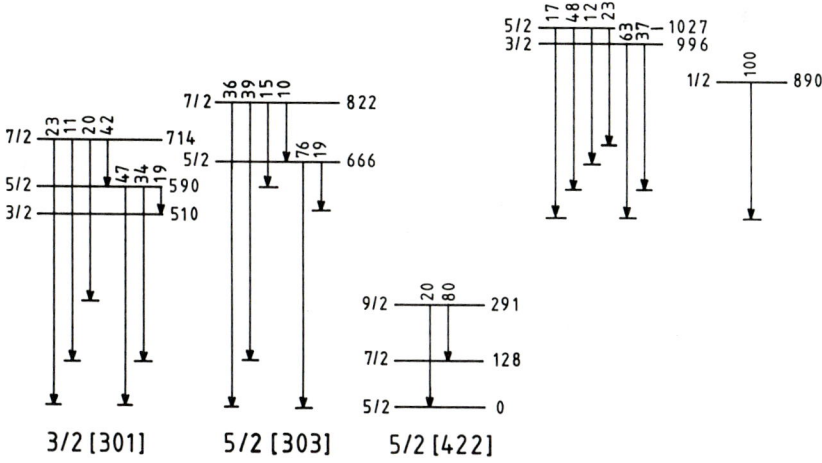

3. Main features of the experimental level energies and transition intensities for ^{101}Y [Pe87].

estsimates. In particular, the suggested procedure [Wo85] of extracting ΔK=0 and ΔK=±1 E1 enhancement factors from the known ^{103}Nb lifetimes for use in subsequent Y calculations seems most dubious.

Transition	3/2[301] → 5/2[422]	5/2[303] → 5/2[422]
Energy (keV)	248	166
Exp. $t_{1/2}$ (ns)	2.0(6)	4.7(5)
Intrinsic matrix element (e fm)	1.45 x 10^{-2} (A)	8.05 x 10^{-3} (A)
	1.85 x 10^{-2} (B)	1.08 x 10^{-2} (B)
uu'-vv'	-0.103 (A)	0.249 (A)
	0.023 (B)	-0.035 (B)
Calc $t^{\gamma}_{1/2}$(ns)	13 (A)	34 (A)
	152 (B)	1000 (B)
F	0.16 (A)	0.14 (A)
	1.3 x 10^{-2} (B)	4.7 x 10^{-4} (B)

Table 2. Comparison of experimental and calculated lifetimes in ^{103}Nb. The hindrance factor F is the ratio of the experimental and calculated halflives.

In view of the uncertainties involved in the E1 calculations, it is worthwhile to review the parity assignment of the K=3/2 band beginning at 536 keV in ^{99}Y. This band was originally assigned [Pe85] as negative parity built on the 3/2[301] orbital, but IBFM/PTQM calculations [Pf86] suggest a positive parity K=3/2 assignment (analogous to the 3/2[431]) instead. It is interesting to note that the g_K factors for these two Nilsson orbitals are very similar (see Table 1), and so it is not possible to distinguish between the two alternatives within a first order analysis. Instead, more detailed calculations are necessary to compute the interband/intraband transition intensities for the two alternatives and compare with experimental results. (Such an analysis has been made previously [Wo87] but with the inclusion of E1 enhancement factors. In the following no such adjustable enhancement factors are included.) The results of such calculations are summarized in Table 3 which lists the relative gamma-ray intensities for transitions to the 5/2+[422] band and to the 5/2[303] band. Both calculations use Δ=0.69 MeV, an effective g_s that is 70% of g_s (free) and a Coriolis attenuation of 0.7 for the positive parity states. The column marked "A" uses parameters taken from [Wo85], i.e. κ=0.067, μ=0.53 and ε=0.34, while column B uses the Nilsson parameters of Fig. 1 with ε=0.32. At this deformation, the computed 3/2+[431] bandhead lies ~580 keV above the 5/2+ groundstate.

I_i	E_i	I_f	E_f	I_γ(A)	I_γ(B)	Expt.
3/2	536	5/2+	0	100 E1	100 M1	100
5/2	654	5/2+	0	26 E1	37 M1	6
		7/2+	125	51 E1	62 M1	53
		5/2-	487	5 E1	<<1 E1	9
		3/2	536	18 M1	<1 M1	32
7/2	814	5/2+	0	2 E1	10 M1	(<1)
		7/2+	125	19 E1	44 M1	(<6)
		9/2+	283	22 E1	45 M1	38
		5/2-	487	15 E1	<<1 E1	20
		5/2	654	32 M1	1 M1	41

Table 3. Comparison of experimental and calculated transition intensities for the K=3/2 band in ^{99}Y, for a 3/2+[431] assignment ("A") or a 3/2-[301] assignment ("B").

At first glance, the agreement between experiment and theory may seem comparable in the two cases. The agreement can be made essentially perfect for the 3/2[301] assignment if E1 enhancement factors are included, but that is not relevant. The crucial difference is that the observed intense transitions within the K=3/2 band cannot be explained with the 3/2+[431] assignment, and that this failure is independent of the details of the calculation, e.g. the strength of the Coriolis mixing. The fast interband M1 transitions are driven by the large j_+ matrix elements between the 3/2+[431] and 5/2+[422] orbitals, and thus the interband transitions dominate the intraband transitions even without Coriolis mixing. This result is also independent of the deformation, since both Nilsson orbitals are essentially pure $g_{9/2}$ states (high-j unique parity orbitals) and remain so for reasonable deformations. Thus, the 3/2+[431] assignment is ruled out, leaving the 3/2-[301] as the only plausible candidate. (This result agrees with the conclusions of Ref. [Wo87].) For the latter assignment, the resemblance between the calculated (E1 and M1) and experimental transition intensities is reassuring, though not convincing due to the uncertainties in the E1 estimates. The total transition rates differ considerably for the two assignments; e.g. for the 536 keV bandhead, the partial halflife is estimated at 0.3 ps for the positive parity assignment and 22 ps for negative parity. The most direct test of course, is to measure transition multipolarities.

Within a symmetric rotor description for ^{99}Y, the assignments of the 3/2[301] and 5/2[303] negative parity bandheads (at 536 and 487 keV respectively) exhaust the negative parity Nilsson orbitals expected at low energy. For parameters near those of Ref. [Wo85], the next available negative parity orbital, the 1/2[301], should produce a 1/2-bandhead around 1.1 MeV above the 5/2-[303], or about 1.6 MeV in the experimental spectrum. For the positive parity states, the 3/2+[431] and 1/2+[431] bandheads are expected around 1 MeV or so above the 5/2+ ground state. Of course, these estimates are sensitive to the Nilsson parameters, deformation, and so on, but the positive parity bandhead positions show a much greater sensitivity in the calculations than do the negative parity bandheads. This is easily understood from Fig. 1 since the relevant negative parity orbitals have very similar slopes while the positive parity orbitals of interest have differing slopes. Deviations from the rigid symmetric rotor description assumed thus far should result in extra states appearing at low energy. Returning to the experimental data for ^{99}Y (Fig. 2), the 1/2 and 3/2 states placed at 1011 and 1009 keV respectively, are each observed to decay by a single transition as indicated. This is consistent with the 1/2[431] assignment made in Ref.[Wo85], and higher lying members of this band are also observed. The 3/2 and 5/2 states at 1119 and 1213 keV cannot be so easily explained, however. These states decay exclusively to the 3/2[301] and 5/2[303] negative parity bands, which is completely inconsistent with a 3/2[431] band. These are most naturally explained as negative parity states, but then imply a departure from the rigid symmetric shape. The effects of triaxiality have been investigated and are summarized in Fig. 4, which shows the energies and calculated decay intensities for the lowest few levels in each band together with their main K components. The calculated E1 intensities (without any enhancement factors) are expected to be rough estimates, and so only the strongest calculated E1 transition is shown for each level. Thus, for example, a 19% E1 branch from the 5/2-(567 keV) to the 1/2+ (125 keV) state is not shown. All calculated M1 and E2 intensities greater than 1% of the total for each level are shown. The significance of the calculation is that an effective triaxiality as small as $\gamma=10°$ can mix in enough of the 1/2-[301] orbital with those nearer the Fermi level to bring an extra negative parity band near the observed position (relative to the other negative parity states; no attempt has been made to adjust the overall separation between the positive and negative parity states). The decay intensities of the 3/2 member of this band are well reproduced, but the 5/2 member is not so well described. The 1/2 bandhead might correspond to the experimental 1/2 level at 907 keV. This small asymmetry also does not destroy the good agreement for the lower states; in particular, the strong intraband transitions in the $\bar{K}$=3/2 band remain. The position of the extra band is sensitive to both γ and the Coriolis attenuation; for Fig. 4 an attenuation of 0.90 was used. If no attenuation is introduced, a slightly larger γ is required (~13°), but the results remain essentially unchanged. Furthermore, similar structures have very recently been observed in ^{101}Y [Pe87], and calculations with $\gamma=10°$ and moderate Coriolis attenuation are shown in Fig. 5. The agreement with the data is very good for this $\bar{K}$;=1/2 band, again suggesting a small effective triaxiality. Also it is interesting to note that the reversal of the 3/2[301] and 5/2[303] bandheads between

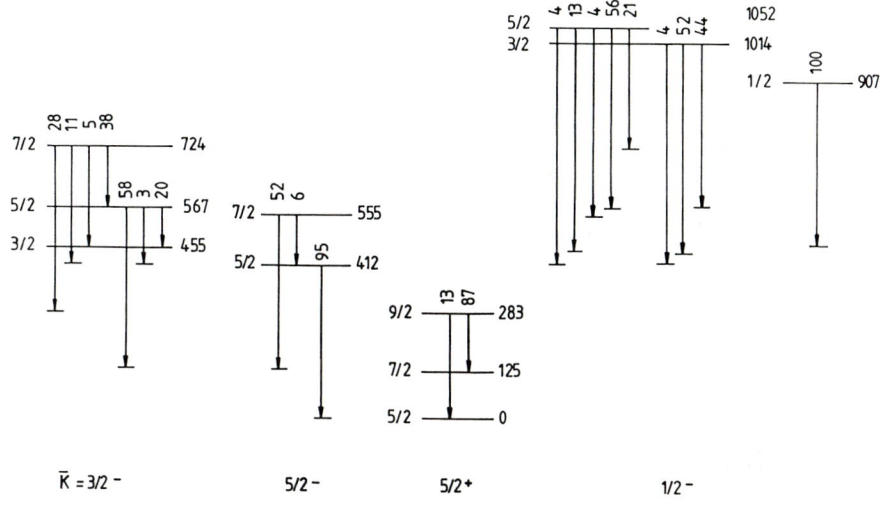

4. Calculated level energies and transition rates for ^{99}Y using $\kappa=0.067$, $\mu=0.51$, $\varepsilon=0.34$, $\gamma=10°$, $\Delta=0.71$ MeV, $\chi=0.90$ and $E_{2+}=0.131$ MeV. Only the strongest E1 transition is shown from each level. The dominant K component is shown for each band.

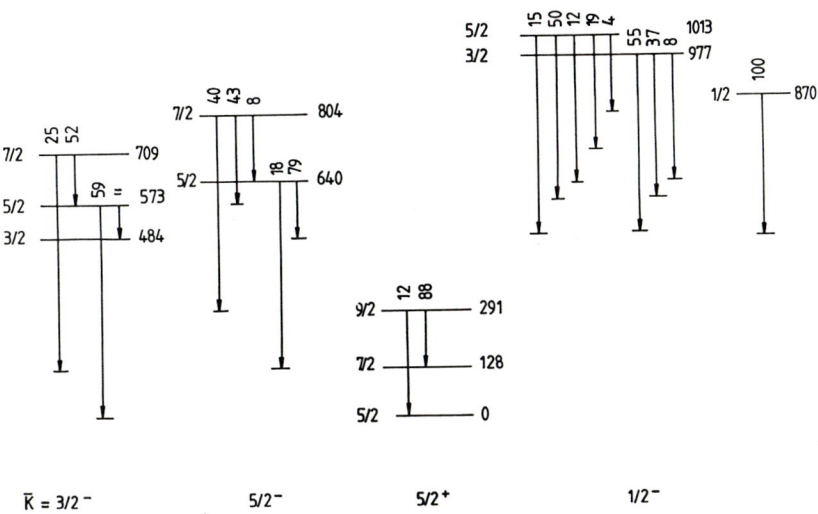

5. Similar to Fig. 4, but for ^{101}Y using $\kappa=0.068$, $\mu=0.47$ and all other parameters are the same.

^{99}Y and ^{101}Y cannot be explained by small deformation charges but requires a decrease in μ which inverts the $f_{5/2}$ and $p_{3/2}$ spherical states. Finally, the author notes that a revised version of Ref. [Pe87] received after completion of this work, also considers a 1/2[301] assignment for these states.

In conclusion, E1 transition rate calculations are subject to great uncertainties, and cannot be relied upon to make multipolarity assignments. The 3/2-[301] assignment in ^{99}Y is confirmed, and possible evidence for a small effective triaxiality in ^{99}Y and ^{101}Y is presented. More experimental measurements of ground states spins and transition multipolarities are badly needed in this mass region.

APPENDIX

For states expanded in the strong coupling basis of ref. [La78] with expansion coefficients $a_K^{I\nu}$, the reduced matrix element of the E1 operator is given by

$$<I'\|M(E1)\|I> = \sqrt{2I+1} \sum_{K'K} a_{K'}^{I'\nu'} a_K^{I\nu}(uu'-vv')$$

$$[\delta_{K'K} <IK\ 1\ 0|I'K'> <\chi_\nu'|e'r\ Y_{10}|\chi_\nu>$$

$$+ (-1)^{I'+1/2} \sum_{m=\pm 1} <IK\ 1\ -m|I'\ -K'> <\chi_\nu'|e'r\ Y_{1m}|\tilde{\chi}_\nu>] \quad \text{(A-1)}$$

where the pairing factor $uu'-vv'$ is calculated in the standard BCS theory without considering blocking effects. The effective charge e' includes the recoil correction and is given by $(\Delta-Z/A)e$, $\Delta=1$ for odd Z, $\Delta=0$ for odd N. The adiabatic single-particle states are eigenfunctions of the triaxially deformed Nilsson Hamiltonian, and are written as expansions

$$\chi_\nu = \sum_{\alpha j\Omega} c_{\alpha j\Omega}^\nu \psi_\Omega^{\alpha j} \qquad \alpha = N, \ell \quad \text{(A-2)}$$

subject to the restriction $\Omega = \ldots, -3/2, 1/2, 5/2, \ldots$. Their conjugate states are

$$\tilde{\chi}_\nu = \sum_{\alpha j\Omega} (-1)^{j-\Omega} c_{\alpha j\Omega}^\nu \psi_{-\Omega}^{\alpha j} . \quad \text{(A-3)}$$

If these single-particle states are written in a spherical basis, then the intrinsic matrix elements are easily shown to be

$$<\chi_\nu'|e'r\ Y_{10}|\chi_\nu> = \sqrt{\frac{3}{4\pi}} e' \sum_{\alpha'j'} \sum_{\alpha j\Omega} c_{\alpha'j'\Omega}^{\nu'} c_{\alpha j\Omega}^\nu <N'\ell'|r|N\ell>$$

$$\sqrt{(2j+1)/(2j'+1)} <j\ \Omega\ 1\ 0|j'\ \Omega> <j\ -1/2\ 1\ 0|j'\ -1/2> \quad \text{(A-4)}$$

and, for $m=\pm 1$,

$$<\chi_\nu'|e'r\ Y_{1m}|\tilde{\chi}_\nu> = \sqrt{\frac{3}{4\pi}} e' \sum_{\alpha'j'} \sum_{\alpha j\Omega} c_{\alpha'j'-(\Omega-m)}^{\nu'} c_{\alpha j\Omega}^\nu$$

$$(-1)^{j'+1/2} <N'\ell'|r|N\ell> \sqrt{(2j+1)/(2j'+1)} <j\ \Omega\ 1-m|j'\Omega-m>$$

$$<j\ -1/2\ 1\ 0|j'\ -1/2> \quad \text{(A-5)}$$

If the single-particle states are defined in a stretched coordinate system, then the E1 operator must likewise be transformed. In terms of the oscillator frequencies ω_x, ω_y, ω_z [La73] this gives

$$rY_{10}(\text{spherical}) = b\sqrt{\frac{\omega_0}{\omega_z}} \rho Y_{10}(\text{stretched}) \quad \text{(A-6)}$$

and for m=±1

$$rY_{1m} = b\sqrt{\frac{\omega_0}{\omega_x}}\, \rho Y_{1m} + b\left(\sqrt{\frac{\omega_0}{\omega_y}} - \sqrt{\frac{\omega_0}{\omega_x}}\right)\frac{1}{2}\rho(Y_{11}+Y_{1-1}) \quad (A-7)$$

where b is the oscillator length $\sqrt{\hbar/m\omega_0}$, and the $Y_{\ell m}$ on the left hand side refer to the spherical coordinate system while those on the right refer to the stretched coordinate system. The matrix elements of the E1 operator in the stretched single-particle basis are obtained directly by combining the above results.

References

[Bo75] A. Bohr and B. Mottelson, Nuclear Structure, vol. 2.

[La73] S.E. Larsson, Physica Scripta $\underline{8}$ (1973) 17.

[La78] S.E. Larsson, G. Leander and I. Ragnarsson, Nucl. Phys. $\underline{A307}$ (1978) 189.

[Me85a] R.A. Meyer, Hyperfine Interactions $\underline{22}$ (1985) 385.

[Me85b] R.A. Meyer et al., Nucl. Phys. $\underline{A439}$ (1985) 510.

[Pe85] R.F. Petry et al., Phys Rev. $\underline{C31}$ (1985) 621.

[Pe87] R.F. Petry et al., preprint (1987).

[Pf86] B. Pfeiffer et al., Z. Phys. $\underline{A325}$ (1986) 487.

[Se84] T. Seo et al., Z. Phys. $\underline{A315}$ (1984) 251.

[Sh84] K. Shizuma et al., Z. Phys. $\underline{A315}$ (1984) 65.

[Wo83] F.K. Wohn et al., Phys. Rev. Lett. $\underline{51}$ (1983) 873.

[Wo85] F.K. Wohn, John C. Hill and R.F. Petry, Phys. Rev. $\underline{C31}$ (1985) 634.

[Wo87] F.K. Wohn, Phys. Rev. $\underline{C36}$ (1987) 1204.

Large Inertial Moments, Pairing Reduction and the Effects of Coriolis Mixing on Rotational Bands in Odd-Z Deformed A ≃ 100 Nuclei

F.K. Wohn[1], J.C. Hill[1], R.F. Petry[2], and R.L. Gill[3]

[1] Ames Laboratory, Iowa State University, Ames, IA 50011, USA
[2] University of Oklahoma, Norman, OK 73019, USA
[3] Brookhaven National Laboratory, Upton, NY 11973, USA

Highly deformed A≃100 nuclei with A=99-102 and N⩾60 are being studied using the TRISTAN facility at BNL. The structures of the rotational bands reveal some unusual features, such as the anomalously large moments of inertia that suggest reduced pairing correlations among the valence nucleons. The apparent reduction in the strength of the pairing interaction could be a result of p-n interactions among the valence nucleons. However, it would be inappropriate to form any firm conclusions concerning either the amount of pairing reduction or the importance of p-n interactions without first characterizing the proton and neutron orbitals occupied by the valence nucleons in the odd-A deformed nuclei.

The main thrust of the present paper is the characterization of the valence protons in the odd-A deformed A≃100 nuclei. Bands have been proposed in ^{99,101}Y [1,2] and ^{101,103}Nb [3,4]. For the Y nuclei studied at TRISTAN, assignments of appropriate Nilsson orbitals for observed rotational bands were made in a manner consistent with experimental results on the observed rotational bands, including β-feeding and γ-decay patterns. The ^{101}Y levels are highlighted here due to our more extensive results for ^{101}Y. The importance of the Coriolis interaction in determining band characteristics is stressed. Its effects on the core inertial moments and deductions of pairing reduction are discussed. Finally, a comparison of the ordering of Nilsson orbitals of the four known deformed odd-A Y and Nb nuclei suggests an influence of the p-n interaction on the ordering of Nilsson orbitals for nuclei in this region.

THE ROTATIONAL STRUCTURE OF ^{101}Y

Our study of the β decay of ^{101}Sr resulted in a level scheme for ^{101}Y with 96 γ rays placed among 34 levels. The first 14 levels of ^{101}Y, which lie below 1.1 MeV, are identified as members of the four rotational bands based on Nilsson orbitals 5/2[422], 3/2[301], 5/2[303] and 1/2[301]. These 4 bands are shown in Fig. 1. Above 1.2 MeV lies a group of 6 levels with I⩽7/2 that have properties expected for the levels of 1/2[431] or 3/2[431] bands, but it is not possible to

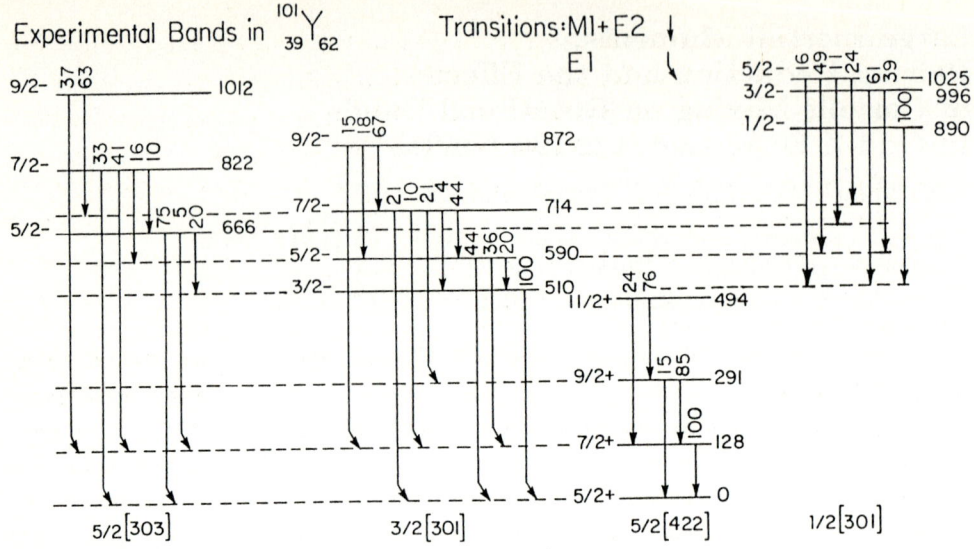

Fig. 1. Experimental bands and relative γ-ray intensities in ^{101}Y.

make unambiguous assignments of all 6 levels. Fig. 2 presents two ways to group these levels, depending upon the decoupling parameter a_1 of the K=1/2 band.

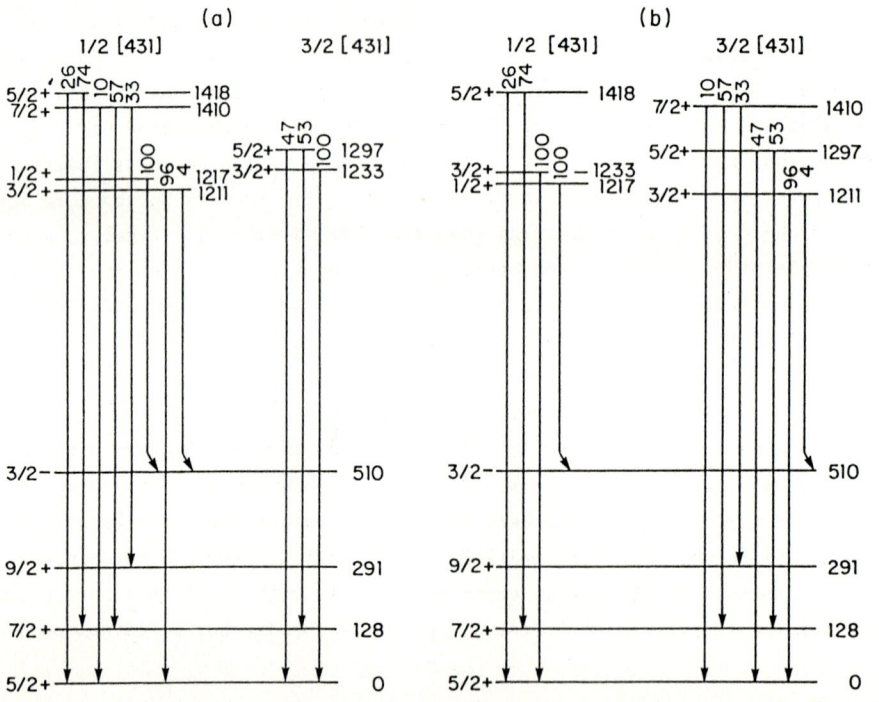

Fig. 2. Two choices for assigning experimental levels to 1/2[431] and 3/2[431] bands in ^{101}Y: (a) with a_1=-1.10; (b) with a_1=-0.74.

Detailed properties of the bands are given in [2]. Only a few features are noted here. E2 γ rays are weak compared to competing E1 and/or M1 γ rays. Only ΔI=2 collective intraband E2 γ rays are seen. Low logft, allowed β decay occurs only to the $5/2^+$ ground state, as expected for β decay of a $3/2^+$ 3/2[411] ground state of ^{101}Sr. γ-decay patterns for the 5/2[422], 1/2[301], and 3/2[431] bands are determined mainly by dipole Alaga rules. However, the 5/2[303], 3/2[301] and 1/2[431] bands have γ-decay patterns with large deviations from Alaga rules. For the latter two, the deviations are very large but, as is shown in [2], these deviations are consequences of Coriolis mixing. The most striking effect occurs for the 1/2[431] band. In this band, a Coriolis-admixed component of 3/2[431] as small as 0.5% is still large enough to give a γ-decay pattern in the K=1/2 band that is dominated by strong ΔK=1 M1 γ rays to the ground band.

As discussed in detail in [2], the simple particle-rotor model provides an excellent reproduction of the level energies, intraband M1+E2 and interband M1 and E1 γ rays. The amount of Coriolis mixing was determined by fitting energy levels. Subsequent calculations of γ-decay patterns confirmed the deduced mixing. Fig. 3 gives results of calculations of the four lowest bands in ^{101}Y.

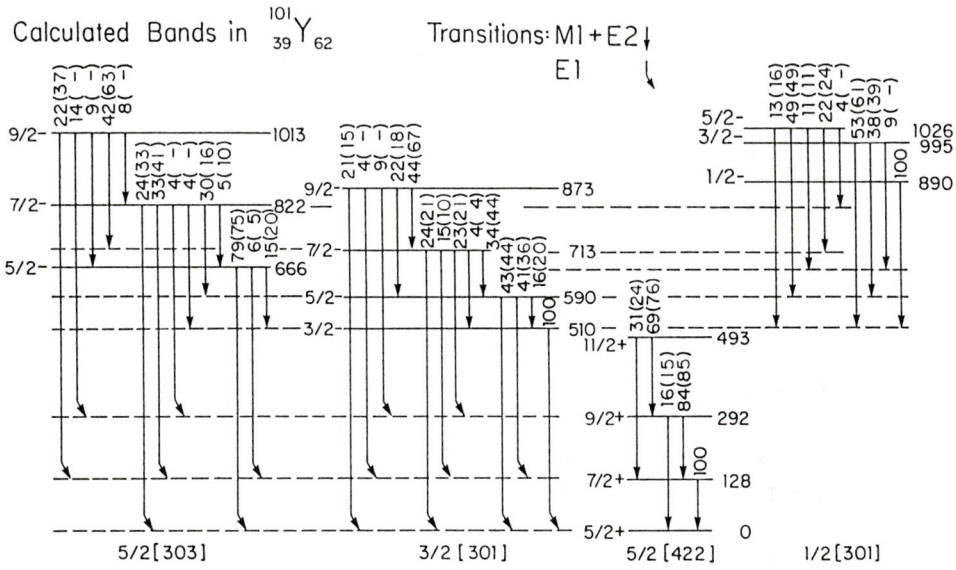

Fig. 3. Calculated bands and relative γ-ray intensities in ^{101}Y. For ease of comparison, the experimental intensities from Fig. 1 are given in parentheses.

Calculations of the observed energy and γ-decay patterns were found to be very insensitive to the choice of the Nilsson parameters κ, μ and ε. Failure to include Coriolis mixing, however, resulted in an inability to reproduce the patterns observed. Furthermore, the γ-decay patterns were found to play a crucial role in deducing a viable assignment of a Nilsson orbital to a band. Detailed discussions of these points are given in [2].

COMPARISON OF BANDS IN ^{99}Y AND ^{101}Y

Nilsson orbital	^{99}Y				^{101}Y			
	Bandhead (keV)	$a=\hbar^2/2\mathcal{J}$ (keV)	a_1	γ-decay pattern	Bandhead (keV)	$a=\hbar^2/2\mathcal{J}$ (keV)	a_1	γ-decay pattern
5/2[422]	0	17.9		M1>>E2	0	18.3		M1>>E2
3/2[301]	536	23.6		E1~M1>E2	510	16.0		E1~M1>E2
5/2[303]	487	19.6		E1>E2>M1	666	22.3		E1~M1>E2
1/2[301]	–	–	–	–	890	20.8	0.70	M1>>E1,E2
1/2[431]	1011	17.8	–1.05	M1>>E1,E2	1217	20.3 21.2	–1.10 –0.74	M1>>E1,E2

Table I. Experimental band properties for ^{99}Y [1] and ^{101}Y [2]. The decoupling parameter a_1 is given only for K=1/2. The 1/2[301] band is not known for ^{99}Y. Two sets of a and a_1 values are given for the 1/2[431] band in ^{101}Y.

Table I shows that ^{99}Y and ^{101}Y have similar general features. 5/2[303] is the second band in ^{99}Y but 3/2[301] is second in ^{101}Y. Coriolis matrix elements for normal-parity bands are weak and other normal-parity bands are well separated in energy, thus these two bands mix mainly with each other. The lower band of the two thus has the smaller energy spacing, as is evident from the empirically deduced inertial parameters a of Table I. This dependence of a on energy order provides unambiguous evidence of the Coriolis mixing of these bands. This mixing is greater in ^{101}Y since its levels of like I^π are closer together. The 1/2[301] band in ^{101}Y has little mixing, thus its a value is essentially the a_0 value for the even-even core. Further comparisons of the bands are given in [2].

3/2[431] bands are not shown in Table I. Fig. 2 shows that members of this band are uncertain for ^{101}Y. As for ^{99}Y, it now appears that no members of this band have been identified. In [1], levels at 1119 and 1213 keV were proposed as 3/2$^+$ and 5/2$^+$ members of a 3/2[431] band whose γ-decay patterns required a very large enhancement of E1 γ rays. From our results for ^{101}Y, it seems likely that these levels are instead 3/2$^-$ and 5/2$^-$ members of a 1/2[301] band. The proposal [6] of 3/2[431] for the 536-keV bandhead in ^{99}Y is also unacceptable [7], thus a 3/2[431] orbital remains unplaced in ^{99}Y. A restudy of ^{99}Y levels is planned at TRISTAN, focusing on bands above the three well-established lower bands in ^{99}Y.

MOMENTS OF INERTIA AND PAIRING REDUCTION

For both ^{99}Y and ^{101}Y the variations in the empirical inertial parameters a of Table I for each nucleus is readily explained by including Coriolis mixing in the particle-rotor model [1,2]. Then it is possible for each band (in a given Y nucleus) to have the same moment of inertia. The Coriolis interaction produces changes in the a values for the observed bands. The analysis of [1] gave a core inertial parameter a_0=21.8 keV for ^{99}Y, while that of [2] (with more bands) gave a_0=20.4 keV for ^{101}Y. These core values are essentially the same as those of the neighboring even-even nuclei. With Coriolis mixing taken into account, the band structures of the odd-A Y nuclei imply no change in the moment of inertia of the core due to the odd proton. This can be interpreted as indicating that there is no additional reduction in pairing correlations due to the odd proton.

In an earlier report [5], effects of the Coriolis interaction were assumed to be negligible. Observed increases of 10-15% in $\underline{a}$ values (of the lowest bands in deformed odd-A and odd-odd A≃100 nuclei) for each additional unpaired nucleon were interpreted as implying pairing reduction due to unpaired nucleons. For two unpaired nucleons, the $\underline{a}$ values are nearly equal to rigid-rotor values, suggesting that these bands could be nearly "pairing free" bands [5]. The present analysis shows, however, that neglect of Coriolis effects is invalid.

Even-even nuclei in this region do, however, appear to have weaker pairing correlations. The same even-even pairing, which is about 50-60% of the global Δ value of $12/\sqrt{A}$ MeV, is adequate for the core of ^{101}Y [2] as well as for ^{99}Y [1]. This may also be true for other odd-A A≃100 nuclei, but similar deductions (with Coriolis mixing included) of core inertial parameters have not yet been reported for odd-A deformed nuclei other than Y (nor for any odd-odd deformed nuclei).

NILSSON-ORBITAL ORDERING IN ^{99,101}Y AND ^{101,103}Nb

An unusual situation occurs here, as the ordering of the three lowest bands in ^{101,103}Nb [3,4] are the same as in ^{99,101}Y. For all four nuclei, the ground band is 5/2[422]. Bands two and three are the nearly degenerate bands 5/2[303] and 3/2[301]. The average bandhead energy of the latter two bands is 550 keV in ^{99,101}Y and 207 keV in ^{101,103}Nb. Thus the addition of a proton pair to Y merely causes a compression in the spacing of the three Nilsson bandheads. Since there is an energy gap at Z=38 in the Nilsson (or Woods-Saxon) scheme for deformations ε of 0.3-0.4, the location of the Fermi level for Z=39 or 41 is straightforward: for Y the Fermi level is closest to the first orbital above Z=38, whereas for Nb the Fermi level is closest to the second orbital above Z=38. The observed band orderings in these nuclei are quite different from this expected behavior.

In analyses of odd-A Y and Nb nuclei, observed band orderings were reproduced by using quite different values of the Nilsson parameter μ. A μ value of ≃0.5 was deduced [1,2] for ^{99,101}Y. (Such a μ value makes 3/2[301] and 5/2[303] orbitals nearly degenerate, and changes of less than ±0.02 suffice to give the observed ^{99,101}Y bandhead energies.) In the ^{101,103}Nb analyses [3,4], a μ value of ≃0.4 puts the 3/2[301] orbit below 5/2[422], giving a Fermi level nearest the 5/2[422] orbital, with the orbitals 5/2[303] and 3/2[301] about equidistant from the Fermi level. Such adjustments of μ are needed to explain the level ordering using the "standard" particle-rotor model, based upon a Hamiltonian that retains only the rotational core and the rotational interaction between the particle and core. BCS pairing (with or without blocking), the Coriolis interaction, and the "recoil" term (essentially $\langle j^2 \rangle$ matrix elements) can be used in the model. However, none of these terms can cause the unusual ordering observed for Y and Nb, thus a drastic change in μ was needed to reproduce the observed orderings. Such large changes in μ are not reasonable and they do not occur for the well-studied deformed rare earths [8]. There are a few rare earths for which the gound state is unchanged by adding a nucleon pair, but this invariably involves two nearly degenerate orbitals near the Fermi level that exchange places (or cross) as the nucleon pair is added. An example is $_{71}$Lu and $_{73}$Ta, where 9/2[514] and 7/2[404] orbitals cross, resulting in a 7/2[404] ground state for both Lu and Ta.

The p-n interaction appears to play an important role for A≈100 nuclei [9]. In the following an oversimplified, schematic calculation illustrates how a p-n interaction and a single set of Nilsson parameters could explain the ordering of these orbitals in Y and Nb. Using BCS with blocking, we postulate an energy E_{pn} that lowers the "chosen" 5/2[422] bandhead due to its attractive p-n interaction with positive-parity neutrons beyond N=50. To obtain average bandhead energies, we use μ=0.51 to make the 3/2[301] and 5/2[303] orbitals degenerate. V_+^2 is the BCS proton-pair occupation of the 5/2[422] orbital, taken to be 0.50 for the odd proton in the orbital. The effect is illustrated in Table II, where the single-particle, quasiparticle and level energies are given in keV. The level energies are just $E_{q.p.} - 2V_+^2 E_{pn}$, and the level energies shown are excitation energies.

Bandhead	99,101Y				101,103Nb			
	$E_{s.p.}$	$E_{q.p.}$	$2V_+^2$	E_{level}	$E_{s.p.}$	$E_{q.p.}$	$2V_+^2$	E_{level}
5/2[422]	203	167	1.000	0	252	51	1.000	0
3/2[301],5/2[303]	0	0	0.063	550	0	0	0.662	207

Table II. Illustration of bandhead "rearrangement" with an attractive p-n energy E_{pn} of 765 keV acting only on protons in the 5/2[422] orbital.

Thus it is possible for a fixed set of Nilsson parameters plus a p-n interaction to explain the observed Y and Nb orderings, as the preceeding schematic calculation shows. It would also be of interest to relate the above p-n energy to that deduced for $g_{9/2}$ protons and $g_{7/2}$ neutrons [10]. The purpose of this illustration is to focus attention on two important questions: Are odd-A A≈100 deformed nuclei affected by a p-n interaction? If so, how do we deduce its effects?

REFERENCES
1. F.K. Wohn, J.C. Hill and R.F. Petry, Phys. Rev. C 31, 634 (1985).
2. R.F. Petry et al., Phys. Rev. C (accepted).
3. A.-M. Schmitt et al., 1983 Annual Report, IKP, KFA Jülich.
4. T. Seo et al., Z. Phys. A315, 251 (1984). (See also these proceedings.)
5. L.K. Peker et al., Phys. Lett. 169B, 323 (1986).
6. B. Pfeiffer et al., Z. Phys. A325, 487 (1986).
7. F.K. Wohn, Phys. Rev. C 36, 1204 (1987).
8. W. Ogle et al., Rev. Mod. Phys. 43, 424 (1971).
9. P. Federmann and S. Pittel, Phys. Rev. C 20, 820 (1979).
10. G. Lhersonneau et al., in "Nuclei Far from Stability," ed. by I.S. Towner, AIP Conf. Proceedings 164, 393 (1988).

The Nuclear Structure of ^{101,103}Nb

T. Seo[1], R.A. Meyer[2], and K. Sistemich[3]

[1]Research Reactor Institute, Kyoto University,
 Kumatori-cho, Sennan-gun, Osaka, 590-04, Japan
[2]Lawrence Livermore National Laboratory, Livermore, CA 94550, USA
[3]Institut für Kernphysik, Kernforschungsanlage Jülich,
 Postfach 1913, D-5170 Jülich, Fed. Rep. of Germany

The lowest state energies of ^{101}Nb and ^{103}Nb with the odd proton occupying the orbitals of [422 5/2], [303 5/2] and [301 3/2] have been calculated and found to have deep minima at $\delta=0.26$ and 0.28, respectively. The low-lying level structures of the nuclei have been well reproduced by new Nilsson model + pairing + Coriolis coupling calculations. Reasonable agreement with experiments has also been obtained for the γ-ray transitions.

1. Introduction

These years special interest has been shown to the new region of nuclear deformation around $A=100$, and several rotational bands have been observed in odd-mass nuclei being assigned to new Nilsson orbitals [1]. One important problem in the theoretical analyses of these nuclei is how to choose the parameters in the Nilsson potential and we first pointed out the experimental basis of choosing the parameter μ for neutrons through the investigation of ^{103,105}Mo [2]. Subsequently, one of us proposed a new method [3] of describing the Nilsson potential with only three parameters for all mass-number regions. The usefulness of this method was corroborated by the agreement of the Nilsson diagram calculated by this model and the one [4] for protons independently obtained by fitting the experimental level energies of ^{99}Y.

In this report, we present the result of theoretical analyses obtained by applying the new method to the isotopes ^{101,103}Nb.

2. Particle-rotor model calculations

2.1. Static deformation

In the mass number region $A \approx 100$, we used values slightly different from those in the previous work [3] as to the parameters κ_0, κ_1 and ν_0:

$$\kappa_0 = 0.020, \quad \kappa_1 = 0.85, \quad \nu_0 = 0.060.$$

The lowest energy $\mathcal{E}_i$ of the system with the orbital i occupied by the last nucleon was calculated using the equation,

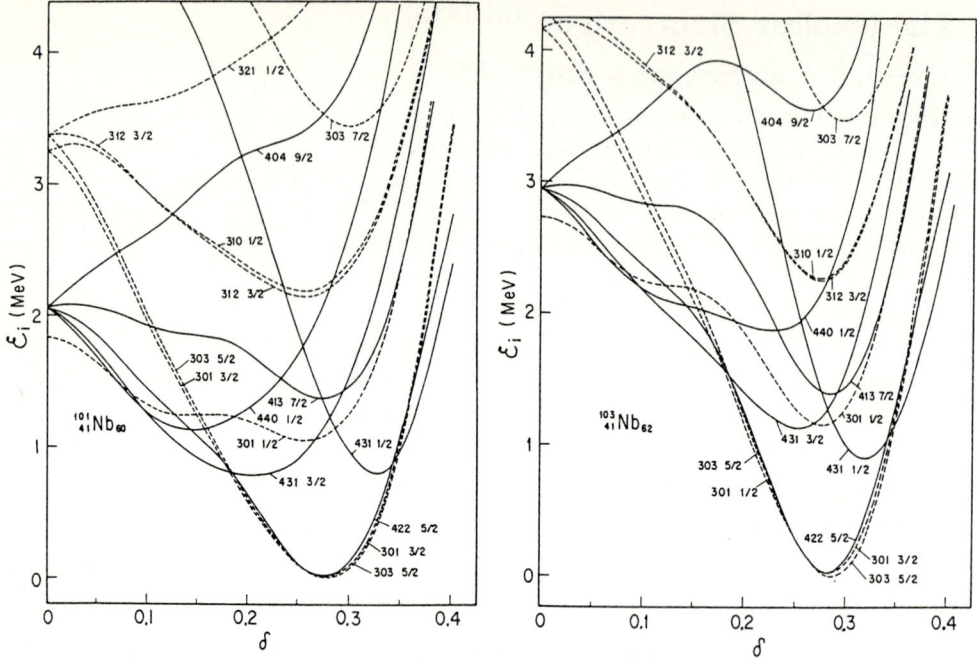

Fig. 1. The lowest energy of the system with an orbital [i] occupied by the last proton plotted vs. deformation.

$$\mathcal{E}_i = e_i + 2 \sum_{k \neq i} V_k^{(i)2}(e_k - \frac{G}{2} V_k^{(i)2}) - \Delta_i^2/G. \quad (1)$$

where e_i is the energy of the Nilsson orbital i, G the pairing strength, $V_k^{(i)}$ and Δ_i the occupation amplitude of the orbital k and the gap parameter with the orbital i blocked. Usually, the Strutinsky renormalization should be taken into account in such calculations, but for lighter nuclei ($A \lesssim 100$) eq. (1) would be reliable enough for the qualitative investigation of the nuclear deformability.

The calculated energies $\mathcal{E}_i$ vs. deformation are plotted in Fig. 1. The pairing parameters used in this calculation are $G_p=(25.9-23.2(N-Z)/A)/A$ and $G_n=(20.8-30.8(N-Z)/A)/A$ which yield the gap energies $\Delta_{p[404\ 9/2]}=0.966$ MeV and $\Delta_n=0.733$ MeV at $\delta=0.26$ and $\Delta_{p[404\ 9/2]}=0.847$ MeV and $\Delta_n=0.573$ MeV at $\delta=0.28$ for ^{101}Nb and ^{103}Nb, respectively. As seen in Fig. 1, energy curves for the orbitals of [422 5/2], [303 5/2] and [301 3/2] have sufficiently deep minima at $\delta=0.25-0.30$, indicating static deformation for both the nuclei.

2.2. Level energies

Level energies were calculated by diagonalizing the rotational Hamiltonian including the Coriolis coupling. Assumed deformation is $\delta=0.26$ and 0.28 for ^{101}Nb

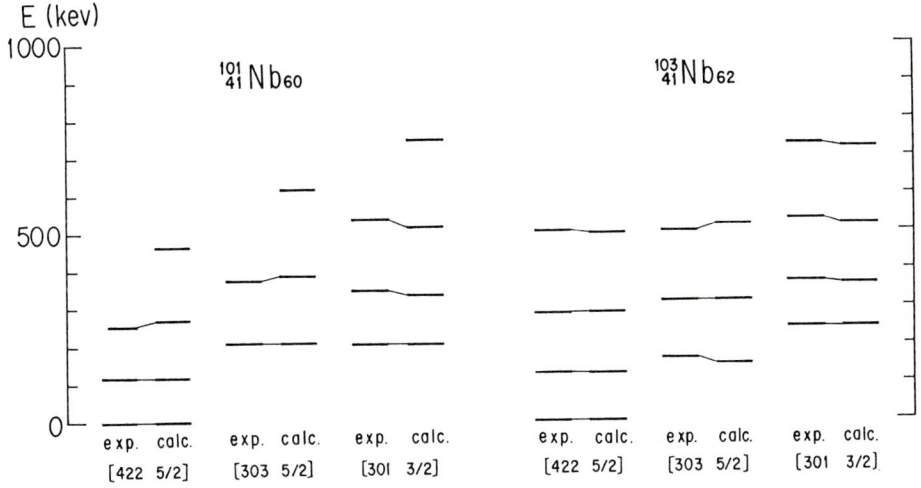

Fig.2. Comparison of the experimental and the calculated level schemes.

and ^{103}Nb, respectively. Although, in principle, the total energy for each orbital $\mathcal{E}_i$ in eq.(1) should be used in the Coriolis coupling calculation, it was necessary to adjust it in order to obtain reasonable agreement with the experimental level energies, by shifting downward all $N=4$ orbitals by 142 keV. Especially, for ^{101}Nb we made further shift of 19 keV down and 64 keV up for the orbital of [301 3/2] and [303 5/2] orbital, respectively. Orbitals included in the diagonalization are [440 1/2], [431 3/2], [422 5/2], [413 7/2], [404 9/2], [431 1/2], [301 1/2], [312 3/2], [303 7/2], [301 3/2], [303 5/2], [301 1/2] and [321 1/2].

Different rotational parameters for $N=3$ and 4 orbitals were used; for ^{101}Nb $a=$ 26.3 keV and 23.4 keV, and for ^{103}Nb $a=23.0$ keV and 25.5 keV, each for $N=3$ and 4 orbitals, respectively. We used the rotational recoil energy $-2K^2$ omitting the $\mathbf{j}^2$-term from the original one, $\mathbf{j}^2-K^2-\Omega^2$, and did not take into account the attenuation of the Coriolis matrix elements.

The calculated level energies are compared with experiments in Fig. 2.

2.3. Gamma-ray transitions

Gamma-ray transition probabilities were calculated using the formulae given in [5] with the inclusion of the pairing effects and the Coriolis mixing. Used parameters are as follows; effective g-factors $g_s=0.6g_s^{free}$ and $g_R=Z/A$, intrinsic quadrupole moment, $Q_0 \approx 0.8ZR_0^2\delta$, and the E1 enhancement for $\Delta K=0$ transitions $H=5$.

In Table 1 the calculated values for the intraband transitions from the first excited states are compared with experiments. In Table 2, γ-ray transitions from the band heads of [303 5/2] and [301 3/2] to the ground-state band are summarized. The patterns of γ-transitions from other levels are illustrated in Fig. 3 for ^{101}Nb and in Fig. 4 for ^{103}Nb, where the calculated transition intensities are normalized

Table 1. Intraband transition rates for [422 5/2].

Nucleus	Transition (keV)	Transition prob.(ns^{-1}) M1	Transition prob.(ns^{-1}) E2	$T_{1/2}$ (ns)	$T_{1/2}^{exp}$ (ns)
^{101}Nb	119.3	6.51	0.09	0.092	-
^{103}Nb	126.3	8.25	0.14	0.074	0.1±0.3

to the experimental one of the biggest transition depopulating each level.

3. Discussion

As seen in Fig. 2, the present Coriolis calculation reproduced well the experimental level energies. Although this good agreement is partly owing to the downward shift of the band head energies for the $N=4$ orbitals, the shifts are small enough to regard the present calculation as reasonable.

The intraband transitions for the [422 5/2] band is M1 predominant as seen in Table 1. The reasonable agreement of the calculated half-life value and the experimental value for ^{103}Nb indicates the validity of the [422 5/2] assignment and of the choice of the effective g-factor, $g_s=0.6g_s^{free}$; the value of $g_s=0.63g_s^{free}$ gives a perfect agreement. As for other intraband transitions, their characteristics can be seen in Figs. 3 and 4. It should be noted that for the [303 5/2] band in ^{103}Nb the crossover transition of 338.4 keV exceeds the cascading one of 187.9 keV, and this tendency is well reproduced by the present calculation. This is related to a large cancelation of the term g_K-g_R involved in the M1 matrix element. Thus, the branching ratio is so sensitive to the effective g-factor g_s that to measure the intensity of the 187.9 keV transition provides key information for

Table 2. Transitions from the band heads.

Nucleus	Initial level E(keV)	Initial level I	Initial level Orbital	Final level E(keV)	Final level I	Final level Orbital	Transition E(keV)	Transition Mul.	Transition Prob.(ns^{-1})	$T_{1/2}$ (ns)	$T_{1/2}^{exp}$ (ns)
^{101}Nb	208.6	5/2	[303 5/2]	0.0	5/2	[422 5/2]	208.6	E1	0.00113	71	1.7±0.2
				119.3	7/2	[422 5/2]	89.3	E1	0.00739		
	205.7	3/2	[301 3/2]	0.0	5/2	[422 5/2]	205.7	E1	0.058	12	0.7±0.2
^{103}Nb	163.9	5/2	[303 5/2]	0.0	5/2	[422 5/2]	163.9	E1	0.160	4.2*	4.7±0.5
	247.6	3/2	[301 3/2]	0.0	5/2	[422 5/2]	247.6	E1	0.105	3.7	2.0±0.6
				163.9	5/2	[303 5/2]	83.7	M1	0.058		

* Fitted to the experimental value by adjusting the E1 enhancement for ΔK=0 transitions; i.e. H=5. Data on ^{101}Nb are from a recent β-γ measurement [6].

353

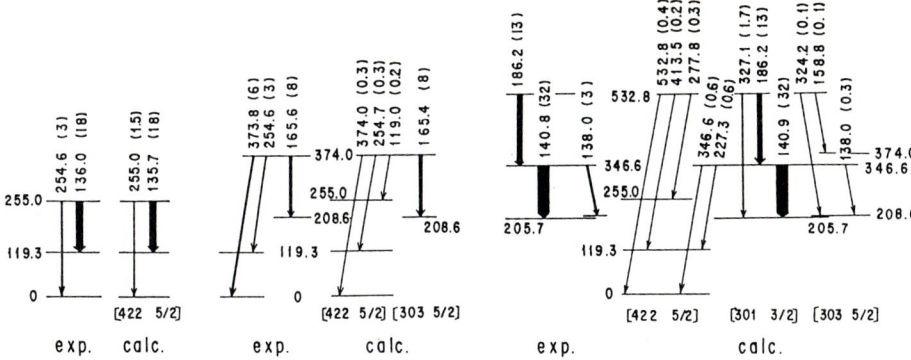

Fig. 3. Gamma-transition pattern for ^{101}Nb.

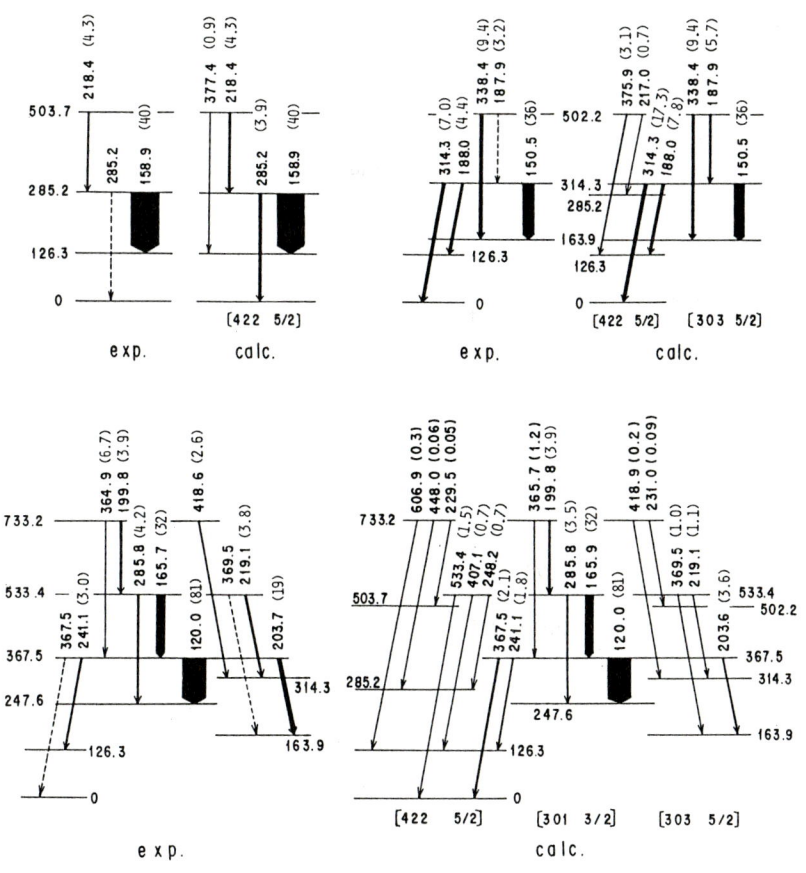

Fig. 4. Gamma transition patterns for ^{103}Nb.

Table 3. Pairing factors $P^- = UU' - VV'$.

Orbital pair	P^-	
	^{101}Nb	^{103}Nb
[303 5/2] [422 5/2]	0.005	-0.099
[301 3/2] [422 5/2]	0.064	-0.030

determining g_s.

As seen in Table 2, the reproduction of the interband transition rates for ^{101}Nb is not very good. The calculated half-lives are more than ten times as large as the experimental ones. Moreover, the E1 enhancement for $\Delta K=0$ transitions $H=5$ which has been determined by fitting the experimental half-life of the 163.9 keV level in ^{103}Nb seems to be high. E1 transitions between such low-lying levels are highly sensitive to the pairing U and V coefficients through $P^- = UU' - VV'$, and the U and V coefficients are subject to the energies of the Nilsson orbitals near the Fermi level. In the case of the present calculation, the values of U and V in Table 3 lose their pronounced meaning because we used the changed values of $\check{\varepsilon}_i$ for the Coriolis coupling calculation. A fine re-adjustment of the values of U and V would improve the agreement.

As for transitions from other levels, the present calculation generally reproduced the experimental transition patterns (Figs. 3 and 4). Some quantitative discrepancies in details would also be removed by the fine re-adjustment of the U- and V-values.

As is seen above, the low-lying levels in ^{101,103}Nb can be interpreted by the rotational bands of [422 5/2], [303 5/2] and [301 3/2]. The Nilsson-orbital assignment given above, however, is not a unique one; the possibility of an alternative assignment for ^{101}Nb, i. e. [301 3/2] to the ground state band and [422 5/2] to the 205.7 keV band, has not been completely ruled out. In order to establish the Nilsson-orbital assignment, it is necessary to determine experimentally the multipolarities of γ-ray transitions as well as the level spins.

References

[1] Sistemich, K., Lhersonneau, G., Meyer, R.A., Seo, T.: Proc. Int. Symp. on In-Beam Nuclear Spectroscopy, Debrecen, Hungary, 51 (1984)
[2] Seo, T., Lawin, H., Lhersonneau, G., Meyer, R.A., Menzen, G., Sistemich, K.: Z. Phys. A-Atoms and Nuclei 320, 393 (1985)
[3] Seo, T.: Z. Phys. A-Atomic Nuclei 324, 43 (1986)
[4] Wohn, F.K., Hill, J.C., Petry, R.F.: Phys. Rev. C31, 634 (1985)
[5] Nilsson, S.G.: K. Dan. Vidensk. Selsk. Mat. Fys. Medd. 29, No. 16 (1955)
[6] Kawade, K., Nagoya University, Japan: private communication

Part VII

Proton-Neutron Interactions and Pairing Reductions

Part VII.

Influence of the Proton-Neutron Interaction in the A ≃ 100 Region

P. Federman

Departamento de Fisica, Facultad de Ciencias Exactas y Naturales,
Universidad de Buenos Aires, Ciudad Universitaria, Nunez,
1428 Buenos Aires, Argentina

This talk is divided in two parts. In the first one I will present some new Shell Model results for the Zr isotopes, obtained with A. Etchegoyen at the TANDAR laboratory. In the second part I will present a weak-coupling description of intruder states and its application to the case of the Zr isotopes, carried out with G. E. Arenas Peris at the University of Buenos Aires.

I. Shell model calculations for the zirconium isotopes

I.1 *Introduction*

It was pointed out in the early 50's that the effective interaction among valence *neutrons* and *protons* is instrumental in the build-up of nuclear collectivity[1]. This is also clear when we compare the experimental spectra of ^{20}O and ^{20}Ne. While ^{20}Ne is deformed and rotational, ^{20}O is a typical spherical Shell Model nucleus. The strong isoscalar interaction makes ^{20}Ne 9.3 MeV more bound than ^{20}O, in spite of the extra Coulomb repulsion of about 8 MeV.

The n-p interaction is also responsible for the dramatic lowering of 4p-4h configurations in ^{16}O[2] (and in ^{40}Ca) that lead to the rotational band built upon the first excited $J^{\pi}=0^{+}$ state.

In the early 60's, the crucial role played by the interaction between neutrons and protons in the $1d_{5/2}$ and $1d_{3/2}$ orbits for light (2s,1d) shell nucleus was emphasized[3]. Moreover, it leads to the coexistence of intruder two-proton core-excitations and valence neutron states in the low-energy spectra of Oxygen and Calcium isotopes[4].

From the perspective of the Collective Model, both collectivity and deformation originate in a long-range quadrupole force that acts among *all* valence nucleons, alike or not. Since heavy nuclei are usually not amenable to Shell Model calculations we are left in a rather uncomfortable situation, with two very different descriptions of nuclear collectivity, depending on the weight of the nucleus considered.

Fortunately Nature (and the experimentalists) were kind, and a new region of deformation was discovered in the early 70's around A ≃ 100 region[5]. In this new region, valence neutrons and protons occupy different shells, as they do for heavy nuclei. Morover, the Zr isotopes up to ^{96}Zr have been described in terms of the spherical Shell Model[6], raising the possibility of extending such a description to the neutron-rich deformed isotopes. This is in my opinion the importance of the Zr isotopes for Nuclear Structure.

They present us for the first time with the opportunity -and the challenge- of microscopically following the shape transition in a case similar to the one of heavy deformed nuclei. Hopefully this will help to bridge the gap between the descriptions of collectivity for light and heavy nuclei.

In the late 70´s the shape transition for the Zr isotopes was studied using restricted Shell Model spaces with active protons in the $2p_{1/2}$, $1g_{9/2}$ and $2d_{5/2}$ orbits, and active neutrons in the $3s_{1/2}$, $2d_{3/2}$ and $1g_{7/2}$ orbits. Results were obtained for ^{98}Zr and only $J^{\pi}=0^+$ states of ^{100}Zr, using a Yukawa force with Rosenfeld admixture[7]. The dramatic lowering of the first excited $J^{\pi}=0^+$ state was well reproduced, and could be understood in terms of the strong isoscalar interaction between nucleons occupying the spin-orbit partner $g_{9/2}$ and $g_{7/2}$ orbitals. The appearance of deformation coincides with -and is probably triggered by- the strong polarization of neutrons and protons into the partner orbits.

The importance of the above results is that they raised the possibility of describing rotational states in terms of restricted Shell Model spaces. Nevertheless, the calculations were too incomplete to draw any definite conclusions. First, the shape transition is completed only for ^{102}Zr, which could not be treated at the time. Results were obtained only for $J^{\pi}=0^+$ states in ^{100}Zr. Thus, the ability of the model to describe rotational bands (including the necessary drastic lowering of the first $J^{\pi}=2^+$ state) remained an open question. And last but not least, B(E2)'s could not be considered either.

I.2 Calculations

We have now been able to begin adressing the above questions, thanks to the invaluable collaboration of Dr. A. Etchegoyen and his OXBASH Shell Model code. We were able to carry out calculations for all isotopes up to ^{104}Zr and for most J values of interest, in the same restricted space mentioned above and using the same single-particle energies, listed in Table I. The proton $2d_{5/2}$ orbit is not included, its influence being neglegible[7]. We used the effective interaction

TABLE I. Single-particle energies for Zr Shell-Model calculations

Protons		Neutrons	
Orbital	Energy (MeV)	Orbital	Energy (MeV)
$2p_{1/2}$	0.00	$3s_{1/2}$	0.00
$1g_{9/2}$	0.89	$2d_{5/2}$	1.13
		$1g_{7/2}$	1.64

recently obtained by fitting the G-matrix elements derived the Paris potential with elements of OBEP-type funcional forms[8]. The matrix elements were kindly supplied to us by E. Vergini. Although the calculations are still in progress, the results obtained until now are interesting enough to be presented here.

In Table II we compare experimental[9] and calculated energies for ^{100}Zr and ^{102}Zr, and list results for ^{104}Zr. Calculated occupation numbers for all active orbits are also included for each level. The spectra of the light Zr isotopes are not well reproduced with the interaction used. For example, the first excited 0^+ state in ^{98}Zr is obtained at around 2.3 MeV. This can be traced back to some shortcomings of T=1 interaction matrix elements, e.g. the pairing matrix element for $1g_{9/2}^2$ is only about a third of the value obtained in effective Shell Model calculations[6]. Nevertheless, the dramatic lowering of the first excited 0^+ level in ^{100}Zr is generously reproduced. When the neutron-proton interaction takes over, the interaction does very well. The calculations were performed changing $V(1g_{9/2}^2\ J=0)$ from its original value of -0.64 MeV to the effective value of -1.7 MeV. This has the effect of lowering the first excited 0^+ state in ^{98}Zr by about 0.8 MeV, without significantly affecting the results for the heavier isotopes.

I.3 *Discussion of Results*

As can be seen in Table II, the drastic lowering of the first excited $J^\pi = 2^+$ level is very well reproduced. Rotational bands are obtained for ^{102}Zr and ^{104}Zr, though not so for ^{100}Zr.

The analysis of the wave functions reveals interesting features. Although the ground state of ^{100}Zr contains a large component of the polarized (or "deformed") configurations (about 60%), this is not the case for the lowest 2^+ state. In ^{102}Zr the situation is well defined, and the rotational band fully developed. The wave functions of all yrast levels are made up of practically 100% polarized components.

TABLE II. Experimental and calculated energies (in MeV) for ^{100}Zr, ^{102}Zr and ^{104}Zr. No experimental data are available for ^{104}Zr. Calculated occupation numbers of the active orbits are also included for each level.

A	J^π	E_{expt}	E_{calc}	$2p_{1/2}$	$1g_{9/2}$	$3s_{1/2}$	$2d_{3/2}$	$1g_{7/2}$
100	0^+_1	0.00	0.00	0.5	1.5	1.9	0.6	1.5
	0^+_2	0.33	0.19	1.4	0.6	2.0	1.5	0.5
	2^+_1	0.21	0.18	1.6	0.4	2.0	1.8	0.2
	2^+_2	—	0.39	0.4	1.6	1.9	0.5	1.6
	4^+_1	0.56	0.84	0.6	1.4	1.9	0.6	1.5
	4^+_2	—	1.04	1.0	1.0	1.9	0.6	1.5
	6^+_1	1.06	1.28	0.3	1.7	1.9	0.2	1.9
	8^+_1	(1.68)	2.06	0.0	2.0	1.9	0.3	1.8
	10^+_1	—	2.40	0.0	2.0	1.9	0.8	1.3
102	0^+_1	0.00	0.00	0.0	2.0	1.9	1.9	2.2
	0^+_2	—	0.57	1.8	0.2	2.0	3.9	0.1
	2^+_1	0.15	0.18	0.1	1.9	1.9	1.8	2.3
	2^+_2	—	0.69	0.1	1.9	2.0	1.5	2.5
	4^+_1	(0.48)	0.59	0.1	1.9	1.9	1.6	2.5
	4^+_2	—	0.79	0.1	1.9	1.9	1.0	3.1
	6^+_1	(0.96)	1.18	0.1	1.9	1.9	1.3	2.8
	8^+_1	(1.55)	1.94	0.1	1.9	1.9	1.0	3.1
104	0^+_1	—	0.00	0.1	1.9	2.0	2.8	3.2
	0^+_2	—	0.13	0.0	2.0	1.9	2.2	3.9
	2^+_1	—	0.07	0.0	2.0	1.9	2.5	3.6
	2^+_2	—	0.15	0.1	1.9	1.9	2.3	3.8
	4^+_1	—	0.23	0.0	2.0	2.0	2.3	3.7
	4^+_2	—	0.44	0.0	2.0	1.9	2.1	4.0
	6^+_1	—	0.64	0.0	2.0	1.9	2.1	4.0
	8^+_1	—	1.22	0.0	2.0	1.9	2.0	4.1

It has been recently suggested[10], and also obtained as a result of HFB calculations[11], that the occupancies of the $1g_{7/2}$ and $1g_{9/2}$ orbits may be larger than two. Indeed, our results show a strong influence of $1g_{7/2}^3$ configurations for ^{102}Zr and also $1g_{7/2}^4$ configurations for ^{104}Zr. Nevertheless, no large components are obtained with four or more neutrons in the $1g_{7/2}$ orbit for ^{102}Zr. Unfortunately, it was not possible to allow for the opening of the Z=38 core, leading to the possibility of more than two $1g_{9/2}$ protons.

We have also calculated B(E2) values. The ratio for the $2_1^+ \rightarrow 0_1^+$ transitions in ^{102}Zr and ^{100}Zr is obtained to be about 4, in good agreement with the experimental value[5] of around 3. Absolute values are off by about an order of magnitude if we use effective charges of 0.5 for neutrons and 1.5 for protons, which is not surprising in view of the highly restricted Shell Model spaces utilized. The inclusion of more orbits leading to small but numerous admixtures of other configurations may be crucial to improve the calculated B(E2) absolute values.

II. Weak-Coupling Description of Intruder States

II.1 Introduction

Intruder states are "core-excited" states that coexist with valence nucleon states in the low energy spectra of nuclei near magic N and/or Z, throughout the Periodic Table. They usually involve the excitation of one or two protons (neutrons) across a major shell, although for light doubly-magic nuclei two neutrons and two protons are involved.

A lot of both experimental and theoretical effort has been devoted in the last years to study intruder states in medium and heavy nuclei[12]. The microscopic description of intruder states in such nuclei poses a rather interesting challenge for theorists. Shell Model approaches rapidly lead to matrices of astronomic dimensions. Several models and approximations have been used in different cases and regions[12]. Perharps the two main motivations for so much interest in intruder states are their puzzling energy systematics (sometimes they even become the ground states) and some measured electromagnetic transition rates which seem to indicate a collective nature for the states involved[13].

II.2 The model

For the sake of clarity, in the following we will treat only the case of proton excitations. The neutron case is completely analogous, and requires only minor modifications.

Proton intruder states in a (Z,N) nucleus are obtained by raising ΔZ protons from the core into the orbits of the valence neutrons (particles or holes). Consider the nucleus consisting of the doubly magic core (Z_M, N_M), the valence proton-particles (*only for* $Z > Z_M$), the valence neutrons and the ΔZ excited protons, and let H_1 be its Hamiltonian.

Let H_2 be the Hamiltonian of the proton-holes, *i.e.* the interaction among them plus their single-hole energies relative to the doubly-magic (Z_M, N_M) core, and consider

$$H = H_1 + H_2 + V_{12}, \tag{1}$$

where V_{12} includes only the interaction of the proton-holes with the valence nucleons. Note that H_1 is the Hamiltonian of the nucleus with $Z_M + Z - Z_R$ protons ($Z_R = \min\{Z_M, Z\} - \Delta Z$) and N neutrons, so that its eigenvalues can be obtained from experiment when available. The eigenvalues of H_2 can likewise be extracted from the spectrum of the nucleus (Z_R, N_M) and the binding energy of the doubly magic core.

Let $|\alpha_{1(2)} J_{1(2)}\rangle$ denote the eigenstates of $H_{1(2)}$ with eigenvalues $E(\alpha_{1(2)} J_{1(2)})$. To the extent that V_{12} is "small", a zero order approximation to the binding energies of the intruder states in the nucleus (Z,N) is given by:

$$\begin{aligned}E_0(\alpha_1 J_1 \alpha_2 J_2; J) &= E(\alpha_1 J_1) + E(\alpha_2 J_2) \\ &= BE(Z_M + Z - Z_R, N; \alpha_1 J_1) \\ &\quad + BE(Z_R, N_M; \alpha_2 J_2) - BE(Z_M, N_M; g.s.).\end{aligned} \tag{2}$$

The first-order correction to E_0 is given by

$$\Delta E(\alpha_1 J_1 \alpha_2 J_2; J) = \langle \alpha_1 J_1 \alpha_2 J_2; J | V_{12} | \alpha_1 J_1 \alpha_2 J_2; J \rangle. \tag{3}$$

The calculation of ΔE becomes particulary simple when J_1 and/or J_2 are 0, since then only the monopole component of V_{12} contributes. Let j'_π, j_π and j_ν denote respectively the orbits occupied by the proton-particles and the valence neutrons. Then,

$$\Delta E = \sum_{j'_\pi} n'_\pi \left[\sum_{j_\pi} n_\pi \bar{V}(j'^{-1}_\pi; j_\pi) + \sum_{j_\nu} n_\nu \bar{V}(j'^{-1}_\pi; j_\nu) \right], \tag{4}$$

where the n's are the occupations of the respective orbits (K. Heyde et al. use a similar expression in their IBA-2 treatment[14]).

The average or monopole interactions are given by

$$\overline{V}(j_\pi'^{-1};j_{\pi(\nu)}) = \frac{\sum_J (2J+1) \, V(j_\pi'^{-1} j_{\pi(\nu)};J)}{\sum_J (2J+1)} = -\overline{V}(j_\pi' ;j_{\pi(\nu)}). \tag{5}$$

For proton intruder states in isotope chains, n_π' and n_π vary little with N as compared to n_ν. It is thus convenient in this case to rewrite ΔE as

$$\Delta E = -(Z_M-Z_R)\left[(Z-Z_R) \, \overline{V}_{\pi\pi} + (N-N_M) \, \overline{V}_{\nu\pi}\right], \tag{6}$$

where

$$\overline{V}_{\pi\pi} = \sum_{j_\pi'} \sum_{j_\pi} \frac{n_\pi' \, n_\pi \, \overline{V}(j_\pi';j_\pi)}{(Z_M-Z_R)(Z-Z_R)} \tag{7}$$

will be nearly N-independent, while

$$\overline{V}_{\pi\nu} = \sum_{j_\pi'} \sum_{j_\nu} \frac{n_\pi' \, n_\nu \, \overline{V}(j_\pi';j_\nu)}{(Z_M-Z_R)(N-N_M)} \tag{8}$$

may depend on N. It is the differences in the $\overline{V}(j_\pi';j_\nu)$ (i.e. in the monopole components) for different orbits j_ν weighted with the relative occupancies $n_\nu/|N-N_M|$ that are responsible for this dependence.

However, the

$$\frac{\sum_{j_\pi'} n_\pi' \, \overline{V}(j_\pi';j_\nu)}{\sum_{j_\pi'} n_\pi'} = \overline{V}_\nu \tag{9}$$

are expected to be nearly constant for series of isotopes. Moreover, they are expected to be the same even for different values of Z, provided the number of proton holes is the same, as occurs for example in the cases of Z=81 (1p-2h $J^\pi = \frac{9^-}{2}$) and Z=82 (2p-2h $J^\pi = 0^+$) intruder states.

To calculate ΔE with Eq.(5) requires $\bar{V}_{\pi\pi}$ and $\bar{V}_{\nu\pi}$. We take $\bar{V}_{\pi\pi}$ to be constant over a given region, since $\bar{V}(j'_\pi;j_\pi)$ is generally a small and repulsive quantity nearly independent of the orbits involved, and use values extracted from 1p-1h multiplets in (Z_M,N_M), or obtained from effective interactions.

Since ΔE is very sensitive to small variations of $\bar{V}_{\nu\pi}$ (the "amplification" factors sometimes are of order 100), its determination requires more care. This is achieved self-consistently within the model by requiring that also the ground states of nuclei with $Z=Z_R$ be described by Eqs.(2)-(6) with $\Delta Z=0$. We then obtain:

$$BE(Z_R,N;g.s.) = BE(Z_M,N;g.s.) + BE(Z_R,N_M;g.s.)$$
$$- BE(Z_M,N_M;g.s.) - (Z_M-Z_R) \sum_{j_\nu} n_\nu \bar{V}_\nu. \qquad (10)$$

For a given value of Z_R the $\bar{V}_\nu$'s can be determined to best fit Eq.(10) for different values of N.

Calculations were performed for series of isotopes with Z near 28, 50 and 82, and series of isotones with N close to 82[15,16]. The excellent agreements obtained in general are striking, particularly in view of the simplicity of the model and the lack of free parameters. It is worth stressing that the average interactions $\bar{V}_\nu$ are not free parameters. For example, the values used to calculate the excitation energies of 1p-4h and 2p-4h intruder states for respectively Au and Hg isotopes ($Z_R=78$ in both cases) are obtained by fitting through Eq.(10) the ground states of Pt isotopes ($Z=Z_R$).

In the case of core-excited protons, intruder intra-band quadrupole transition rates in the nucleus (Z,N) are predicted within the framework of the present model to be approximately the same as for the ground state band of the (Z_M+Z-Z_R,N) isotone. While this seems to be the case in the tin region[16], lack of experimental information makes comparison impossible in other regions.

II.3 Application to the Zr isotopes

In the following we apply the above model to the Zr isotopes, assuming two-proton excitations. A value of -260 keV is obtained for $\bar{V}_{\pi\pi}$ by fitting the first excited 0^+ level of ^{90}Zr. Available experimental data allow the determination of $\bar{V}_\nu$ through Eq.(10) only for $j_\nu=2d_{5/2}$ and $3s_{1/2}$, for which we obtain 380 keV and 270 keV respectively. For $j_\nu=1g_{7/2}$ we use 370 keV as obtained previously for the Z=50 region[15].

Using those values we calculated the excitation energies of the intruder $J^\pi = 0^+$ band heads for even Zr isotopes, shown and compared to experiment in Fig. 1. The detailed agreement is striking. The change in the structure of the ground state during the shape transition is correctly predicted to occur between A=98 and A=100.

The same parameters yield much higher energies for the four-proton excitations, although in this case the ΔE corrections are indeed larger than for the two-proton excitations. Nevertheless they cannot overcome the higher E_o values, leading to higher excitation energies, as also shown in Fig. 1.

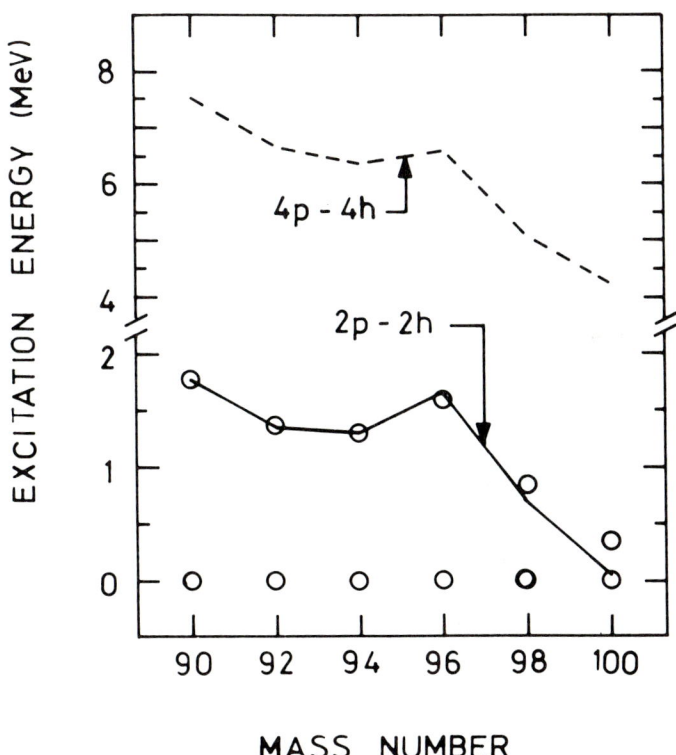

Fig. 1. Experimental excitation energies (in MeV) of the lowest 0^+ states in the even Zr isotopes with $90 \leq A \leq 100$ (open circles) compared with calculated values assuming two-proton (full line) and four-proton (dashed line) excitations.

III. Conclusions

Summing up, the present results indicate that it may be possible to describe the rotational bands of the neutron-rich Zr isotopes in terms of the effective Shell Model in restricted spaces, provided the relevant orbits -i.e. those that maximize the n-p interaction- are included. In the present case, they are mainly the $1g_{7/2}$ and $1g_{9/2}$ spin-orbit partners. A more complete calculation should also include the $h_{11/2}$ orbit of neutrons, since the n-p interaction may also be relatively strong when the orbitals have large values of ℓ that differ in one unit[1]. The opening of the Z=38 core would also be of interest. Having both the $2p_{3/2}$ and $2p_{1/2}$ orbits active should help deformation, as observed for the Sr isotopes. Also the $1g_{7/2}$ proton orbits may be of interest, regarding the nature of the Nilsson-like proton orbitals.

The question on the effective occupation of the $g_{7/2}$ and $g_{9/2}$ partner orbits is partially answered by the present Shell Model calculations. Indeed $g_{7/2}^3$-type configurations play a significant role, and to a lesser degree also $g_{7/2}^4$ configurations are important. About the occupancy of the $g_{9/2}$ proton orbit, the weak-coupling results suggest that $g_{9/2}^4$-type configurations should contribute little to the deformed ground states of the heavier Zr isotopes.

In heavier nuclei more than one pair of orbits may be relevant. That may also be the case for the recently observed super-deformed excited bands[17]. In addition to the partners $h_{9/2}$-$h_{11/2}$, the $i_{13/2}$ proton orbit may provide the "extra" deformation.

Acknowledgments

This work has been partially supported by CONICET and UBA, Argentina.

References

1) A. de Shalit and M. Goldhaber, Phys. Rev. **92** (1953) 1211.
2) I. Unna and I. Talmi, Phys. Rev. **112** (1958) 452.
3) I. Talmi, Rev. Mod. Phys. **34** (1962) 704.
   I. Unna, Phys. Rev. **132** (1963) 2225.
4) P. Federman, Cargèse Lect. Phys., Ed. M. Jean (Gordon and Breach, New York, 1969) Vol. 3, p. 21.
5) E. Cheifetz, R. C. Jared, S. G. Thompson and J. B. Wilhelmy, Phys. Rev. Lett. **25** (1970) 38.
6) N. Auerbach and I. Talmi, Nucl. Phys. **64** (1965) 458.
   D. H. Gloeckner, Nucl. Phys. **A253** (1975) 301.
7) P. Federman and S. Pittel, Phys. Rev. **C20** (1979) 820.
8) A. Hosaka, K. I. Kubo and H. Toki, Nucl. Phys. **A444** (1985) 76.
9) C. M. Lederer and V. S. Shirley (eds.), Table of Isotopes (John Wiley & Sons Inc., New York, 1978).
10) R. A. Meyer, E. A. Henry, L. G. Mann and K. Heyde, Phys. Lett. **B177**

(1986) 271.
11) A. Kumar and M. R. Gunye, Phys. Rev. **C32** (1985) 2116.
12) See K. Heyde, P. Van Isacker, M. Waroquier, J. L. Wodd and R. A. Meyer, Phys. Reports **102** (1983) 291 for a comprehensive review until December 1983.
13) G. Wenes, P. Van Isacker, M. Waroquier, K. Heyde and J. Van Maldeghem, Phys. Rev. C23 (1981) 2291.
14) K. Heyde, P. Van Isacker, R. F. Casten and J. L. Wood, Phys. Lett. **B155** (1985) 303.
15) G. E. Arenas Peris and P. Federman, Phys. Lett. **B173** (1986) 359.
16) G. E. Arenas Peris and P. Federman, Proc. IX Workshop on Nuclear Physics, eds. A. O. Macchiavelli et al. (World Scientific, Singapore, 1987) p. 428.
17) B. Haas et al., Phys. Rev. Lett. **60** (1988) 503.

The Proton-Neutron Interaction in Neutron-Rich A ≃ 100 Nuclei

B. Pfeiffer and K.-L. Kratz

Institut für Kernchemie, Universität Mainz,
D-6500 Mainz, Fed. Rep. of Germany

Abstract

The possible factors responsible for the sudden onset of nuclear deformation in neutron-rich A≅100 nuclei are reviewed from the systematics of subshell occupation numbers for neutron and proton orbitals. The agreement between our results from a Nilsson model with existing selfconsistent HFB calculations strengthens the role of a cumulative neutron-proton interaction, but does not substantiate the conjecture concerning a selective involvement of the $\pi g_{9/2}$-$\nu g_{7/2}$ spin-orbit-partner orbitals in producing deformation as was suggested in earlier work.

Introduction

One of the most characteristic features of neutron-rich A≅100 nuclei is the vary rapid phase transition from spherical to strongly deformed ground-state shapes at N=60. With reference to shell-model and Hartree-Fock-Bogolyubov (HFB) studies performed by Federman and Pittel [1-4] in the late 1970's, it has become widely accepted to attribute this onset of deformation to a particularly strong isoscalar residual interaction between protons and neutrons occupying the spin-orbit partner (SOP) orbitals $\pi g_{9/2}$ and $\nu g_{7/2}$ which have large spatial overlap. Although already in the beginning of the 1980's the exclusive role played by the above SOP orbitals was questioned by the selfconsistent HFB calculations of Khosa et al. [5] and later by those of Kumar and Gunye [6], the arguments about a selective $\pi g_{9/2}$-$\nu g_{7/2}$ interaction as driving force for nuclear deformation are still used by many authors. In order to help clarifying the situation, we have performed Nilsson-model calculations using the BCS pairing part of the quasi-particle RPA shell-model of Krumlinde and Möller [7] to derive occupancies of proton and neutron single-particle orbitals (SP) for neutron-rich even-even $_{38}$Sr, $_{40}$Zr and $_{42}$Mo isotopes.

The 'SOP' Idea for Describing Deformation

Nuclear deformation has successfully been described within collective models by introducing a phenomenological long-range residual quadrupole-quadrupole interaction between all nucleons [8]. On the other hand, from a shell-model point of view De-Shalit and Goldhaber [9] suggested that collective effects could be produced by the isoscalar residual interaction between neutrons and protons leading to strong mixing of shell model configurations. This effect should be most pronounced for highly overlapping orbitals, in particular when they satisfy the conditions $n_n = n_p$ and $l_n \cong l_p$.

Based on these ideas, Federman and Pittel [1-4] presented a "unified shell model description of nuclear deformation valid throughout the periodic table". Shell-model calculations for ^{98,100}Zr were performed assuming ^{94}Sr as an inert core and restricting the valence neutrons to the $2s_{1/2}$, $1d_{3/2}$ and $0g_{7/2}$ orbitals and the valence protons to the $1p_{1/2}$, $0g_{9/2}$ and $1d_{5/2}$ orbitals. The calculated occupation numbers for the ground states (see Tables 4 and 5 in [4], and Fig. 2 of this article) seemed to confirm the crucial role of the $\pi g_{9/2}$-$\nu g_{7/2}$ partner orbitals: whereas the spherical ground state in ^{98}Zr is mainly composed of $(\pi p_{1/2})^2$ and $(\nu s_{1/2})^2$, the deformed ground state of ^{100}Zr is dominated by $g_{9/2}$ protons and $g_{7/2}$ neutrons.

Due to the increased number of valence protons relative to the Zr isotopes, the same authors treated the Mo isotopes in the framework of the HFB with inclusion of the $\nu h_{11/2}$ orbit in the configuration space. The occupation numbers (see Tab. 8 in [4], and Fig. 3 of this article) demonstrated an important influence of the $\nu h_{11/2}$ orbital on deformation, diminishing strongly the importance of the SOP $\nu g_{7/2}$. These results already contradicted the early concept of an exclusive role of the SOP's $\pi g_{9/2}$-$\nu g_{7/2}$. Nevertheless, in the following years many authors adopted the original SOP arguments of Federman and Pittel without considering their modifications for the Mo case.

HFB Calculations

Already in the early 1980's, it was questioned whether the predominant role attributed to the SOP coupling as deformation driving force was not merely an artefact of the extremely restricted valence space used by Federman and Pittel. Khosa et al. [5] and Kumar and Gunye [6] performed calculations in the framework of the HFB employing configuration spaces large enough to permit systematic and unbiased studies of the phase transition in isotopes of Sr, Zr and Mo around N=60. Khosa et al. [5] used ^{76}Sr as an inert core and included all shells up to $0h_{11/2}$ for neutrons and protons. In the second paper [6], the valence space was even chosen so large that no subshell closures, such as $0f_{5/2}$ for protons or $0g_{9/2}$ and $1d_{5/2}$ for neutrons, had to be assumed. Both calculations employed the usual pairing-plus-quadrupole-quadrupole effective interaction and were carried out in a selfconsistent way to determine equilibrium deformation, intrinsic quadrupole moments, pairing gaps, occupation numbers and B(E2) values, simultaneously. The experimentally observed sudden onset of large deformation in Sr, Zr, Mo isotopes was reproduced by both calculations which indicated a sharp increase of deformation parameters and quadrupole moments, as well as a decrease of the pairing gaps for proton and neutrons at N=60. This shape transition was shown to be correlated with the sudden depletion of protons from the $0f_{5/2}$ orbital and of neutrons from the $1d_{5/2}$ and $0g_{9/2}$ orbitals. The reduction of nucleon numbers in these lower orbitals was observed to be mainly compensated by a substantial increase in the occupancies of the high-j orbitals $\pi g_{9/2}$ and $\nu h_{11/2}$. The simultaneous increase in the occupancies of $\pi g_{9/2}$ and of its SOP $\nu g_{7/2}$ around N=60 was found to contribute also to some extent to the onset of deformation, but it could not be confirmed being the most important factor as proposed by Federman and Pittel [1-4]. In addition, the calculations of Kumar and Gunye [6] clearly indicated that ^{94}Sr or even ^{88}Sr cannot be used as inert cores for structure calculations at A≅100.

The Nilsson Model - BCS Pairing Approach

In order to help clarifying the discrepancies, we have calculated subshell occupation numbers in the framework of the Nilsson model - BCS pairing approach of Krumlinde and Möller [7]. Based on the early work of Halbleib and Sorensen [10], the model calculates both the SP energy levels and wave functions using a modified oscillator potential of the Nilsson model. The κ- and μ-parameters for proton and neutron SP potentials recommended by Ragnarsson and Sheline [11] have been applied. This set contains different values for different N-shells, obtained from adjustment to experimental data over the whole region of nuclei. It yields a similar SP level scheme for A≅100 in the harmonic oscillator potential as is obtained with Woods-Saxon [12] and Folded-

Yukawa [13] potentials. Further input parameters of the model are the deformation parameters ε_2 and ε_4 which were either taken from experiment or from the most recent Möller and Nix mass tables [14]. Pairing is treated in the BCS approximation with the pairing gap Δ as input variable. The code then calculates the pairing strength G, the location of the Fermi surface λ and the occupation probabilities v_κ^2. This method of solving the pairing equations assures that the BCS method does not collapse for near-spherical nuclei, as obviously occurs in other BCS models that prescribe a value of G and calculate Δ, λ and v_κ^2. In order to account for the influence of the structure of the level spectrum reflected in specific values of Δ, pairing gaps according to the recent model of Madland and Nix [15] were used.

In the framework of the Nilsson model, changes of occupation numbers with deformation can qualitatively be understood as an interplay

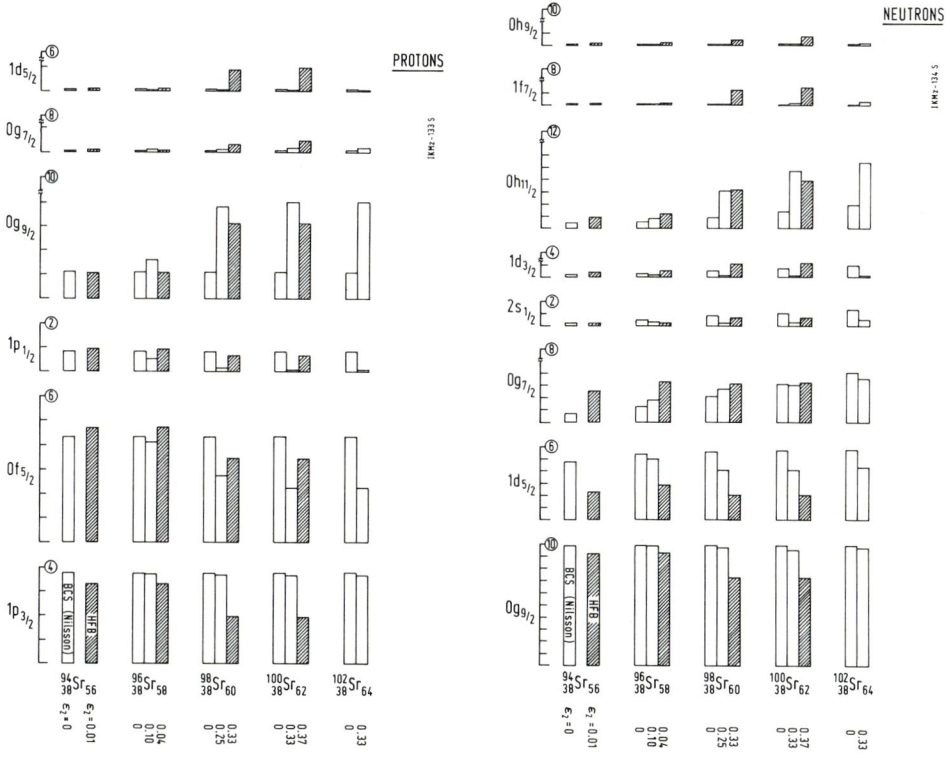

Fig. 1: Influence of the deformation parameter ε_2 on subshell occupation numbers for even-even $_{38}$Sr isotopes around N=60. Nilsson model - BCS pairing results from this work (open squares) are compared to HFB calculations (hatched areas) from Ref. [6].

between upbending high-Ω and downsloping low-Ω orbitals crossing the Fermi level. In Fig. 1, the influence of the deformation parameter ε_2 on proton and neutron occupancies for $_{38}$Sr isotopes with N=56-64 obtained from our calculations are displayed and compared to the HFB results of [6]. The overall agreement between the two approaches is quite good.

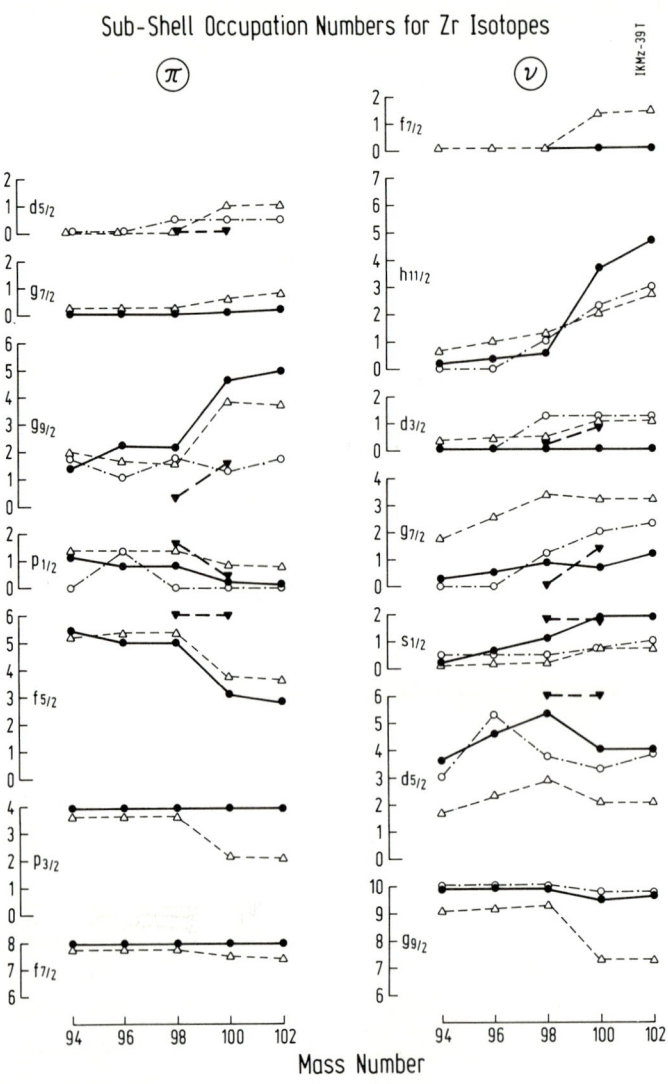

Fig. 2: Subshell occupation numbers for even-even $_{40}$Zr isotopes with N=54-62 obtained in this work (●), compared to shell model calculations (▼: Ref.[4]) and two HFB approaches (o: Ref.[5] and △: Ref.[6], respectively).

The addition of two neutrons to the spherical isotope $^{96}Sr_{58}$ leads to drastic rearrangements of neutron as well as proton occupation numbers. The sudden depletion of the nearly filled $\nu d_{5/2}$ orbital (making it a semi-magic subshell in the spherical case) is compensated by a sharp increase in the population of the Nilsson orbitals $\nu[550\ 1/2]$ and $\nu[541\ 3/2]$ of $h_{11/2}$ parentage. The high intrinsic quadrupole

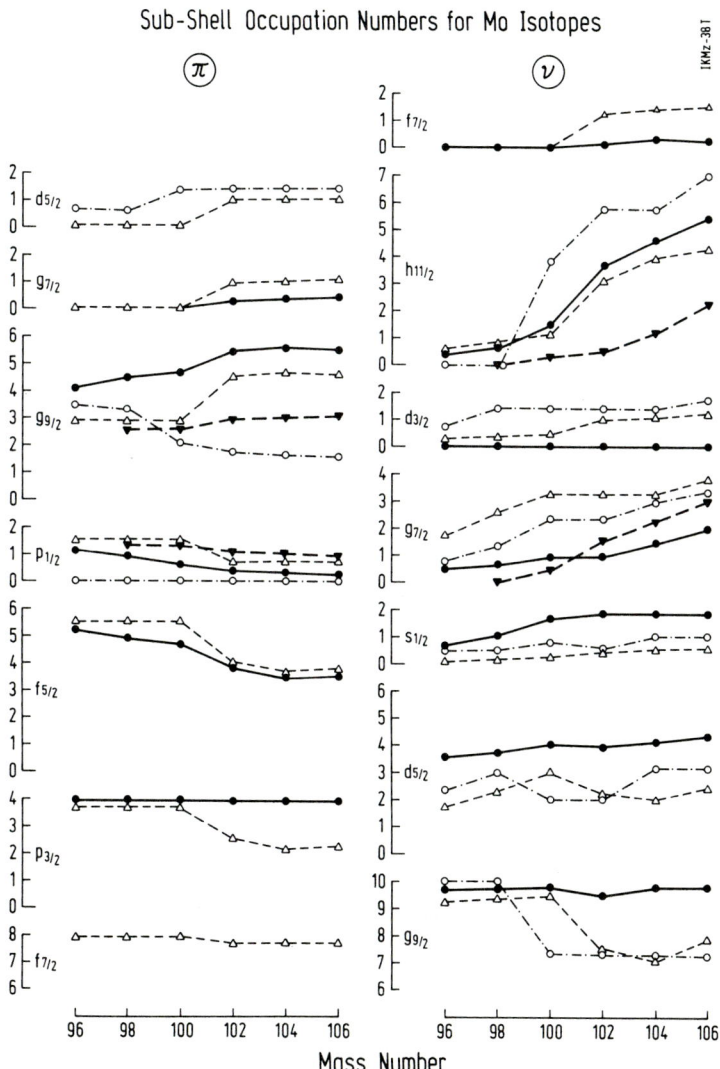

Fig. 3: Occupation numbers from the Nilsson model - BCS pairing approach for $^{96-106}$Mo isotopes (●), compared to HFB results from Refs. [2,4] (▼), [5] (o), and [6] (Δ), respectively.

moments of these orbitals then induce a shift in the proton occupation numbers. Protons from the orbitals $0f_{5/2}$ and $1p_{1/2}$ which are highly filled in the spherical case are lifted to low-Ω members of $0g_{9/2}$. This "polarization effect" [4] seems to be more pronounced in the case of the Nilsson model than in the HFB calculations of [6]. In the harmonic-oscillator approach, the range of the pairing interaction (as given by the pairing gap Δ) influences only quasi-particles near the Fermi surface, such as $0f_{5/2}$, $1p_{1/2}$ and $0g_{9/2}$ in the proton case, whereas in the HFB method [6] the whole valence space is affected, even strongly weakening major magic shells such as Z=28 and N=50.

The same picture as for the $_{38}$Sr nuclei also holds for the series of $_{40}$Zr and $_{42}$Mo isotopes. From Figs. 2 and 3 it is furthermore evident that both the shell-model study for Zr and the HFB calculations for Mo performed by Federman and Pittel [1-4] are inconclusive due to far too restricted valence spaces chosen. In general, results from the HFB methods of [5,6] and our Nilsson model agree remarkably well. The a priori disadvantage of the latter approach of not being 'selfconsistent', seems to be compensated by the fact that experimental information on nuclear structure can be included explicitly via an appropriate choice of certain input parameters. For example, the κ- and μ-parameters of [11] yield a sufficiently large spherical N=56 subshell gap and give - in contrast to the neutron SP levels chosen in [1-6] - the 'correct' SP sequence above N=50, i.e. $1d_{5/2}$, $2s_{1/2}$, $0g_{7/2}$, $1d_{3/2}$, $0h_{11/2}$. Moreover, the faster onset of strong deformation in the $_{38}$Sr and $_{40}$Zr isotopes compared to the respective $_{42}$Mo isotones could be taken into account by applying measured quadrupole-deformation parameters (as far as available) instead of model predictions [14].

Conclusions

The fair agreement between HFB calculations in large valence spaces [5,6] and our Nilsson model - BCS pairing approach strengthens the dominant role of the cumulative np-quadrupole interaction for producing nuclear deformation at A$\cong$100, but does not support the widely accepted exclusive role of the $\pi g_{9/2}$-$\nu g_{7/2}$ SOP coupling, originally suggested by Federman and Pittel [1-4]. As was already indicated by the latter authors for the Mo isotopes, the occupation of the $\nu 0h_{11/2}$ orbital not only stabilizes deformation beyond N=62 [4], but seems rather to be the main mechanism for the sudden onset of deforamtion in Sr, Zr and Mo isotopes at N=60. This neutron shell is practically empty at spherical shape but becomes occupied at large deformation due to its large SP quadrupole moment. Such unnatural parity high-j intruders are observed in all regions of medium to heavy nuclei just at the onset of shape transitions.

Further support for the concept of a <u>cumulative</u> np-interaction rather than a <u>selective</u> $\pi g_{9/2}$-$\nu g_{7/2}$ coupling comes from the successful application of $N_\pi N_\nu$ type models [16,17,18] to describe the onset of deformation at $A \cong 100$, and from a recent determination of the $[\pi g_{9/2}$-$\nu g_{7/2}]_{1^+}$ interaction matrix element which did not show any exceptional strength compared to similar configurations in other nuclear regions [19].

Acknowledgements

We would like to acknowledge valuable discussion with P. Möller and K. Heyde. This work has been funded by the German Federal Minister for Research and Technology (BMFT) under contract number 06MZ552.

References

[1] P. Federman and S. Pittel, Phys. Lett. <u>25</u>, 38 (1970)
[2] P. Federman and S. Pittel, Phys. Lett. <u>69B</u>, 385 (1977)
[3] P. Federman et al., Phys. Lett. <u>82B</u>, 9 (1979)
[4] P. Federman and S. Pittel, Phys. Rev. <u>C20</u>, 820 (1979)
[5] S.K. Khosa et al., Phys. Lett. <u>119B</u>, 257 (1982)
[6] A. Kumar and M.R. Gunye, Phys. Rev. <u>C32</u>, 2116 (1985)
[7] J. Krumlinde and P. Möller, Nucl. Phys. <u>A417</u>, 419 (1984)
[8] A. Bohr and B.R. Mottelson, Nuclear Structure (Benjamin, New York, 1975)
[9] A. De-Shalit and M. Goldhaber, Phys. Rev. <u>92</u>, 1211 (1953)
[10] J.A. Halbleib,Sr., and R.A. Sorensen, Nucl. Phys. <u>A98</u>, 542 (1967)
[11] I. Ragnarsson and R.K. Sheline, Physica Scripta <u>29</u>, 385 (1984)
[12] W. Nazarewicz et al., Nucl. Phys. <u>A435</u>, 397 (1985)
[13] R. Bengtsson et al., Physica Scripta <u>29</u>, 402 (1984)
[14] P. Möller and J.R. Nix, Rep. LA-UR-86-3983 and At. Data Nucl. Data Tables, in print
[15] D.G. Madland and J.R. Nix, Nucl. Phys. <u>A476</u>, 1 (1988)
[16] I. Hamamoto, Nucl. Phys. <u>73</u>, 225 (1965)
[17] R.F. Casten, Nucl. Phys. <u>A443</u>, 1 (1985)
[18] H.J. Daley et al., Phys. Rev. Lett. <u>57</u>, 198 (1986)
[19] G. Lhersonneau et al., Proc. Int. Conf. on Nuclei far from Stability, Sep. 87, Lake Rosseau, Canada;
AIP Conf. Ser. <u>164</u>, 393 (1988)

The P-Factor and Atomic Mass Systematics: Application to Medium Mass Nuclei

D.S. Brenner[1], *P.E. Haustein*[2], *and R.F. Casten*[2]

[1] Clark University, Worcester, MA 01610, USA
[2] Brookhaven National Laboratory, Upton, NY 11973, USA

Recently it has been shown[1] that a measure of the average number of interactions of each valence nucleon with those of the other type given by the parameter P = $N_p N_n/(N_p+N_n)$ [where $N_p(N_n)$ are the number of valence protons (neutrons)] provides a general and physically meaningful explanation for the development of collectivity and the onset of deformation in medium and heavy nuclei. For example, as shown in Fig. 1, the energy ratio $E_{4^+_1}/E_{2^+_1}$ plotted versus P exhibits remarkably similar behavior independent of mass region and, in all cases indicates a transition to deformation ($E_{4^+_1}/E_{2^+_1} \approx 3.0$) when P $\approx$ 4-5. This rule provides a necessary condition for deformation that the nuclear ground state cannot be deformed for medium or heavy nuclei if the number of valence nucleons of either type is less than 4. Furthermore, this result was seen as a manifestation of the competition between the like nucleon pairing and p-n quadrupole interactions since the pairing strength, which is about 1 MeV and which supports sphericity, is approximately off-set by the quadrupole deformation-driving term when 4-5 valence p-n interactions (each with strength $\approx$ 200 keV) are present.

It is of interest to investigate whether the success of the P formalism can be extended to systematics of the mass surface. It has long been known that mass data reflect variation in stability which accompany changes in nuclear structure and shape. Such quantities as S_{2n} and S_{2p}, the nucleon pair separation energies, have been used to illustrate the effects of shell closures and the onset of stable ground state deformation. An example of such systematics is shown in Fig. 2 where S_{2n}, the two-neutron separation energy, is plotted for isotopic chains Z=30-49. The decreased binding beyond the N=50 shell closure and the increased binding associated with the onset of deformation around N=60 in the Zr region are clearly visible. While S_{2n} systematics have been extremely useful for illustrating and identifying shell closures and new regions of deformation, they are not well suited to quantitative applications such as the prediction of masses of unknown nuclides since extrapolation of non-linear systematics is involved.

In applying the P formalism to masses it is essential to separate two components which together comprise the atomic mass. The overwhelmingly dominant one is a strong, basically structure independent (macroscopic) component, that scales with A, Z or N. The second (microscopic) component contains the structure and deformation dependence and it is in this regime where the influence of the valence p-n interaction might be expected to manifest itself. In order to isolate

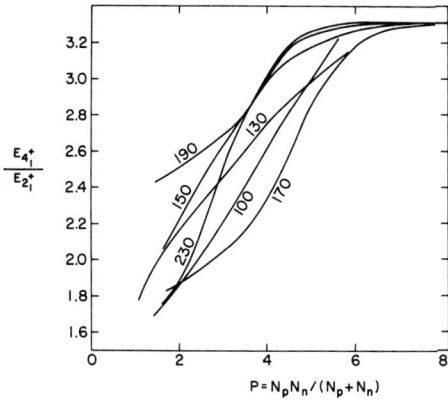
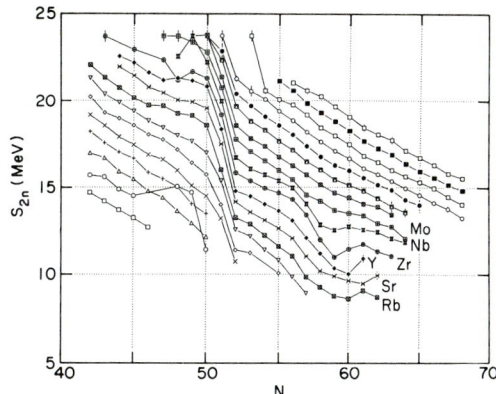

Figure 1: Smoothed P plots of $E_{4^+_1}/E_{2^+_1}$ for six mass regions.

Figure 2: Two-neutron separation energies for isotopic chains in the Zr region.

this latter component we have in each case subtracted the calculated spherical macroscopic mass from the experimental mass to arrive at a semi-empirical microscopic (SEM) mass which is presumed to be a measure of the structure dependent part of the mass. The spherical masses were taken from the model calculations of Möller and Nix[2] and the experimental masses from the compilation of Wapstra et al.[3].

Figure 3 shows plots of SEM-masses versus N_n and the P-factor for the actinide region (Z=84-104 with N>128). Note that points lying higher on the plots represent lower binding energies and that the open circles in Fig. 3b are for data with N>146. Data are shown up to midshell with the assumption that the next shell closures occur at Z=126 and N=184. These choices of shell definition, which in the case of the actinides may be altered depending on the exact locations of the next magic numbers, are important aspects of our approach. We shall elaborate on this point when we consider the effects of subshell closures in the Zr region. At first glance there appears to be no overwhelming simplification in the mass systematics from using the P formalism. However, on closer examination of separate isotopic and isotonic sequences, the P-factor plots exhibit some interesting and useful features. Figure 4 shows such sequences for the actinide region. There are three noteworthy results: first, the SEM masses for individual isotopic or isotonic sequences fall on <u>straight lines</u>. Second, the isotopic sequence (e.g., the Po nuclei with Z=84 (N_p=2)) falls on the <u>same</u> straight line as the corresponding isotonic sequence (the N=128 isotones (N_n=2)). This fact provides a means for predicting unknown masses for nuclei far from stability in selected regions of the chart of the nuclides by <u>interpolation</u> rather than by traditional methods involving extrapolation[4]. Third, the slopes of these straight lines decrease <u>smoothly</u> as a function of increasing values of the constant N_p or N_n.

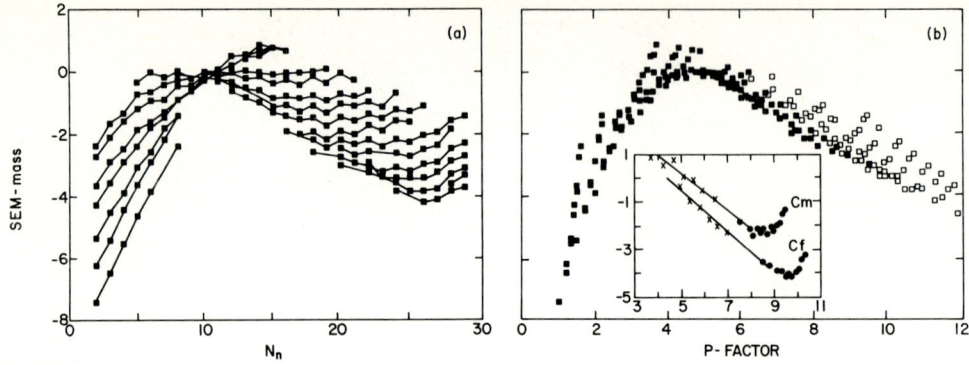

Figure 3: N_n and P plots for the actinides. In the N_n plot the solid lines connect isotopic chains. In part (b) the SEM masses for N>146 are represented by open squares. The inset illustrates the upturn for N>146 for two high Z sequences.

The linear dependence of SEM masses with P is striking but not yet fully understood theoretically. A simple model of SEM masses in terms of the p-n interaction would lead to isotopic and isotonic mass sequences proportional to $(A-BN_p)N_n$ and $(A'-B'N_n)N_p$, respectively, whereas, the best linearity is observed in P-factor rather than in $N_p N_n$ plots. Nevertheless, since the P-factor embodies the concept of the p-n interaction the implication that the structure part of the mass is somehow related to the p-n interaction among valence nucleons seems plausible.

Before leaving the actinides a closer look at Fig. 3 is warranted. Note that the open squares on the global P-plot (Fig. 3b) show a different feature which is illustrated in more detail in the inset. Starting from N=146 (the leftmost open square) the data level off relative to the descending line defined at lower P-values. Then as N increases further, an up-bend occurs. A simple explanation for this behavior is possible. There is some evidence in the actinide region for proton and neutron shells or subshells at Z=114 and N=152. If these are assumed valid instead of Z=126 and N=184 then points for N≥140 are plotted at incorrect P values in Fig. 3 and should be omitted from Fig. 4 since they fall in a different quadrant of p-n space. Another explanation is also possible in terms of the Fermion Dynamical Symmetry Model. We will not go into this explanation here but instead refer the reader to Refs. 4 and 5 for details.

We have extended our investigation of P-factor mass systematics to other regions of medium and heavy nuclei. In particular, a discussion of the rare earth region can be found in a forthcoming publication[4]. A feature of this region that complicates the analysis is the presence of a strong subshell closure at Z=64. A strict application of the P-formalism in this case dictates that the number of valence protons be counted as particles or as holes (whichever is appropriate) relative to Z=64 for N<90. It is well known that for N≥90 the traditional shell definitions (Z=50-82, N=82-126) are operable. Thus the P-formalism as currently

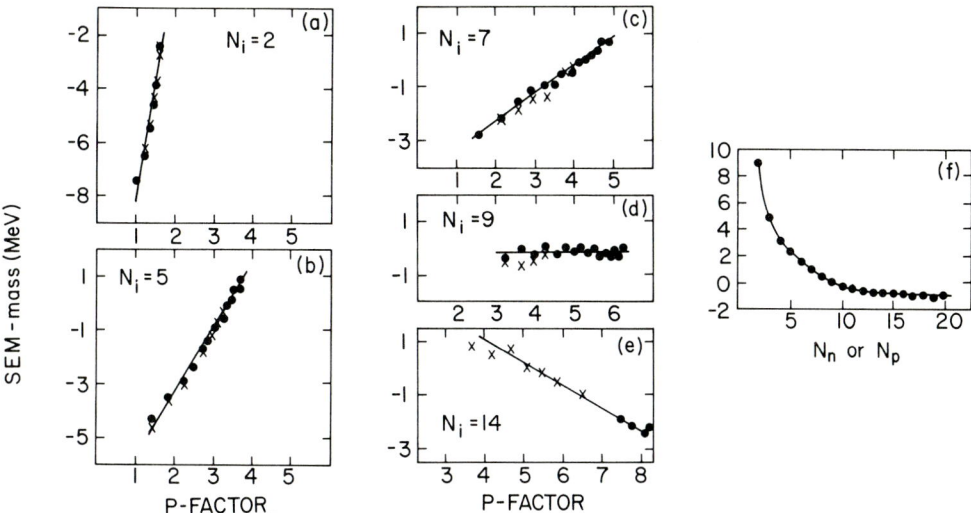

Figure 4: P factor plots for selected N_p and N_n values for actinide nuclei. Dots represent isotopic sequences and crosses are for isotonic chains. Panel (f) summarizes the slopes (per unit P) of the fitted straight lines as a function of N_p or N_n.

prescribed requires an abrupt transition in the number of valence protons for elements with 58<Z<72 when N=90. Recent studies of experimental g-factors and B(E2) values of 2^+_1 states in this region[6] have shown that one should more appropriately use N_p^{eff} values which reflect a smoothed transition in shell definition in the region of the Z=64 subshell. It is not surprising, therefore, to find that the presence of subshells causes some difficulties for any scheme (including the P-factor method) which relies solely on counting valence nucleons. This will become more evident as we examine mass systematics in the Zr region.

Figure 5 shows isotopic and isotonic SEM masses as a function of P for part of the "Zr region" which for our purposes we designate as either the mass space confined within the Z=28-50, N=50-82 shells or within the Z=38-50, N=50-82 region depending on whether one ignores or invokes the shell closure at Z=38. Systematics for the former are shown to the left (panels a-e) in Fig. 5 and those which presume a strong subshell closure at Z=38 are shown to the right (panels f-h). In the former case, we see that certain general features which were seen in the actinide region are again present here, namely, that the isotopic and isotonic data fall on more-or-less straight lines when plotted versus P and that the slope decreases as the number of valence nucleons of either types increases. A closer look at Fig. 5, however, reveals two discrepant features. In each case the isotonic data (crosses) depart from linearity by "rolling over" for higher values of P. In addition, as N_i increases there is a separation of the isotopic and isotonic systematics which require two, nearly parallel curves, instead of one, to represent the data. If we

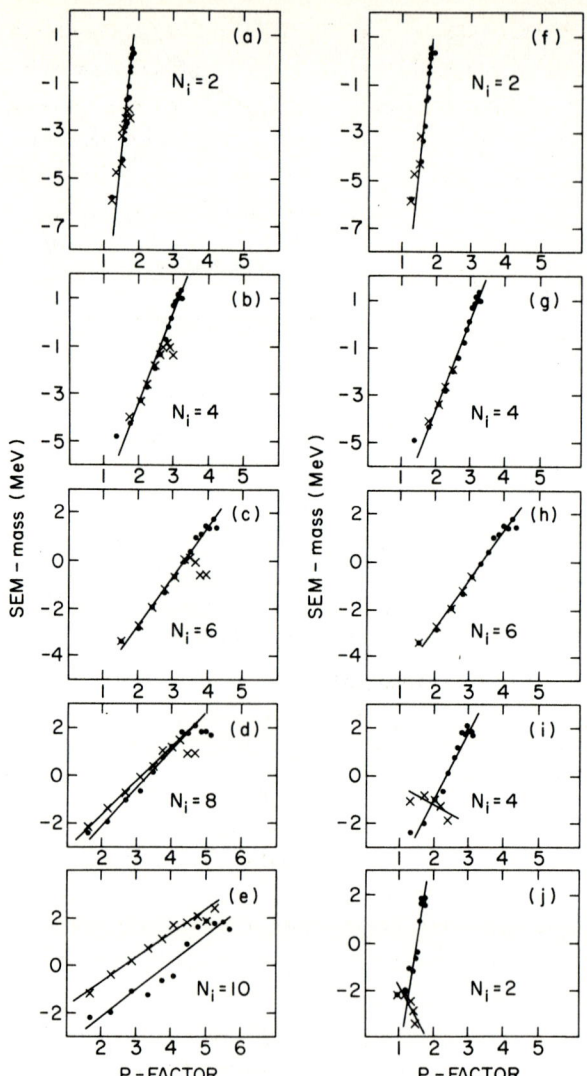

Figure 5: P factor plots for selected N_p and N_n values in the Zr region. Dots represent isotopic data and crosses are for isotonic chains. Panels (a-e) display data for p-hole, n-particle nuclei in the region defined by the shells Z=28-50, N=50-82 while panels (f-h) assume a proton shell closure at Z=38. Panels (i-j) display data for the p-particle, n-particle quadrant defined by the Z=38-50, N=50-82 shells.

instead incorporate a Z=38 subshell in valence particle counting we see (right hand side of Fig. 5) that the systematics are much improved for N_i = 2, 4 and 6 (the N_i = 8 and 10 cases no longer fall in the neutron-particle, proton-hole mass quadrant when the Z=38 subshell is invoked) and a very linear systematics in P results, again suggesting that the p-n interaction, as manifest in the P-factor, seems to describe to a remarkable extent the structure-dependent part of the mass. One must be cautious, however, because this agreement has actually been achieved by shifting the offending isotonic data to another mass quadrant (proton-particle, neutron-particle) rather than by bringing them onto a common line consistent with the other points. Examples of these data are shown in panels i-j. Neither the isotopic nor the shifted isotonic data display convincing linearity in the particle-particle quadrant nor is there any congruence. The best that can be said is that the isotonic data are more linear in P when the Z=38 subshell is invoked.

In summary, the SEM masses display remarkable linearity when plotted vs P in the neutron-particle, proton-hole space when a strong Z=38 subshell is invoked (a similar statement would apply were Z=40 assumed as the subshell closure), and the decrease in slope with increasing N_i is consistent with the expectation of binding effects between valence protons and neutrons. To the extent that deviations from linearity occur, we may be observing effects arising from the simplicity of our assumptions about the abrupt termination of shells and subshells.

In conclusion, we have applied the P formalism to atomic mass systematics for medium and heavy nuclei. The P-factor linearizes the structure dependent part of the nuclear mass in those regions which are free from subshell effects indicating that the attractive quadrupole p-n force plays an important role in determining the binding of valence nucleons. Where marked non-linearities occur, the P-factor provides a means for recognizing subshell closures and/or other structural features not embodied in the simple assumptions of abrupt shell or subshell changes. These are thought to be regions where the monopole part of the p-n interaction is highly orbit dependent and alters the underlying single particle structure as a function of A, N or Z. Finally, in those regions where the systematics are smooth and subshells are absent, the P-factor provides a means for predicting masses of some nuclei far-from-stability by <u>interpolation</u> rather than by extrapolation.

Work supported by the United States Department of Energy under contract DE-AC02-76CH00016 and grant DE-FG02-88ER40417.

References

1. R. F. Casten, D. S. Brenner, P. E. Haustein, Phys. Rev. Lett. <u>58</u>, 658 (1987).
2. P. Moller and J. R. Nix, Los Alamos National Laboratory Report LA-UR-90-1996; At. Data and Nuc. Data Tables <u>26</u>, 165 (1981).
3. A. H. Wapstra, G. Audi, and R. Hoekstra, Midstream (1986) Atomic Mass Evaluation, available from the NNDC, Brookhaven National Laboratory.
4. P. E. Haustein, D. S. Brenner, and R. F. Casten, Phys. Rev. C, in press (1988).
5. C.-L. Wu et al., Phys. Lett. <u>194B</u>, 447 (1987).
6. A. Wolf and R. F. Casten, Phys. Rev. C <u>36</u>, 851 (1987).

The Spin Dipole Interaction in Odd-Odd Nuclides

W.B. Walters

Department of Chemistry*, University of Maryland,
College Park, MD 20742, USA

In this paper, I wish to discuss some of the systematics of the spin-dipole term in the splitting of multiplets in odd-odd nuclides. Since the introduction of the Paar model for treating odd-odd nuclides in 1979, most of the attention has been focussed on the quadrupole term which produces the now-familiar parabolic splitting.[1] While the need for yet a third cubic term has been obvious for nuclides that are quite near double closed shells, the dimensions of the effects of the spin dipole term have not been explored in a general and systematic way. Inasmuch as some of the cleanest doublets are found in the mass 100 region in the Nb nuclides, this discussion would seem appropriate to this conference.

The position of any level in any multiplet is given by the sum of 6 terms, two arise from the single particle energies of the states in the adjacent odd-mass nuclide, two in the quadrupole term, and two from the spin-dipole term. For the quadrupole terms, one is a constant that has a sign opposite to the term with the $I(I+1)$ dependence in it. That is, if the ends of the parabola are pushed down (a tragic parabola), the whole parabola is pushed up, and if the ends of the parabola are pushed up (a comic parabola), the whole parabola is pushed down. For the dipole term which is linear in $I(I+1)$, the constant term is always positive and pushes up the whole line. In a general way, the quadrupole term works most strongly on the ends of the multiplet while the dipole term works most strongly on the higher spin states. Because many radioactive decay studies deal only with the low spin states, the effects of the dipole term are usually not significant.

On the other hand, for the doublets involving spin 1/2 levels, it is the dipole term that produces the entire splitting as there is no quadrupole interaction for doublets involving spin 1/2 levels. It is therefore possible to examine doublet splitting to determine the degree to which the estimates proposed by Paar in 1979 account for the observed splitting.

The actual splitting is dependent upon various constants associated with the spins of the single particle levels involved and the strength of the interaction. That strength has been given as $\alpha = 4a_1^2/\hbar\omega \sim 40/A$. In addition there is a multiplier for occupancy of $U_n V_n U_p V_p$. Because one of the particles is 1/2, the U and V are equal at 0.7 and the other particle is typically near a closed shell where UV is at a minimum relative and can be as low as 0.3 as contrasted with near midshell where $UV = 0.5$. But, the total variation produced by the UV term is less than a factor of 2.

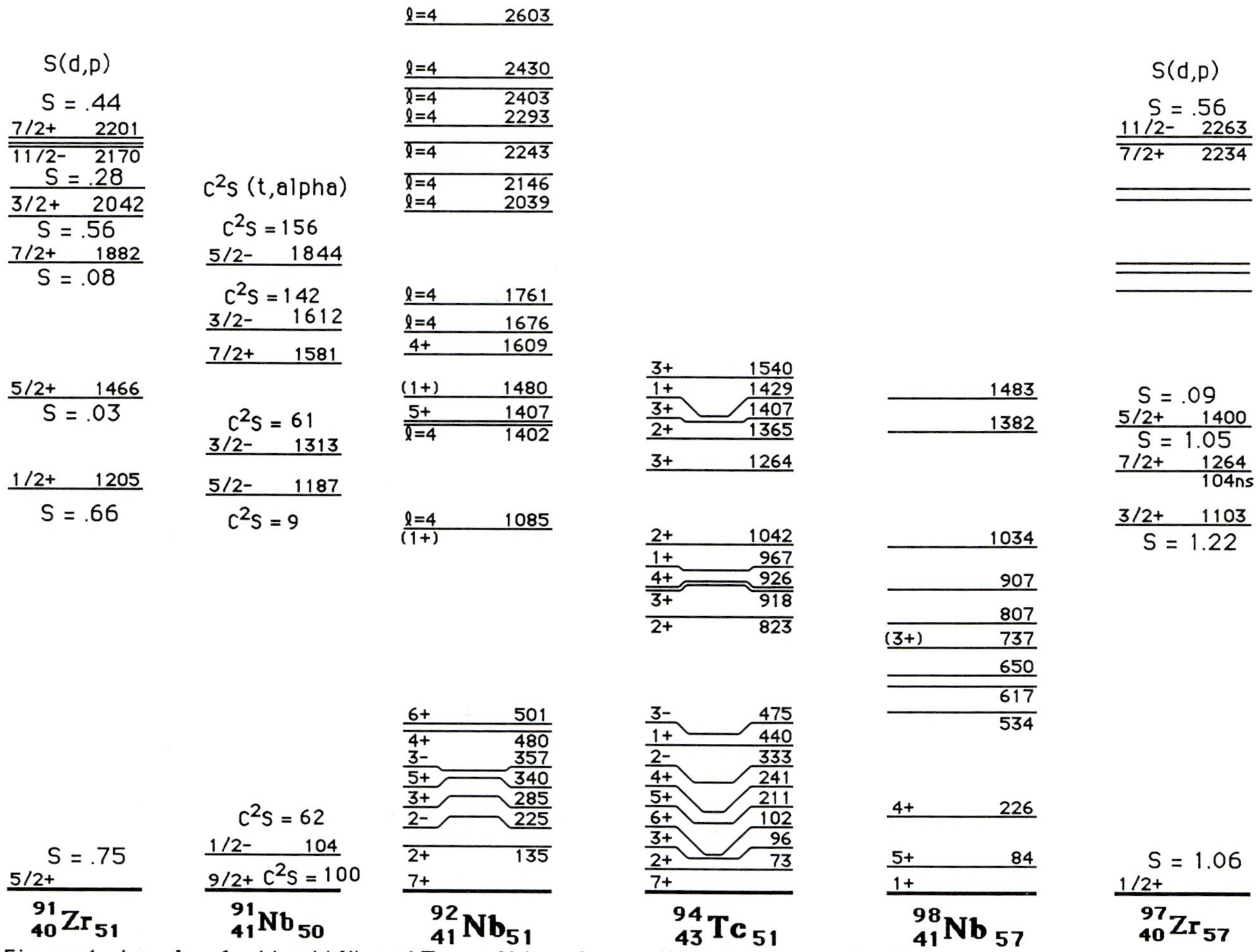

Figure 1. Levels of odd-odd Nb and Tc nuclides along with adjacent single particle levels

The least perturbed set of spin 1/2 doublets are present in the odd-odd Nb nuclides where both $\Pi g_{9/2} V s_{1/2}$ and $\Pi p_{1/2} V d_{5/2}$ doublets are present along with the $\Pi g_{9/2} V d_{5/2}$ multiplet and the spin-orbit partner $\Pi g_{9/2} V g_{7/2}$ multiplet. The low-lying proton orbitals are the $p_{1/2}$ and $g_{9/2}$ levels. The neutron structure includes the low-lying $d_{5/2}$ orbital with the $s_{1/2}$ and $g_{7/2}$ levels lying at considerably higher energies. Because of this separation, considerable attention has been previously focussed on the gd multiplet and its inversion with neutron occupancy change and quenching with proton occupancy change. The inversion is well known and an outstanding example of the effects of the $U^2 - V^2$ term. The size of the multiplet, shown in Figure 1, is about 500 keV in ^{92}Nb where $U^2 - V^2$ is proportional to 8. The quenching can be seen to be nearly quantitative in ^{94}Tc where $U^2 - V^2$ is proportional to 4 and the size of the multiplet has been reduced to 241 keV. At higher excitation, transfer reactions from ^{93}Nb have permitted considerable insight into the spin-orbit partner gg multiplet in ^{92}Nb. The 1$^+$ level is the lowest level and it eventually becomes the ground state in ^{98}Nb and in many other nuclides in this region. It can be seen that the stronger interaction between these 2 spin-orbit partner orbitals that was suggested by Federman and Pittel[2] is readily manifested by the nearly 1500-keV size of this multiplet. The size of the 2$^-$ - 3$^-$ doublet can be seen to remain virtually unchanged between ^{92}Nb and ^{94}Tc, as would be expected.

In Table 1 are presented the results of the most straightforward calculation of this doublet splitting using the formulae and constants given by Paar for the doublets in the Nb region and in In and light Sb nuclides as well. The critical constant is the spin-dipole interaction strength which was taken from Bohr and Mottelson[3] and is estimated at 40/A. The adjustable parameter is the 40 and the splitting is linear in this constant. There is, of course, the effect of the Nordheim number as shown in the parameter ξ. That calculation determines the sign of the $I(I+1)$ term, and whether the lower spin member of the doublet is down or up. For both of these doublets, the order is predicted correctly and the actual magnitude is not in bad agreement as shown in Table 1. This table is a printout of a general calculation which also has provision for the quadrupole contribution as well. As the proton occupancies do not change in this region and as the neutron 1/2 occupancy is always 0.7, one calculation suffices for all of the nuclides for the gs doublet. A slight broadening should be expected for the pd doublet as the d occupancy values change. Also shown are some data from the In nuclides. The interesting doublet is the $\Pi p_{1/2} v s_{1/2}$ multiplet identified in ^{112}In by the (^{3}He, d) reaction on ^{111}Cd which has an $s_{1/2}$ ground state. In Table 1 is the calculation for this doublet which is in modest agreement with the observed splitting. In this mass region, we have shown several examples of doublet splitting that are reasonably well described by the spin-dipole interaction.

Just across the shell are the levels of the light odd-odd Sb nuclides. The one that has the most interesting information is ^{116}Sb. Present are 2 multiplets that have been identified in stripping on ^{115}Sn which has a 1/2$^+$ ground state. Thus, the ds and gs doublets are readily identified.

Table 1. Doublet splitting in Nb, In, and Sb nuclides.

I	Jn	Jp	I(I+1)	Exp	Enorm	E1	C1o	C1	A1	Zeta	exp diff	calc diff	Up	Vp	Un	Vn

Calculation for the d 5/2 - p 1/2 multiplet in Nb-92

| 2 | 5/2 | 1/2 | 6 | 225 | 225 | 276 | 40 | 0.147 | 0.324 | -1.40 | 132 | 130 | 0.71 | 0.71 | 0.9 | 0.4 |
| 3 | | | 12 | 357 | 355 | 406 | 40 | 0.147 | 0.324 | -1.40 | | | 0.71 | 0.71 | 0.9 | 0.4 |

Calculation for the s 1/2 - g 9/2 multiplet in Nb-92

| 4 | 1/2 | 9/2 | 20 | 1609 | 1697 | -378 | 40 | 0.202 | 0.261 | 3.67 | 202 | 290 | 0.95 | 0.32 | 0.7 | 0.7 |
| 5 | | | 30 | 1407 | 1407 | -668 | 40 | 0.202 | 0.261 | 3.67 | | | 0.95 | 0.32 | 0.7 | 0.7 |

Calculation for the s 1/2 - p 1/2 split in In-112

| 0 | 1/2 | 1/2 | 0 | 915 | 903 | 60 | 40 | 0.060 | 0.357 | -3.00 | 300 | 238 | 0.71 | 0.71 | 0.7 | 0.7 |
| 1 | | | 2 | 1215 | 1141 | 298 | 40 | 0.060 | 0.357 | -3.00 | | | 0.71 | 0.71 | 0.7 | 0.7 |

Calculation for the s1/2 - g 7/2 multiplet in Sb-116

| 3 | 1/2 | 7/2 | 12 | 772 | 772 | 443 | 40 | 0.139 | 0.228 | -3.00 | 176 | 203 | 0.94 | 0.35 | 0.7 | 0.7 |
| 4 | | | 20 | 948 | 975 | 646 | 40 | 0.139 | 0.228 | -3.00 | | | 0.94 | 0.35 | 0.7 | 0.7 |

Calculation for the s1/2 - d 5/2 multiplet in Sb-116

| 2 | 1/2 | 5/2 | 6 | 103 | 309 | -192 | 40 | 0.116 | 0.257 | 4.20 | 103 | 308 | 0.91 | 0.41 | 0.7 | 0.7 |
| 3 | | | 12 | 0 | 0 | -501 | 40 | 0.116 | 0.257 | 4.20 | | | 0.91 | 0.41 | 0.7 | 0.7 |

Calculation for the s1/2 - g 7/2 multiplet in Sb-132

| 3 | 1/2 | 7/2 | 12 | 484 | 484 | 390 | 40 | 0.122 | 0.200 | -3.00 | 192 | 178 | 0.94 | 0.35 | 0.7 | 0.7 |
| 4 | | | 20 | 676 | 662 | 568 | 40 | 0.122 | 0.200 | -3.00 | | | 0.94 | 0.35 | 0.7 | 0.7 |

Calculation for the s1/2 - g 7/2 multiplet in Sb-122

| 3 | 1/2 | 7/2 | 12 | 61 | 61 | 422 | 40 | 0.133 | 0.217 | -3.00 | 192 | 193 | 0.94 | 0.35 | 0.7 | 0.7 |
| 4 | | | 20 | 253 | 253 | 614 | 40 | 0.133 | 0.217 | -3.00 | | | 0.94 | 0.35 | 0.7 | 0.7 |

Calculation for the s1/2 - d 5/2 multiplet in Sb-122

| 2 | 1/2 | 5/2 | 6 | 323 | 323 | -183 | 40 | 0.111 | 0.244 | 4.20 | 293 | 293 | 0.91 | 0.41 | 0.7 | 0.7 |
| 3 | | | 12 | 30 | 30 | -476 | 40 | 0.111 | 0.244 | 4.20 | | | 0.91 | 0.41 | 0.7 | 0.7 |

Calculation for the s1/2 - g 7/2 multiplet in Sb-132 using a larger interaction strength

| 3 | 1/2 | 7/2 | 12 | 484 | 484 | 487 | 50 | 0.153 | 0.251 | -3.00 | 192 | 223 | 0.94 | 0.35 | 0.7 | 0.7 |
| 4 | | | 20 | 676 | 707 | 710 | 50 | 0.153 | 0.251 | -3.00 | | | 0.94 | 0.35 | 0.7 | 0.7 |

The interaction strength is C1o/A, C1 is the constant term.

In view of these examples from the Zr region, it is possible to attempt to identify the spin-dipole term in the higher Sb nuclides and extend this study to more complex structures of ^{132}Sb and ^{130}Sb. Our initial reason for studying these nuclides was the identification of the 5+ member of the gd multiplet, the 4+ member of the gs doublet, and the 3+ member of the dd multiplet.[4] We succeeded in establishing clear identification of the 5+ level in both ^{132}Sb and ^{130}Sb and have some evidence for the other 2 levels in ^{132}Sb and for the 3+ level in ^{130}Sb. The importance was that we could now set about following the inversion of the gd multiplet as neutrons are removed from ^{132}Sb as the occupancy of the $d_{3/2}$ orbital changes. While the levels are quite complex and there is some mixing, transfer reactions on ^{123}Sb which has a $g_{7/2}$ ground state and on ^{121}Sb which has a $d_{5/2}$ ground state make it possible to identify the appropriate multiplets in ^{122}Sb. The fits are shown in Table 2.

The most serious problem has been the structure of ^{128}Sb which appears to have a linear structure. It became apparent that this nuclide was showing the effects of the linear dipole term in a nuclide where the quadrupole term was nearly zero because of the $U^2 - V^2$ term being nearly zero. Somewhat better fits were obtained with the larger spin-dipole interaction constant of 50, rather than the original estimate of 40. One additional important change was made for the spin-orbit partner multiplet, the proton $d_{5/2}$-neutron $d_{3/2}$ multiplet. The interaction strength was doubled by lowering the effective energy of the 2+ level, which is in the denominator of the formula for the energy splitting. With this doubled strength, both dd multiplets in ^{132}Sb and ^{130}Sb were much better fit.

From this investigation, it may be concluded that the spin-dipole term does play a role in describing the interaction between the single odd proton and the single odd neutron in an odd-odd nuclide. When coupled with an appropriately chosen quadrupole interaction strength, the parabola formulation of interactions in odd-odd nuclides can give an approximate picture of the expected structure. While the development of more comprehensive calculational procedures that take into account many of the terms missing in this simple formulation and that provide for calculation of the effects of level mixing will surely result in much better descriptions of odd-odd nuclides, this simple model has proved effective at predicting the order of doublets and the systematic inversion of both large and small multiplets in nuclides that are near closed shells.

The author is grateful for many fruitful discussions with numerous persons on the subject of odd-odd nuclides. Particularly helpful have been those with Vladimir Paar, Kris Heyde, Chien Chung, Richard Meyer, and Craig Stone from whose thesis the new data for the Sb nuclides is drawn.

---

* Work supported by the U. S. Dept. of Energy under Grant DE-FG05-88ER40418.
1   V. Paar, Nucl. Phys. **A331** (1979) 16.
2   P. Federman and S. Pittel, Phys. Rev. C **20**, (1979) 820.
3   A. Bohr and B. R. Mottelson, Nuclear Structure **II**, W. A. Benjamin, New York (1975).
4   C. A. Stone, Ph. D. Thesis, University of Maryland, 1987 (unpublished).

Table 2. Full calculations for odd-odd Sb nuclides

I	I(I+1)	Enorm	Exp	E1+E2	E1	E2	C1o	B(E2)	E2+	A2	C2	A1	C1	Up	Vp	UUp-VVp	UUn-VVn	Un	Vn

Calculation for Sb-132 7/2-3/2 multiplet
2	6	344	426	210	56	153	40	0.3	1.2	1.61	-0.13	0.19	0.08	0.94	0.35	0.75	0.5	0.9	0.5
3	12	114	85	-20	31	-51	40	0.3	1.2	1.61	-0.13	0.19	0.08	0.94	0.35	0.75	0.5	0.9	0.5
4	20	-1	0	-135	-2	-133	40	0.3	1.2	1.61	-0.13	0.19	0.08	0.94	0.35	0.75	0.5	0.9	0.5
5	30	162	162	28	-44	71	40	0.3	1.2	1.61	-0.13	0.19	0.08	0.94	0.35	0.75	0.5	0.9	0.5

Calculation for Sb-122 7/2 - 3/2 multiplet using same parameters except for the signs of the quadrupole terms
2	6	156	167	-97	56	-153	40	0.3	1.2	1.61	0.13	0.19	0.08	0.94	0.35	0.75	0.5	0.9	0.5
3	12	335	334	82	31	51	40	0.3	1.2	1.61	0.13	0.19	0.08	0.94	0.35	0.75	0.5	0.9	0.5
4	20	384	481	131	-2	133	40	0.3	1.2	1.61	0.13	0.19	0.08	0.94	0.35	0.75	0.5	0.9	0.5
5	30	138	137	-115	-44	-71	40	0.3	1.2	1.61	0.13	0.19	0.08	0.94	0.35	0.75	0.5	0.9	0.5

Calculation for Sb-132 for 5/2 - 3/2 multiplet
1	2	1342	1325	245	93	153	40	0.3	1.2	1.43	-0.12	0.21	0.08	0.91	0.41	0.667	0.5	0.9	0.5
2	6	1202	1078	105	126	-22	40	0.3	1.2	1.43	-0.12	0.21	0.08	0.91	0.41	0.667	0.5	0.9	0.5
3	12	1154	?955	57	177	-120	40	0.3	1.2	1.43	-0.12	0.21	0.08	0.91	0.41	0.667	0.5	0.9	0.5
4	20	1396	?	299	245	54	40	0.3	1.2	1.43	-0.12	0.21	0.08	0.91	0.41	0.667	0.5	0.9	0.5

Calculation for Sb-130 for 5/2 - 3/2 multiplet with slightly changed occupancies
1	2	877	1042	232	125	107	50	0.3	1.2	1.00	-0.08	0.29	0.10	0.91	0.41	0.667	0.35	0.8	0.6
2	6	801	813	156	171	-15	50	0.3	1.2	1.00	-0.08	0.29	0.10	0.91	0.41	0.667	0.35	0.8	0.6
3	12	801	744	156	240	-84	50	0.3	1.2	1.00	-0.08	0.29	0.10	0.91	0.41	0.667	0.35	0.8	0.6
4	20	1014	?	369	331	38	50	0.3	1.2	1.00	-0.08	0.29	0.10	0.91	0.41	0.667	0.35	0.8	0.6

Calculation for Sb-122 for 5/2 - 3/2 multiplet with the signs of the quadrupole terms changed
1	2	121	121	-60	93	-153	40	0.3	1.2	1.43	0.12	0.21	0.08	0.91	0.41	0.667	0.5	0.9	0.5
2	6	329	397	148	126	22	40	0.3	1.2	1.43	0.12	0.21	0.08	0.91	0.41	0.667	0.5	0.9	0.5
3	12	478	484	297	177	120	40	0.3	1.2	1.43	0.12	0.21	0.08	0.91	0.41	0.667	0.5	0.9	0.5
4	20	371	394	190	245	-54	40	0.3	1.2	1.43	0.12	0.21	0.08	0.91	0.41	0.667	0.5	0.9	0.5

Calculation for Sb-132 7/2-3/2 multiplet using C1o = 50
2	6	358	426	224	70	153	50	0.3	1.2	1.61	-0.13	0.23	0.10	0.94	0.35	0.75	0.5	0.9	0.5
3	12	122	85	-12	39	-51	50	0.3	1.2	1.61	-0.13	0.23	0.10	0.94	0.35	0.75	0.5	0.9	0.5
4	20	-1	0	-135	-3	-133	50	0.3	1.2	1.61	-0.13	0.23	0.10	0.94	0.35	0.75	0.5	0.9	0.5
5	30	151	162	17	-55	71	50	0.3	1.2	1.61	-0.13	0.23	0.10	0.94	0.35	0.75	0.5	0.9	0.5

Calculation for Sb-130 7/2-3/2 multiplet using C1o = 50
2	6	263	262	183	76	107	50	0.3	1.2	1.13	-0.09	0.25	0.11	0.94	0.35	0.75	0.35	0.8	0.6
3	12	87	70	7	42	-36	50	0.3	1.2	1.13	-0.09	0.25	0.11	0.94	0.35	0.75	0.35	0.8	0.6
4	20	-16	0	-96	-3	-93	50	0.3	1.2	1.13	-0.09	0.25	0.11	0.94	0.35	0.75	0.35	0.8	0.6
5	30	71	63	-9	-59	50	50	0.3	1.2	1.13	-0.09	0.25	0.11	0.94	0.35	0.75	0.35	0.8	0.6

Calculation for Sb-128 7/2-3/2 multiplet
2	6	153	152	97	81	15	50	0.3	1.2	-0.16	-0.01	0.27	0.12	0.94	0.35	0.75	-0.05	0.7	0.7
3	12	96	79	40	45	-5	50	0.3	1.2	-0.16	-0.01	0.27	0.12	0.94	0.35	0.75	-0.05	0.7	0.7
4	20	40	46	-16	-3	-13	50	0.3	1.2	-0.16	-0.01	0.27	0.12	0.94	0.35	0.75	-0.05	0.7	0.7
5	30	0	0	-56	-63	7	50	0.3	1.2	-0.16	-0.01	0.27	0.12	0.94	0.35	0.75	-0.05	0.7	0.7

Calculation for Sb-132 for 5/2 - 3/2 multiplet using double interaction strength
1	2	1325	1325	421	116	305	50	0.3	0.6	2.86	-0.24	0.26	0.09	0.91	0.41	0.667	0.5	0.9	0.5
2	6	1018	1078	114	158	-44	50	0.3	0.6	2.86	-0.24	0.26	0.09	0.91	0.41	0.667	0.5	0.9	0.5
3	12	886	?955	-18	221	-240	50	0.3	0.6	2.86	-0.24	0.26	0.09	0.91	0.41	0.667	0.5	0.9	0.5
4	20	1319	?	415	306	109	50	0.3	0.6	2.86	-0.24	0.26	0.09	0.91	0.41	0.667	0.5	0.9	0.5

Calculation for Sb-130 for 5/2 - 3/2 multiplet using double interaction strength
1	2	1042	1042	339	125	214	50	0.3	0.6	2.00	-0.17	0.29	0.10	0.91	0.41	0.667	0.35	0.8	0.6
2	6	843	813	140	171	-31	50	0.3	0.6	2.00	-0.17	0.29	0.10	0.91	0.41	0.667	0.35	0.8	0.6
3	12	775	744	72	240	-168	50	0.3	0.6	2.00	-0.17	0.29	0.10	0.91	0.41	0.667	0.35	0.8	0.6
4	20	1110	?	407	331	76	50	0.3	0.6	2.00	-0.17	0.29	0.10	0.91	0.41	0.667	0.35	0.8	0.6

Part VIII

Beta Decay at A~100

Part VIII

Fast First-Forbidden Transitions and Subshell Closures in the Region of ^{96}Zr

H. Mach[1,2], E.K. Warburton[2], R.L. Gill[2], R.F. Casten[2], A. Wolf[2,3],
Z. Berant[2,3], J.A. Winger[4], K. Sistemich[5], G. Molnár[6,7], and S.W. Yates[6]

[1] Clark University, Worcester, MA 01610, USA
[2] Brookhaven National Laboratory, Upton, NY 11973, USA
[3] Physics Department, Nuclear Research Centre, Beer Shiva, Israel
[4] Ames Laboratory, Iowa State University, Ames, IA 50011, USA
[5] Institut für Kernphysik, Kernforschungsanlage Jülich,
 Postfach 1913, D-5170 Jülich, Fed. Rep. of Germany
[6] University of Kentucky, Lexington, KY 40505, USA
[7] Institute of Isotopes, H-1525 Budapest, Hungary

Abstract

The low-spin β-decay of ^{96,98}Y has been reinvestigated to obtain detailed information on the fast first-forbidden transitions to the ground states of ^{96,98}Zr. A considerably more detailed decay scheme of ^{96}Y confirms the general features of a previous study. The spin of the ground state of ^{98}Y has been measured to be 0^- from $\gamma\gamma(\Theta)$ angular correlations in ^{98}Sr$\rightarrow^{98}$Y decay and from Q_β measurements by $\beta-\gamma$ coincidences in the decay of ^{98}Y$\rightarrow^{98}$Zr. From β-decay data on $^{90-100}$Y nuclei, mixing of the "spherical" and "collective" 0^+ states is deduced, and the strength of the mixing matrix elements is found to vary greatly across the Zr nuclei.

Introduction

One of the unique features of the A=100 region is the existence of a nearly doubly magic ^{96}Zr nucleus in close vicinity to deformed heavier Zr nuclei. Strong subshell closures at Z=40 and N=56 are manifested [1] by a high-lying 2_1^+ state at 1750 keV and by the purity of the 0^+ ground state (g.s.) configuration. A previous study [1] clarified the existence of a fast first-forbidden $0^-\rightarrow 0^+$ β transition in the decay of ^{96}Y.

Fast $0^-\rightarrow 0^+$ first-forbidden β transitions have been observed [1-3] in the vicinity of doubly-closed shell nuclei such as ^{16}O, ^{96}Zr, and ^{208}Pb, and attributed to the $\nu s_{1/2}\rightarrow\pi p_{1/2}$ transition. The purity of the shell model configurations and the high overlap between the configurations involved provide enhancement [4] of the decay rates over typical limits [5]. Further enhancement is caused by meson-exchange effects [6]. In fact, shell model calculations based on the data from Ref. 1 indicate a ~70% enhancement of the ^{96}Y decay rate due to meson-exchange effects [6].

The study of Ref. 1 did not elucidate the β feeding to high-lying levels in ^{96}Zr or high energy γ transitions feeding the g.s. Although this feeding is generally weak and does not significantly alter $\log ft$ values deduced for low-lying levels, evidence of strongly β-fed (~50%) 1^- levels at excitation energies of ~4.2 MeV has been reported [7] for the decay of ^{98}Y$\rightarrow^{98}$Zr. Thus the possibility that the decay of 0^- ^{96}Y→^{96}Zr has similar features is opened. Motivated by the importance of a correct decay scheme to the study of meson-exchange effects we have reinvestigated the low-spin β decay of ^{96}Y in order to clarify the β feeding to the high-lying levels and to confirm other quantities pertinent to the assignment of logft values. We have also clarified the spin assignment of the ^{98}Y g.s. which feeds the 0^+ g.s. of ^{98}Zr by a rather fast first-forbidden β transition [7].

The experiments were performed at the TRISTAN fission product mass separator at BNL. Details of the experimental procedures will be published elsewhere [8]. We have measured γ singles, γ-multispectral scaling, and $\gamma - \gamma$ and $e^- - \gamma$ coincidences in the decay of the 0^- ^{96}Y isomer. For A=96,98 nuclei Q_β was measured using a $\beta - \gamma$ coincidence technique and absolute E0 and γ-ray intensities were obtained from singles conversion electron and γ spectra measured at beam saturation. The $\gamma\gamma(\Theta)$ angular correlation experiments for the decay of ^{98}Sr→^{98}Y were performed using a four-detector system [9] which allowed simultaneous measurements at angles of 90°, 105°, 120°, 135°, 150°, and 165°.

^{96}Y→^{96}Zr

Our more precise measurements of this decay confirm the findings of Ref. 1. Although 35 new levels were found at the excitation energies from 3.5 to 6.5 MeV, their total β feeding is only ~0.5%. The new decay scheme will not be presented here. It is sufficient to note for the present purpose that no new energy levels are found below 3.0 MeV and that the new values for the half-life of 5.34(5) s, Q_β=7.07(3) MeV and the g.s. β feeding of 95.5(5)% for ^{96}Y→^{96}Zr compare well with the previous values [1,10] of 5.4(1) s, 7.14(4) MeV and 95.2(9)%. The important β feeding [1] of the 0^+ 1581-keV level is confirmed to be 1.2(2)% in agreement with the value of 1.3(5)% reported previously [1]. The β feeding to the 2^+ state at 1750 keV was found to be 1.9(3)% which is slightly lower than the previously obtained [1] value of 2.4(9)%. Absolute intensities of the E0 and γ-ray transitions reported [1,7] for the A=96 and 98 mass chains are confirmed. In particular, the absolute intensity of the 1581-keV E0 transition in ^{96}Zr was found to be 1.40(15)% in agreement with the value of 1.4(5)% of Ref. 1.

^{98}Y→^{98}Zr.

Similar to the decay of ^{96}Y, the decay of ^{98}Y has been found [7] to proceed via a rather fast first-forbidden β transition to the g.s. of ^{98}Zr, logft=5.8. The negative parity of the g.s. of ^{98}Y depends [7] on the J^π=1^+ assignments for the 547 and 600-keV levels in this nucleus. These assignments were deduced [7] from logft values of 4.7 and 4.3, respectively, observed in the β decay of the 0^+ g.s. of ^{98}Sr and electron conversion coefficients for the γ transitions deexciting these 1^+ levels and feeding the g.s. New conversion coefficients of α_K=0.08(2), 0.0017(3), 0.0037(4), 0.0013(8), and 0.0025(7) for the 119, 428, 444, 481 and 564-keV γ rays in ^{98}Y support the multipolarity assignments used in Ref. 7.

The spin J=1 was assigned [7] for the g.s. of ^{98}Y based on β-decay rates of logft~6 to the known 0^+ and 2^+ levels in ^{98}Zr that were deduced from the absolute E0 and γ-ray intensities of Ref. 7 and the decay scheme of Ref. 11. A fast first-forbidden β transition [7], logft=5.8, between the ground states of ^{98}Y and ^{98}Zr implies [4] rather pure shell model configurations for these states. $J^\pi=1^-$ can be formed in the ^{98}Y g.s. by coupling either the $s_{1/2}$ or $d_{3/2}$ neutron to a valence $p_{1/2}$ proton. However, the parabolic rule suggests [12] the g.s. is 0^- for $\nu s_{1/2}$ or 2^- for $\nu d_{3/2}$ coupling.

The angular correlation data rule out the J=1 assignment for the ^{98}Y g.s. but support either a J=0 or 2 value. The 428–119 keV γ cascade in ^{98}Y has $a_2=-0.33(5)$ and $a_4=0.09(8)$ and is consistent with $a_2=-0.23(6)$ and $a_4=0.07(10)$ measured by Becker [13]. The correlation data were fitted to various spin sequences (see Fig. 1) restricted by $J^\pi=1^+$ for the 547-keV level and δ(L+1/L)-mixing ratios allowed by the internal conversion coefficients listed above. Two spin sequences, 1→1→0 and 1→2→2, were found to be consistent with the data and give $J^\pi=0^-$ or 2^- for the g.s. of ^{98}Y. This assignment is in conflict with the apparent strong β feeding, logft~6, to levels of spin 0^+ and 2^+ in ^{98}Zr [7,11]. This issue was resolved by the Q_β measurements.

Table 1 presents results from Q_β measurements by $\beta-\gamma$ coincidences for three groups of levels in ^{98}Zr. In the first group, which consists of high-lying levels ($E_x \geq 4.0$ MeV), β-feeding energies, E_β, provide a Q_β value of 8963(41) keV which is consistent with the adopted value of Q_β=8890(70) keV [10]. Similar Q_β values are deduced for the group of 0^+ levels and confirm direct β-feeding to these levels. However, the Q_β values for the group of 2^+ levels are lower than the accepted Q_β energy and show no consistent pattern. The values imply an indirect β feeding to these levels while the lack of a pattern rules out the possibility of a second isomer in ^{98}Y. The decay scheme of Ref. 11 is thus incomplete. Strong β feeding to the 0^+ states in ^{98}Zr and weak feeding to the 2^+ levels rules out $J^\pi=2^-$ for the g.s. of ^{98}Y but supports the 0^- spin assignment.

Mixing matrix elements for 0_1^+ and 0_2^+ states in even-even Zr nuclei

The mixing of the 0^+ g.s. and the low-lying 0^+ intruder state has been previously deduced for $^{90-94}$Zr from particle transfer reaction data [14] under the assumption that the dominant proton component of the 0_2^+ states in Zr arises from a proton pair excitation across the Z=40 subshell gap from the $\pi p_{1/2}$ to the $\pi g_{9/2}$ orbit. The g.s. and 0_2^+ states are assumed to have admixed wave functions $\Psi_{gs}=A\varphi_1+B\varphi_2$ and $\Psi_{0_2^+}=-B\varphi_1+A\varphi_2$, where φ_1 has a $(\pi p_{1/2})^2$ and φ_2 a $(\pi g_{9/2})^2$ configuration. The mixing of these states can also be deduced from the β^- decay of Y→Zr nuclei [1,7]. Since the β-decay matrix element connecting the odd $d_{5/2}$ neutron in $^{90-94}$Y (and $s_{1/2}$ neutron in ^{96,98}Y) with a $p_{1/2}$ proton is large, while the transition to a $g_{9/2}$ proton is highly hindered, the β feeding of the 0_2^+ state is attributed to admixtures of the g.s. configuration. Consequently, from the ratio of the comparative half-lives, which is approximately equal to the square of the amount of mixing, one can deduce the B^2 value for these levels. Table 2 (columns 3 and 5) gives B^2 values deduced from particle transfer [14] and β-decay data [1,7,16], respectively. In columns 4 and 6 the strength of the mixing matrix element, $M=\langle\varphi_1 \mid V \mid \varphi_2\rangle$, is deduced assuming two-level mixing and using the data in columns 2, 3 and 5, respectively.

Table 2 illustrates two characteristic features of the heavy Zr nuclei: a) the mixing amplitude of the 0_1^+ and 0_2^+ states, B^2, shows a sharp minimum at N=56, where it has a much lower value than at N=50, and b) the value of the mixing matrix element, M, is gradually reduced with increased neutron number (there may be a slight increase for N=58) and shows a sudden drop at N=60. In comparison to ^{90}Zr, the very low mixing amplitude in ^{96}Zr arises from the much lower mixing matrix element; while in comparison to the N=52-60 nuclei, it is mainly due to the higher 0_2^+ excitation energy in ^{96}Zr which reflects the strong subshell closure at Z=40 and N=56. The striking difference observed [1] in the mixing amplitudes for two otherwise similar nuclei, ^{90}Zr and ^{96}Zr, can now be explained by following the argument of Heyde et al. [17] discussed next.

Heyde et al. [17] have pointed out that a large reduction in the mixing matrix element connecting the 0_1^+ and 0_2^+ states is needed (from ~800 keV at N=50 to ~200 keV at N=60) to explain the very low 0_2^+ excitation energy in ^{98,100}Zr. This reduction has been calculated [17] as arising from a diminished overlap between spherical and Nilsson model deformed neutron wave functions, which describe the neutron component in spherical and collective (intruder) configurations, respectively. As deformation sets in the configuration mixing characterizing the Nilsson wave functions reduces their overlap with spherical shell model configurations.

Discussion

The present study confirmed the general features of the decay of the 0^- isomer of ^{96}Y reported in Ref. 1 including the logft=5.6 for the crucial $0^- \to 0^+$ fast first-forbidden β transition to g.s. of ^{96}Zr. Moreover, it established $J^\pi=0^-$ for the g.s. of ^{98}Y with an expected shell model configuration of $(\nu s_{1/2} \pi p_{1/2})0^-$. Consequently, the $0^- \to 0^+$ transitions in ^{96,98}Y$\to^{96,98}$Zr similarly arise predominantly from the $\nu s_{1/2} \to \pi p_{1/2}$ decay. Furthermore, we postulate that the g.s. of ^{96}Sr, which is expected to be characterized by a strong $(\nu s_{1/2})^2 0^+$ component [18], β decays to the 0^- g.s. of ^{96}Y via a fast $\nu s_{1/2} \to \pi p_{1/2}$ first-forbidden transition. This transition is still to be verified as it is masked by an intense GT $0^+ \to 1^+$ β transition observed in this decay. These three transitions would form a region of fast $0^- \leftrightarrow 0^+$ β transitions in the vicinity of ^{96}Zr.

The fast first-forbidden $0^- \to 0^+$ β transitions near ^{96}Zr are related to the strong subshell closures at Z=40 and N=56 [1]. However, the $\nu s_{1/2}$ subshell closure at N=58 expected [7,19] from the similarity of ^{96}Zr and ^{98}Zr level schemes is not as strong as the $\nu d_{5/2}$ subshell closure at N=56 in ^{96}Zr. This is evident from much lower 0_2^+ energy in ^{98}Zr and disruption of normal filling of the shell model orbits at N=59. At N=58 the $\nu s_{1/2}$ orbit should be filled and the next neutron should go to the $\nu g_{7/2}$ orbit. However, $J^\pi=0^-$ for the g.s. of ^{98}Y (Z=39, N=59) indicates that the odd valence neutron occupies the $s_{1/2}$ orbit. A similar effect is observed [18] in ^{97}Sr (Z=38, N=59) where the odd N=59 neutron also resides in the spherical $s_{1/2}$ orbit.

Large mixing of the intruder deformed configurations with the g.s. wave functions can significantly alter the composition of the β-transition matrix element. However, studies of the intruder bands in ^{96,98}Zr imply [1,19,20] a rather weak deformation, $\beta_2 \sim 0.2$, for the 0_2^+ intruder states. The same conclusion can be drawn from the calculations of Heyde et al. [17] and the relatively large values of the mixing matrix elements for ^{96,98}Zr listed in Table 2. As a consequence, one does not expect strong admixtures of other configurations (like $\nu d_{3/2} \to \pi p_{3/2}$ [6]) to the main component $\nu s_{1/2} \to \pi p_{1/2}$ in the $0^- \to 0^+$ β decay of ^{96}Y and, perhaps to a lesser extent, of ^{98}Y, as indicated by its larger logft value.

Research has been performed under the contracts No. DE-AC02-76CH00016 and DE-AC02-79ER10493 with the U.S. Department of Energy.

[1] H.Mach et al., Phys.Rev. **C37**, 254 (1988).

[2] C.A.Galiardi, G.T.Garvey, and J.R.Wroubel, Phys.Rev. **C28**, 2423 (1983).

[3] M.P.Webb, Nucl.Data Sheets **26**, 145 (1979).

[4] J.Damgaard, R.Broglia, and C.Riedel, Nucl.Phys. **A135**, 310 (1969).

[5] S.Raman and N.B.Gove, Phys.Rev. **C7**, 1995 (1973).

[6] J.A.Becker, E.K.Warburton, and B.A.Brown, *Proceedings of Interactions and Structures in Nuclei*, Eds. R.Blin-Stoyle and W.Hamilton (Institute of Physics, London, 1987) to be published.

[7] H.Mach and R.L.Gill, Phys.Rev. **C36**, 2721 (1987).

[8] H.Mach et al., to be published.

[9] A.Wolf et al., Nucl.Instrum.Methods **206**, 397 (1983).

[10] A.H.Wapstra and G.Audi, Nucl.Phys. **A432**, 1 (1985).

[11] H.-W. Muller, Nucl.Data Sheets **39**, 467 (1983).

[12] S.Brant et al., Z.Phys. **A329**, 301 (1988), see also V.Paar Nucl.Phys. **A331**, 16 (1979).

[13] K.Becker, Thesis Universität Giessen, F.R.Germany, 1983.

[14] M.R.Cates, J.B.Ball, and E.Newman, Phys.Rev. **187**, 1682 (1969).

[15] Table of Isotopes, ed. by C.M.Lederer and V.S.Shirley (Wiley, N.Y., 1978).

[16] F.K.Wohn et al., Phys.Rev. **C33**, 677 (1986).

[17] K.Heyde, E.D.Kirchuk, and P.Federman, to be published.

[18] F.Buchinger et al., Z.Phys. **A327**, 361 (1987).

[19] R.A.Meyer et al., Phys.Lett. **177B**, 271 (1986).

[20] M.L.Stolzenwald et al., this Conference.

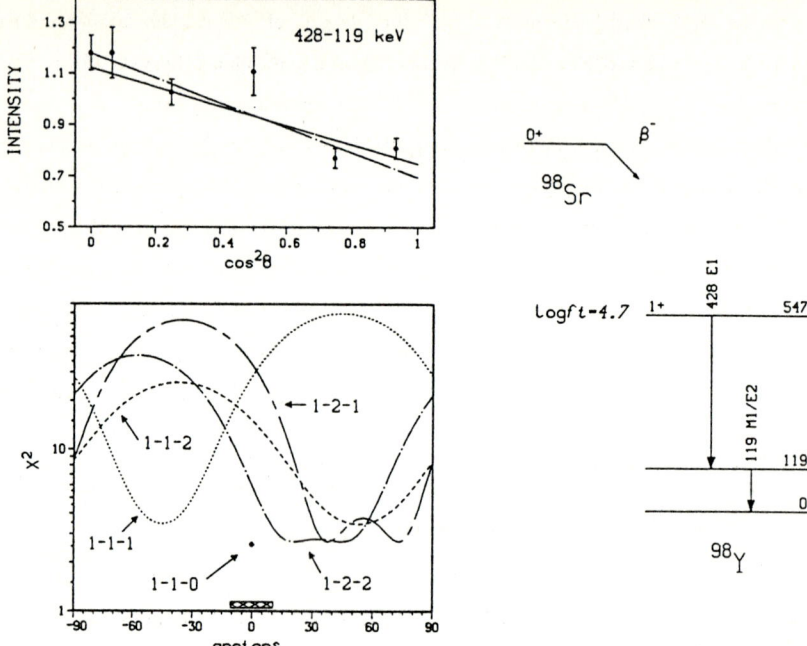

Fig. 1. Angular correlations for 428-119 keV cascade in ^{98}Y. The top panel indicates the best fit of the theoretical coefficients to the experimental data for 1-2-1, 1-2-2 (solid line) and 1-1-0, 1-1-1, and 1-1-2 (broken line) spin sequences. In the bottom panel χ^2 is plotted against $\delta(L+1/L)$ (expressed as $arctan\delta$) for the same sequence of spins. The shaded area illustrates the acceptable δ mixing ratio for the 119-keV transition obtained from conversion coefficients. A partial decay scheme of ^{98}Sr [7] is illustrated on the right.

TABLE 1

E_γ [keV]	E_x [keV]	J	E_β [keV]	Q_β [keV]	$Q'_\beta - E_x - E_\beta$ [keV]
2941	4164		4820(45)	8984(45)	–
3228	4451		4483(88)	8934(88)	–
3310	4164		4854(62)	9018(62)	–
4450	4450		4430(62)	8881(62)	–
			Q'_β =	8963(41)	
213	1437	0$^+$	7437(76)	8874(76)	85(86)
268	1859	0$^+$	7049(46)	8908(46)	52(61)
1223	1223	2$^+$	7248(100)	8471(100)	488(108)
1591	1591	2$^+$	6605(80)	8196(80)	763(89)
1744	1744	2$^+$	4648(64)	6392(64)	2567(75)

TABLE 2

Nucleus	$E_{0_2^+}$ a) [keV]	β^2 b)	M [keV]	β^2 c)	M [keV]
^{90}Zr$_{50}$	1760.7	0.40(4)	863	0.26	772
^{92}Zr$_{52}$	1383.0	0.50(4)	692	0.33	650
^{94}Zr$_{54}$	1300.4	0.35(4)	620	0.23	547
^{96}Zr$_{56}$	1581.4	0.09(3)	453	0.04(2)	310
^{98}Zr$_{58}$	854.0	–	–	0.29	388
^{100}Zr$_{60}$	331.0	–	–	0.24	141

a) from Refs. 1,15;   b) from Ref. 14.;   c) from Refs. 1,7,15,16.

Structure of 1^+ States in Neutron Deficient Odd-Odd Indium Isotopes

L. Kalinowski

Institute of Experimental Physics, Warsaw University,
ul. Hoza 69, 00681 Warsaw, Poland

Abstract

The particle-core coupling and residual neutron-proton charge exchange interaction are applied to the description of excited states in odd-odd indium isotopes. The distribution of the Gamow-Teller strength over a number of 1^+ levels is naturally explained.

1. Introduction

In a series of recent experiments the Gamow-Teller beta decays of even-even nuclei near doubly magic ^{100}Sn were studied (see e.g. [1-3] and references quoted therein). For each of these decays, the single particle shell model predicts only one possibility of the transition from the 0^+ ground state of the parent nucleus to a 1^+ state of the daughter, namely $\pi\, 1g_{9/2} \to \nu\, 1g_{7/2}$. However, experimentally one finds transitions to a few 1^+ states, which is interpreted as a result of configuration mixing. The splitting of the GT strength has been observed in particular for the ^{104,106,108}Sn $\to$ ^{104,106,108}In decays [2]. The present paper attempts to present a simple phenomenological description of the positive parity states in the daughter indium isotopes and to explain experimental energy levels and the GT strength distribution. The description is based on the coupling of neutrons and protons to core vibrations - a model succesfully applied to negative parity levels [4]. An additional residual charge exchange interaction between $1g_{7/2}$ neutrons and $1g_{9/2}$ protons is introduced to obtain appropriate strength splitting.

2. Description of the Model

Basis for the description of the J^π states in indium nuclei consists of single quasiparticle neutron ν nlj and proton π nlj states of the spherical Woods-Saxon potential with pairing [5,6] coupled to quadrupole core vibrations:

$$|(N,L)[\nu\, nlj, \pi\, nlj] I; J^\pi\rangle, \qquad (1)$$

where (N, L) denote the number and angular momentum of quadrupole phonons. Quasiparticle energies and occupation probabilities for subshells above N=50 and below Z=50 in $^{104-108}$In are listed in Table 1 [6]. In fact, only the $1g_{9/2}$ proton and the $2d_{5/2}$, $1g_{7/2}$, $3s_{1/2}$, $2d_{3/2}$ neutrons are considered to obtain low lying positive parity levels. Up to 3 quadrupole phonons are taken into account.

Table 1.
Relative quasiparticle energies (in MeV) and occupation probabilities in the spherical Woods-Saxon potential with pairing [6]

	^{104}In		^{106}In		^{108}In	
Neutrons	E_ν	v_ν^2	E_ν	v_ν^2	E_ν	v_ν^2
$2d_{5/2}$	0.00	0.368	0.00	0.490	0.00	0.583
$1g_{7/2}$	0.55	0.145	0.29	0.234	0.08	0.332
$3s_{1/2}$	1.26	0.074	0.92	0.115	0.60	0.161
$2d_{3/2}$	1.75	0.052	1.34	0.081	0.96	0.115
$1h_{11/2}$	1.78	0.051	1.40	0.078	1.40	0.107
Protons	E_π	v_π^2	E_π	v_π^2	E_π	v_π^2
$1g_{9/2}$	0.00	0.929	0.00	0.933	0.00	0.936
$2p_{1/2}$	1.00	0.970	1.00	0.972	0.97	0.973
$2p_{3/2}$	2.51	0.987	2.51	0.988	2.56	0.989
$1f_{5/2}$	2.89	0.989	2.93	0.990	3.03	0.991

The hamiltonian of the neutron-core interaction reads:

$$H_{\nu-core} = -(\pi/5)^{1/2}\, \xi_\nu\, \hbar\omega \sum_{\nu\nu'\mu} \langle\nu'|Y_{2\mu}|\nu\rangle \times$$
$$\times [b_{2\mu} + (-1)^\mu b^+_{2-\mu}]\, N(a^+_{\nu'} a_\nu), \qquad (2)$$

where $b_{2\mu}$ and a_ν are annihilation operators of quadrupole phonons and neutrons, respectively. The proton-core interaction has the same form and may be obtained by changing the ν, ν' indices to π, π'. Following ref. [4], $\hbar\omega$ is the core excitation energy and ξ_ν, ξ_π are the coupling constants.

The effect of the particle-core interaction is a well known parabolic splitting of neutron-proton multiplets [7]. As far as positive parity indium states are concerned, the interaction reproduces quite properly the position of low lying 2^+ and 3^+ levels which are built of almost pure $[\nu\ 2d_{5/2}, \pi\ 1g_{9/2}]$, $[\nu\ 1g_{7/2}, \pi\ 1g_{9/2}]$ configurations. It also mixes the $|(0,0)[\nu\ 1g_{7/2}, \pi\ 1g_{9/2}]1,1^+\rangle$ state with states containing one quadrupole phonon, giving rise to the distribution of beta decay strength over a number of 1^+ levels. However, the mixing is very weak and almost the entire strength is focused in the lowest 1^+ level.

Stronger configuration mixing is obtained by introducing the residual neutron-proton interaction of Gamow-Teller type [8]:

$$H_{GT} = 2\varkappa \sum_{\nu\nu'\pi\pi'\mu} \langle\pi|\sigma_\mu^+|\nu\rangle \langle\nu'|\sigma_\mu|\pi'\rangle\ N(a^+_\nu, a_\pi, a^+_\pi a_\nu) \qquad (3)$$

In the case of indium nuclei, it affects only the $[\nu\ 1g_{7/2}, \pi\ 1g_{9/2}]$ multiplet. The corresponding matrix element has a simple form:

$$\langle[\nu\ 1g_{7/2}, \pi\ 1g_{9/2}]I|H_{GT}|[\nu\ 1g_{7/2}, \pi\ 1g_{9/2}]I\rangle =$$
$$= 2\varkappa\ (160/9)\ [\ \delta_{1I}(u_\nu^2 v_\pi^2 + v_\nu^2 u_\pi^2) + \qquad (4)$$
$$+ (-1)^{1+I}\ (I(I+1)/240)\ (u_\nu^2 u_\pi^2 + v_\nu^2 v_\pi^2)\ ]$$

For $I>1$, the matrix element reduces to the second term in square brackets, which is relatively small and does not change the spectra significantly. However, for $I=1$ the $|(0,0)[\nu\ 1g_{7/2}, \pi\ 1g_{9/2}]1,1^+\rangle$ state is shifted up in the energy scale, towards 1^+ states containing the wave function of 1-phonon excitation. The decrease of the energetic distance makes the mixing, especially with the lowest 1-phonon $|(1,2)[\nu\ 2d_{5/2}, \pi\ 1g_{9/2}]1,1^+\rangle$ state, more effective.

3. Numerical Results and Comparison with Experiment

Positive energy levels and wave functions were obtained by diagonalization of particle-core and GT interactions in the basis (1). Two parameters were adjusted to experimental data - the energy of the core vibration phonon $\hbar\omega$ and the GT coupling constant $\varkappa$. The phonon energy was found to be 0.8 MeV. This is an intermediate value between 1.2 MeV, the energy of 2^+ level in even-even tin isotopes, and the corresponding energy in even-even cadmium, which is about 0.65 MeV. The coupling constant $\varkappa$ seems to follow N and Z dependence of the form $\varkappa = \varkappa_0 (N-Z)/A$ with $\varkappa_0 \approx 0.5$ MeV. Particle-core interaction strengths ξ_ν, ξ_π for ^{106,108}In were taken from ref. [4] and extrapolated to ^{104}In. All of the parameters are listed in Table 2.

Table 2.
Model parameters used in calculation

	^{104}In	^{106}In	^{108}In
$\hbar\omega$ (MeV)	0.8	0.8	0.8
ξ_ν	4.00	3.75	3.50
ξ_π	3.00	2.75	2.50
$\varkappa$ (MeV)	0.03	0.04	0.05

Calculated levels and relative strength distribution are compared to the relevant experimental data in Fig. 1. Only those theoretical 1^+ levels that contain more than 1% of relative strength are shown. One finds that the experimental splitting of the 1^+ levels observed in the beta decays of even tin isotopes can be naturally explained in the framework of the model. The position of the 1^+ levels and the energy scale of their splitting are reproduced. The distribution of the GT strength and particular 1^+ energies are only in semiquantitative agreement with experiment due to the simplicity of the model (e.g. neglection of deformation).

Weakening of the experimental configuration mixing in the ^{108}In case can be also understood within the model. It is due to the pairing factor $u_\nu, u_\nu - v_\nu, v_\nu$ appearing in the matrix elements of the neutron core interaction. This factor suppresses the coupling of the $1g_{7/2}$ and $2d_{5/2}$ neutrons to the quadrupole phonon, preventing the mixing.

Similar splitting of 1^+ levels in GT transitions would be expected in the region of ^{146}Gd, where the π $1h_{11/2}$ and ν $1h_{9/2}$ shells are candidates for the charge exchange interaction. However, only single 1^+ levels were found experimentally [9]. The model described here might be further applied to investigate this phenomenon.

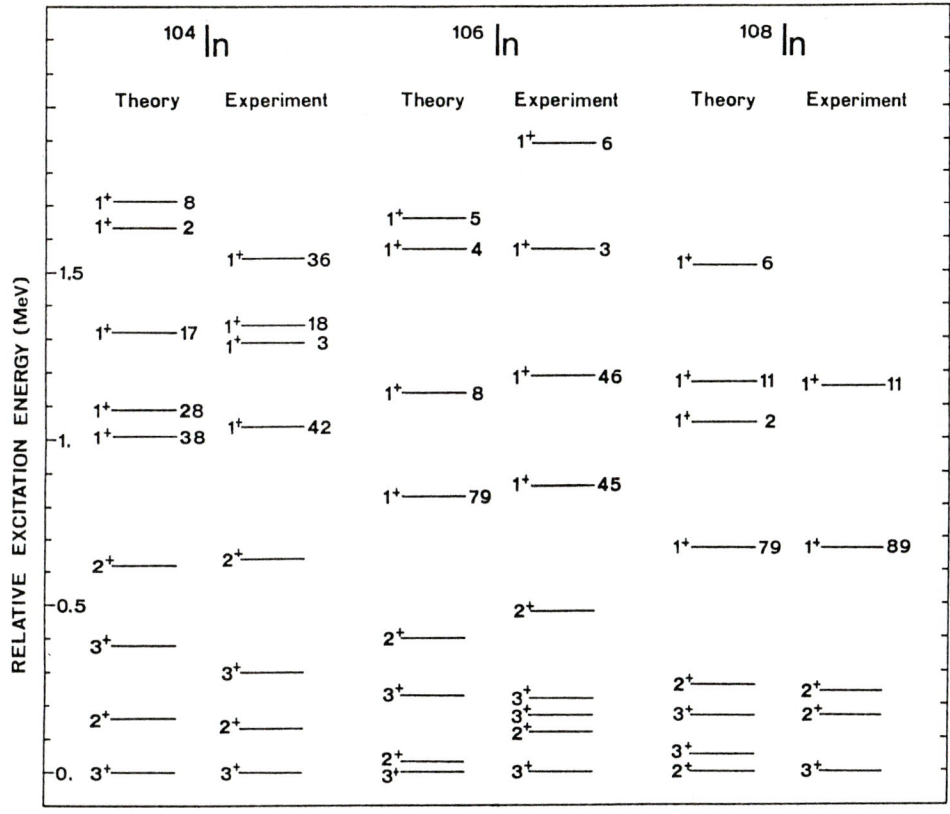

Fig. 1.
Comparison of the calculated and experimental energy levels of
104-108In. The distribution of the relative beta decay strength is
shown on the right of the 1+ levels (in %).

References

[1] K. Rykaczewski et al., Proc. 5th Int. Conf. on Nuclei Far from
    Stability, Rosseau Lake (1987), ed. I. S. Towner, p. 656
[2] R. Barden et al., Z. Phys. A329, 319 (1988)
[3] E. Roeckl, Contribution to this Workshop
[4] J. van Maldeghem, K. Heyde, J. Sau, Phys. Rev. C32, 1067 (1985)
[5] G. A. Leander et al., Phys. Rev. C30, 416 (1984)
    S. Cwiok et al., Comput. Phys. Commun. 46, 379 (1987)
[6] W. Nazarewicz, K. Rykaczewski, private communication

[7] V. Paar, Nucl. Phys. A331, 16 (1979)
[8] J. A. Halbleib, R. A. Sorensen, Nucl. Phys. A98, 542 (1967)
[9] P. Kleinheinz, Proc. Int. Symp. Weak and Electromagnetic Interactions in Nuclei, ed. V. Klapdor, p. 250

Gamow-Teller Beta Decay of Even Nuclides Near ^{100}Sn

R. Barden[1], H. Gabelmann[2], I.S. Grant[2], R. Kirchner[1], O. Klepper[1],
G. Nyman[3], A. Plochocki[4], G.-E. Rathke[1], E. Roeckl[1], R. Rykaczewski[4],
D. Schardt[1], J. Szerypo[1], and J. Zylicz[1,4]

[1] GSI, Postfach 110552, D-6100 Darmstadt 11, Fed. Rep. of Germany
[2] EP Division, CERN, CH-1211 Geneva 23, Switzerland
[3] Department of Physics, Chalmers University of Technology,
    Fack, S-41296 Göteburg 5, Sweden
[4] Department of Physics, University of Warsaw,
    ul. Hoza 69, 00681 Warsaw, Poland

ABSTRACT

Using on-line separation combined with γ-ray and conversion-electron spectroscopy, decay properties of neutron-deficient isotopes were measured near ^{100}Sn. The Q_{EC} values, the excitation energies of 1^+ levels, and the strength for $0^+ \to 1^+$ Gamow-Teller β transitions, as obtained for the decay of the even-even nuclides ^{94}Ru, $^{96-100}$Pd, $^{100-104}$Cd, and $^{104-110}$Sn, are compared with predictions from shell-model calculations including deformation, pairing and core-polarization effects. The quenching of the Gamow-Teller strength, as deduced from this comparison, is presented. Furthermore, low-lying levels in ^{104}In are discussed with particular emphasis on a possibility of observing spin-mixing.

INTRODUCTION

The double shell closure $Z = N = 50$ at ^{100}Sn, being very far from the β stability line, has not yet been reached experimentally. The investigation of nearby nuclei, however, has made progress in recent years, and has met with considerably increased interest in studying nuclear Gamow-Teller (GT) transitions in general. This interest is related to fact that the measured GT strength tends to be retarded (quenched) in comparison to predictions from nuclear models. Both experimental and theoretical efforts aim now at obtaining more precise and more reliable quenching factors in a systematic manner (e.g., as a function of mass number, or for chains of isotopes and isotones), and at subsequently interpreting the remaining deficiency, if there is any, of the nuclear model.

As far as β decay is concerned, an average quenching factor of 0.58, defined as the ratio between measured and calculated strength, is obtained[1] for the sd-shell nuclei (^{16}O to ^{40}Ca). For this region, large-basis shell model calculations and an extended set of experimental data are available. Even though both requirements are not fulfilled quite as well for fp-shell nuclei (^{40}Ca to ^{56}Ni), a quenching factor similar to the one observed for the sd-shell region is suggested here, too (see e.g. ref. 2). When going to heavier nuclei, studies ought to be carried out preferentially near closed shells, since model prescriptions with a few valence nucleons are expected to be comparatively reliable.

In this paper we want to present decay studies of very neutron- deficient isotopes near ^{100}Sn. At first, the experimental technique will be described, which consists in performing γ-ray and conversion-electron measurements of radioactive sources prepared by on-line mass separation at the GSI UNILAC and the CERN SC. Results will then be exemplified by showing measured decay schemes, Q_{EC} values, excitation energies of 1^+ levels, and $0^+ \to 1^+$ β transition probabilities, as obtained for the decay of the even nuclei $^{94}_{44}$Ru, $^{96-100}_{46}$Pd (ref. 3, 4), $^{100-104}_{48}$Cd (ref. 5), $^{104-110}_{50}$Sn (ref. 6), and to present the data with reference to model calculations (ref. 4, 7) low-lying levels in the odd-odd nuclide ^{104}In (refs. 8, 9) will also be discussed, including the identification of a new M3 isomer and a possibility of observing spin-mixing. As these aspects can be covered only briefly here, the reader is refered to refs. 3-9 for more details on the new GSI/CERN results, on data taken from the literature, and on the model calculations. We shall conclude with a few remarks on future possibilities of decay spectroscopy near ^{100}Sn.

EXPERIMENTAL TECHNIQUES AND RESULTS

The nuclei ^{96}Pd, ^{104}In, and $^{104-108}$Sn were produced by fusion-evaporation reactions, using ^{40}Ca and ^{58}Ni beams from the GSI UNILAC, and ^{60}Ni, ^{50}Cr and ^{58}Ni targets, respectively. In case of the ^{104}Sn measurements, the more abundantly produced isobaric contaminants were strongly reduced by applying the technique of bunched beam release[6,10] at the GSI On-line Mass Separator. This novel technique consists in cooling a trap inside the FEBIAD ion source, heating the trap subsequently and setting a time window corresponding to the element-specific release characteristic. The decay of ^{98,100}Cd was investigated at the ISOLDE II facility[11] on-line to the CERN SC. Chemically pure mass-separated beams of cadmium isotopes were produced, using spallation reactions between a molten tin target and a beam of 600 MeV protons.

Both for the GSI and the CERN experiments, mass-separated beams were implanted into tapes, and the resulting point-like radioactive sources were periodically transported to detector stations. Germanium detectors and a miniorange spectrometer were used to perform X-ray, γ-ray, and conversion-electron measurements. Singles data were taken in multispectrum mode, while coincidences were stored for γγ and γ conversion-electron events. From singles and coincidence data measured for the decay of the 0^+ nuclides ^{96}Pd, ^{100}Cd, and $^{104-108}$Sn, partial level schemes were constructed for the daughter nuclei. The relative β decay feeding of individual levels was obtained from the intensity balance using measured or estimated multipolarities for internal-conversion corrections. The Q_{EC} values, deduced from experimental $β^+/(EC+β^+)$ probability ratios, together with the measured β-decay half-lives allowed then to determine ft values from relative branching ratios.

As an example for the experimental results, $0^+ \rightarrow 1^+$ transitions are shown in Fig. 1 as observed for the decay of ^{94}Ru, ^{96}Pd, ^{100}Cd, and ^{104}Sn (the latter three nuclides have the same isospin-projection quantum number $T_z = (N-Z)/2 = 2$). We shall return below to a discussion of number and energy spacing of 1^+ levels, but note here that 4 such levels were observed for the decays of ^{94}Ru, ^{96}Pd and ^{104}Sn, while 7 where identified in the case of ^{100}Cd. In this context it is important to realize that the quality of the spectroscopic data for ^{100}Cd is higher than for the other nuclides. Leaving aside differences in Q value, chemical purity of radioactive sources, detector geometry, and counting time, this quality is connected to the relevant (maximum) intensities of mass-separated beams, which amounted to 2.8×10^4 atoms/s for ^{100}Cd, 3.5×10^3 atoms/s for ^{96}Pd, 10^3 and 200 atoms/s for ^{104}Sn in DC mode and for bunched beam release, respectively. It is thus more probable that 1^+ states may have remained unobserved for the ^{96}Pd and ^{104}Sn decay than it is for the ^{100}Cd case.

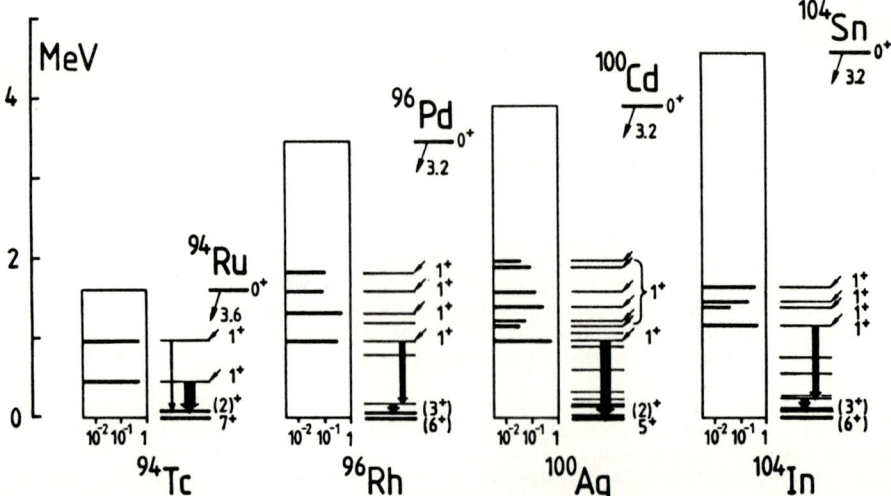

Fig.1. Simplified level schemes for the β decay of ^{94}Ru, ^{96}Pd, ^{100}Cd, and ^{104}Sn. The excited states in the daughter nuclides and the parent states' Q_{EC} values are displayed on scale with reference to the respective daughter ground-state. For each parent 0^+ ground-state, the total log ft value is indicated and the relative GT strength distribution is given on a logarithmic scale. Additional data, e.g., on excited daughter levels, half-lives, Q_{EC} values, or individual log ft values, can be taken from refs. 3, 5, 6.

While the search for decay properties of ^{102}Sn remained unsuccessful so far, some progress has been made recently in studying ^{98}Cd, this nucleus having two protons less than ^{100}Sn. From measuring KX rays, γ-rays, and conversion electrons, 60 and 107 keV transitions in ^{98}Ag were found and the half-life of ^{98}Cd was determined preliminarily to be 8.1(5) s.

In the course of the ^{104}Sn→^{104}In decay studies, low-lying ^{104}In levels were investigated in addition to the above-mentioned 1^+ states. Of particular interest are two close-lying levels, one of them representing a M3 isomer with a half-life of 15.7(5)s, an excitation energy of 93.5(1) keV, and tentative spin/parity values of 3^+; the other one is short-lived ($T_{1/2} < 0.1$ μs), lies 290(20) eV below the isomer, is depopulated by M1 radiation, and has a tentative spin/parity assignment of 5^+.

NUCLEAR STRUCTURE AND GAMOW-TELLER DECAY

In this section, we compare the new experimental results with model predictions. In order to calculate the $0^+ \rightarrow 1^+$ GT transition strength, we start out with an extreme single-particle shell model (ESPSM) and add then corrections for pairing as obtained by Dobaczewski et al.[4], and for core polarization as given by Towner[7].

Before discussing GT transition probabilities in detail, it is interesting to consider results from Q_{EC} values and 1^+ excitation energies as calculated by Dobaczewski et al.[4]. This macroscopic-microscopic model is based on an average deformed Woods-Saxon potential, includes a residual interaction of the monopole pairing form, and uses the Strutinsky shell-correction method with a droplet-model mass formula for calculating total binding energies[4]. The Q_{EC} values of ^{94}Ru, $^{96-100}$Pd, $^{100-104}$Cd, $^{104-110}$Sn, and the 1^+ excitation energies of the corresponding daughter nuclei were calculated. Good agreement with experiment was found by including deformation effects, that is by treating quadrupole (β_2) and hexadecapole (β_4) deformation as free parameters, while spherical symmetry both for this model and for a self-consistent Hartree-Fock-Bogolyubov approach yields poorer agreement. The quadrupole-deformation parameters β_2, calculated for the ground-states of odd-odd nuclides, increase from 0.067 for ^{104}In to 0.12 for ^{110}In in fair agreement with recent results from collinear laser spectroscopy[12]. However, opposite to the ground-states, the calculations yield non-deformed shapes for the ($\pi g_{9/2}^{-1}$, $vg_{7/2}$) 1^+ configurations, and hence cannot explain the observed spread of the β strength over several GT transitions. On the other hand, recent calculation, performed by Kalinowski[13], indicates that this fragmentation may be related to residual proton-neutron interaction and particle-core coupling.

Turning now to a quantitive comparison between measured GT strength and model predictions, we take from experiment the reduced transition probability B_Σ(GT), summed over all identified 1^+ levels in a β-decay daughter according to

$$\frac{6160s}{(g_A/g_V)^2} \sum_i \left(\frac{1}{(ft)_i}\right) = B_\Sigma(\text{GT}) \tag{1}$$

B_Σ(GT) as well as the calculated values of B(GT), to be discussed below, are given in units of $g_A^2/4\pi$ (compare ref. 14). The constant in the numerator, and the free-neutron ratio of the weak coupling constants $|g_A/g_V| = 1.263$, are taken from ref. 7; we omit experimental errors of these constants in the discussion here, since the corresponding uncertainties are far below those from ft-determination.

According to the ESPSM without residual interaction, the GT strength for the decay of neutron-deficient even isotopes in the ^{100}Sn region is concentrated in a single $\pi g_{9/2} \rightarrow vg_{7/2}$ transition and is given by

$$B(\text{GT}) = \frac{160}{9} v_\pi^2 (1 - v_\nu^2) \tag{2}$$

v_π^2 and v_ν^2 being the occupation numbers of protons and neutrons in the $\pi g_{9/2}$ and $vg_{7/2}$ orbits, respectively. Within the extreme model ESPSM it is assumed, that the valence protons between niobium (Z = 41) and tin (Z = 50) occupy $g_{9/2}$ orbits ($v_\pi^2 = (Z-40)/10$), and that the valence neutrons above N = 50 occupy first $d_{5/2}$ and then $g_{7/2}$ orbits ($v_\nu^2 = 0$ for N ≤ 56; $v_\nu^2 = (N-56)/8$ for 57 ≤ N ≤ 64). In order to include pairing effects, the occupation numbers in equ. 2 were taken from the macroscopic-microscopic model calculations of Dobaczewski et al.[4]. As can be seen from Fig.2, the resulting quenching factors $B_\Sigma(GT)/B(GT)$ are of the order of 0.1 to 0.2. The pairing corrections shift the calculated strength values slightly towards the experimental ones except for the decay of ^{94}Ru, ^{96}Pd and ^{110}Sn where the opposite effect is observed.

An important reduction of the predicted GT strength is obtained when taking first-order core polarization[7] into account. The resulting quenching factors, displayed also in Fig.2, are approximately 0.5 for the decays of ^{94}Ru, ^{96,98}Pd, this value being in agreement, e.g., with β-decay data obtained for sd-shell nuclei[1], and with results from (p,n) reaction studies[15]. However, For ^{102,104}Cd and $^{104-110}$Sn the GT strength reduction appears to be stronger, the quenching factor being roughly 0.2 to 0.3.

The origin of such GT strength quenching is searched for in higher-order effects such as configuration mixing due to tensor forces or mixing with Δ-isobar nucleon-hole configurations. In this context, one should bear in mind that part of the strength may be missed experimentally in cases where the 1^+ state(s) is close or even outside the β-decay window. This represents a problem for studying the decays of ^{94}Ru, ^{100}Pd, and ^{110}Sn and hence the quenching factors obtained for these nuclei are presumably not the most meaningful ones. On the other hand, even for large Q_{EC} values weak β-decay branches to high-lying 1^+ states may remain unobserved. In this respect, higher intensities and higher purity of the radioactive sources are clearly desirable.

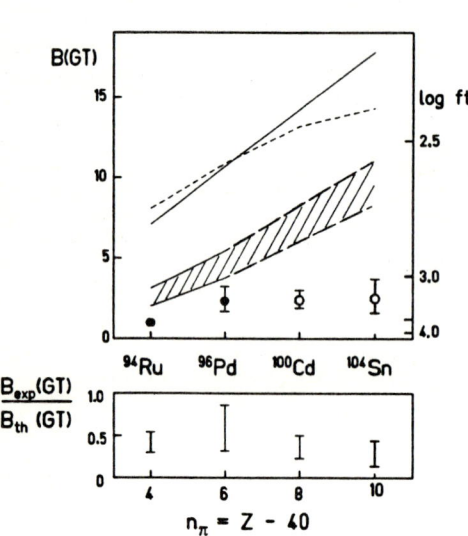

Fig. 2. **Top:** Experimentally observed total GT transition strength B(GT) of even-mass N = 50 isotones (full dots) or the closest available isotope in the case of ^{98}Cd and ^{100}Sn (open dots) and predictions. The B(GT) value are given in units of $g_A^2/4\pi$, the corresponding log ft values are indicated at the right-hand scale. The full line represent the ESPSM prediction as function of the number n_π of protons in the $g_{9/2}$ shell, assuming the free-neutron value of g_A. This prediction is modified by taking into account, at first, pairing correlations of protons and neutrons (dashed line; from model II of ref. 4) and then first-order core polarization from particle-hole interactions. For ^{100}Cd and ^{104}Sn the effect of the latter interaction was assumed to be the same as in the corresponding isotopes with N = 50. The shaded region is obtained for different residual forces[7].
Bottom: The fraction of strength ("quenching factor") observed experimentally, relative to the calculated B(GT) in the shaded areas, is displayed by vertical bars. Their lengths include also the experimental error.

LOW-LYING LEVELS IN ^{104}In

The experimentally identified low-lying levels in ^{104}In, with tentative spin/parity assignments between 2^+ and 6^+, can be discussed in the framework of a $\pi g_{9/2} \nu d_{5/2}$ multiplet-structure systematics together with ^{92,96}Nb and ^{100}In. This systematics seem to indicate[8] that, in contrast to the Z = 40 region, around Z = 50 there is no evidence for a subshell closure at N = 56, and that the $\nu d_{5/2}$ and $\nu g_{7/2}$ shells are rather filled simultaneously. In this context, unambigious spin determination from collinear laser measurements or from low-temperature nuclear orientation data are highly desirable. This holds for the discussion of π-ν multiplets, for the question of spin-mixing to be discussed in the following, but also for the interesting results on the ^{104}In decay obtained recently at Leuven[16,17]

Of particular interest are the 15.7-s isomeric level at 93.5 keV and close-lying short-lived one, with the presently available data suggesting a spin difference of $\Delta I = 2$ between the two states (see Fig. 3). It is tempting to speculate about the possibility of observing spin-mixing of these two levels, a phenomenon, which to our knowledge has not been observed in nuclear physics so far. In the system of three quantum levels sketched in Fig. 4, the wave functions $|1>$ and $|2>$, as well as the transition probabilities P_1 and P_2, change due to mixing in case of a weak perturbation. For $T_1 \gg T_2$, the value of T_1 decreases due to this perturbation. The relevant admixture coefficient is calculated with first-order perturbation theory to be $\beta = <2|\hat{H}'|1>/\Delta E$. $\hat{H}'$ denotes the perturbation hamiltonian which takes the form $-\hat{\mu}B$ and $-0.25\, e\, \hat{Q}\, (\delta E/\delta Z)_o$, respectively, if one considers the magnetic field B and the electric field gradient $(\delta E/\delta Z)_o$ caused at a nucleus by a single electron. $\hat{\mu}$ and $\hat{Q}$ denote the magnetic-dipole and electric-quadrupole operators. A change in half-life for 15.7-s ^{104m}In may be arise from admixture of the 93.2-keV level (see Fig. 3) assuming hydrogen-like ^{104}In^{46+}, or ^{104}In^{42+} with one $2p_{3/2}$ electron. The corresponding values of B and $(\delta E/\delta Z)_o$ are 2 MT and 2×10^{25} V/m^2, respectively. The latter gradient is relevant for mixing in case of a spin difference $\Delta l = 2$ deduced from measured multipolarities. Assuming 20% for the beta-decay branching of ^{104m}In, and the Weisskopf estimate of $\approx 10^{-11}$ s for the half-life of the 93.2-keV level, rough estimates[9] seem to indicate a possible decrease of the ^{104m}In^{42+} half-life due to level mixing. Whether such an effect can be observed at the SIS/ESR facility at GSI, is being investigated.

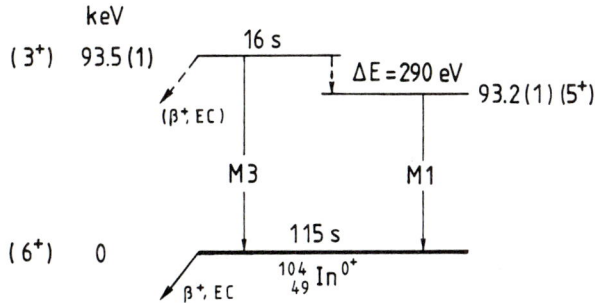

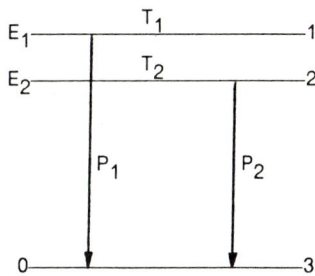

Fig. 3. Lowest-lying energy levels of ^{104}In established experimentally. The half-lives given correspond to the ones for neutral atoms.

Fig. 4. Level scheme for the discussion of of level-mixing phenomena. Levels 1 and 2 lie very close to each other. The symbols E_i, T_i and P_i indicate level energies, half-lives and corresponding transition probabilities.

CONCLUDING REMARKS

We presented results from recent measurements and from literature for $0^+ \rightarrow 1^+$ GT β decays in the ^{100}Sn region, and showed that a comparison with model predictions suggests GT quenching factors of approximately 0.5 for the decays of ^{94}Ru and $^{96, 98}$Pd, and approximately 0.2 to 0.3 for the decays of $^{102, 104}$Cd and $^{104-110}$Sn.

In order to get beyond the approximate characters of these quenching factors, the model calculations ought to be improved, e.g., with respect to the residual proton-neutron interaction. For experiments, a twofold task seems to emerge. On the one hand side, the decays studied already in this area of the nuclidic chart should be re-investigated by "more complete" spectroscopic means which allow to identify also weak GT transitions to 1^+ levels at high excitations energies and/or to derive a quantitative detection limit for such transitions. In this context, spectroscopy of β-delayed protons and application of a sum spectrometer for γ-rays have to be considered, which may be particularly interesting techniques also for further studying the ^{104}In decay. On the one hand, it is important to systematically extend these measurements to nuclei even further away from the β-stability line, where due to higher Q_{EC} values the probability of having 1^+ levels outside the observation window is smaller. The properties of such new neutron-deficient isotopes can be predicted on the basis of the results presented in this paper. For the decays of ^{98}Cd, ^{100}Sn, and ^{102}Sn, for example, a log-ft value of 3.1 follows from the GT quenching systematics[6] which, together with calculated $0^+ \rightarrow 1^+$ energies[4], yield half-lives of 6 s, 0.6 s, and 2 s, respectively. The measured ^{98}Cd half-live of 8.1(5) s is in good agreement with this estimate. An experiment planned at the ISOLDE facility aims at identifying $0^+ \rightarrow 1^+$ β decays of ^{98}Cd which ought to be feasible with a beam intensity of 100 atoms/s expected for this isotope. In the cases of ^{102}Sn and ^{100}Sn, perspectives are not quite as optimistic concerning the available proton or heavy-ion beams. We thus may have to wait for the Projectile Fragment Separator[18] at the SIS/ESR facility where, on the basis of admittedly rather tentative estimates of production cross-sections, such exotic nuclei may be produced and separated in sufficient source strength.

1. B.A. Brown and B. H. Wildenthal, At. Data and Nucl. 33, 347 (1985).
2. T. Sekine et al., Nucl. Phys. A467, 93 (1987).
3. K. Rykaczewski et al., Z. Phys. A322, 263 (1985).
4. J. Dobaczewski et al., Z. Phys. A329, 267 (1988).
5. K. Rykaczewski et al., Gamow-Teller Transitions in the Beta Decay of ^{100}Cd, to be published.
6. R. Barden et al., Z. Phys. A329, 319 (1988).
7. I.S. Towner, Nucl. Phys. A444, 402 (1985).
8. R. Barden et al., Gamow-Teller Beta Decay of Even-Even Nuclides near ^{100}Sn, contribution to the XXIII Zakopane School on Physics, April 1988.
9. J. Szerypo et al., Low-Lying Levels in ^{104}In and a Possibility of Observing Spin-Mixing, contribution to the XXIII Zakopane School on Physics, April 1988.
10. R. Kirchner, Nucl. Instr. and Meth. in Phys. Res. B26, 204 (1987).
11. H. L. Ravn et al., Nucl. Instr. and Meth. 123, 131 (1975).
12. J. Eberz et al., Nucl. Phys. A464, 9 (1987).
13. L. Kalinowski, contribution to this Workshop.
14. A. Bohr and B. Mottelson, Nuclear Structure, Vol. 1 (Benjamin, New York, 1969), p. 349.
15. C. Gaarde, in Proc. Int. Conf. Weak and Electromagnetic Interactions in Nuclei, H.V. Klapdor, ed. (Springer, Berlin-Heidelberg, 1986), p. 260.
16. V.R. Bom et al., The Beta Decay Energies of ^{103}In and ^{104}In, Z. Phys. A, in print.
17. M. Huyse et al., Giant Gamow-Teller Excitation in ^{104}Cd, to be published.
18. H. Geissel et al., Projectile Fragment Separator, A Proposal for the SIS-ESR Experimental Programme, GSI Int. Rep. (1987), unpublished.

Gamow-Teller Beta Decay of Neutron Rich Tc, Ru, Rh and Pd Isotopes

J. Äystö[1], P. Jauho[1], V. Koponen[1], H. Penttilä[1], K. Rykaczewski[2], P. Taskinen[1], and J. Zylicz[2]

[1]Department of Physics, University of Jyväskylä,
 Seminaarinkatu 15, SF-40100 Jyväskylä, Finland
[2]Department of Physics, University of Warsaw,
 ul. Hoza 69, 00681 Warsaw, Poland

ABSTRACT

During recent experiments at the IGISOL-facility several new neutron rich nuclides have been discovered, including the first direct observations of the beta decays of ^{111}Tc, ^{112}Ru, ^{113}Ru, ^{114}Ru, ^{113}Rh, ^{115}Rh, ^{116}Rh and ^{118}Pd. The role of the $\nu g_{7/2} \rightarrow \pi g_{9/2}$ transformation in the beta decays of the odd-odd $^{110-116}$Rh and the even-even $^{114-118}$Pd isotopes is shown to be dominant. The experimental results are compared with shell model calculations including deformation and pairing.

INTRODUCTION

High energy beta decay of nuclides far from the valley of beta stability provides a unique means to study the distribution and the quenching of the Gamow-Teller (GT) strength. In several recent experimental and theoretical studies of the beta decays of neutron deficient nuclides in the sd-shell, in the fp-shell, close to ^{100}Sn and in the vicinity of ^{146}Gd quenching similar to that observed in the (p,n)-reactions has been obtained [1]. Due to limitation in the energy window of the beta decay only a limited group of transitions are available for such studies. A good example of such transitions are those originating from the $\pi g_{9/2} \rightarrow \nu g_{7/2}$ and the $\pi h_{11/2} \rightarrow \nu h_{9/2}$ transitions in the ^{100}Sn and ^{146}Gd regions, respectively.

In this work we present first tentative results of the GT-decays of a group of neutron-rich nuclides with 42<Z<48 and with 64<N<72. According to the shell model picture the allowed beta decay in this region takes place via the transformation of a $g_{7/2}$ neutron into a proton in the spin-partner $g_{9/2}$ orbit. In the present work a special emphasis is on the decays of even-mass nuclides, whose GT-decays selectively populate mainly two-quasiparticle structures of neutron-neutron or neutron-proton type.

Up to now the experimental investigation of the transitional nuclides has not been possible because of their low production rates in thermal neutron induced fission as well as because of low ionization efficiency for these elements in conventional on-line isotope separators. Much of the earlier knowledge of these transitional nuclides has been obtained by radiochemical methods. We have employed a novel Ion Guide Isotope Separator On-Line, IGISOL, method in combination with a charged particle induced fission to produce these isotopes [2]. This technique is based on the separation of the primary thermalized fission fragments and thus has very little selectivity with respect to chemical

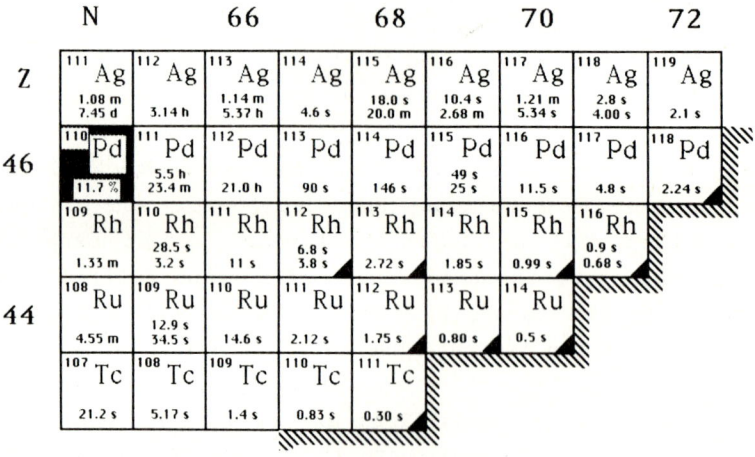

Fig.1 Summary of the known n-rich nuclides from A=110 to 118. The squares labeled by a black triangle are new or earlier erroneously assigned isotopes.

or physical properties. It is also very fast; separation times below one millisecond have been observed for activities produced in light ion reactions. Figure 1. displays a summary of the present experimental situation in this part of the nuclide chart [3-5]. The most exotic and shortest-lived activities detected are the 300 ms ^{111}Tc and 500 ms ^{114}Ru.

EXPERIMENTAL

The activities are produced by the 20 MeV proton-induced fission of ^{238}U, which has a sizeable symmetric component in its product mass distribution resulting in useful production rates for A=110-120 nuclides. The IGISOL technique currently produces a mass separated beam of nearly 10^4 ions/s per isobar with a mass resolving power of about 400 (FWHM). To insure the observation of the shortest-lived activities, decay of the fission products is detected using a close geometry at the point of implantation in the transport tape. Gamma-ray spectra in the energy range from 10 keV to 2 MeV are recorded using Ge detectors of

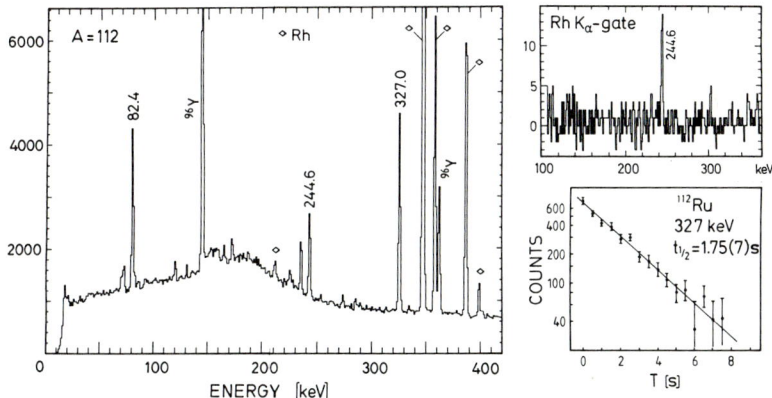

Fig. 2. A gamma-ray spectrum observed at mass 112 in coincidence with a thin plastic scintillator. The lines labeled by energy are associated with the decay of ^{112}Ru through coincidence relations.

about 20 % efficiency. Low energy spectra, including K x-rays, are measured with a 1.4 cm^3 planar HPGe detector. Beta-rays are observed with a 0.9 mm thick NE102 plastic scintillator. As an example a part of the gamma-ray spectrum observed at the mass 112 is shown in Fig.2. The half-life of ^{112}Ru disagree clearly with the earlier values suggesting erroneous assignment in these experiments.

GAMOW-TELLER DECAYS OF EVEN Rh AND Pd ISOTOPES

The systematic study of the beta decays of odd-odd ^{110,112,114,116}Rh isotopes showed that their exhibit isomerism and consist of slightly hindered decays to collective low-lying states and of considerably faster decays to a few states at or above 2.5 MeV excitation [4]. The summary of the beta-strength for these nuclides is given in Fig.3. It shows the existence of at least two beta-decaying states in each case. One is associated with a spin of 1^+ feeding mainly the low-lying structures and the other(s) with a spin value of 4 or more feeding the higher-lying levels. The 1^+ isomer could be the member of the $\pi g_{9/2} \nu g_{7/2}$ multiplet. These isomers could have their origin in the multiplets formed by coupling a $g_{9/2}$ proton to a $d_{5/2}$, $d_{3/2}$, $s_{1/2}$ or $h_{11/2}$ neutron.

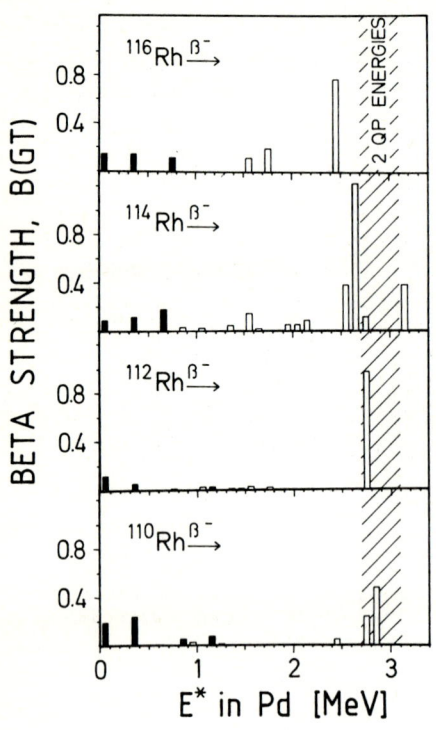

Fig.3. ß-srength for $^{110-116}$Rh.

The ß$^-$ decay of the isomer via the $\nu g_{7/2} \rightarrow \pi g_{9/2}$ transformation would result in an odd $g_{7/2}$ neutron, which then couples with the spectator neutron to form two-quasiparticle multiplets. It is known that these Pd isotopes have a ground state deformation of $\beta_2 \geq 0.2$ and therefore a simple shell model interpretation is certainly too naive. The zeroth order energies of the lowest two-quasiparticle configurations are estimated to be near 3 MeV excitation, above the pairing gap [4]. The main strength of the beta decay is indeed concentrated in this region. This can be taken as an evidence of the dominant role of the g-orbitals in the beta decay of nuclides in this mass region. Even-even neutron-rich nuclei decay typically by few fast $0^+ \rightarrow 1^+$ Gamow-Teller beta transitions.

In the studied region the origin of these 1^+ final states is expected to be the $\pi g_{9/2}, \nu g_{7/2}^{-1}$ multiplet. The Gamow-Teller strength is split over several states due to a slight deformation and residual interaction. The total strength is then obtained by summing over all the observed 1^+ states. We have recently initiated measurements of such transitions in the case of slightly deformed neutron-rich ^{114,116,118}Pd isotopes. Many new transitions have been observed and the decay of ^{118}Pd was observed directly for the first time. The total observed log ft -value for all these cases is close to 4. As compared with the prediction of the spherical single particle shell model with pairing for the pure $\nu g_{7/2} \rightarrow$

$\pi g_{9/2}$ transition, a hindrance of the order of 20 is tentatively obtained. A significant improvement is obtained by the use of the spherical proton-neutron quasiparticle random-phase approximation calculation [6]. The resulting quenching is now of the order of 5-10 [7].

In order to see the effect of deformation on both the decay energy and on the average energy of the unperturbed $(\pi g_{9/2}, \nu g_{7/2}^{-1})1^+$ state and its splitting, we have performed the calculation by using a deformed shell model with pairing based on a macroscopic-microscopic approach. The macroscopic part of the calculation is based on the droplet model and it employs the average deformed Woods-Saxon potential and the Strutinsky shell-correction method for calculating the two-quasiparticle states. The residual interaction is assumed to be of the monopole pairing form. A more detailed discussion of the calculation and its parametrization is described by Dobaczewski et al. [8]. Fig.4 shows the predicted

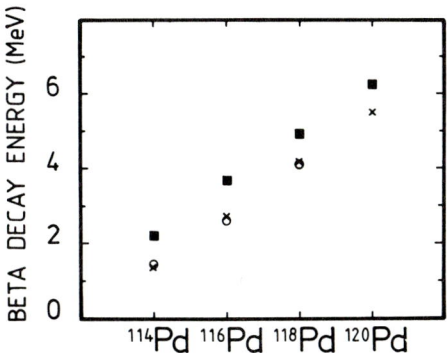

Fig.4. Q_β values obtained from (o) the experiment and the calculation employing as a macroscopic part the approaches of (x) Myers and Swiatecki and (■) Möller and Nix. See ref. [8].

and the experimental Q_β- values for the isotopes of interest. It suggest that especially for the lightest isotopes the missing experimental strength is due to small Q_β. A good agreement of the experimental points with the prediction based on the macroscopic approach of Myers and Swiatecki may be accidental. However, both theoretical models seem to have a good predictive power for estimating the decay energy of so far unobserved isotope ^{120}Pd.

Fig. 5 shows the excitation energy range of the $|\Omega_\pi - \Omega_\nu| = 1$ states originating from the $(\pi g_{9/2}, \nu g_{7/2})$ configuration. The corresponding ground state deformations are of the order of $\beta_2 = 0.2$ The black bars show the tentative experimental range of 1+ states. Fig. 5 indicates that a substantial part of the GT-strength may

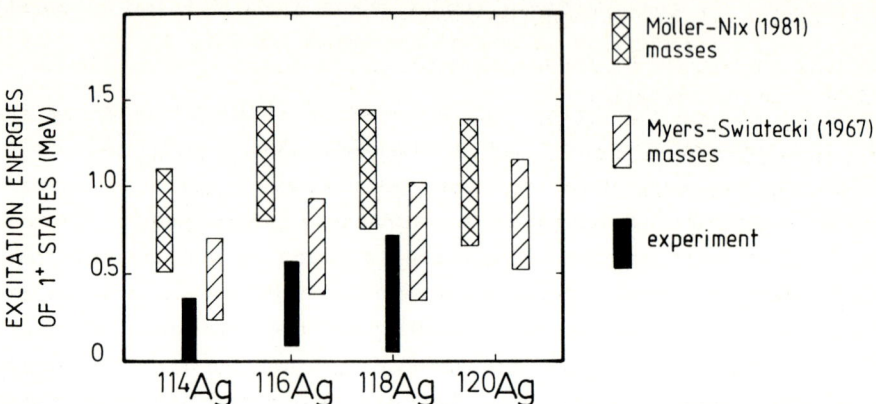

Fig. 5. Excitation energies of the $(\pi g_{9/2}, \nu g_{7/2}^{-1})1^+$ states from tentative experiments and from calculations based on two different macroscopic approach.

have been missed, especially for the lighter Pd isotopes. More detailed calculation to extract the actual GT-strength is in progress. It is clear that future experiments need improved experimental statistics and they should be extended to nuclei with higher Q_β values such as ^{120}Pd and ^{114}Ru.

This work was financially supported by the Academy of Finland.

REFERENCES

1. Proc. of the 5th Int. Conf. on Nuclei Far From Stability, AIP Conf.Proc. 164, I. S. Towner (ed.), New York, 1988.

2. J. Äystö, P. Taskinen, M. Yoshii, J. Honkanen, P. Jauho, J. Ärje and K. Valli, Nucl. Instr. Meth. B26, 394 (1987).

3. J. Äystö, P. Taskinen, M. Yoshii, J. Honkanen, P. Jauho, H. Penttilä and C. N. Davids, Phys. Lett. B201, 211 (1988).

4. J. Äystö et al., Nucl. Phys. A480, 104 (1988).

5. H. Penttilä, P. Taskinen, P. Jauho, V. Koponen, C. N. Davids and J. Äystö, Preprint No. 5/88, Dept. Physics, Univ. Jyväskylä, Finland

6. J. Suhonen, A. Faessler, T. Taigel and T. Tomoda, Phys. Lett. B202, 174 (1988).

7. J. Suhonen, private communication, 1988.

8. J. Dobaczewski, W. Nazarewicz, A. Plochocki and J. Zylicz, Z. Phys. A329, 267 (1988).

Giant Gamow-Teller Excitations in ^{104}Cd

M. Huyse[1], V.R. Bom[2], P. Dendooven[1], R.W. Hollander[2],
P. Van Duppen[1], and J. Vanhorenbeeck[3]

[1]LISOL, University of Leuven, Celestijnenlaan 200 D,
  B-3030 Leuven, Belgium
[2]Technische Universiteit Delft,
  Faculteit der Technische Natuurkunde, Delft, The Netherlands
[3]Maître de Recherche FNRS, Université Libre de Bruxelles,
  CP 165, Avenue F.D. Roosevelt 50, B-1050 Bruxelles, Belgium

1. Introduction.

The decay of ^{104}In is since long a puzzle, not only because of the significant deviation of the measured Q_{EC} value [1] from the mass tables of Wapstra and Audi [2] but also because of the inconsistenties in the decay scheme: the β^+/EC decay of the high-spin groundstate (I = 5, 6 or 7) feeds, according to the published decay scheme [3], levels with spins ranging from spin 2 up till 8 with log ft values between 5.5 and 6.9. By studying the β spectrum of ^{104}In with a solid-state beta detector and by taking extensive γ-γ coincidences at the LISOL on-line isotope separator it becomes clear that the published endpoint energy [1] and level scheme [3] are quite correct but that the deduced Q_{EC} value, β feedings and log ft values are completely wrong. This is due to a strong β^+/EC feeding to a group of levels around 5.5 MeV: in this region far off stability near the mirror nucleus ^{100}Sn the β^+/EC decay is strongly influenced by the Gamow-Teller Resonance.

2. Experimental procedure.

Mass-separated yields of ^{104}In in the order of 5×10^4 atoms per sec are obtained at the LISOL separator [4] in the reaction of a ^{20}Ne beam with an enriched ^{92}Mo target. Time-sequential singles β and γ spectra are simultanously taken with a solid-state beta detector. It is the same set-up that has been used to measure the Q_{EC} values of the heavier In isotopes [5]. The experimental conditions for the β and γ spectra taken with the beta detector are identical and this makes it possible to use the time-sequential γ spectra to decompose the beta spectra for the various isobars.

* Technische Universiteit Delft, Faculteit der Technische Natuurkunde, Delft,
  Nederland
+ Maître de Recherche FNRS, Université Libre de Bruxelles, CP 165, Avenue F.D.
  Roosevelt 50, B-1050 Bruxelles, Belgium

Time-sequential singles x and γ spectra are taken with high purity germanium detectors: special attention has been paid to measure the relative efficiencies, in order to deduce EC/β⁺ ratios. Finally a total of 14 million 2 MeV by 3.5 MeV γ-γ coincidence events are collected: the spread in the intensity of the isobars is 60% ^{104}In, 5% ^{104}Cd and 35% ^{104}Ag.

3. Results.

3.1. The β-spectrum.

Figure 1 gives a comparison between the β spectra of ^{104}In and ^{107}In. In the case of ^{107}In (already discussed in ref. 5) it is possible to fit the total β spectrum down to 0.5 MeV, by using as input the 8 β branches known from literature and as parameters the Q value and the total spectrum intensity. Such a fit procedure is completely impossible for the β spectrum in fig. 1b.

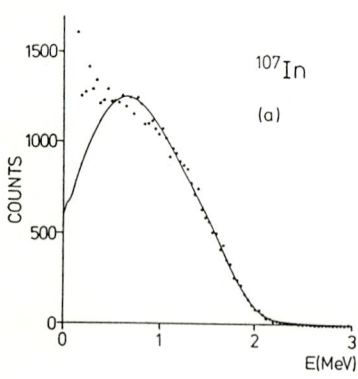

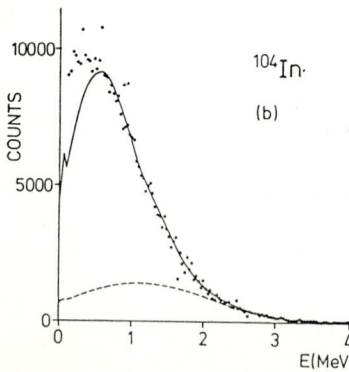

Figure 1. Comparison between the β spectrum obtained in the decay of mass-separated ^{107}In (a) and ^{104}In (b).

In the published decay scheme of ^{104}In [3] 73% of the total β+/EC intensity goes to levels below 2.5 MeV. The dashed line in fig. 1b fits this high-energy portion of the β spectrum, explaining only a small fraction of the total β spectrum. Clearly a strong feeding to higher-lying levels is present. The full line in fig. 1b is a fit through the total spectrum with 8 β branches, having their energy and intensity as fit parameters. Table 2 gives the fit results: although this is not the only possible fit, it displays the main features of the β spectrum. The most important point here is that, in conflict with the published level scheme, 67% of the total β⁺/EC intensity goes to levels above 5 MeV.

It is also possible to deduce out of the β and γ spectra taken with the beta detector the mean EC/β⁺ ratio and this in two independent ways (for more details see ref. 6). The obtained results (1.07 and 0.92) disagree completely with the EC/β⁺ ratio deduced from the published level scheme (0.14) [3].

E-level (MeV)	Endpoint (MeV)	intensity β⁺	intensity β⁺/EC %	log ft	$S_\beta(E)$ $s^{-1}MeV^{-1}$
3.0	4.0	3.7	4.4	6.4	4.0 10^{-7}
3.75	3.2	2.85	3.3	6.0	1.0 10^{-6}
4.25	2.8	5.4	6.6	5.5	3.2 10^{-6}
4.5	2.5	2.4	3.1	5.6	2.5 10^{-6}
4.75	2.2	10.5	15.9	4.7	2.0 10^{-5}
5.25	1.8	9.3	16.8	4.5	3.2 10^{-5}
5.5	1.5	2.4	6.2	4.7	2.0 10^{-5}
5.75	1.3	11.7	44.1	3.6	2.5 10^{-4}

Table 1: Results from the full-line fit from figure 1b. The upper value of the energy interval (0.25 MeV) which is fed directly in the β⁺/EC decay is given in the first column (the Q_{EC} value used here is 8.0 ± 0.2 MeV [2])). The β strength $S_\beta(E)$ corresponds with 1/ft summed over the energy interval.

As it becomes clear that the β decay is wrongly described in ref. 3, a Q_{EC} value deduced from an endpoint-energy determination and this level scheme will also be wrong. Therefore we extracted the Q_{EC} value out of our data without using a level scheme, this by determining the mean β and γ energy (see ref. 6). The obtained value (7.80 ± 0.25 MeV) rejects the value of Wouters et al [1]) (7.26 ± 0.25 MeV) but agrees with the Wapstra and Audi tables [2]) (8.0 ± 0.2 MeV).

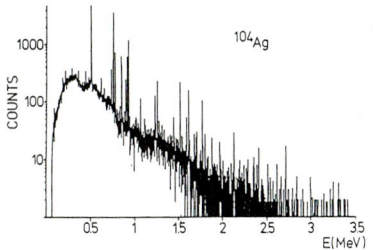

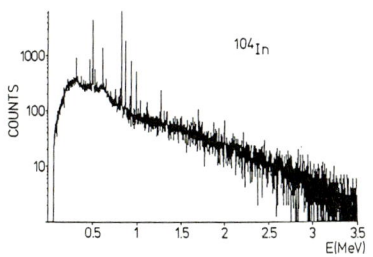

Figure 2. Two projections out of the mass 104 γ-γ coincidence matrix. The ^{104}Ag spectrum consists of γ rays coincident with the 556 keV 2⁺-0⁺ transition in ^{104}Pd. The ^{104}In spectrum is the coincidence with the 658 keV 2⁺-0⁺ transition in ^{104}Cd.

3.2. The γ-γ coincidences.

Figure 2 shows two projected γ-ray spectra from the A=104 γ-γ coincidence data set. The spectrum connected to the ^{104}Ag decay consists mainly of discrete lines superimposed on a background that can be explained as Compton scattering from higher-lying discrete lines. A completely different situation is present for ^{104}In: here a high γ-ray continuum is present with only a small number of discrete lines visible on top of it.

Our conclusion from these experimental observations, is that most of the β^+/EC decay of ^{104}In feeds high-lying levels in ^{104}Cd which on their turn decay by statistical γ-ray deexcitation. Only when the lower-lying levels are reached discrete γ rays become visible. Constructing a level scheme in the "classical" way by deducing the β feeding out of the intensity balance of the discrete γ rays feeding and leaving a certain level is doomed to failure as most of the γ-ray intensity can not be resolved. A level scheme has been constructed, taking this into account and more details can be found in ref. 7.

3.3. The β-strength function.

Figure 3 shows in open boxes the strength function $S_\beta(E)$ (the energy distribution of the reciprocal ft value) as deduced from the previously published decay scheme [3]. This is based on the intensity balance of the discrete γ rays. Taking the statistical γ-ray deexcitation into account is a hard job as this imply a excellent knowledge of γ-detection response function.

An alternative is to deduce out of the singles β spectrum a crude estimation of the β feeding to certain groups of levels. We can use the fit results from fig. 1b that are displayed in table 1 and extract summed log ft values over energy windows. The obtained log ft values are also given in table 1 and figure 3 shows the deduced β-strength function.

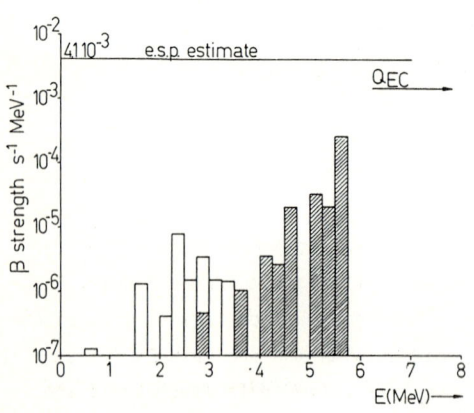

Figure 3. Comparison between the β-strength deduced from ref. 3 (open boxes) and this work (filled boxes).

4. Discussion.

In the region around $N \simeq Z \simeq 50$ most of the β decay proceeds via the $\pi g_{9/2}^n \rightarrow \pi g_{9/2}^{n-1} \nu g_{7/2}$ GT transition. This leads in the decay of the even-even Sn isotopes to very fast 0^+-1^+ β decay with log ft values around 3.5 [8]. In contrast, the decay of the $\pi g_{9/2}^{-1}$ groundstate in the odd In nuclei towards the lowest-lying $7/2^+$ state (of $\nu g_{7/2}$ character) in the odd Cd daughter is much slower and has log ft values around 5.6. P.G. Hansen discussed this phenomenon [9] as coming from the enhancement of the decay of one of the paired nucleons (towards seniority v=3 states) over the decay of the odd nucleon (towards the v=1 state). If we apply this to the odd-odd ^{104}In decay, the two following decay modes are possible (we start from a ^{90}Zr core and take the simplest configuration)

$(\pi g_{9/2}^8)^{0^+} (\nu d_{5/2}^4)^{0^+} (\pi g_{9/2} \times \nu d_{5/2})^{6^+,7^+}$

$\rightarrow v_\pi = 0, v_\nu = 2 \; (\pi g_{9/2}^8)^{0^+} (\nu d_{5/2}^4)^{0^+} (\nu g_{7/2} \times \nu d_{5/2})^{5^+,6^+}$

$\rightarrow v_\pi = 2, v_\nu = 2 \; (\pi g_{9/2}^6)^{0^+} (\nu d_{5/2}^4)^{0^+} (\pi g_{9/2} \times \nu g_{7/2})^{1^+} (\pi g_{9/2} \times \nu d_{5/2})^{6^+,7^+}$

The $v_\pi = 0$, $v_\nu = 2$ configuration will lie at the neutron pairing gap around 2.5MeV. As can be seen in fig. 3 and table 1, only a small fraction of the β^+/EC intensity feeds these levels directly. The $v_\pi = 2$, $v_\nu = 2$ configuration involves the breaking of a neutron and proton pair, which brings these levels around 5 MeV of excitation energy. It is around this energy that most of the β^+/EC intensity can be placed and the log ft values indeed indicate that it involves very fast decay.

The v = 2 proton configurations coupled to the broken neutron pair states give raise to a multitude of levels and the β strength is spread in this resonant type of structure. Evidence of this spreading comes from the fact that a strong component in the γ-decay intensity is not resolved into discrete γ-lines but is present as a statistical continuum.

A more detailed discussion can be found in ref. 10.

5. Conclusion.

In conclusion we can say that due to the specific orbitals that are active around ^{100}Sn (a nearly full $g_{9/2}$ shell for the protons, an empty $g_{7/2}$ shell for the neutrons) and due to the high Q_{EC} value, the β decay of ^{104}In is strongly influenced by the Giant Gamow-Teller Resonance. This leads to an enhancement factor of more than 500 in the β strength feeding the resonance around 5.5 MeV compared to the β strength feeding the "normal" states around 2.5 MeV. Such an enhancement is one order of magnitude greater than is calculated by theory [11,12].

Out of EC/β^+ ratios we observe similar effects in ^{102}In and in the odd nuclei ^{105,103}In and this means that the spectroscopy in this region has to be redone with special emphasis on the β- and γ-strength functions.

References.

1. J.M. Wouters et al., Phys. Rev. C27 (1983) 1745
2. A.H. Wapstra and G. Audi, Nucl. Phys. A432 (1985) 1
3. J. Blachot et al., Nucl. Data Sheets 41 (1984) 325
4. M. Huyse et al., Nucl. Instr. and Meth. B26 (1987) 105
5. V.R. Bom et al., Z. Phys. A325 (1986) 149
6. V.R. Bom et al., to be published
7. J. Vanhorenbeeck et al., to be published
8. R. Barden et al., Z. Phys. A329 (1988) 319
9. P.G. Hansen, Advances in Nuclear Physics, Vol. 7 (1973) 159
10. M. Huyse et al., to be published
11. K.L. Kratz, private communication
12. I.S. Towner, Nucl. Phys. A444 (1985) 402

Subject Index

Reference is made to the first page of relevant articels

Alaga rules 343
Angular distributions
— β particles 245
— γ rays 172, 215, 227, 245, 309, 315
— γ-γ angular correlations 239, 391

Band structures
— backbending 131, 320
— band termination 193
— decoupling parameter 343
— intruder bands 3, 45, 76, 82, 88, 215, 233, 239, 303
— rotational bands 45, 58, 76, 101, 120, 127, 199, 263, 303, 315, 332, 343, 349, 357
— β and γ bands 64, 157
Beta decay
— GT transitions 58, 239, 397, 403, 409, 415
— giant resonance influence 415
— ground state feeding 391
— isomers 64, 239
— positron-to-electron capture values 403
— Q_β values 251, 257, 391, 409
— quenching factor 403
— shape hindrance 58
— strength function 403, 415
— Way-Wood diagrams 257
— $0^- \to 0^+$ transitions 215, 391

Centrifugal forces 193
Complete spectroscopy 157, 172, 176
Conversion coefficients 45, 58, 88, 391
Core polarization 277, 303, 403
Cranking approximation 45, 127, 277

Deformation
— energies 143, 277
— hexadecapole 215
— hyperdeformation intruder states 143
— oblate 17, 45, 143, 298
— octupole 277
— parameters 52, 58, 127, 326, 368
— quadrupole 17, 39, 45, 76, 127, 143, 298, 315
— superdeformation 143, 193, 277
— triaxial 17, 45, 143, 193, 277, 303, 309, 326, 332

EXCITED VAMPIR 32
Experimental devices
— Beta-γ-γ array 251
— Daresbury recoil separator 101, 298, 320
— ESSA 30 detector array 101

— DOLIS COLD 309
— GAMS 120
— GSI separator 403
— IGISOL 409
— INS (inel. neutron scatt. fac.) 215
— ISOLDE 52, 257
— JOSEF 239
— KOOL 245
— Laser-spectroscopy 52
— LISOL 245, 415
— LOHENGRIN 251
— n-γ-γ arrays 45, 101, 127
— OSIRIS 45, 157, 172, 269, 315
— OSTIS 58
— TRISTAN 64, 215, 343, 391

Folded Yukawa potential 277

Gamma softness 17, 45, 277, 303
g factors
— effective boson factors g_π, g_ν 70
— experimental results 52, 70, 131, 245, 269, 309, 326
— intrinsic factor g_K 58, 332
— transient field method 131
Giant Resonance 176, 415

Hartree-Fock calculations 39, 143
High-spin states
— aligned orbitals 45, 115, 127, 277, 303, 320
— experimental information 45, 70, 131, 303, 320
— HI induced fission 101
— shape changes 17, 45, 263

Interacting boson approximation
— effective boson numbers N_π, N_ν 70, 376
— effective charges e_π, e_ν 70
— effective g factors 70
— IBA-1 157, 184
— IBA-2 70
— IBFM 26, 157, 184, 199
— IBFFM 184, 199, 239
Intruder states 3, 88, 184, 277, 357, 391
Isotope shift 52

Large amplitude collective motion 293
Level lifetimes
— delayed coincidences 45, 58, 64, 76, 115, 137, 239, 349
— DSAM in neutron scattering 215, 227

— DSAM in neutron capture 120
— DSAM in α or HI induced reactions 115, 120
— RDM 120
— sidefeeding times 120

Meson-exchange effects 391
Modified oscillator potential 193, 368
Moments of inertia 45, 127, 277, 298, 343
Monopole
— degrees of freedom 88
— strength 3, 64, 88, 233

N = Z nuclei 298
Nilsson model
— Coriolis interaction 277, 332, 343, 349
— orbitals 58, 76, 193, 199, 309, 343, 349
— parameters 3, 332, 343, 349
— RPA 58, 368
Nuclear masses
— binding energies 251, 257
— mass surfaces 251, 376
— predictions 376
— semiempirical microscopic masses 376
— 2-nucleon decay/separation energies 251
Nuclear orientation 245, 309
Nuclear radii 52

Occupation numbers 199, 349, 357, 368, 397
Octupole strength 215, 277
Onset of deformation 3, 52, 64, 251, 357, 368, 376

Pairing
— gap 3, 277, 332, 343
— interaction 17, 39, 277, 368
— reduction 3, 277, 343, 368
— strength 332, 343, 349, 376
Parabolic rule 199, 382, 397
Particle-rotor calculations 315, 326, 332, 343, 349
P-factor 26, 298, 376
Potential-energy surfaces 17, 45, 277, 293
Proton-proton interaction 88, 269
Proton-neutron charge exchange 397
Proton-neutron interaction
— general 3, 26, 269, 277, 326, 343, 368, 382, 397
— $\pi d_{3/2} - \nu d_{5/2}$ 82
— $\pi g_{9/2} - \nu g_{7/2}$ 199, 277, 357, 368, 382
— $\pi g_{9/2} - \nu h_{11/2}$ 277, 357, 368

Quadrupole moments 52, 303
Quantum chaos 176, 184

Rearrangement of pairs 245
Rotation (noncollective) 143
Routhians 45, 277, 293, 303

Shell closures
— major shells 39, 251, 263
— subshells 3, 45, 64, 70, 193, 215, 245, 251, 257, 263, 277, 298, 376, 391
Shape coexistence 17, 32, 39, 45, 58, 64, 76, 82, 88, 193, 233, 263, 277, 303
Shell-model calculations
— shell-model orbitals 58, 263, 368
— single-particle level density 3, 277
Shell model configurations
— pair scattering 269, 397
— particle-hole excitations 88, 215, 233, 269, 326, 357
— 2 proton excitations 239, 263, 357
— 4 proton excitations 357
— proton-neutron multiplets 199, 263, 382
Spin mixing 403
Splitting of 1^+ states 397, 403, 409
Statistical-model calculations 120, 131, 157, 176
Symmetric rotors 45, 315, 320, 332

Transition strength
— anomalies 315
— core related effects 115, 137, 303, 326
— E0 64, 88, 233
— E1 215, 227, 332, 349
— E2 39, 58, 64, 70, 76, 82, 115, 137, 157, 172, 199, 215, 227, 315, 326, 357
— E3 215, 239
— E2/M1-mixing ratio 45, 76, 172, 233, 245
— M1 76, 115, 137, 199, 227, 349
— X(E2/E0) ratio 233
Two-phonon states 137, 215, 233

Woods-Saxon potential 17, 45, 193, 277, 397

0^+ states
— experimental information 45, 64, 88, 215, 239, 326
— intruder configuration 88, 215, 326, 357
— mixing 64, 391

Index of Contributors

Abdelrahman, Y. 101
Äystö, J. 409
Alber, D. 269
Alstad, J. 76
Andrejtscheff, W. 137
Aprahamian, A. 233
Arnold, E. 52
Ashworth, C.J. 309

Barden, R. 403
Batsch, T. 263
Belgya, T. 215,227
Bengtsson, R. 17
Bengtsson, T. 193
Berant, Z. 391
Billowes, J. 131
Bom, V.R. 415
Brant, S. 184,199,239
Brenner, D.S. 376
Brentano, P., von 45,157, 172
Buchinger, F. 52
Bucurescu, D. 26
Busch, J. 45,315

Cǎta, G. 26
Campbell, M. 320
Casten, R.F. 70,176,376, 391
Chishti, A.A. 101,127,298, 320
Cottle, P.D. 303
Cristancho, F. 120
Cutoiu, D. 26

Daniels, W.R. 76
De Frenne, D. 76
Dendooven, P. 415
Dewald, A. 157,172
Dönau, F. 293
Durell, J.L. 101

Eberth, J. 45,315
Egidy, T., v. 176

Fazekas, B. 215,227
Federman, P. 357
Fieber, W. 120

Fitzgerald, J. 101
Flocard, H.C. 143
Funke, L. 115

Gabelmann, H. 403
Gatenby, R.A. 215,227
Gelberg, A. 172
Gelletly, W. 101,127,298,320
Gill, R.L. 64,343,391
Goettig, L. 320
Graefenstedt, M. 251,257
Grant, I.S. 309,403
Grawe, H. 269
Griffiths, A.G. 309,326
Gross, C. 120
Gross, C.J. 127,303
Grümmer, F. 32
Guttormsen, M. 263

Haas, H. 269
Harakeh, M.N. 76
Haustein, P.E. 376
Headley, D.M. 303
Heese, J. 120,127
Henry, E.A. 233
Heyde, K. 3,76
Hill, J.C. 64,343
Hlawatsch, G. 176
Hollander, R.W. 415
Hoyler, F. 176
Hüttmeier, U.J. 303
Huyse, M. 415

Ivaşcu, M. 26

Jacobs, E. 76
James, A.N. 298,320
Jauho, P. 409
Julin, R. 215

Kaffrell, N. 76
Kalinowski, L. 397
Keyser, U. 251,257
Kirchner, R. 403
Klepper, O. 403
Kleppinger, E.W. 215
Kluge, H. 269
Koponen, V. 409

Kostov, L.K. 137
Kownacki, J. 263
Kratz, K.-L. 58,199,368
Krusche, B. 176
Kucharska, A.I. 131
Kumpulainen, J. 215
Kusnezov, D.F. 82

Lhersonneau, G. 58,199,239
Liang, C.F. 26
Lieb, K.P. 120,127,176
Liebchen, M. 45,315
Lieberz, W. 157,172
Lievens, P. 52
Lister, C.J. 101,127,131, 298,320
Lopac, V. 184,199
Love, D.J.G. 320

Mach, H. 64,215,391
Maier, K.H. 233,269
Mann, L.G. 233
McNeill, J.H. 101,320
Meyer, R.A. 82,199,233,349
Molnár, G. 215,227,391
Moore, E.F. 303
Moscrop, R. 320
Moszynski, M. 64
Münnich, F. 251,257
Myläus, T. 315
Mylaeus, T. 45

Nazarewicz, W. 45,277,303
Neu, W. 52
Neugart, R. 52
Nyman, G. 403

Ohm, H. 58,199,239
Ohya, S. 309
Osipowicz, T. 120

Paar, V. 184,199
Pandya, S.P. 39
Paris, P. 26
Passoja, A. 215
Penttilä, H. 409
Petkov, P. 137
Petrovici, A. 32

Petry, R.F. 64,343
Pfeiffer, B. 58,199,368
Phillips, W.R. 101
Plochocki, A. 403

Ragnarsson, I. 193
Ramsay, E.B. 52
Ramsøy, T. 263
Rathke, G.-E. 403
Reinhardt, R. 157,172
Rekstad, J. 263
Riedinger, L.L. 293
Rikovska, J. 309,326
Roeckl, E. 403
Rogowski, J. 76
Roy, N. 233
Rykaczewski, K. 409
Rykaczewski, R. 403

Sahu, R. 39
Schardt, D. 403
Schippers, J.M. 76
Schmal, N. 45,315
Schmid, K.W. 32
Schmittgen, K.P. 172
Schreiber, F. 251,257
Sefzig, R. 45,315
Seiffert, F. 172
Semmes, P.B. 332
Seo, T. 349
Severijns, N. 245
Seyfarth, H. 184,199

Silverans, R.E. 52
Sistemich, K. 58,199,239, 349,391
Skarnemark, G. 76
Skeppstedt, O. 320
Skoda, S. 45,315
Spellmeyer, B. 269
Stolzenwald, M.L. 199,239
Stone, N.J. 309
Stoyer, M.A. 82
Sun, X. 269
Sy Savane, Y. 137
Szerypo, J. 403

Tabor, S.L. 303
Taskinen, P. 409
Teichert, W. 45,315
Tetzlaff, H. 76
Trautmann, N. 76

Ulbig, S. 120
Ulm, G. 52

Van Duppen, P. 415
Vandeplassche, D. 245
Vanderpoorten, W. 245
Vanhaverbeke, J. 245
Vanhorenbeeck, J. 415
Vanneste, L. 245
Varley, B.J. 101,127,298,320
Veres, A. 215,227
Verho, E. 215

Vermeeren, L. 52
Vorkapić, D. 184,199
Vretenar, D. 199

Walker, P.M. 309
Walle, E., van 245
Walters, W.B. 382
Warburton, E.K. 391
Wendt, K. 52
Werf, S.Y., van der 76
Werner, T. 277
Winger, J.A. 391
Winter, C. 176
Winter, G. 115
Wiosna, M. 45,315
Wörmann, B. 120
Wohn, F.K. 64,343
Wolf, A. 70,391
Wolfsberg, K. 76
Wood, J.L. 88
Wouters, J. 245
Wrzal, R. 172

Yaffe, R.P. 82
Yates, S.W. 215,227,391

Zamfir, N.V. 26
Zelazny, Z. 263
Zell, K.O. 157,172
Zganjar, E.F. 88
Zhang, Jing-ye 293
Zipper, V. 157
Zylicz, J. 403,409

Research Reports in Physics

The categories of camera-ready manuscripts (e.g., written in TEX; preferably hard plus soft copy) considered for publication in the **Research Reports** include:

1. Reports of meetings of particular interest that are devoted to a single topic (provided that the camera-ready manuscript is received within four weeks of the meeting's close!).
2. Preliminary drafts of original papers and monographs.
3. Seminar notes on topics of current interest.
4. Reviews of new fields.

Should a manuscript appear better suited to another series, consent will be sought from the author for its transfer to the other series.

Research Reports in Physics are divided into numerous subseries, e.g., nonlinear dynamics or nuclear and particle physics. Besides covering material of general interest, the series provides an opening for topics that are too specialized or controversial to be published within the traditional context. The implied small print runs make a consistent price structure impossible and will sometimes have to presuppose a financial contribution from the author (or a sponsor). In particular, in the case of proceedings the organizers are expected to place a bulk order and/or provide some funding.

Within **Research Reports** the timeliness of a manuscript is more important than its form, which may be unfinished or tentative. Thus in some instances, proofs may be merely outlined and results presented that will be published in full elsewhere later. Since the manuscripts are directly reproduced, the responsibility for form and content is mainly the author's, implying that special care has to be taken in the preparation of the manuscripts.

Springer-Verlag
Berlin Heidelberg New York
London Paris Tokyo

Research Reports in Physics

Manuscripts should be no less than 100 and no more than 400 pages in length. They are reproduced by a photographic process and must therefore be typed with extreme care. Corrections to the typescript should be made by pasting in the new text or painting out errors with white correction fluid. The typescript is reduced slightly in size during reproduction; the text on every page has to be kept within a frame of 18 × 26.5 cm (7 × 10.5 inches). On request, the publisher will supply special stationary with the typing area outlined.

Editors or authors (of complete volumes) receive 5 complimentary copies and are free to use individual parts of the material in other publications later on.

All manuscripts, including proceedings, must contain a subject index. In the case of many-author books and proceedings an index of contributors is also required. Proceedings should also contain a list of participants, with complete addresses.

Our Instructions for the Preparation of Camera-Ready Manuscripts and further details are available on request.

Manuscripts (in English) or inquiries should be directed to

Dr. Ernst F. Hefter,
Physics Editorial 4,
Springer-Verlag, Tiergartenstrasse 17,
D-6900 Heidelberg, FRG,

Springer-Verlag
Berlin Heidelberg New York
London Paris Tokyo

(Tel. [0]6221-487495;
Telex 461723; Telefax 06221-43982).

QC 793.3 .S8 N834 1988

Nuclear structure of the
 zirconium region

APR 1 2 1989